Carbohydrate Photochemistry

Carbohydrate Photochemistry

Edith R. Binkley
Oberlin Center for Carbohydrate Study

Roger W. Binkley
Cleveland State University
Oberlin Center for Carbohydrate Study

ACS Monograph 191

American Chemical Society, Washington, DC

Library of Congress Cataloging-in-Publication Data

Binkley, Edith R., 1940–
Carbohydrate photochemistry / Edith R. Binkley and Roger W. Binkley.
p. cm. — (ACS monograph, ISSN 0065–7719 ; 191)
Includes bibliographical references (p. –) and index.
ISBN 0–8412–3539–2 (alk. paper)
1. Carbohydrates—Congresses. 2. Photochemistry—Congresses. I. Binkley, Roger W., 1941– . II. Title. III. Series.
QD321.B588 1998
547'.780455—dc21 98–8007
CIP

The paper used in this publication meets the minimum requirements of American National Standard for Information Sciences—Permanence of Paper for Printed Library Materials, ANSI Z39.48-1984.

PRINTED IN THE UNITED STATES OF AMERICA

About the Authors

EDITH R. BINKLEY is a chemist who has dedicated her career to public school teaching. They both retired from teaching in 1994 to devote their efforts to writing about chemistry.

ROGER W. BINKLEY is an emeritus professor of chemistry at Cleveland State University and an affiliate scholar at Oberlin College. He has been engaged in research and writing in the areas of carbohydrate chemistry and photochemistry for over thirty years. Much of his research has involved mechanistic and synthetic photochemistry of carbohydrates.

Foreword

ACS MONOGRAPH SERIES was started by arrangements with the interallied Conference of Pure and Applied Chemistry, which met in London and Brussels in July 1919, when the American Chemical Society undertook the production and publication of Scientific and Technological Monographs on chemical subjects. At the same time, it was agreed that the National Research Council, in cooperation with the American Chemical Society and the American Physical Society, should undertake the production and publication of Critical Tables of Chemical and Physical Constants. The American Chemical Society and the National Research Council mutually agreed to care for these two fields of chemical progress.

The Council of the American Chemical Society, acting through its Committee on National Policy, appointed editors and associates to select authors of competent authority in their respective fields and to consider critically the manuscripts submitted. The first Monograph appeared in 1921. Since 1944, the Scientific and Technological Monographs have been combined in the series.

These Monographs are intended to serve two principal purposes: first, to make available to chemists a thorough treatment of a selected area in a form usable by person working in more or less unrelated fields so that they may correlate their own work with a larger area of physical science; and second, to stimulate further research in the specific field treated. To implement this purpose, the authors of Monographs give extended references to the literature.

Contents

Preface

Broad-based interest in the chemical reactions of carbohydrates has grown dramatically in recent years as the diverse biological roles of these compounds have become better understood. This interest has stimulated the discovery of new photochemical reactions of carbohydrates and caused further study of existing reactions of these compounds. To appreciate the scope of recent discoveries one need only consider the variety of new synthetic reactions that involve photochemical generation of carbohydrate radicals. To understand the further development of existing reactions one need only turn to recent studies of release of synthetic oligonucleotides from solid-phase supports to which they are linked by a photolabile group. These types of reaction serve not only as examples of the importance of photochemical transformations of carbohydrates to synthetic chemistry, but they also are indicators that other photochemical reactions of these compounds may have similar, as yet unknown, synthetic value. We believe that one factor that will stimulate the discovery of new reactions and assist development of existing ones is for researchers to have readily available, complete information about the photochemistry of carbohydrates. Our goal in writing this book is to provide such information.

Carbohydrate Photochemistry is a book of twenty chapters in which the photochemical reactions of carbohydrates are discussed, analyzed, and comprehensively reviewed. This book is intended to cover the published literature as fully as possible through 1996. A typical chapter begins with an introduction that documents the discovery and development of a group of related reactions. This introductory section is followed by one containing a mechanistic discussion that draws on information from all areas of chemistry. Attention next focuses specifically on carbohydrates with the major portion of each chapter consisting of a detailed, amply illustrated, discussion of pertinent photochemical reactions. The chapter then closes with a summary, the references, and a group of tables that contain a comprehensive listing of reported reactions.

In addition to carbohydrate chemists, the information in this book should be of interest to persons specializing in synthetic organic chemistry, biochemistry, and photochemistry. This book provides its readers with an in-depth discussion of the photochemical reactions of carbohydrates and allows them to easily determine what has been accomplished in the field. Such information should be valuable to most carbohydrate chemists and many biochemists. *Carbohydrate Photochemistry* offers organic chemists, who use carbohydrates in synthesis, an understanding of the photochemical reactions of these compounds that can be the basis for solving particular synthetic problems. Photochemists interested in the opportunities for study of reaction mechanisms provided by investigation of enantiomerically pure compounds with one or more chiral centers should find the information in this book worthwhile.

Several persons and organizations have helped us in this endeavor. Professor Thomas W. Flechtner carefully read the manuscript and provided us with many valuable suggestions. Four individuals from the American Chemical Society provided timely guidance. Cheryl Shanks first expressed interest in our manuscript and worked with us during the review process. Cheryl Wurzbacher directed the production process and enabled it to proceed smoothly. Jay Cherniak edited the manuscript, and Kathy Strum monitored the quality of the final product. Finally, the libraries at Cleveland State University and Oberlin College made it possible for us to obtain the primary literature upon which our writing depended.

EDITH R. BINKLEY
ROGER W. BINKLEY

Dedication

This book is dedicated to the early leaders in the field of carbohydrate photochemistry.

K. Araki

P. M. Collins

G. Descotes

B. Fraser-Reid

D. Horton

Y. Ishido

G. O. Phillips

W. A. Szarek

R. L. Whistler

U. Zehavi

1

An Introduction to Carbohydrate Photochemistry

Introduction

A Long-Standing Relationship

Photochemistry and carbohydrate chemistry are two branches of chemical science that are joined by the process of photosynthesis. (Photosynthesis has been defined as "the system of reactions by which higher plants, some algae, and bacteria capture solar energy, convert it to chemical energy, and use this energy in the reduction of carbon dioxide to sugars."[1]) Photochemistry and carbohydrate chemistry have more in common, however, than the vital, photochemically based process of photosynthesis. Photochemical reactions have been used by carbohydrate chemists for many years to perform a variety of molecular transformations, including removal of protecting groups, oxidation of alcohols, reduction of esters, and formation of unique ring systems. Although the list of applications has gradually expanded over the years, the synthetic potential of these reactions will not be realized until they are more routinely incorporated into reaction sequences involving carbohydrates. Many applications of photochemical reactions in carbohydrate synthesis are the result of efforts of a group of carbohydrate chemists who have a special interest in photochemistry. A more general appreciation for the advantages and synthetic value of photochemical reactions is needed in order for these reactions to be more widely used in carbohydrate chemistry. Continued discovery of new reactions and, perhaps more important, further optimization and simplification of existing ones, are elements that will foster this appreciation.

Carbohydrate chemistry has been a field for application of photochemical reactions but, with few exceptions, not a place where new understanding of photochemistry has occurred. One of these exceptions is in the area of chirality transfer, where the study of carbohydrates has made an important contribution.[2–4] Another exception is the photodeoxygenation of esters, where carbohydrate photolysis also has had a central role in elucidating the reaction mechanism.[5] These exceptions serve as reminders that, in some situations, the study of carbohydrates provides an excellent pathway to understanding basic photochemistry. Whenever investigation of enantiomerically pure compounds (particularly those with multiple chiral centers) is desirable, use of carbohydrates is a logical choice. One helpful step, we believe, in stimulating greater appreciation of carbohydrate photochemistry and in fostering more interaction between the areas of carbohydrate chemistry and photochemistry is to bring together and analyze comprehensively the information that already exists about photochemical reactions of carbohydrates. Also, this information will be useful to carbohydrate chemists who are interested in reactions that offer valuable, sometimes unique, synthetic possibilities and to photochemists who are attracted to the study of compounds that offer different possibilities for understanding photochemical reactions. Stimulating interest in the opportunities that carbohydrate chemistry and photochemistry hold for each other is one rationale for writing this book.

The Purpose of an Introductory Chapter

The primary goal of this first chapter is to offer an introduction to the terminology, definitions, and concepts of photochemistry that are necessary to understand the photochemical reactions of carbohydrates. When we considered the amount of information available about photochemical reactions, we needed to decide what topics should be included in this first chapter and what should be the depth of presentation. Our decision was to include only that information required to understand the reactions and ideas presented in the remaining chapters. This principle also guided our decisions about the depth of the treatment of topics. We understand that some readers need none of this introductory material, and they may wish to go immediately to the next chapter. We realize also that others may wish to read this chapter and have additional information about the topics discussed. For those in the latter group, we have tried to provide the needed information by making reference at appropriate points to books and review articles that contain much more extensive treatment of basic photochemical topics than that found here.

Because this chapter contains an introduction to photochemistry, one might wonder whether an introductory chapter on carbohydrate chemistry also is needed. We don't think so. We have presented material in such a fashion that a knowledge of organic chemistry, but not the details of carbohydrate chemistry, is necessary to understand the information discussed. For example, carbohydrate nomenclature, as important as it is, is sometimes a "stumbling block" for noncarbohydrate (and carbohydrate) chemists; consequently, we always include structural formulas in presenting and discussing information.

Ideas and terminology in use in carbohydrate chemistry but not in other areas of organic chemistry (there are very few) are explained when encountered.

Two Definitions

A Photochemical Reaction

A photochemical reaction is a process that begins with a molecule in its ground state being converted into one in an electronically excited state by absorbing a photon of light.* This excitation process can be viewed as the first step of a three-part sequence.[6] The second step involves some fundamental change in bonding for the excited molecule. This change can include forming bonds, breaking bonds, donating electrons, accepting electrons, and sometimes combinations of these events. The energy for this second step comes largely, sometimes totally, from the energy supplied by excitation. In step 3, any reactive intermediate (e.g., radicals, ions, and radical ions) formed in step 2 undergoes further reaction to produce a stable product or products.

A Photochemical Reaction of a Carbohydrate

Once the definition of a photochemical reaction has been determined, the meaning of "a photochemical reaction of a carbohydrate" needs to be established. Because carbohydrates have been defined as "polyhydroxy aldehydes, ketones, alcohols, acids, and their derivatives, and the polymers derived from these compounds",[7] it seems that the definition of a photochemical reaction of a carbohydrate merely requires combining the definitions of "photochemical reaction" and "carbohydrate". Difficulties arise, however, for compounds in which the carbohydrate portion† of a molecule represents only a part of its structure. Nucleosides, for example, are compounds for which many (but not all) photochemical reactions actually involve only the nitrogenous base portion of the molecule. Even though a typical nucleoside is a derivative of a polyhydroxy aldehyde and, therefore, is a carbohydrate, its photochemical reactions may have little, if anything, to do with the carbohydrate portion of the molecule. An example of such a reaction is given in eq 1.[8] A reaction that

*Another important way of forming an excited state, discussed in the section "Electronic Energy Transfer" later in this chapter, is by energy transfer. There are still other ways of producing excited states (e.g., chemical reaction and electrical discharge), but these are not of concern to the photochemistry discussed in this book.

†The "carbohydrate portion" of a molecule is not particularly well defined but generally can be considered to be that part of the molecule having a structure resembling a polyhydroxy alcohol, aldehyde, or ketone.

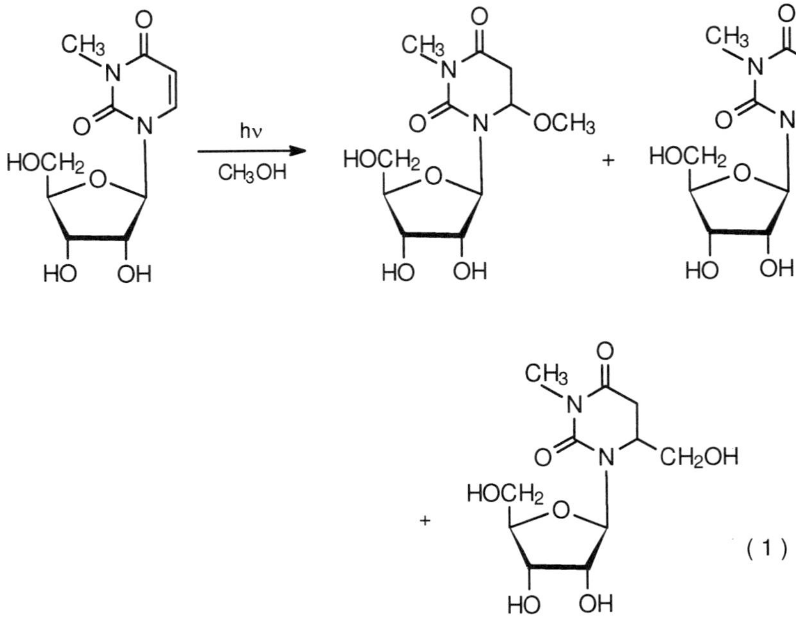

(1)

Equation 1

involves only that part of a molecule classified as noncarbohydrate logically does not belong with photochemical reactions of carbohydrates.

The challenge then is to arrive at a definition that includes all appropriate reactions and only those reactions that reasonably do belong. With this thought in mind, we adopt a definition that is somewhat cumbersome but meets the challenge. A photochemical reaction of a carbohydrate is a photochemical reaction of any molecule defined as a carbohydrate; however, for those compounds with both carbohydrate and noncarbohydrate portions, a reaction is included only if the photochemical process involves or is clearly influenced by the carbohydrate part of the molecule.

Electronically Excited States

Singlet States

Because a photochemical reaction begins with the formation of an excited state,[9] it is useful to consider the possible states that might be involved. The three most important states in the solution-phase photochemistry of organic compounds are the ground state (S_0), the lowest triplet state (T_1), and the lowest excited singlet state (S_1). Singlet states will be discussed in this section and triplet states in the next.

Excitation promotes an electron from a bonding or nonbonding molecular orbital of the ground state (S_0) to an antibonding molecular orbital to create an excited state. The lowest excited singlet state (S_1) forms when the orbital receiving the electron has the lowest energy of the antibonding orbitals. (In an excited singlet state the spins of the two electrons in partially filled molecular orbitals remain paired, as they were in the ground state.) Any change between states in a molecule is a photophysical process, and in state diagrams, such as that shown in Figure 1.1, is represented by an arrow connecting the states involved. Straight arrows indicate radiative processes, such as absorption of a photon, and wavy arrows represent nonradiative processes, such as collisional loss of vibrational energy. Absorption occurs in such a short time that the positions of the nuclei in a molecule remain effectively unchanged during this period; therefore, in the instant after absorption, nuclei occupy essentially the same relative positions as they did in the ground state. (The idea that nuclear positions do not change during excitation is known as the Franck–Condon principle.) Because the equilibrium geometry for an excited state is rarely that of the ground state, excited states typically are created in a vibrationally excited form. In solution, collisions with solvent molecules rapidly remove this excess vibrational energy so that a molecule relaxes to the lowest vibrational level of S_1 (Figure 1.1).

A higher excited state (e.g., S_2) forms when the antibonding orbital receiving an electron does not correspond to the lowest-energy antibonding level. This state rapidly converts to the lowest-energy excited state (S_1) by a nonradiative process known as internal conversion (Figure 1.2). Internal conversion, therefore, is a change from one electronic state into another without emission of light (or spin change). The internal conversion of higher excited states (S_2,

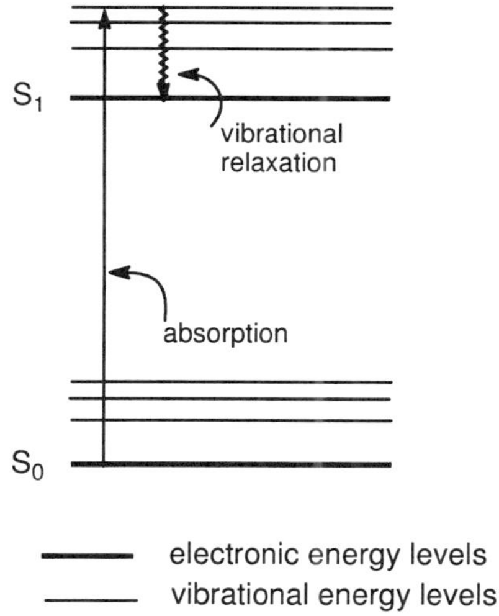

Figure 1.1. Formation and vibrational relaxation of a singlet excited state.

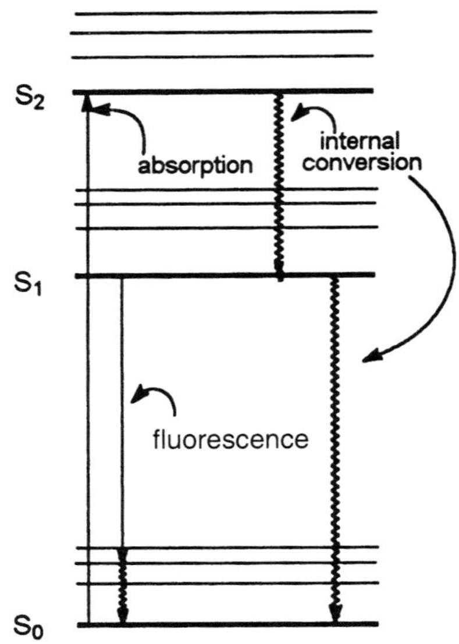

Figure 1.2. Internal conversion and fluorescence of a singlet excited state.

S_3, ...) to S_1 is so rapid that this process generally occurs before emission of light (fluorescence) or photochemical reaction can take place. The observation that only the lowest excited singlet state (S_1) [or, when triplet states are involved, the lowest triplet state (T_1)] is a probable candidate for reaction is an expression of Kasha's rule.[10,11] Instances of emission or reaction from higher excited states are known but are not common.[12] When an excited singlet state returns to the ground state, it can do so either by emitting a photon (fluorescence) or by internal conversion (Figure 1.2).

Triplet States

One possible change for a molecule in an excited singlet state is for it to undergo conversion into a triplet state; that is, the excited molecule can experience intersystem crossing (Figure 1.3). The triplet state, which has two unpaired electrons and is always lower in energy than the corresponding singlet state, cannot be readily produced by direct absorption of a photon because the probability of a spin inversion during absorption is extremely low. To be formed efficiently, triplet states must be produced indirectly either from the singlet state or by energy transfer. (Energy transfer is discussed in the section "Electronic Energy Transfer" later in this chapter.) The rate constant for intersystem crossing (e.g., $S_1 \rightarrow T_1$) is inversely related to the difference in energy between singlet and triplet states. This means that when a small energy difference exists, as it does in many aldehydes and ketones, intersystem crossing to produce a triplet state is fast and efficient. Intersystem crossing is slower and

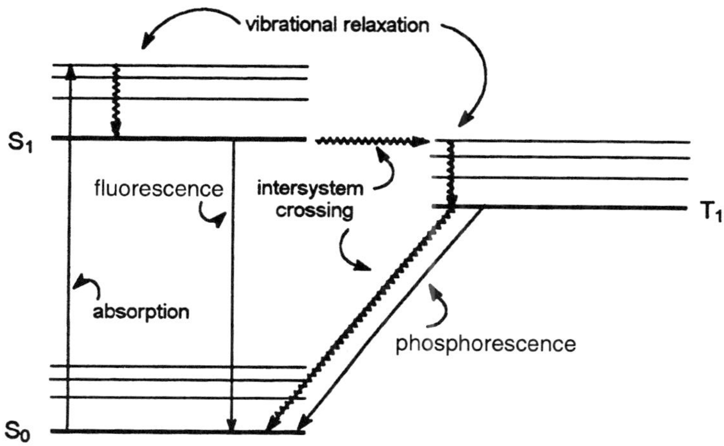

Figure 1.3. Intersystem crossing of a singlet excited state to a triplet state.

less efficient for molecules, such as alkenes, that have a larger singlet–triplet energy difference.

Efficiency of Reaction

The electronic configuration of an excited state and the excited-state reaction efficiency are two critical factors in determining product formation in a photochemical reaction. The electronic configuration controls the type of reaction that is possible for an excited molecule, and reaction efficiency determines the extent to which product formation will occur, that is, the extent to which product formation is competitive with the other processes (e.g., internal conversion and intersystem crossing) available to an excited molecule. Reaction efficiency is expressed in terms of the quantum yield (Φ), which is defined as the number of product molecules formed divided by the quanta of light absorbed.*

The quantum yield for a reaction also can be expressed in terms of rate constants. As an illustration, consider the process outlined in Scheme 1.1. The pathways (and associated rate constants) open to compound A in its excited singlet state (^{1}A) are internal conversion (k_{ic}), product formation (k_r), and fluorescence (k_f). Because every photon absorbed produces one ^{1}A, and every ^{1}A must undergo one of the processes shown in Scheme 1.1, the quantum yield for product formation can be expressed in terms of rate constants, as shown in eq 2.

*It is possible to define other types of quantum yields. For example, the quantum yield for fluorescence is the number of photons emitted by a group of molecules each in a singlet state divided by the quanta of light absorbed, and the triplet quantum yield is the number of triplet states formed divided by the number of photons absorbed.

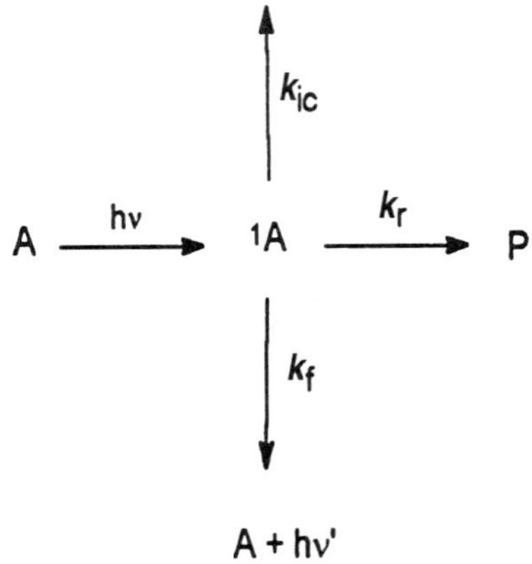

Scheme 1.1

$$\Phi = \frac{\text{number of product molecules formed}}{\text{number of photons absorbed}} = \frac{k_r}{k_{ic} + k_r + k_f}$$

Equation 2

One conclusion that can be drawn from the information in Scheme 1.1 and eq 2 is that a low quantum yield will result if either k_{ic} or k_f is much larger than k_r. Because internal conversion and fluorescence both return the excited starting material to its ground state, a low quantum yield does not necessarily translate into a low chemical yield for a reaction. For example, a reaction with a quantum yield of 0.001 may require a long irradiation time to reach completion (99.9% of the photons absorbed are "wasted" by return of ^{1}A to A). If the only reaction pathway open to the excited state is formation of product P, the chemical yield for that product could be high or even quantitative, depending on product stability.

Excited-State Lifetimes

The measured lifetime (τ) of an excited state is defined as the reciprocal of the sum of the first-order (or pseudo first-order) rate constants (k_i) for all processes that the excited molecule can undergo (eq 3).[13] By way of illustration, the lifetime for the triplet excited state (^{3}A) in the reaction shown in Scheme 1.2 (every photon absorbed produces one ^{3}A) is given by eq 4.

If the quantum yield for product formation (Φ) and the lifetime of the reactive excited state (τ) both are known, the rate constant for reaction (k_r) can be calculated (eq 5). Knowledge of reaction rate constants is critical in mechanistic studies where, for example, the effect of changes in structure on the reactivity of a group of compounds is being investigated. It is necessary to compare the rate constants for product formation to know in what way structural change is increasing or decreasing reactivity.

One direct method for measuring excited-state lifetimes uses the technique of flash photolysis. In a flash-photolysis experiment, a pulse of light with high intensity and short duration produces a high concentration of excited molecules. After the pulse has created the excited molecules, observing their disappearance by an appropriate technique enables experimental determination of the excited-state lifetime. If, for example, the excited state formed is a phosphorescing triplet, its disappearance can be followed by monitoring the phosphorescence of the sample. Alternatively, because many excited compounds (and other transients such as radicals and radical ions) can be detected

$$\tau = \frac{1}{\sum_i k_i}$$

Equation 3

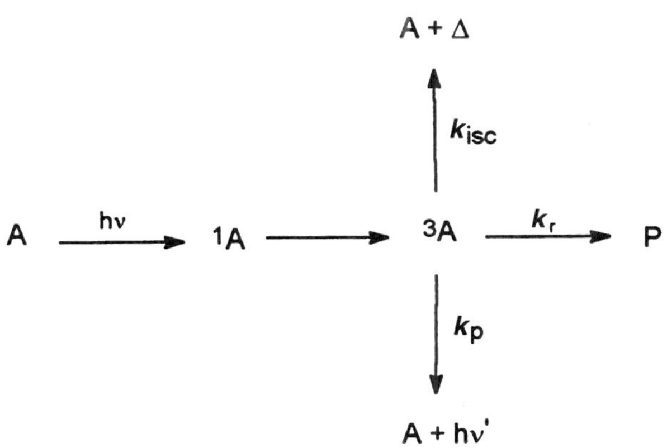

Scheme 1.2

$$\tau = \frac{1}{k_{isc} + k_r + k_p}$$

Equation 4

$$\Phi = k_r \tau = \frac{k_r}{k_{isc} + k_r + k_p}$$

Equation 5

by their visible or UV absorption, a decrease in excited-state concentration can be observed by measuring the amount of light absorbed by the excited compound as a function of time.

As mentioned in the section "Singlet States", higher excited states (S_2, S_3, ..., and T_2, T_3, ...) generally have very short lifetimes because, according to Kasha's rule, they undergo rapid internal conversion ($S_n \rightarrow S_1$, $T_n \rightarrow T_1$). Because two of the major decay pathways for T_1 (phosphorescence and intersystem crossing) involve a spin change, that is, are "spin-forbidden", their rate constants generally are smaller than the comparable decay processes (fluorescence and internal conversion) for S_1. One result of this difference is that T_1 generally has a longer lifetime than the corresponding S_1.

Exciplexes

When an electronically excited molecule A* comes into contact with a polar or polarizable ground-state molecule B, stabilization can occur that leads to formation of the metastable species AB*. Intermediates such as AB* are excited complexes known as exciplexes.[14–16] These complexes can be formed even when A and B have no corresponding ground-state attraction. Exciplexes are involved in many photochemical reactions; for example, when an aldehyde or ketone in an (n,π*) singlet state adds to an alkene to form an oxetane (the Paternò–Büchi reaction, discussed in Chapter 4), exciplex formation is the first step in the process after photon absorption.

Molecules That Contain Heteroatoms

For a saturated compound that contains no heteroatoms, the lowest-energy excited state is (σ,σ*) in nature. If the compound is unsaturated, the lowest excited state is (π,π*). When a heteroatom is incorporated into the molecular framework, the possible lowest excited states increase to include (n,π*) and (n,σ*) states. Identifying the excited state responsible for a reaction can be useful in understanding (and even predicting) the reactivity of the excited compound; however, excited-state identification is sometimes difficult and uncertainty can exist. This type of uncertainty plagued the photochemistry of carbohydrates for many years.

Most of the photochemical reactions of carbohydrates studied since 1965 have involved compounds in which a light-absorbing chromophore has been introduced into the reactant by a chemical reaction such as oxidation or derivatization. Before 1965, however, the photochemical study of carbohydrates concentrated on underivatized compounds. Even though photolysis of these compounds produced photochemical reaction, identification of the light-absorbing chromophore proved to be a controversial task. Phillips has discussed the background information on this controversy in two excellent reviews.[17,18] The basic question concerned whether underivatized carbohydrates have no absorption or very weak absorption at wavelengths greater than 220 nm. A significant problem in resolving this question was that for molecules with very weak absorption, it was difficult to be certain that the absorptivity did not arise from a small amount of a highly absorbing impurity.[19]

For many carbohydrates, the problem was complicated further by their existence in solution as a mixture of equilibrating cyclic and acyclic structures (Scheme 1.3). Even though any acyclic compound containing a carbonyl group would have UV absorption at wavelengths greater than 220 nm, it was not certain that acyclic structures were responsible for significant UV absorption in carbohydrates because the relative amounts of acyclic structures vary widely from compound to compound and can be quite small. D-Glucose, for example, contains 0.024% of a reducible form at 25 °C; this percentage, which would include any aldehydo-D-glucose (**1**, Chart 1.1), is likely to be too small for meaningful light absorption.[20] D-Fructose, on the other hand, contains about 0.8% of the *keto* form (**2**) under comparable conditions, and even though this percentage is low, **2** has been found to be the significant light-absorbing component of the equilibrium mixture at wavelengths longer than 220 nm.[21,22]

Phillips and Barber conducted an extensive study aimed at determining the actual UV absorption of a typical underivatized carbohydrate.[23] As a

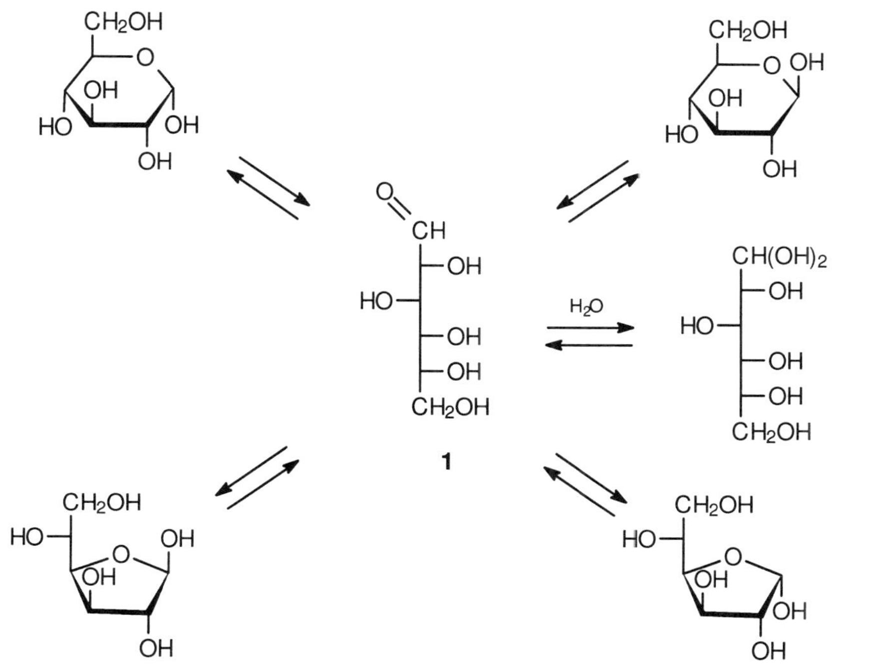

Scheme 1.3

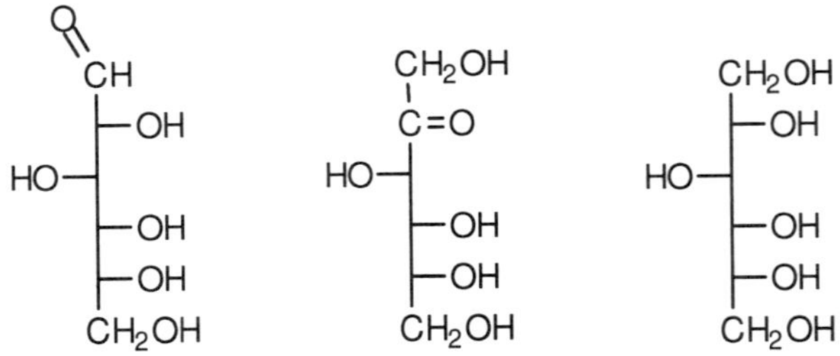

1 2 3

Chart 1.1. Aldehydo, keto, and reduced forms of sugars.

beginning point, they noted that although compounds such as cyclohexane, tetrahydrofuran, and *p*-dioxane do not have UV maxima at wavelengths longer that 200 nm, the presence of an oxygen atom in two of these compounds causes weak but definite absorption extending well into the ultraviolet region. (The $\%T$ (T = transmittance) for cyclohexane, tetrahydrofuran, and *p*-dioxane are 100%, 57%, and 64% at 254 nm, respectively.[24]) This increase in UV absorption from introduction of an oxygen atom into a molecule suggested that saturated, highly oxygenated compounds, such as carbohydrates, could have UV absorption at wavelengths much longer than 220 nm.

Phillips and Barber elected to examine in detail the UV absorption of the sugar alcohol D-glucitol (**3**, Chart 1.1).[23] This compound was selected because it contained no carbonyl group and, therefore, existed as a single structure rather than an equilibrating mixture of compounds. Careful experimentation showed that D-glucitol (**3**) had a definite, although quite weak, absorption in the region from 254 nm to 200 nm. (The molar absorption coefficient was determined to be 0.2 (L mol^{-1} cm^{-1}) at 254 nm, and it rose to a value slightly greater than 3.0 at 200 nm.) Because hexane does not absorb light at all in this region, the conclusion reached was that the absorption of D-glucitol (**3**) was associated with the nonbonded electrons on the oxygen atoms and that the absorption was (n,σ^*) in nature.[23]

Selectivity in Photochemical Reactions

Under the proper conditions, absorption of light can selectively excite one compound from among those present in a reaction mixture. This situation stands in contrast to thermal reactions, where heat normally is supplied indiscriminately to all the compounds present in solution (starting material, solvent, catalysts, products). Although it is difficult to synthesize heat-sensitive compounds in thermal reactions, this difficulty does not extend to photochemical reactions because selective light absorption may be used to activate only the desired reactant. When the product forms, it is in a solution at or below room temperature, where a thermally labile substance is more likely to be stable. If, however, the photoproduct also absorbs the incident light, its yield may be reduced by further photochemical reaction.

Because various groups within a molecule have different wavelengths at which they absorb light, it is possible to selectively excite a particular chromophore within a molecule. Because this excitation often remains highly localized within the absorbing chromophore, selective activation of a particular portion of a molecule can occur. This excitation makes possible selective reaction involving the excited chromophore. Selective excitation and reaction of a particular chromophore within a molecule is the principle upon which photodeprotection reactions are based.

Electronic Energy Transfer

Sensitization of Reactions

Turro has defined electronic energy transfer as "the radiationless one-step transfer of electronic excitation, which is initially localized on a donor (sensitizer) molecule, D^*, to a second qualified acceptor (quencher) molecule, A_0, which is initially in its ground electronic state."[25] This energy exchange is summarized in eq 6. When an excited compound is formed by energy transfer, the process is known as sensitization. Sensitization occurs readily if energy transfer is sufficiently exothermic, and the lifetime of the sensitizer is long enough for the sensitizer to encounter the acceptor molecule.

Because only singlet excited states can be formed efficiently by direct absorption of light, population of triplet states must take place indirectly. The two pathways that can lead efficiently to a triplet state are (1) intersystem crossing from an excited singlet state, and (2) energy transfer from another molecule in its triplet state (i.e., triplet–triplet energy transfer). Although singlet states also can be formed indirectly by energy transfer, singlet–singlet energy transfer is less common than triplet–triplet energy transfer because, in general, singlet states have much shorter lifetimes than triplet states and thus are more likely to return to the ground state before energy transfer can occur.

Because selection of an appropriate sensitizer is essential for effective energy transfer, it is useful to consider some desirable sensitizer characteristics. One way of doing this is to list some attributes of an ideal sensitizer, specifically, one useful for triplet–triplet energy transfer. An ideal triplet sensitizer is one that has a high enough triplet energy so that it can sensitize all compounds of interest. Its rate of intersystem crossing is sufficiently rapid that every singlet state is converted to a triplet state; that is, the formation of triplet states from singlet states is 100% efficient. The ideal sensitizer itself has a long enough lifetime for effective energy transfer to occur even in dilute solutions. Finally, this

$$D^* + A_0 \rightarrow A^* + D_0$$

Equation 6

hypothetical compound only transfers energy and does not otherwise participate in photochemical reactions.

Although no real molecule is an ideal sensitizer, some are very good. Most of the ones that are commonly used are aromatic carbonyl compounds; among these, benzophenone is often selected. Benzophenone, like many aromatic ketones, is converted into its lowest triplet state [E_t = 69 kcal/mole (287 kJ/mole)] rapidly and efficiently following excitation.[25,26] Its lowest triplet state has a relatively high energy and long lifetime, so that it is an effective sensitizer for many compounds. It is far from ideal, however, because there are many molecules that the triplet state of benzophenone is not sufficiently energetic to sensitize. Excited benzophenone also reacts with some compounds (e.g., hydrogen atom donors and some alkenes) rather than functioning as a sensitizer; in fact, benzophenone is used more often in carbohydrate chemistry to generate radicals by hydrogen abstraction than to function as a sensitizer.

Quenching of Reactions

A powerful method for obtaining information about excited states and their reactivity requires deliberately adding to a photochemical reaction mixture a substance that can accept energy from an excited compound. Valuable information can be obtained if a series of experiments is conducted in which the added substance, the quencher, is present in different concentrations. The importance of such experiments can be appreciated by considering an example in which compound A reacts to give a product P by following the pathway outlined in Scheme 1.4.

The process described in Scheme 1.4 assumes a sequence of events in which each absorbed photon generates an excited singlet state (^1A) that then undergoes intersystem crossing with 100% efficiency to give a triplet state (^3A). If no quencher (Q) is present, the quantum yield for reaction (Φ_0) can be expressed, as in eq 7, in terms of the rate constants k_r (product formation), k_p (phosphorescence), and k_{isc} (intersystem crossing). In the presence of the quencher, Q, the quantum yield is given by eq 8, which contains k_q (the rate constant for bimolecular quenching) and [Q] (the quencher concentration). Taking the reciprocal of each side of eq 8 gives a new equation (eq 9), in which there is a linear relationship between $1/\Phi$ and [Q]; therefore, if a triplet state is being quenched, a plot of $1/\Phi$ versus [Q] will produce a straight line with a positive slope (Figure 1.4). The existence of such a linear relationship is excellent evidence for a single reactive species. (Another way in which this information is often presented is by dividing eq 7 by eq 8 to give eq 10, from which a plot of

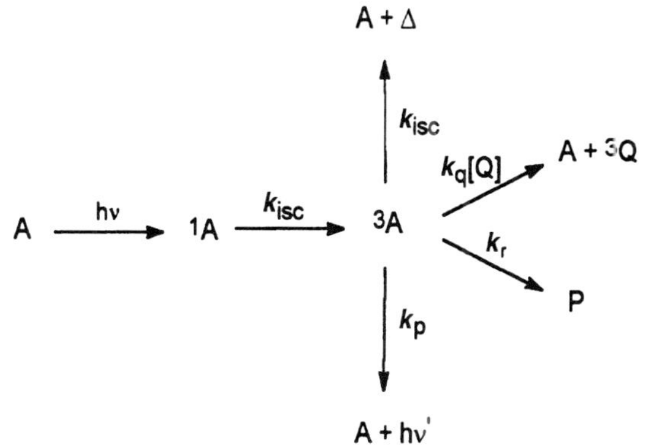

Scheme 1.4

$$\Phi_0 = \frac{k_r}{k_r + k_p + k_{isc}}$$

Equation 7

$$\Phi = \frac{k_r}{k_r + k_p + k_{isc} + k_q[Q]}$$

Equation 8

$$\frac{1}{\Phi} = \frac{k_r + k_p + k_{isc}}{k_r} + \frac{k_q}{k_r}[Q]$$

Equation 9

$$\frac{\Phi_0}{\Phi} = 1 + \frac{k_q}{k_r + k_p + k_{isc}}[Q]$$

Equation 10

Φ_0/Φ versus [Q], known as a Stern–Volmer plot, is linear if a single reactive species is being quenched.) Because the slope of the line in Figure 1.4 is k_q/k_r, the rate constant for product formation can be determined if the slope is measured and if k_q is known. If the quencher has been selected so that energy transfer is exothermic by at least several kilocalories per mole, then every encounter between ^{3}A and Q results in energy transfer. Under these conditions, k_q will be equal to the rate of diffusion in the reaction solvent. Because rates of diffusion in typical reaction solvents are known (e.g., 1.9×10^{10} (acetonitrile), 1.0×10^{10}

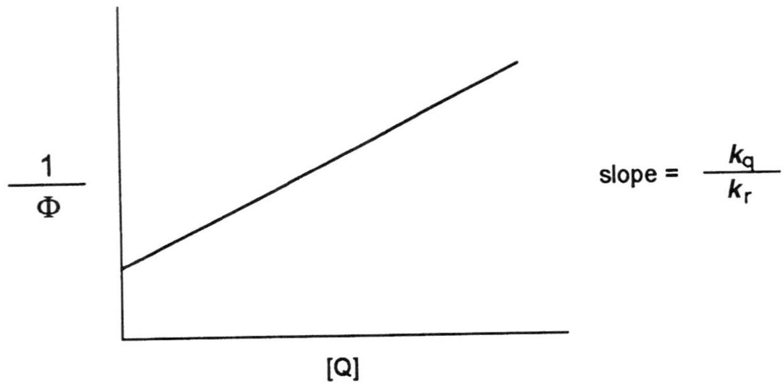

Figure 1.4. A Stern–Volmer plot.

(benzene), 6.5 × 10^9 (water), and 4.6 × 10^6 (glycerol) L mol^{-1} s^{-1}),[27] all the information needed is available for determining the value for the rate constant for product formation (k_r).

As mentioned in the section "Excited-State Lifetimes", determining the rate constant for a reaction is a powerful tool for studying reaction mechanisms. If a mechanistic proposal for a photochemical reaction is to be tested by examining the effect of changes in reaction conditions or reactant structure, the rate constants for reaction are the quantities that need to be compared. Quantum yields and chemical yields are much easier to obtain, but they are not necessarily good guides to the relative reactivity of the excited states involved. As can be seen from eq 7, the relationship between the quantum yield and the rate constant for product formation is determined, in part, by the values of k_p and k_{isc}. Because these rate constants also may change with changes in reactant structure or reaction conditions, no correlation between the quantum yield and the rate constant for product formation necessarily exists.

Electron Transfer

Excitation of an electron in a molecule from a bonding (or nonbonding) to an antibonding molecular orbital creates an excited compound that is both a more powerful oxidizing and reducing agent than the compound in its ground state. The excited molecule is a stronger reducing agent (electron donor) because an electron in an antibonding orbital is more easily transferred to another molecule (Figure 1.5) than an electron in a bonding molecular orbital (Figure 1.6). Likewise, the excited molecule also is a more effective oxidizing agent (electron acceptor) than its ground-state counterpart because the excited molecule has a vacancy in a bonding molecular orbital into which an

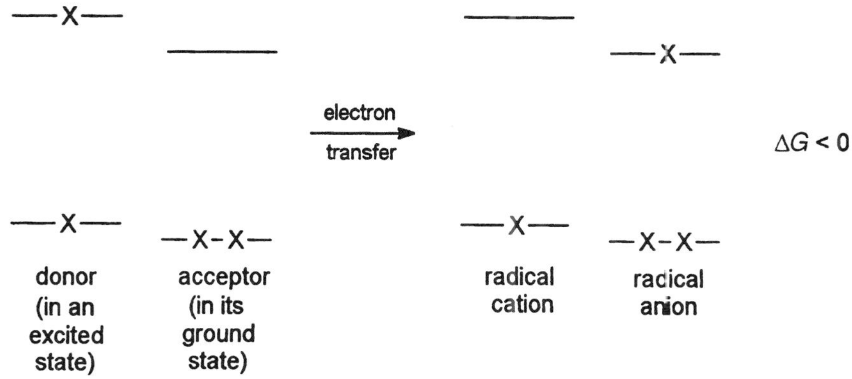

Figure 1.5. An excited molecule acting as an electron donor.

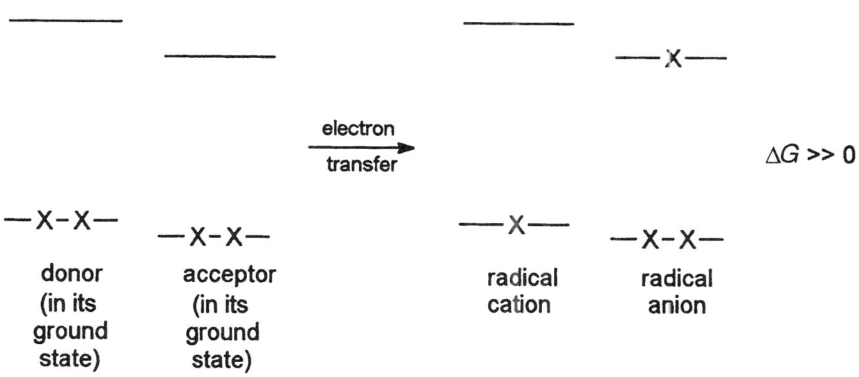

Figure 1.6. A ground-state molecule acting as an electron donor.

electron can be transferred (Figure 1.7). Any electron transferred into a ground-state molecule would need to enter an antibonding molecular orbital (Figure 1.8). When excited states are viewed in this way, photochemical electron transfer can be seen as a likely process.

The feasibility of electron transfer between a molecule in its ground state and one in its excited state is primarily a matter of thermodynamics.[28] The free energy of electron transfer can be estimated from the Weller equation,[29,30] which is given in a simplified form in eq 11.[31] All of the quantities in eq 11, except ΔE_{coul}, can be measured directly (or obtained from the literature). The Coulombic interaction energy, ΔE_{coul}, can be obtained from the Born equation.[30,31] References 32 and 33 contain some examples involving carbohydrates in which the Weller equation has been used to estimate whether electron transfer should occur. Regarding the Weller equation, Mattay has briefly summarized its significance in the following way.[34] "The Weller equation therefore allows an estimate, to a first approximation, as to whether elec-

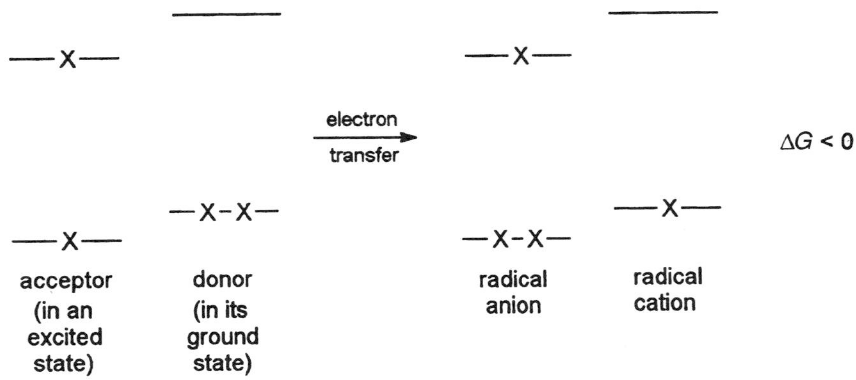

Figure 1.7. An excited molecule acting as an electron acceptor.

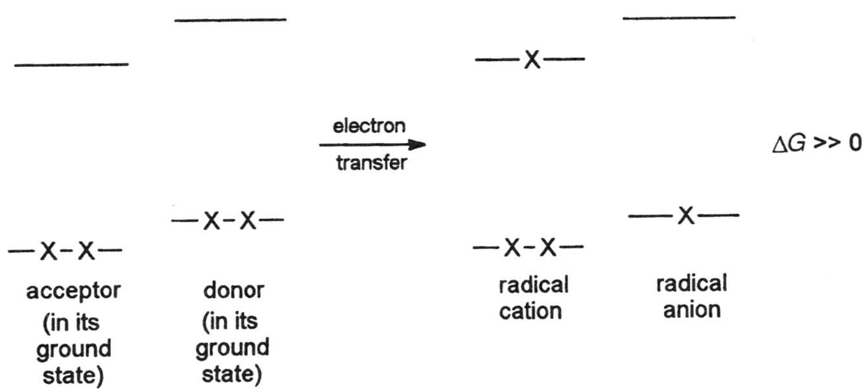

Figure 1.8. A ground-state molecule acting as an electron acceptor.

$$\Delta G = E_{1/2}^{\mathrm{ox}}(\mathrm{D}) - E_{1/2}^{\mathrm{red}}(\mathrm{A}) - \Delta E_{\mathrm{ecit}} + \Delta E_{\mathrm{coul}}$$

$E_{1/2}^{\mathrm{ox}}(\mathrm{D})$ = the oxidation potential of the donor]

$E_{1/2}^{\mathrm{red}}(\mathrm{A})$ = the reduction potential of the acceptor

ΔE_{ecit} = the excitation energy of the electronically excited species

ΔE_{coul} = the Coulombic interaction energy of the radical ions at a defined distance in a given solvent

Equation 11

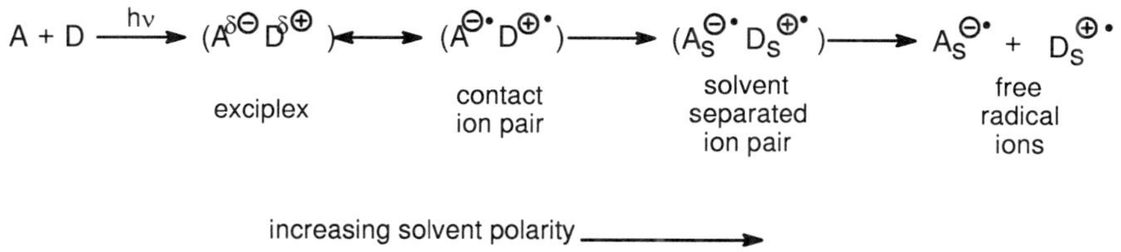

Scheme 1.5

tron transfer within a donor–acceptor system is thermodynamically allowed (ΔG negative, exergonic process) or not (ΔG positive, endergonic process). In addition, it shows that the electron-transfer process can be decisively influenced by the polarity of the solvent as well as by the electronic properties of the reacting species."

Because electron transfer between neutral molecules creates radical ions, this transfer is favored by polar solvents because they stabilize the developing charges. In fact, depending upon the polarity of the solvent, a number of possibilities exist for the extent of electron transfer and separation of the radical ions produced. Mattay has summarized these possibilities as shown in Scheme 1.5.[35] Electron transfer, however, can be a rapidly reversible process; therefore, even if electron transfer takes place, it does not ensure that reaction will be observed because back electron transfer can occur before any other reaction takes place. Considering this fact, Fox has pointed out that because the possibility of back electron transfer exists, rapid reactions are those with the best opportunity to be observed.[36]

Further Reading about Photochemistry

The review articles and books cited in this chapter provide excellent sources of further information about photochemistry. Citations to some additional books that contain very fine explanations of organic photochemistry at several different levels are given in references 37 through 40.

References

1. Bishop, M. B.; Bishop, C. B. *J. Chem. Ed.* **1987**, *64*, 302.
2. Pelzer, R.; Jütten, P.; Scharf, H.-D. *Chem. Ber.* **1989**, *122*, 487.
3. Piva, O.; Pete, J.-P. *Tetrahedron:Asymmetry* **1992**, *3*, 759.
4. Tochtermann, W.; Schlösser, U.; Ott, F.; Popp, B.; Sdunnus, N.; Snatzke, G.; Peters, E.-M.; Peters, K.; von Schnering, H. G. *Chem. Ber.* **1992**, *122*, 1733.
5. Klausener, A.; Beyer, G.; Leismann, H.; Scharf, H.-D.; Müller, E.; Runsink, J.; Görner, H. *Tetrahedron* **1989**, *45*, 4989.
6. Turro, N. J. *Molecular Photochemistry;* W. A. Benjamin: New York, 1965; p 2.

7. Binkley, R. W. *Modern Carbohydrate Chemistry;* Marcel Dekker: New York, 1988; p 7.
8. Fourrey, J.-L.; Jouin, P. *Tetrahedron Lett.* **1977**, 3397.
9. Turro, N. J.; Schuster, G. *Science (Washington, D. C.)* **1975**, *187*, 303.
10. Turro, N. J. *Modern Molecular Photochemistry;* Benjamin/Cummings: Menlo Park, CA, 1978; p 103.
11. Kasha, M. *Disc. Faraday Soc.* **1950**, *9*, 14.
12. Turro, N. J.; Ramamurthy, V.; Cherry, W.; Farneth, W. *Chem. Rev.* **1978**, *78*, 125. This article is concerned with reactions from upper excited states.
13. Turro, N. J. *Molecular Photochemistry;* W. A. Benjamin: New York, 1965; p 55.
14. Mattes, S. L.; Farid, S. *Science (Washington, D. C.)* **1984**, *226*, 917.
15. Caldwell, R. A.; Creed, D. *Acc. Chem. Res.* **1980**, *13*, 45.
16. Mattes, S. L.; Farid, S. *Acc. Chem. Res.* **1982**, *15*, 80.
17. Phillips, G. O. *Adv. Carbohydr. Chem.* **1963**, *18*, 9.
18. Phillips, G. O. In *The Carbohydrates;* Pigman W.; Horton, D., Eds.; Academic: New York, 1980, Vol. IB, p 1217.
19. McKelvey, R. D. *Carbohydr. Res.* **1975**, *42*, 187.
20. Pigman W.; Anet, E. F. L J. In *The Carbohydrates*; Pigman W.; Horton, D., Eds.; Academic: New York, 1972, Vol. IA, p 165.
21. Funcke, W.; von Sonntag, C.; Triantaphylides, C. *Carbohydr. Res.* **1979**, *75*, 305.
22. Hayward, L. D.; Angyal, S. J. *Carbohydr. Res.* **1977**, *53*, 13.
23. Phillips, G. O.; Barber, P. *J. Chem. Soc.* **1963**, 3990.
24. Murov, S. L.; Carmichael, I.; Hug, G. L. *Handbook of Photochemistry;* 2nd ed.; Marcel Dekker: New York, 1993; p 279.
25. Turro, N. J. *J. Chem. Ed.* **1966**, *43*, 13.
26. Murov, S. L.; Carmichael, I.; Hug, G. L. *Handbook of Photochemistry;* 2nd ed.; Marcel Dekker: New York, 1993; p 56. This book contains an extensive listing of triplet state energies.
27. Murov, S. L.; Carmichael, I.; Hug, G. L. *Handbook of Photochemistry;* 2nd ed.; Marcel Dekker: New York, 1993; p 208. This book contains a list of the rates of diffusion in a large number of solvents.
28. Maslak, P. In *Photoinduced Electron Transfer V;* Mattay, J., Ed.; Springer-Verlag: New York, 1993; p 1.
29. Rehm, D.; Weller, A. *Isr. J. Chem.* **1970**, *8*, 259.
30. Weller, A. *Z. Phys. Chem. (Munich)* **1982**, *133*, 93.
31. Mattay, J. *Angew. Chem. Int. Ed. Engl.* **1987**, *26*, 825.
32. Masnovi, J.; Koholic, D. J.; Berki, R. J.; Binkley, R. W. *J. Am. Chem. Soc.* **1987**, *109*, 2851.
33. Nishida, A.; Hamada, T.; Yonemitsu, Y. *Chem. Pharm. Bull.* **1990**, *38*, 2977.
34. Mattay, J. *Angew. Chem. Int. Ed. Engl.* **1987**, *26*, 826.
35. Mattay, J. *Chem. Rev.* **1993**, *93*, 99.
36. Fox, M. A. *Advan. Photochem.* **1986**, *13*, 237.
37. Coyle, J. D. *Introduction to Organic Photochemistry;* John Wiley & Sons: New York, 1986.
38. Kagan, J. *Organic Photochemistry: Principles and Applications;* Academic: San Diego, CA, 1993.
39. Kopecký, J. *Organic Photochemistry: A Visual Approach;* VCH: New York, 1992.
40. Gilbert, A.; Baggott, J. *Essentials of Molecular Photochemistry;* CRC: Boca Raton, FL, 1991.

2

Norrish Type I Reaction: An α-Cleavage Process

Mechanism of the Norrish Type I Reaction

Because carbonyl compounds undergo such a variety of photochemical reactions, they occupy a prominent position in organic photochemistry. It is not surprising, therefore, that considerable effort has been invested in understanding carbonyl photochemistry. One early result of this effort was a proposal by Zimmerman for a valence-bond representation of an (n,π^*) excited state of a carbonyl compound (Figure 2.1).[1] In this picture of the carbonyl group, electrons in different orbitals are represented with different symbols; that is, an electron in either a π or π^* orbital is a dot, an electron in an oxygen orbital with considerable *s* character is an open circle, and an electron in an oxygen orbital with considerable *p* character is a "y". The importance of the Zimmerman proposal is that it allows one to use valence bond structures and reasoning to mechanistically analyze photochemical reactions of various aldehydes and ketones. Because the lowest excited singlet state and lowest triplet state for nonconjugated carbonyl compounds are both (n,π^*) in nature, the Zimmerman picture can be used to advantage in representing the excited states for many carbonyl compounds, particularly those undergoing the type of reaction of interest in this chapter, that is, the Norrish Type I reaction.

One idea that arises from the Zimmerman representation of an excited carbonyl group is that an aldehyde or ketone in its lowest (n,π^*) excited singlet or triplet state should have reactivity in common with an alkoxy radical because both intermediates have an orbital on oxygen containing a single electron (Scheme 2.1).[2] Because both intermediates undergo α-cleavage and hydrogen abstraction reactions, the anticipated similar reactivity does exist.[3,4] (The products from α-cleavage and hydrogen-abstraction reactions (Scheme 2.1) are ground-state species for which the "•, ○, y" notation normally is not used.) The

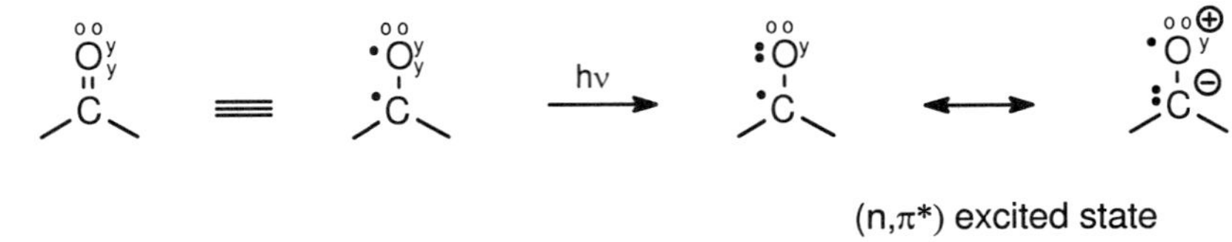

y = electron in an orbital with considerable p character
o = electron in an orbital with considerable s character
• = electron in a π or π* orbital

Figure 2.1. The Zimmerman representation of an excited carbonyl group.

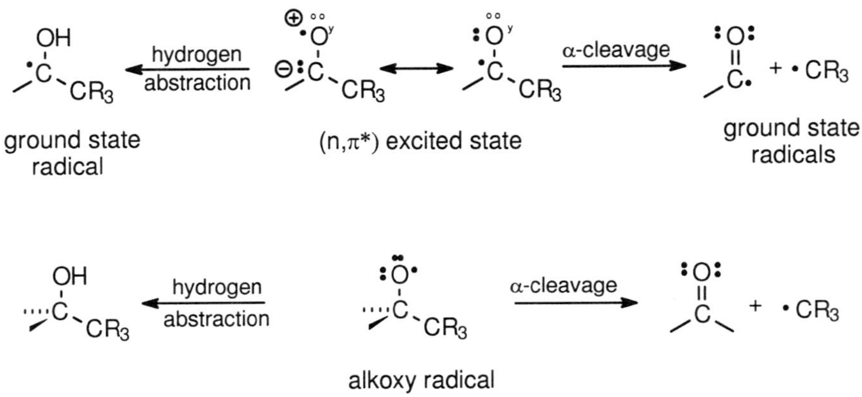

Scheme 2.1

α-cleavage that occurs in an excited carbonyl compound is often called a Type I or a Norrish Type I process.* The radical pair produced by this α-cleavage can undergo the traditional radical reactions of disproportionation, combination, and, because an acyl radical is involved, decarbonylation (Scheme 2.2). When the carbonyl compound is cyclic and, therefore, forms a diradical, this list of possible reactions increases to include ring expansion (Scheme 2.3). Among all these reactions, only decarbonylation necessarily leads to further change because it alone produces a new radical pair that also can undergo disproportionation and combination reactions. Except for the decarbonylation process, the structures of the photoproducts depend upon the direction of α cleavage, that is, which of the two carbonyl-to-α-carbon bonds is broken when the excited compound reacts. This topic is discussed in the next section.

Although Type I reaction has been observed from (n,π*) singlet and triplet states, in general, the triplet state is more reactive than the singlet.[6] The use of correlation diagrams has shown that this difference in reactivity is a natural con-

*Although sometimes this process also is called a Type I reaction, Type I cleavage, or a similar name, a more precise terminology has been proposed by Turro.[5]

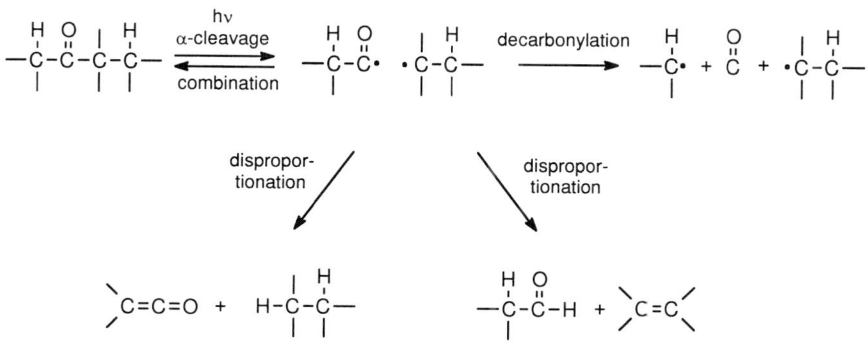

Scheme 2.2

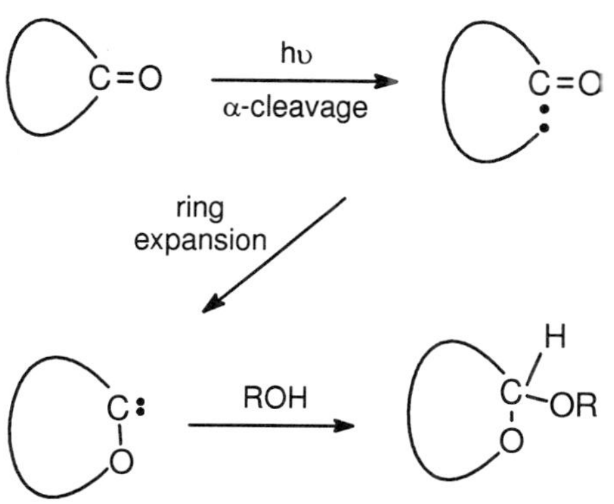

Scheme 2.3

sequence of the need for the singlet state to undergo a forbidden intersystem crossing in order for reaction to occur.[7] Relatively little is known about the identity of reactive excited states of carbohydrates undergoing Type I reaction. In one study, where such information is available, Goodman and Greenberg have shown that α-cleavage occurs from a triplet state.[8] (This work is described in the section "Decarbonylation" later in this chapter.) In another investigation, levoglucosenone was shown by sensitization experiments to react from a triplet state to give the same product as that formed by direct irradiation (see the section "Rearrangement"). These two studies contain the only information available about the identity of excited states involved in Type I reactions of carbohydrates.

Direction of α-Cleavage

Because bond breaking in a Type I reaction will generate the more stable of the two possible radical pairs, the favored direction for α-cleavage depends upon the stability of the radicals produced. In many molecules the direction of cleavage is easily deduced from knowledge of radical stability or by consulting tables of bond dissociation energies. For example, because the *t*-butyl radical is much more stable than the phenyl radical, Type I cleavage of the ketone **1** should be in the direction that produces the *t*-butyl radical (Scheme 2.4).[9] In carbohydrates, however, where nearly every carbon–carbon bond cleavage produces pairs of radicals, each of which is stabilized by an oxygen atom, the direction of cleavage often is not easy to predict. Even deciding the direction of cleavage from the identity of the photoproducts formed can be difficult because, if decarbonylation is taking place, any information about the direction of α-cleavage must come indirectly.

Collins et al. have used epimerization at a chiral α-carbon atom as an indication of the direction of bond breaking; thus, epimerization at C1 in the product **3** indicated cleavage of the C1–C2 bond during photolysis of **2** (Scheme 2.5).[10,11] Although C1 epimerization is a good criterion for cleavage of the C1–C2 bond, lack of epimerization at C3 does not exclude C2–C3 bond breaking because even if the C2–C3 bond were broken, the dioxolane ring containing C3 and C4 could inhibit epimerization.

It is possible to determine the direction of α-cleavage in some compounds by trapping the intermediate diradical (or radical pair) with a spin trap. This technique has been used sometimes to detect carbohydrate radicals,[12] but none of these instances have involved a photochemical α-cleavage reaction. Another possibility for obtaining evidence about α-cleavage reactions is by observation of chemically induced dynamic nuclear polarization (CIDNP). Seifert has used CIDNP to demonstrate that photolysis of D-glyceraldehyde and 1,3-dihydroxyacetone causes α-cleavage followed by recombination.[13]

Reactions Occurring after α-Cleavage

Decarbonylation

Because α-cleavage of an acyl radical is an activated process, radical stability also is instrumental in determining whether carbon monoxide will be lost fol-

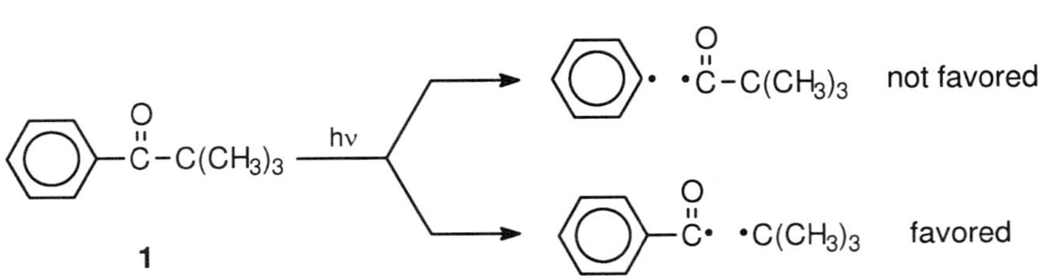

1

Scheme 2.4

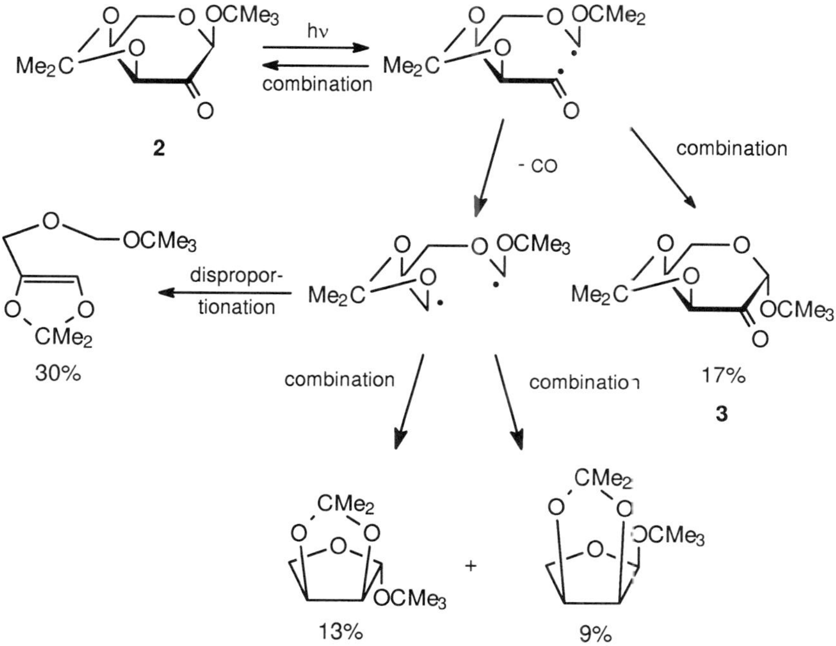

Scheme 2.5

lowing the initial bond homolysis (Scheme 2.2). Photochemical decarbonylation is common in the gas phase, but in solution it is a less likely event.[14,15] Two proposals have been advanced to explain the reason for this difference in reactivity. One possibility is that in solution, where intermediates frequently collide with solvent molecules, the energy needed to break the C–C bond in the acyl radical is transferred to the solvent before reaction can occur.[14,15] Because dissipation of energy by collision takes much longer in the gas phase, time exists for decarbonylation to take place. The second suggestion is that the radical pairs formed by bond homolysis in solution do not immediately separate because they are held together by the solvent molecules ("a cage effect") and, consequently, recombination occurs before CO loss.[15] In contrast, in the gas phase, where there is no solvent to form a cage, the separation of radicals permits the acyl radical to exist long enough to lose carbon monoxide.

Carbon monoxide loss can become an important process in solution if a carbon atom α to the carbonyl group contains a substituent that stabilizes the developing radical center. This condition typically is met easily in carbohydrates (with their multiple hydroxyl or other oxygen-containing groups) and, consequently, photodecarbonylation is a significant reaction pathway for these compounds (Scheme 2.5). The most thorough studies on this topic have been conducted by Collins and co-workers who, besides determining the identity of the compounds formed after carbon monoxide extrusion, identified other Type I reaction products (Tables 2.1 and 2.2). Heyns and co-workers also have investigated the photodecarbonylation of a number of carbohydrates (Tables 2.1 and 2.2); however, in their studies only photoproducts resulting from loss of carbon monoxide were reported.

Photodecarbonylation also is the most important reaction pathway for carbohydrates that exist in an equilibrium between cyclic and acyclic structures (Table 2.3), whenever there is a sufficient concentration of the carbonyl-containing constituent to absorb the light necessary for reaction.* (Equilibrium concentrations of carbonyl-containing constituents are discussed in Chapter 1.) This condition is met by D-fructose, for example, even though the equilibrium concentration of the keto component is only 0.7–0.8% (Scheme 2.6).[36,37] Triantaphylides et al. studied several unprotected sugars, including D-fructose, and found that the photoproducts isolated were formed from CO loss followed by either a disproportionation, combination, or elimination reaction (Scheme 2.7) of the radicals that remained.[26] The elimination reaction, a process that involves loss of a molecule of water from a sugar radical (Scheme 2.8), is a common reaction for radicals derived from unprotected carbohydrates.[38]

Photolysis of acyclic sugars can take a quite different course in the presence of metal ions. Rather than the variety of products produced by irradiation of D-fructose by itself (Scheme 2.7), photolysis of a D-fructose solution containing iron(III) chloride is reported to form only D-erythrose in 80% yield (Table 2.3).[27–30] Although there is conflicting information in the literature[39] about product identity and yield, there is no question that complexation of sugars with metal ions can change the course of their photochemical reactions. Reaction of sugars in the presence of metal ions is discussed in Chapter 11.

Type I reaction can be used for generating radicals for specific purposes. For example, as a part of an investigation of the role of radicals in nucleic acid damage, Goodman and Greenberg needed to generate the radical **5**, a feat accomplished by photolysis of the ketone **4** (Scheme 2.9).[8] These researchers investigated the reactivity of **5** under various conditions and found that it abstracted a hydrogen atom from β-mercaptoethanol to give an anomeric mixture of reduction products (**6α** and **6β**). Also, when molecular oxygen was added to the reaction mixture instead of β-mercaptoethanol, oxidative degradation took place to give the 2′-deoxyribonolactone **7** and uracil (Scheme 2.9). Goodman and Greenberg further determined that piperylene quenching of the reaction of **4** gave a linear Stern–Volmer plot, indicating that the reactive excited state of **4** was a triplet.

---

*Even though most of the compounds listed in Table 2.3 exist as an equilibrium mixture of cyclic and acyclic structures, it is the acyclic structure that is primarily responsible for reaction because it contains the carbonyl chromophore.

Table 2.1. Decarbonylation of 2-keto sugars. The reaction solvent is benzene.

Reactant	*Product and Percent Yield*		*Ref.*
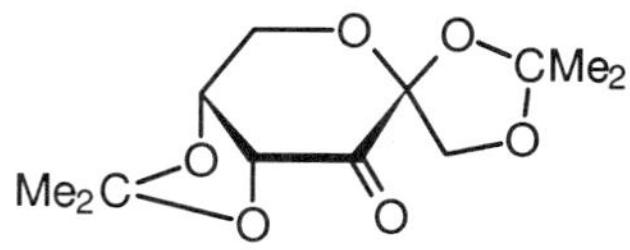	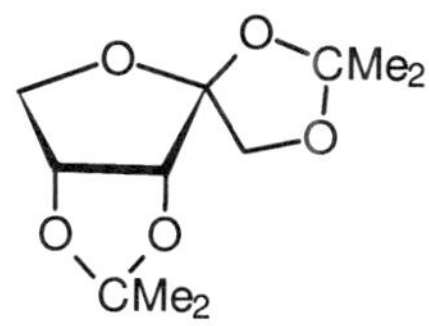 **A** 39%	+ 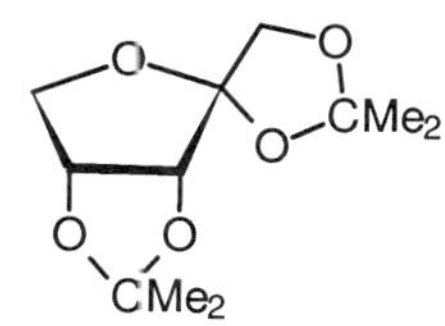 **B** 20%	16
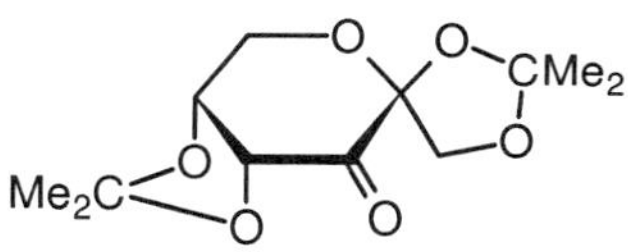	**A** 16%	+ **B** 19%	
	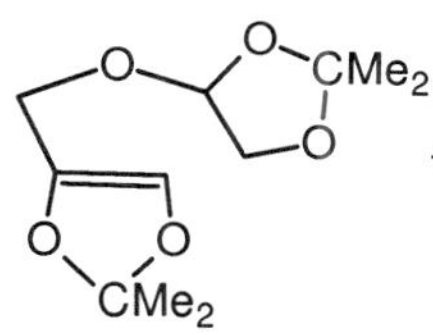 10%	+ 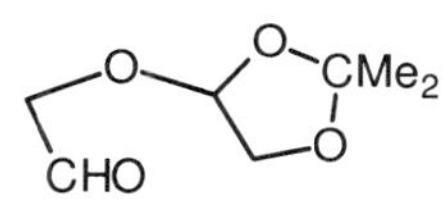 40%	17
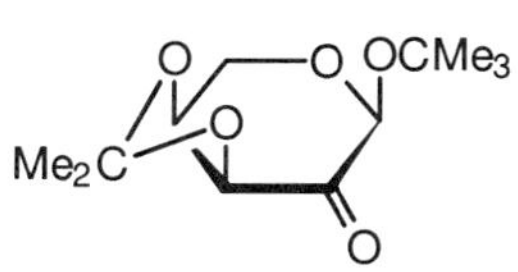	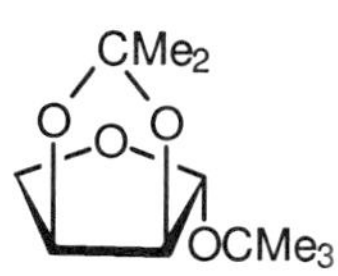 13%	+ 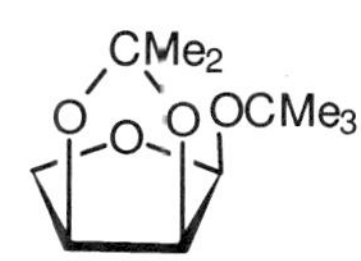 9%	10, 11
	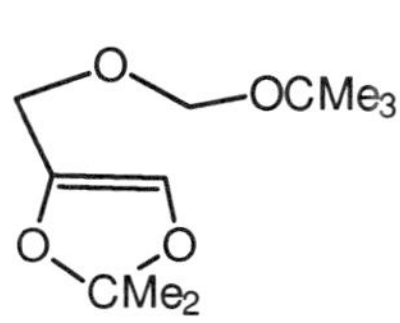 30%	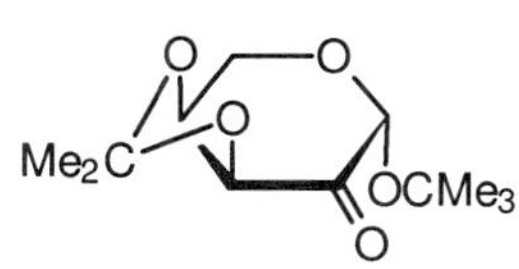 17%	
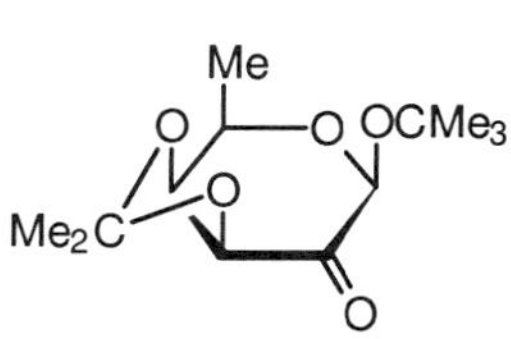	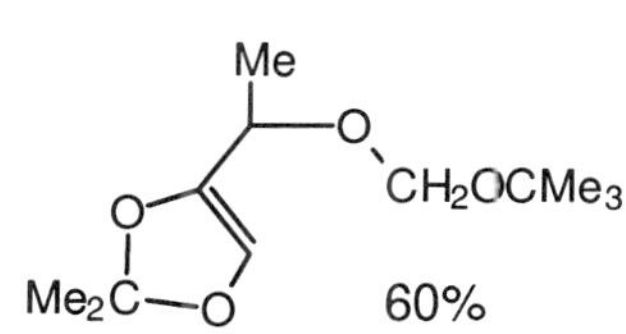 60%		11

Continued on next page

Table 2.1. Decarbonylation of 2-keto sugars—*Continued*

Reactant	*Product and Percent Yield*	*Ref.*
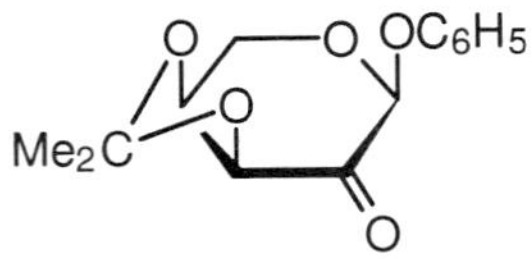	no reaction	11
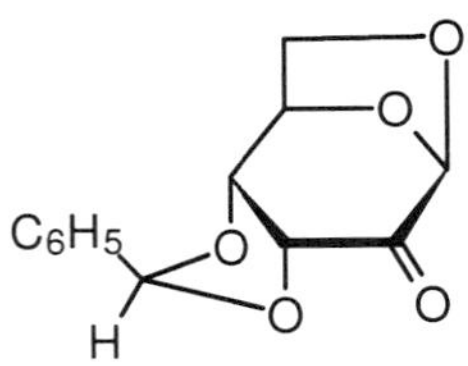	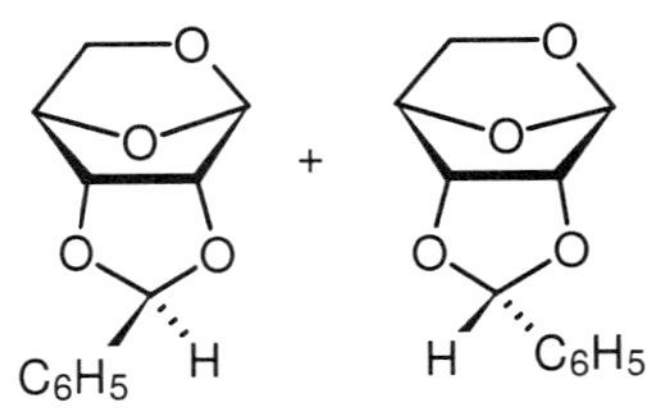 **C** 9% **D** 9%	18[b]
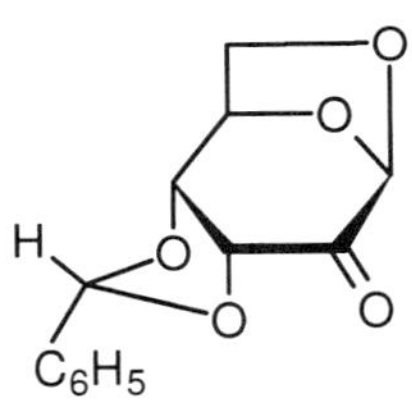	**C** + **D** 8% 8%	18[b]
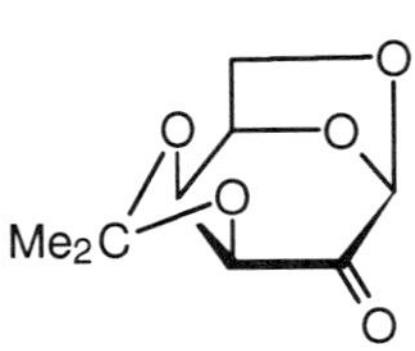	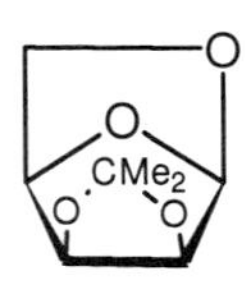 19%	18[b]
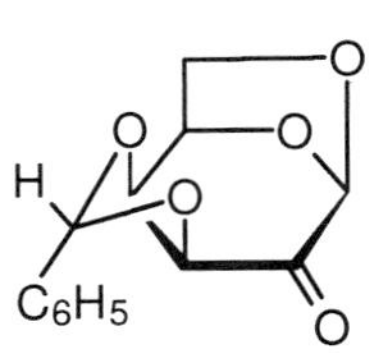	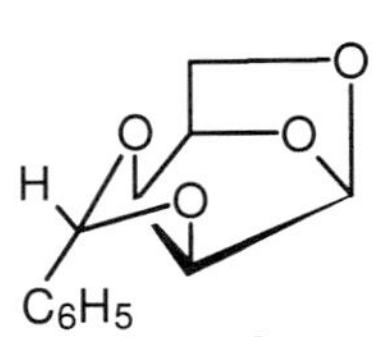 12%	18[b]

Continued on next page

Table 2.1. Decarbonylation of 2-keto sugars—*Continued*

Reactant	*Product*	*Percent Yield*	*Ref.*
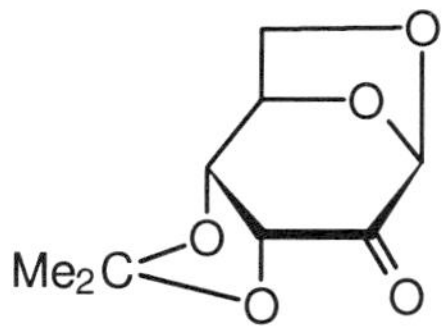	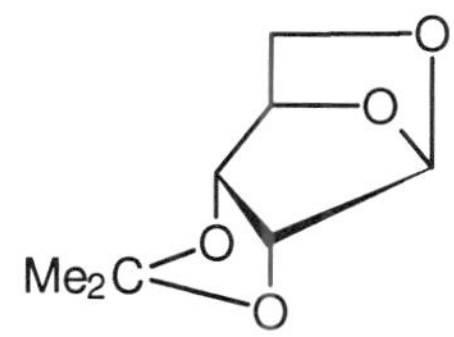	23	18[b]
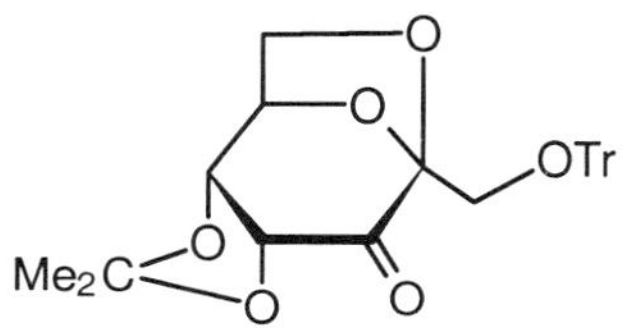	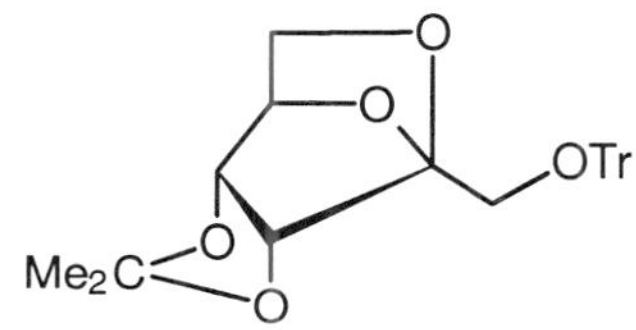	15	18[b]
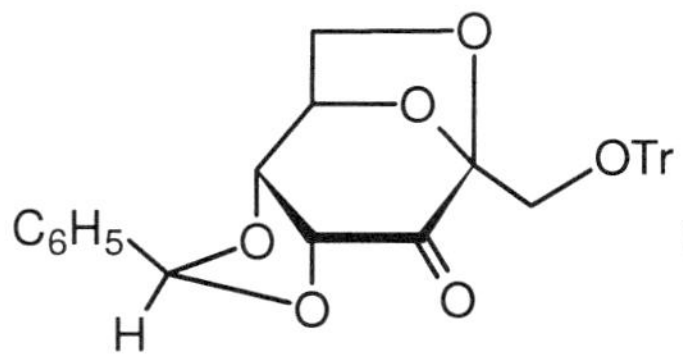	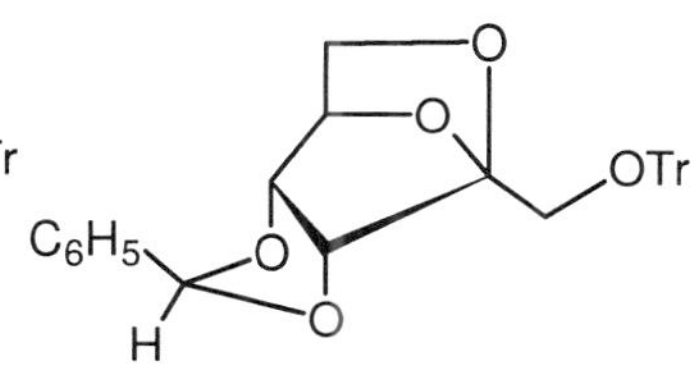	12	18[b]
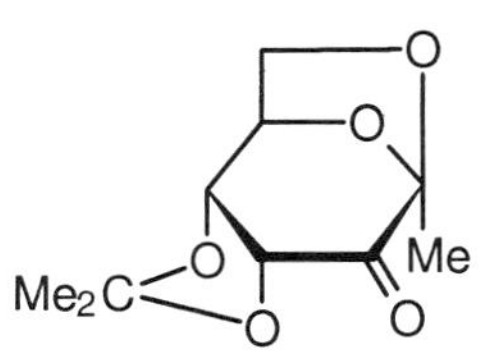	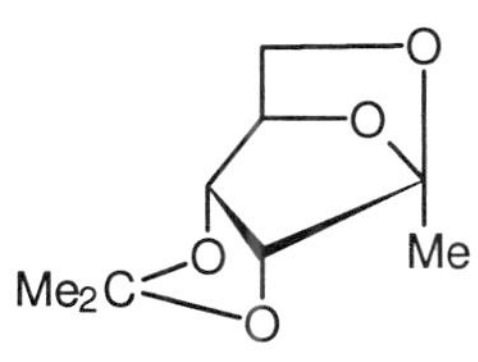	23	18[b]
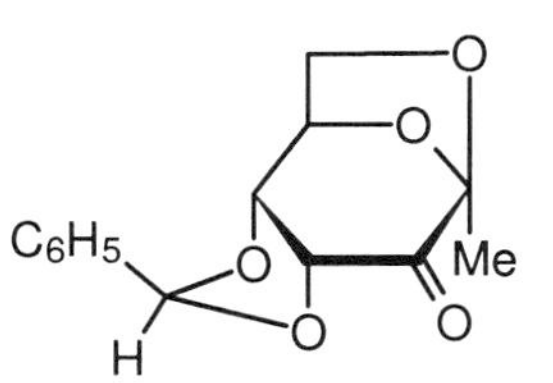	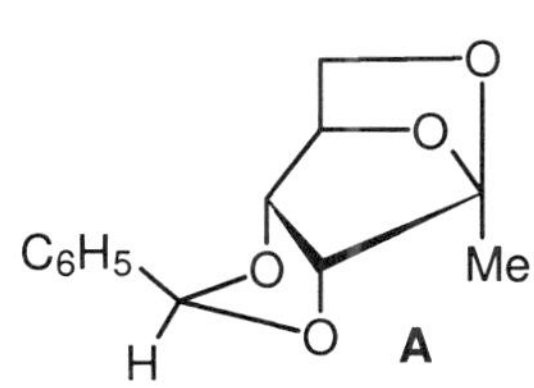 **A**	23	18[b]
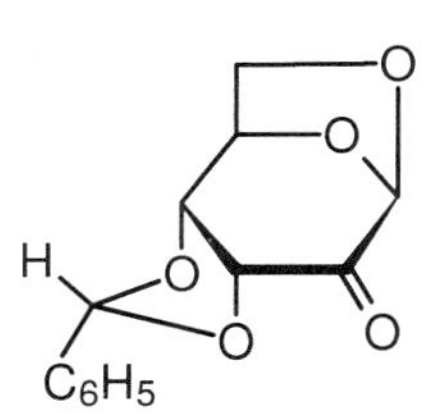	**A** + 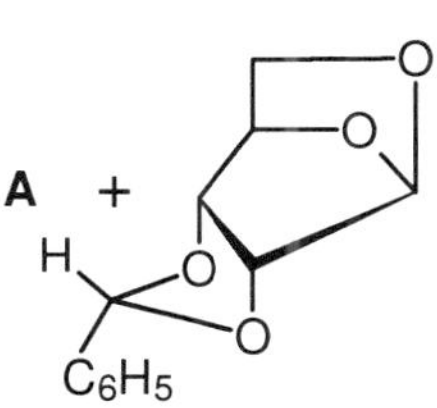	19	18[b]

[a]The reaction solvent was benzene.
[b]No other products were reported.

Table 2.2. Decarbonylation of 4-keto and related sugars. The reaction solvent is benzene unless otherwise noted.[a]

Reactant[a]	*Product and Percent Yield*	*Ref.*
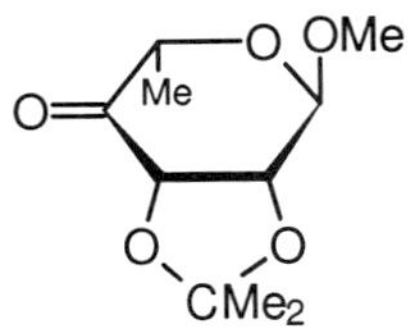	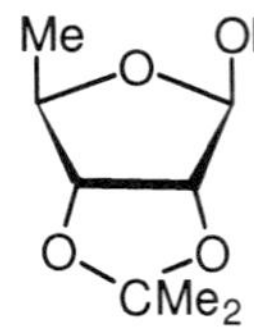 + 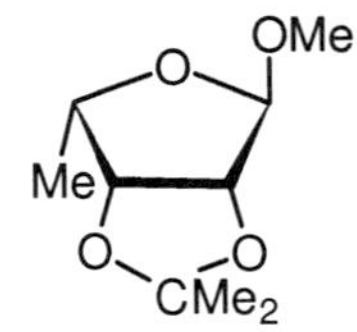 55% 5% **A** **B**	19 20 21
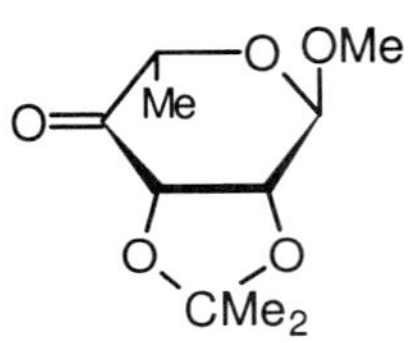	**A** + **B** + 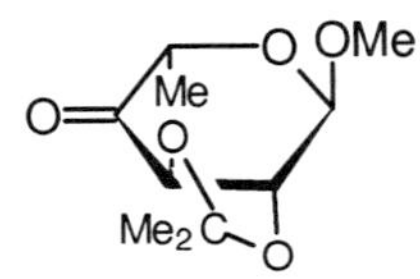 54% 6% 19%	22[b]
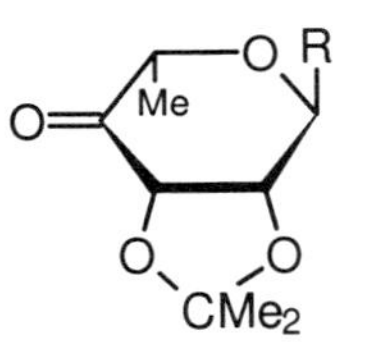 R = $OCH_2C_6H_5$ R = OC_6H_5 R = CH_2CH_3 R = C_6H_5	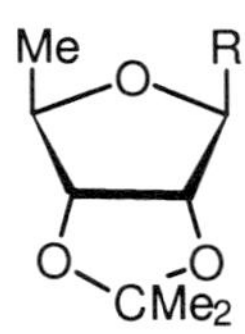 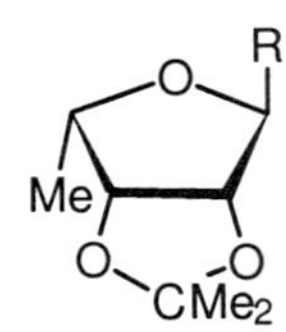 40% c 40% c 30% 40%	23
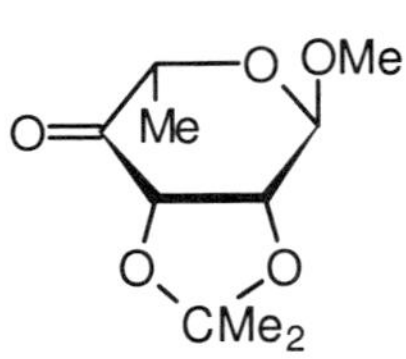	**A** 54% + **B** 6% + 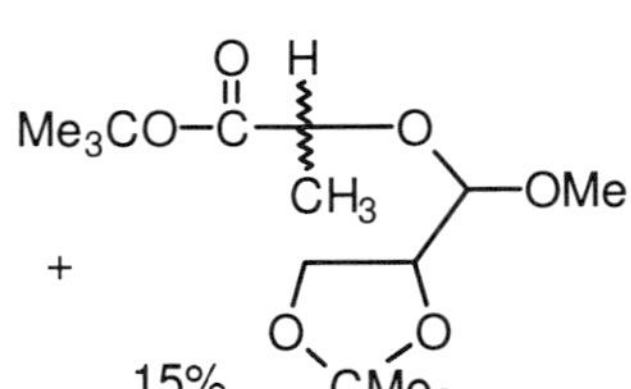 15%	20[d] 22[d] 25[d]

Continued on next page

Table 2.2. Decarbonylation of 4-keto and related sugars—*Continued*

Reactant	*Product*	*Percent Yield*	*Ref.*
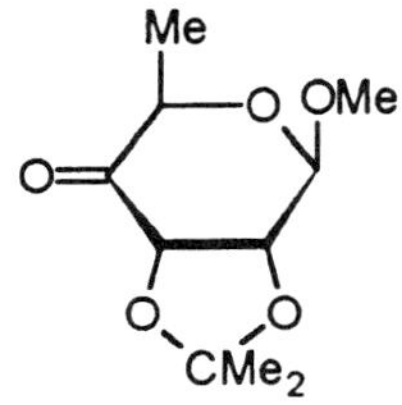	A + B 78% 7%		21
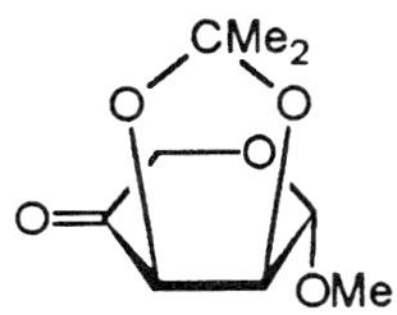	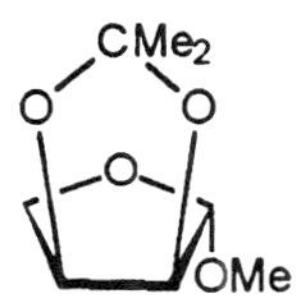	17	21
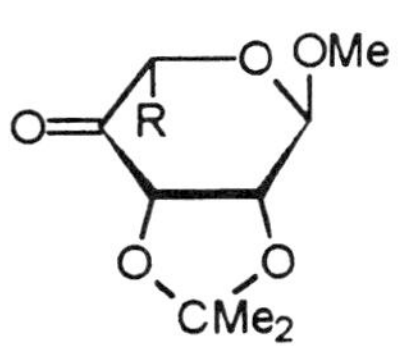 R = CH_2OMs	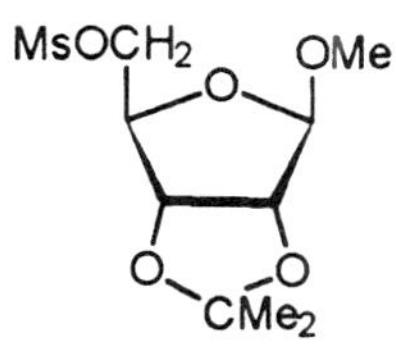	43	23
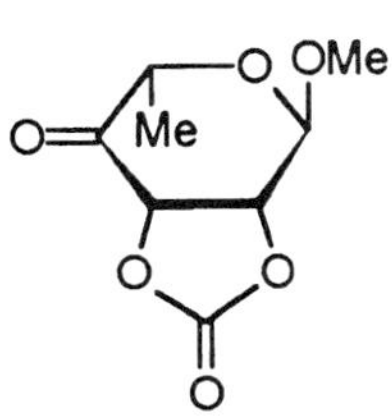	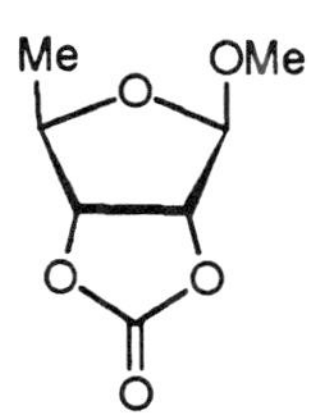	22	23
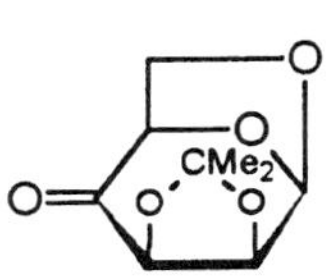	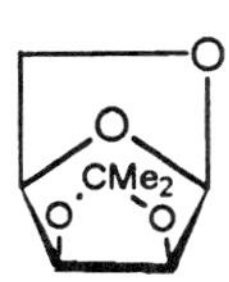	16	18
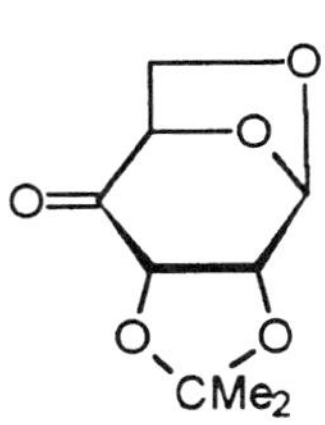	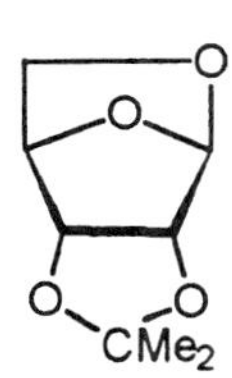	21	18

Continued on next page

Table 2.2. Decarbonylation of 4-keto and related sugars—*Continued*

Reactant	*Product*	*Percent Yield*	*Ref.*
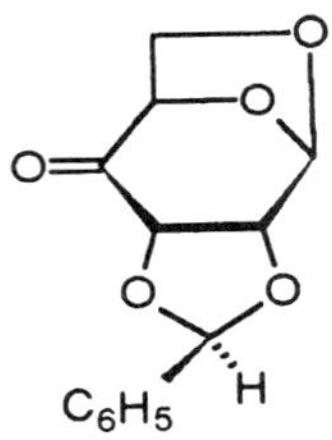	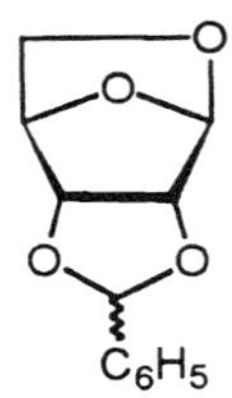	6	18
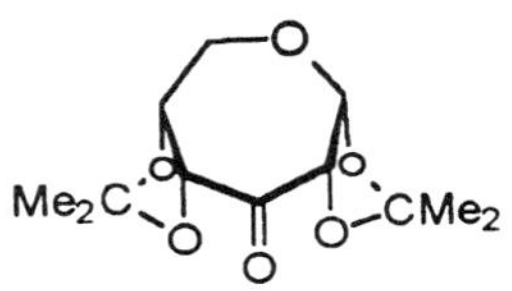	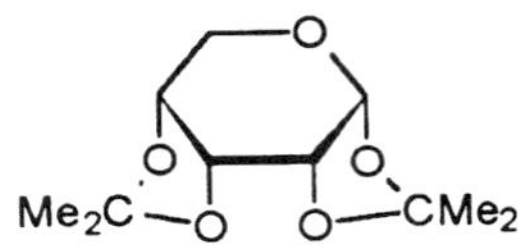	10	24
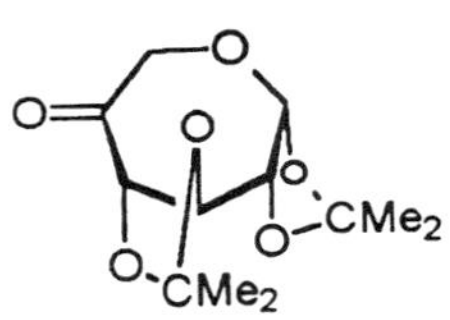	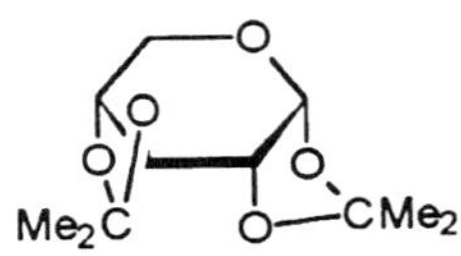	10	24

[a]The reaction solvent was benzene, unless otherwise noted.
[b]The yields of A and B are those reported in reference 20.
[c]These compounds may have been formed, but they were not identified.
[d]$(CH_3)_3COH$ was the reaction solvent.

Table 2.3. Reactions of acyclic carbonyl compounds.
The reaction solvent is water.

Reactant[a]	*Product*	*Ref.*
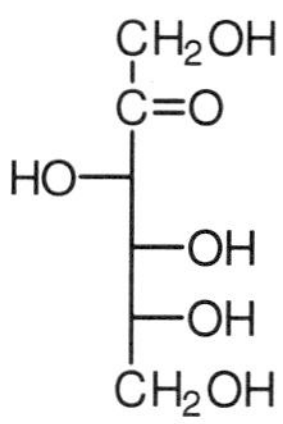	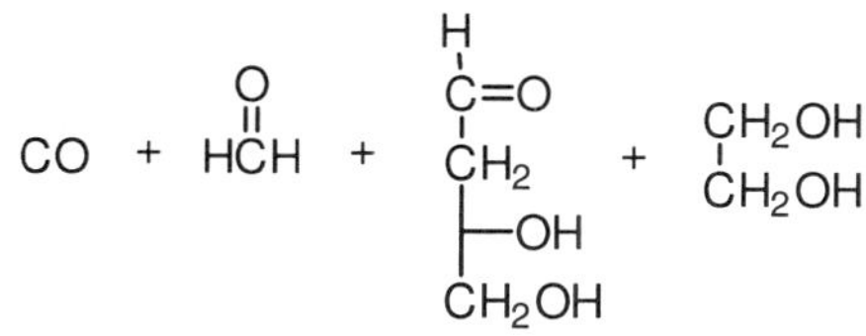 0.02 (0.02) · 0.05 (0.06) · 0.035 (0.00) · 0.03 (0.00)	26[b]
	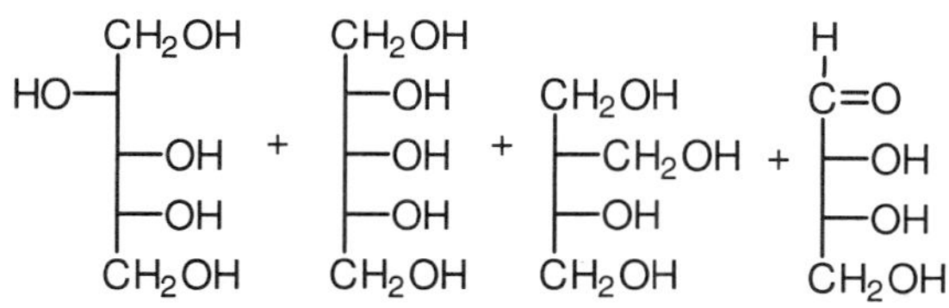 0.02 (0.00) · 0.02 (0.00) · 0.02 (0.00) · 0.03 (0.08)	
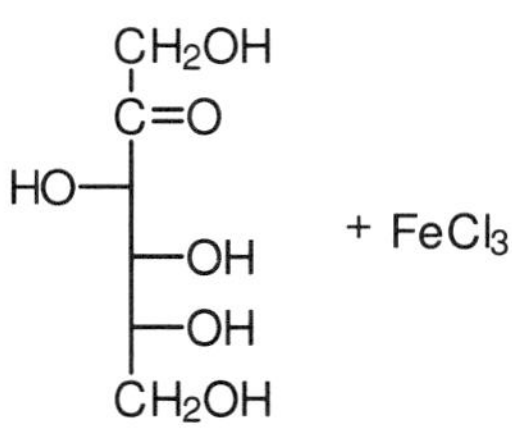	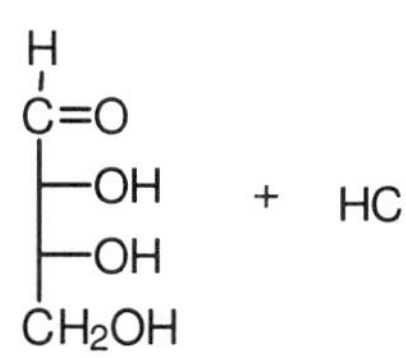 80%	27-30[c]
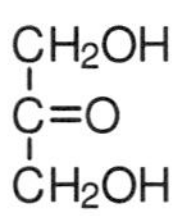	 80%	31[d]
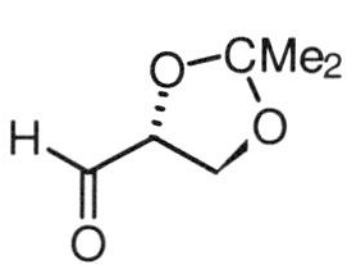	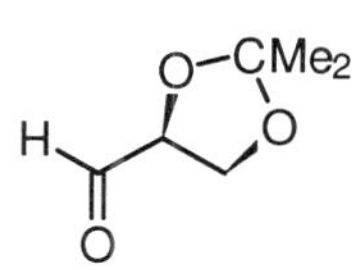	32

Continued on next page

Table 2.3. Reactions of acyclic carbonyl compounds—*Continued*

Reactant	*Product*	*Ref.*
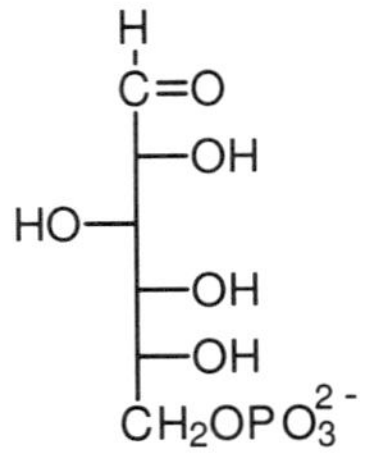	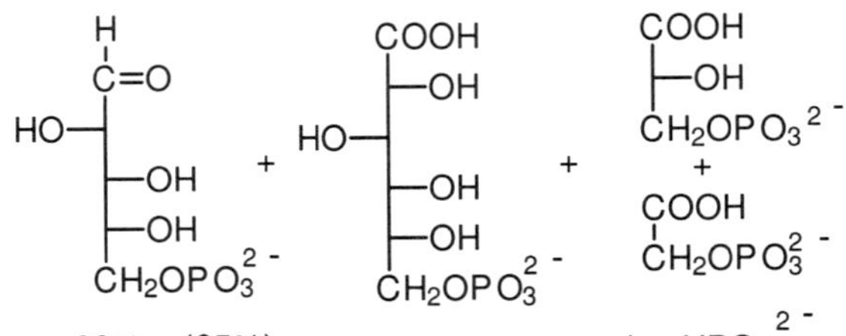 60% (65%)	33[e]
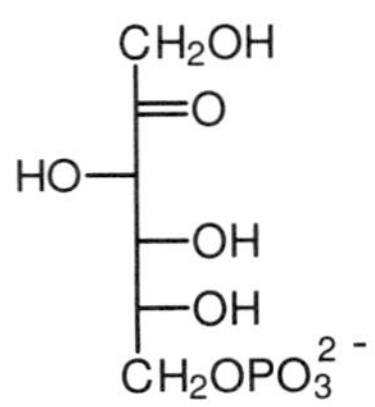	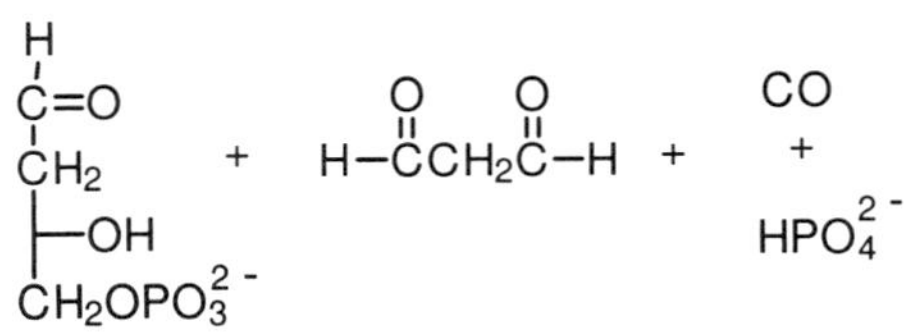	34[f,g]
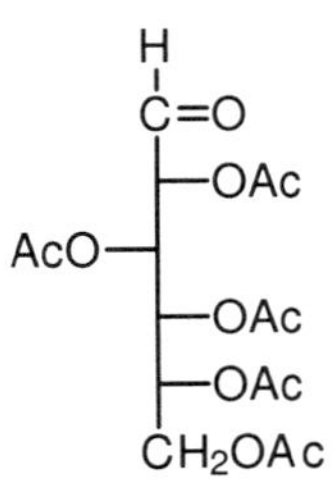	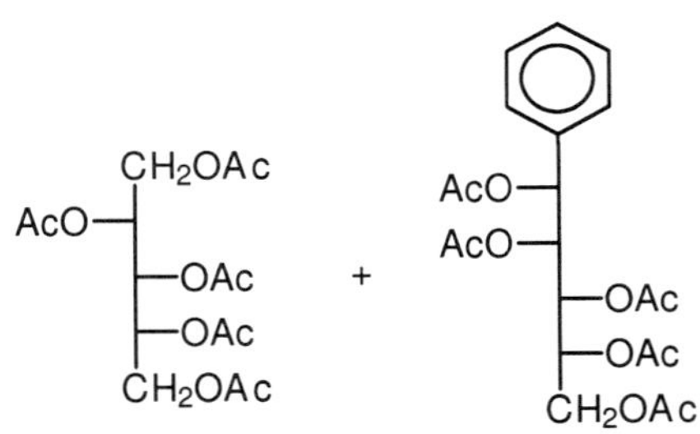 16% 4.5%	35[h]

[a]The reaction solvent was water. [b]Numbers without parentheses are quantum yields determined in an inert atmosphere. [c]Replacing Fe(III) with Mn(II) gives a 56% yield of D-erythrose.[30] [d]1,3-Diacetoxyacetone also was irradiated but only CO was identified. [e]This reaction was conducted in the presence and absence of oxygen. The yield in parenthesis is for reaction in the presence of oxygen. [f]The first photoproduct experiences further reaction. This reaction is discussed with other Type II reactions in Chapter 3. [h]The reaction solvent was benzene. A dimer formed in one percent yield.

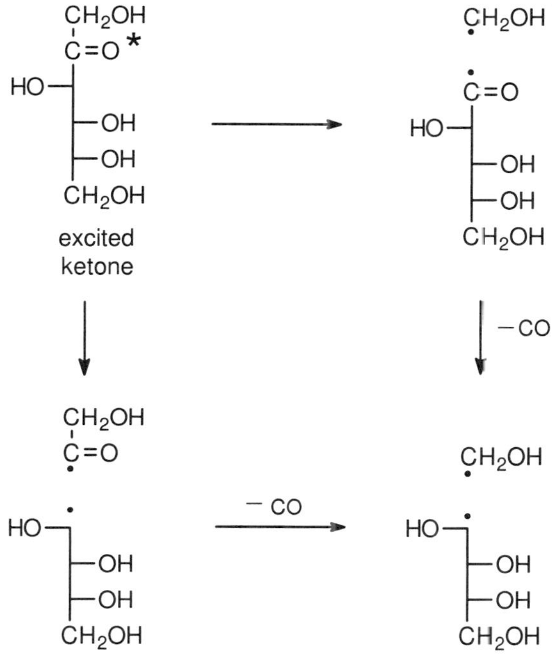

Scheme 2.6

Giese and co-workers also have used Type I reactions of *t*-butyl ketones to generate radicals (Table 2.4).[40] In their work, radical formation was part of a process of cleavage of oligonucleotides (Scheme 2.10).[41,42]

Disproportionation and Combination

Although α-cleavage can be followed by decarbonylation, combination, or disproportionation reactions, only decarbonylation is widely observed. Because combination regenerates the starting carbonyl compound, it too could be a common process but one that often goes undetected. As mentioned in the section "Direction of α-Cleavage", the only type of evidence available for recombination following α-cleavage in most carbohydrates is epimerization at one α-carbon atom (Scheme 2.5). Based upon this evidence, recombination is a relatively uncommon occurrence (Tables 2.1–2.4). Disproportionation is the least frequently observed of the typical reactions that follow α-cleavage in carbohydrates. An example of a disproportionation reaction is shown in Scheme 2.11.[20] It is not certain why disproportionation is so infrequent among Norrish Type I reactions of carbohydrates, but the ease with which decarbonylation takes place may account for the inability of other processes to compete effectively with carbon monoxide loss.

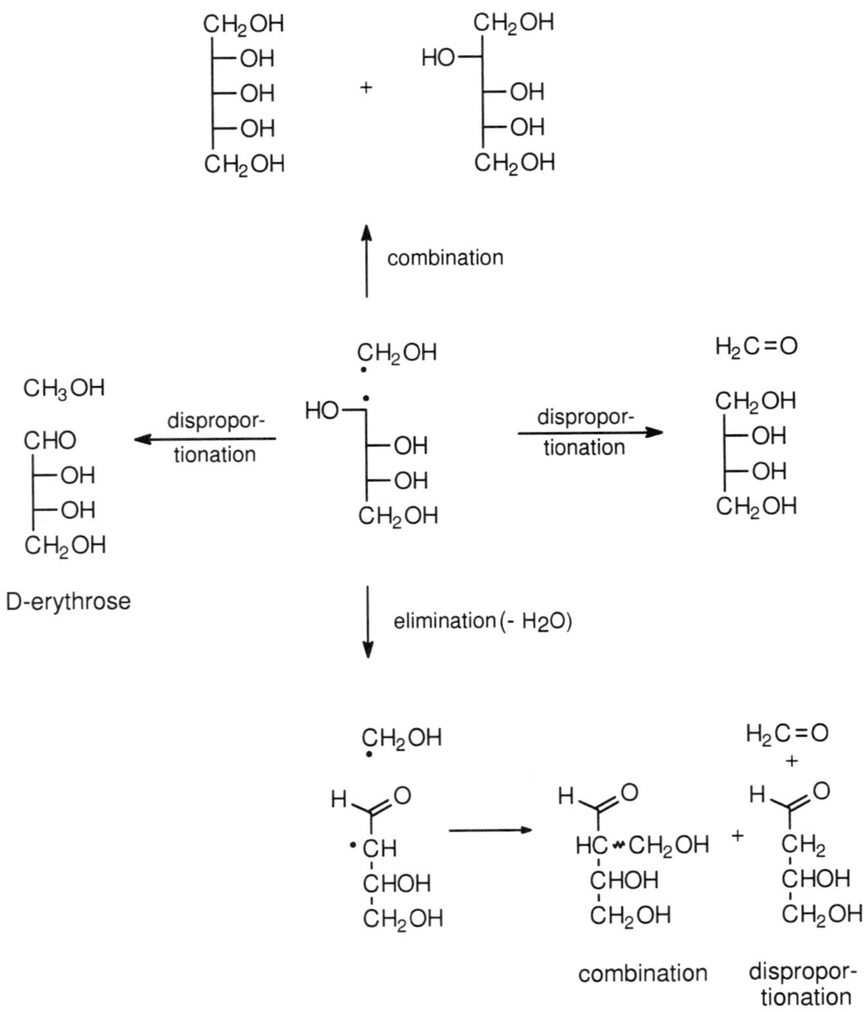
combination
dispropor-
tionation
D-erythrose
elimination (- H2O)
combination
dispropor-
tionation

Scheme 2.7

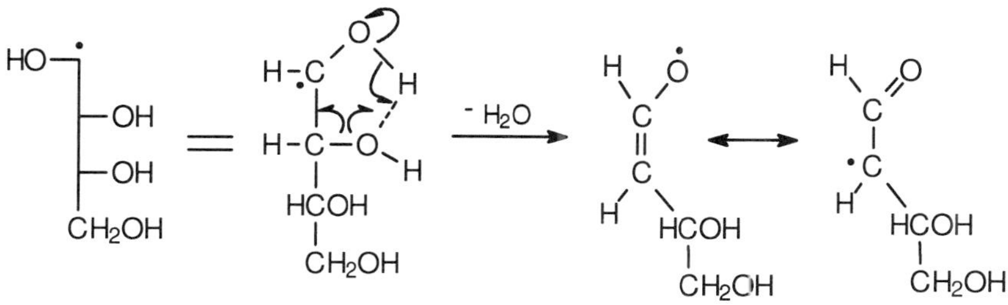

Scheme 2.8

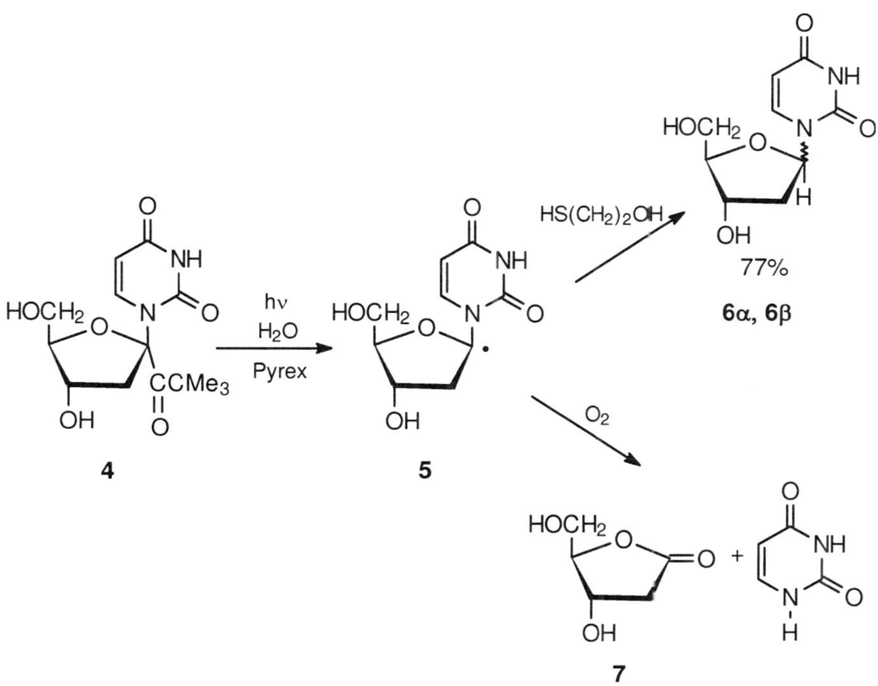

Scheme 2.9

Table 2.4. Reactions of nucleosides, oligonucleotides, and model compounds.

Reactant	*Product*	*Percent Yield*	*Ref.*
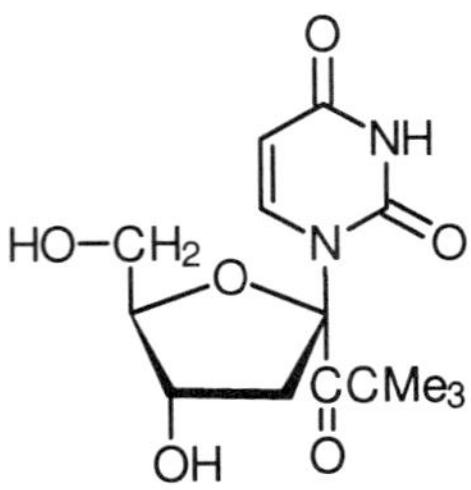 **4** + $HSCH_2CH_2OH$	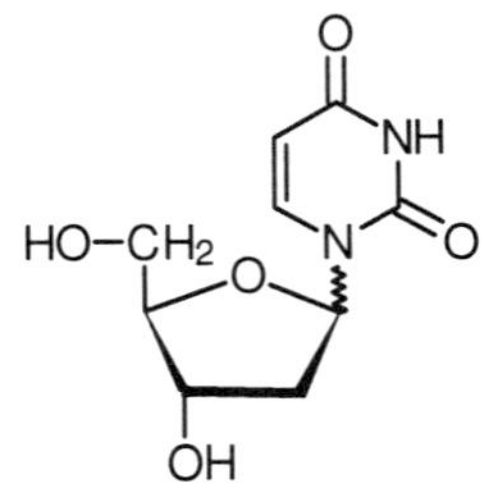 **6α, 6β**	77	8
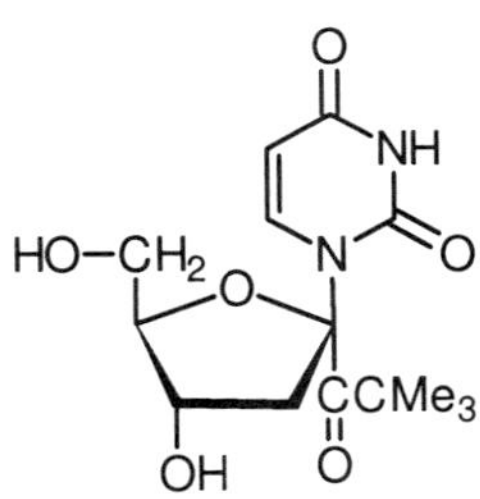 **4** + O_2	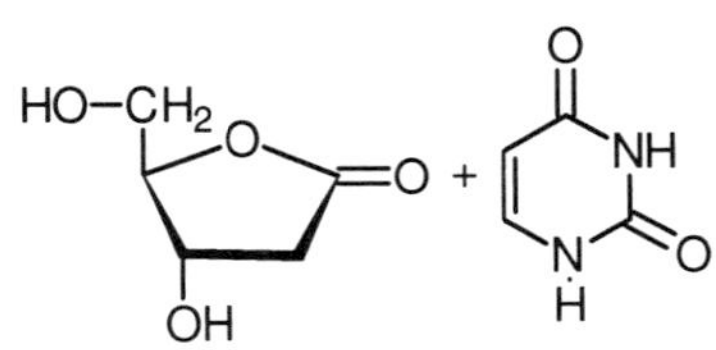 **7**	not reported	8
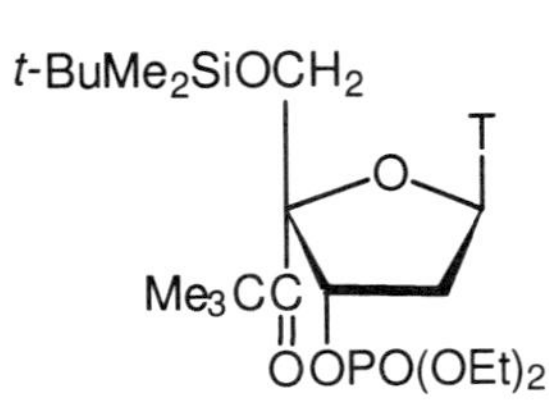 + MeOH + Bu_3SnH	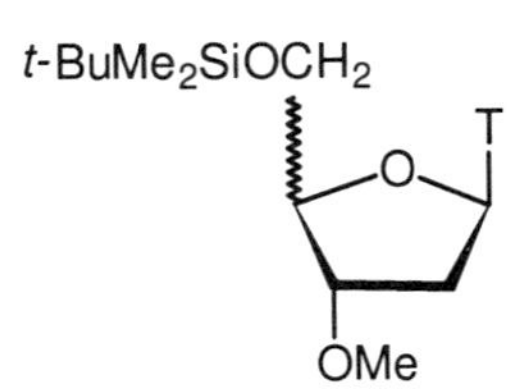	45	42
	+ 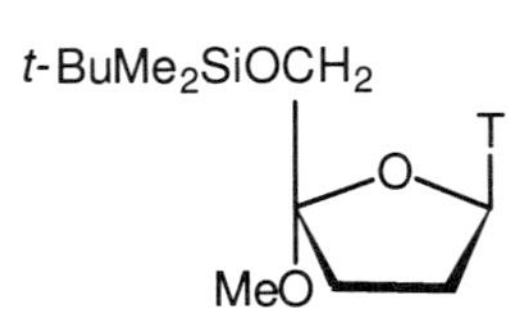	36	42

Continued on next page

Table 2.4. Reactions of nucleosides, oligonucleotides, and model compounds—*Continued*

Reactant	*Product*	*Percent Yield*	*Ref.*
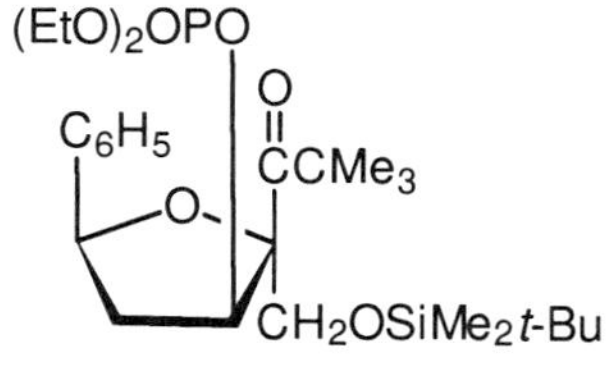 or 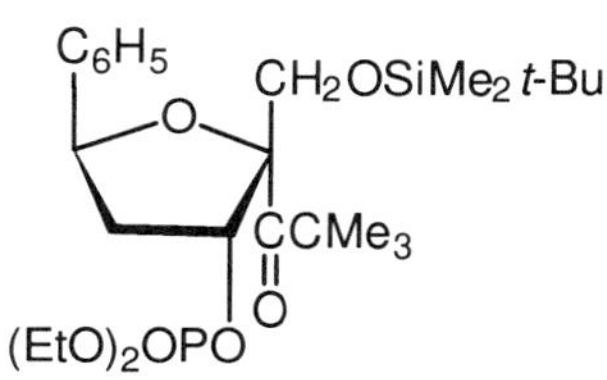 + MeOH + Bu_3SnH	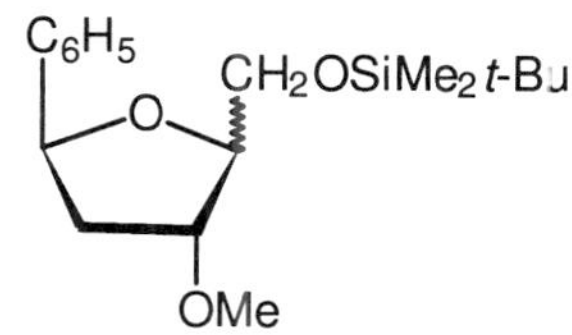	34	42
	+ 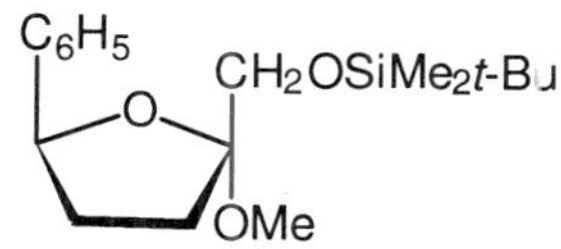	28	
	+ 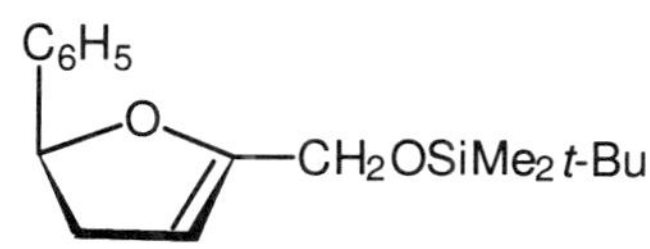	8	
	+ 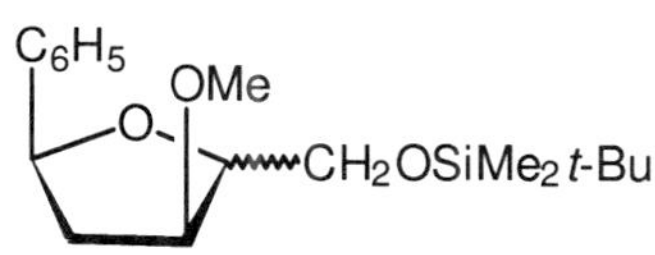	5	
	+ 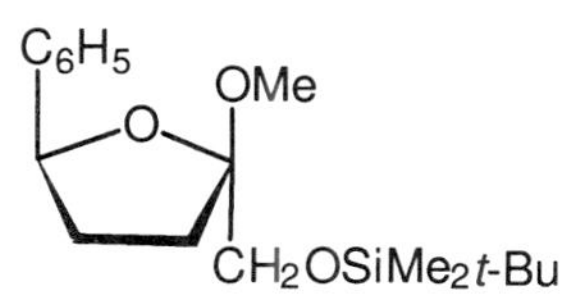	13	
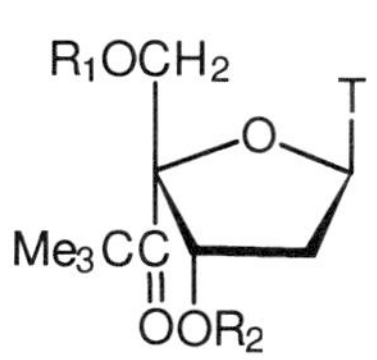 R_2 = 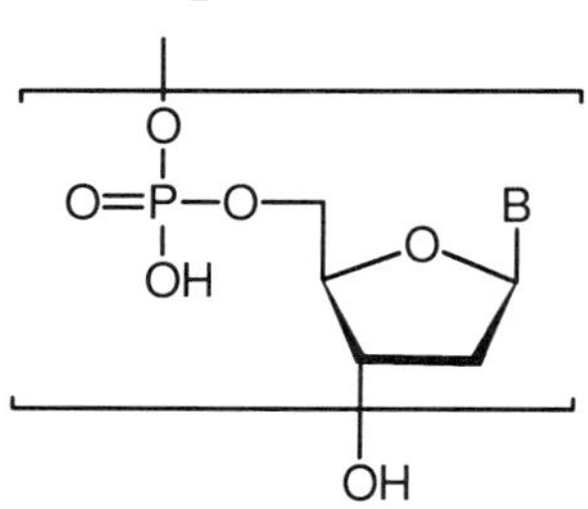 7	R_1H + R_2H R_1 = 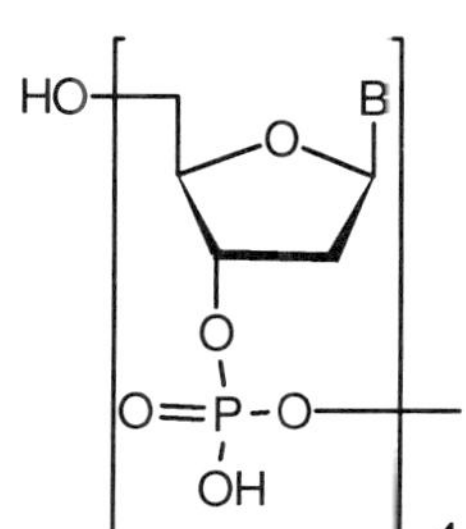 4		40, 41

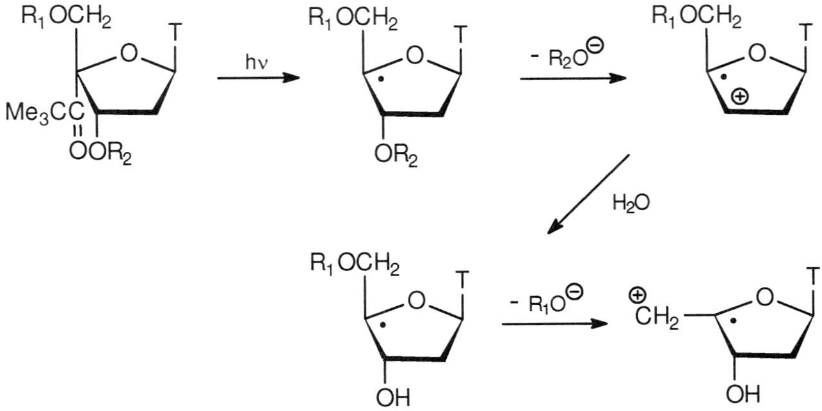

Scheme 2.10. See Table 2.4 for the identities of R_1 and R_2.

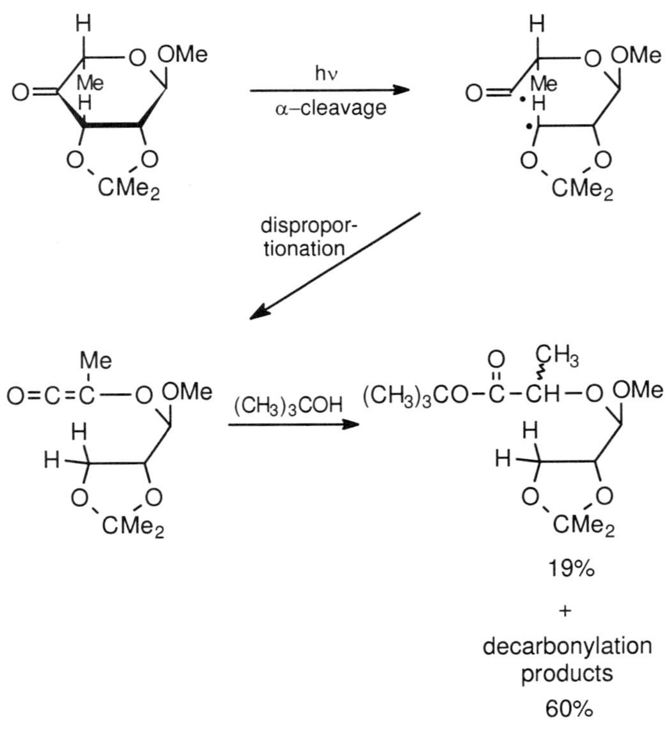

Scheme 2.11

Rearrangement

Ring Expansion

In 1964 Yates and Kilmurry reported a new photochemical rearrangement for cyclic carbonyl compounds.[43] This reaction began with a characteristic α-cleavage to form a diradical, but rather than undergoing typical further reactions (Scheme 2.2), this diradical cyclized to form an oxacarbene (Scheme 2.12). The oxacarbene then reacted with the solvent (methanol) to form a cyclic acetal. Following the original report, a variety of additional examples of this ring-expansion reaction were discovered.[44,45] These additional discoveries demonstrated that oxacarbene formation is not a general photochemical process for ketones, but it is common for certain structural types. Cyclobutanones and cyclopentanones, but in the latter group only those that are incorporated into a bridged ring system, are likely to undergo ring expansion via an oxacarbene.[46] As far as six-membered rings are concerned, oxacarbene formation was unknown until it was observed from photolysis of a carbohydrate (Scheme 2.13).[47]

Collins et al. in 1974 reported that compound **8** underwent a ring expansion upon photolysis to give the cyclic acetal **9**.[47] These authors proposed that an oxacarbene was an intermediate in the process (Scheme 2.13).[25,47] An interesting feature of this reaction is that it only occurred at low temperature (–70 °C); irradiation at room temperature afforded carbon monoxide.[25,47] (No other products were mentioned as being formed from room-temperature photolysis.) Collins and co-workers reported two other examples of this ring-expansion reaction, but these additional examples were less noteworthy because the product yields ranged from low to very low (Table 2.5).[25] Lee-

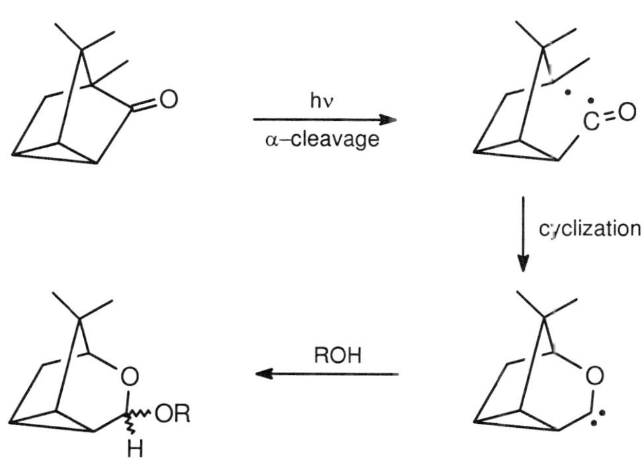

Scheme 2.12

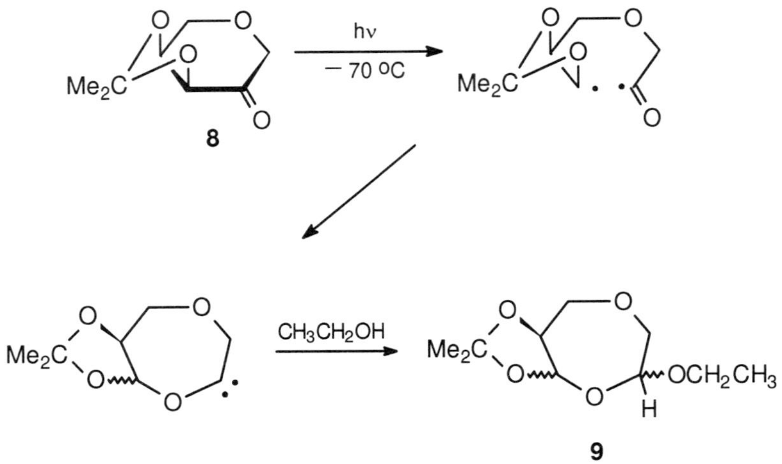

Scheme 2.13

Ruff and co-workers, however, showed that ring expansion has synthetic potential when used to synthesize 2′,3′-dideoxynucleosides (Table 2.5).[48–50]

Lactone Formation

As a result of considerable effort by the Collins and Heyns groups, the photochemical reactions of sugars with carbonyl groups at the 2- and 4-positions of pyranoid rings have been extensively investigated (Tables 2.1 and 2.2). As described in the section "Decarbonylation", the most common photochemical reaction for these compounds is Type I cleavage followed by decarbonylation. When Collins et al. investigated the 3-uloses (Table 2.6), however, they met with an unexpected finding, namely, the photochemical rearrangement (with loss of the isopropylidene group) shown in eq 1.[51] At the time of the discovery of this reaction, no mechanism was proposed; however, the loss of an isopropylidene group provided suggestive evidence for formation of an intermediate that was being intercepted by the water present in the reaction mixture.[51,52] (The isopropylidene group remained in place during photolyses when water was absent.)

Eight years after their original study, Collins et al. returned to the investigation of this process and discovered information that "shed new light" on the rearrangement.[52] In a particularly informative experiment, they demonstrated that the deuterium atom at C1 became attached to C3 as a result of reaction (eq 2). From this information, Collins proposed that the overall process occurred in two distinct photochemical steps. The first of these, a typical α-cleavage followed by intramolecular disproportionation, gave the enal **10** (Scheme 2.14). Because **10** could not be detected in the photolysis mixture, Collins proposed that it reacted by an efficient, second photochemical reaction to give the

Table 2.5. Ring-expansion reactions.

Reactant	*Product*	*Percent Yield*	*Temp (°C)*	*Ref.*
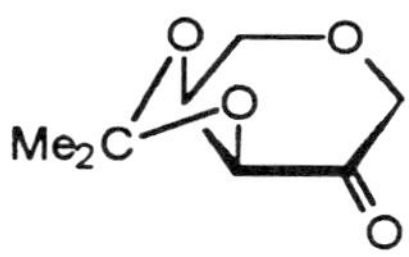	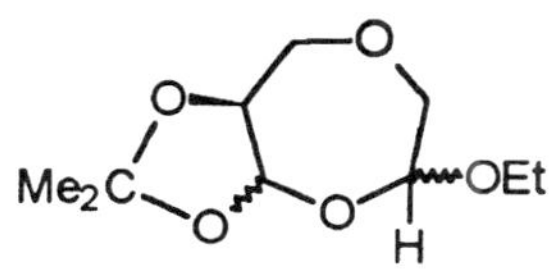	65[a]	-70	25, 47
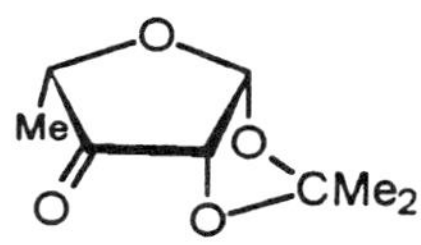	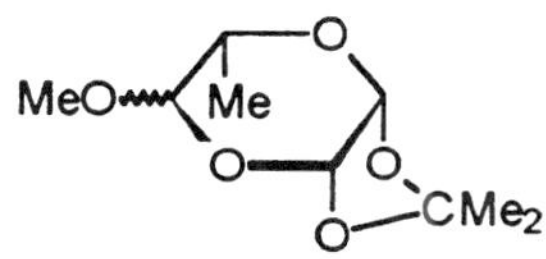	5	+20[b]	25, 47
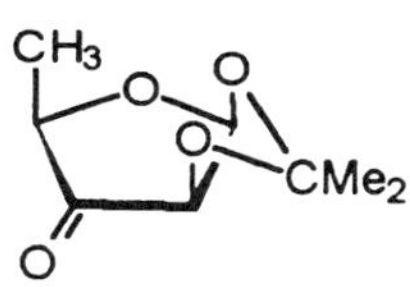	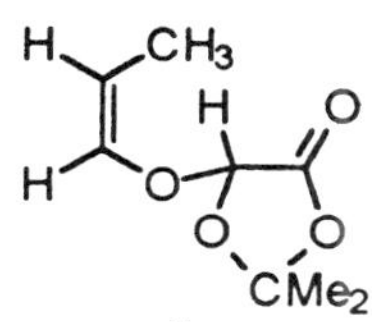 **A**	A+B = 5 C = 27	-60[b]	25
	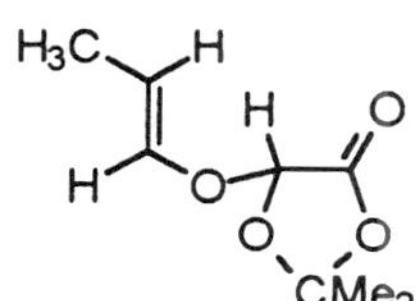 **B**	A+B > 7 No C	+20	
	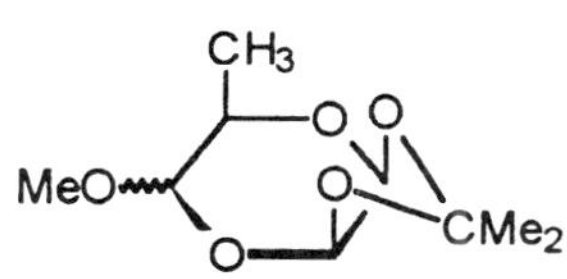 **C**	C>7 C = 14	+20[b] -60[b]	
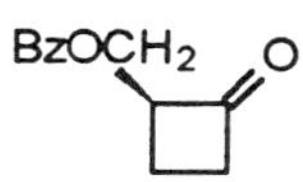 + ROH	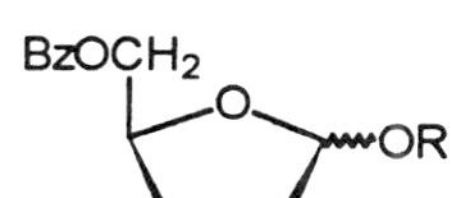			50
R = H		55		
R = OCH₃		77		

Continued on next page

Table 2.5. Ring-expansion reactions—*Continued*

Reactant	*Product*	*Percent Yield*	*Ref.*
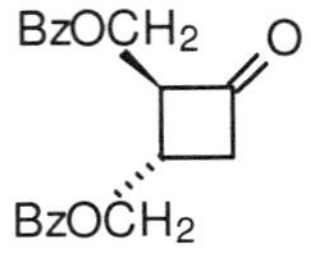	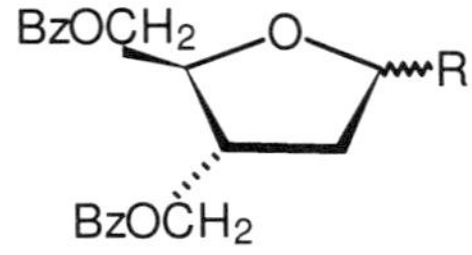		48, 49
R = OH		58	
R = OC(CH$_3$)$_3$		9	
R = OCH$_3$		62	
R = 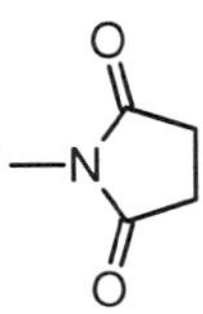		17	
R = 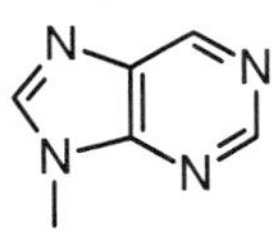		19	
R = 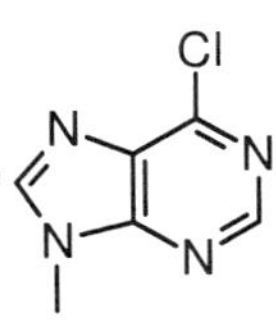		43	
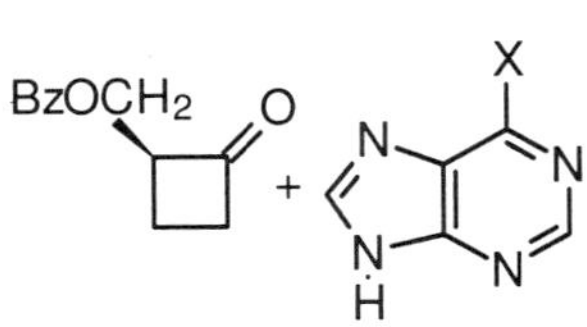	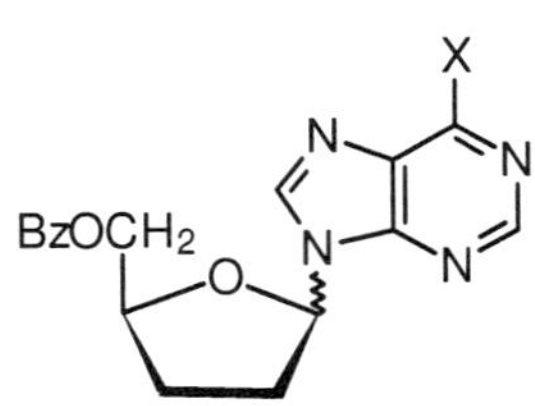		50
X = H		36	
X = Cl		39	
X = OCH$_3$		69	
X = O(CH$_2$)$_5$CH$_3$		45	

[a]Two products were formed.
[b]Methanol was the solvent.

Table 2.6. Lactone formation.

Reactant	*Product*	*Percent Yield*	*Ref.*
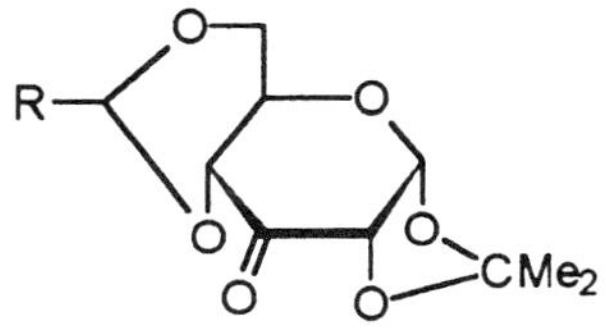	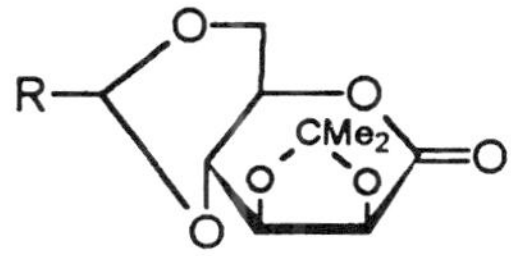		51, 52
R = CH_3		45	
R = C_6H_5		30[a]	
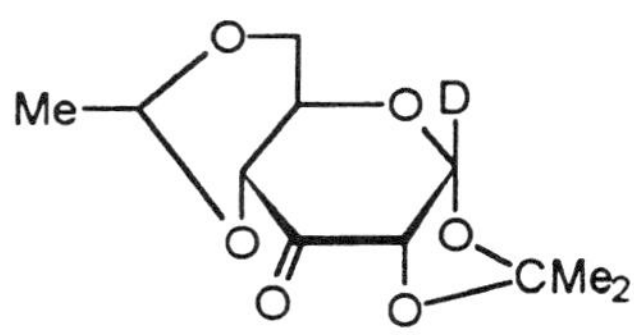	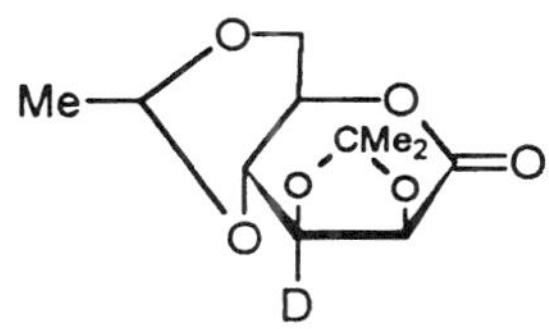	45	51
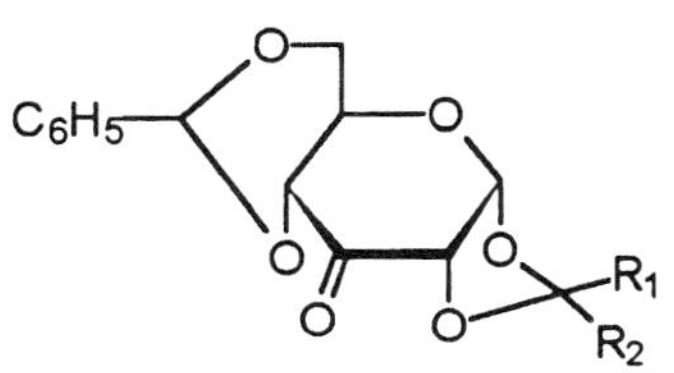	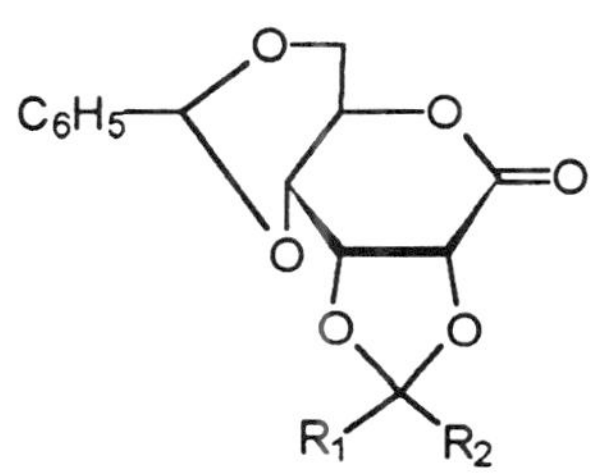		
R_1 = OMe, R_2 = Me		b	51
R_1 = Et, R_2 = H		b	51
R_1 = OEt, R_2 = Me		b	51
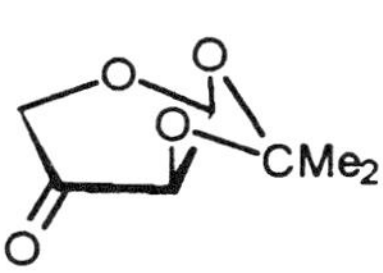	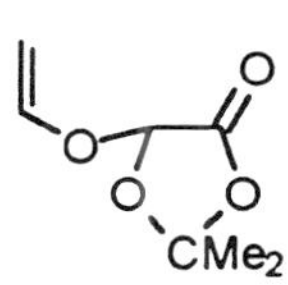	8[d]	25

Continued on next page

Table 2.6. Lactone formation—*Continued*

Reactant	*Product*	*Percent Yield*	*Ref.*
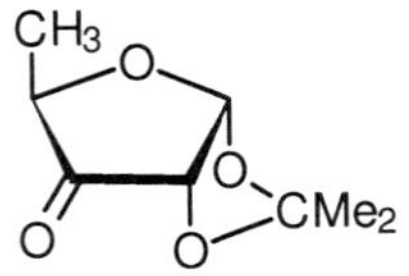	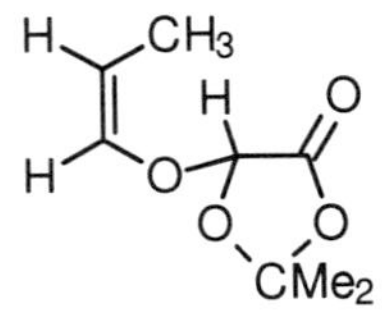 **A** 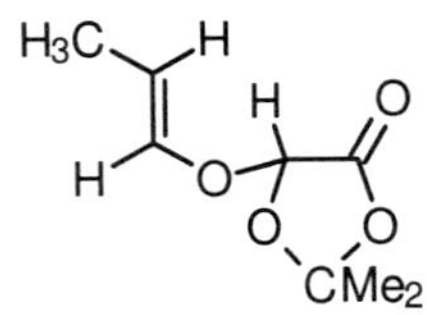 **B**	A + B > 3[c] A + B = 19[d]	25

[a]Other products were formed by reaction with methanol.
[b]Yields were not reported.
[c]The temperature was -60 °C.
[d]The temperature was +20 °C.

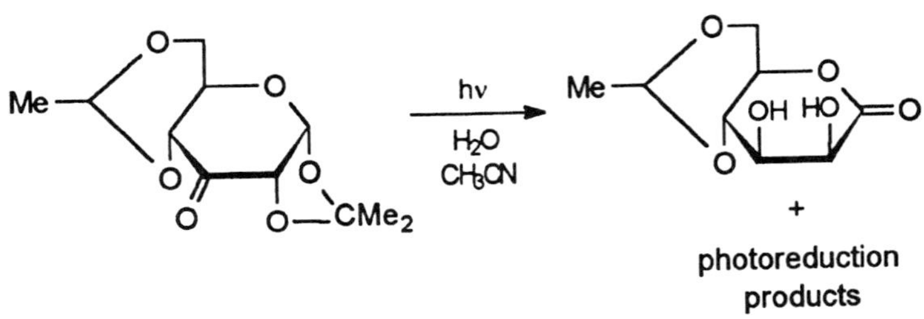

(1)

Equation 1

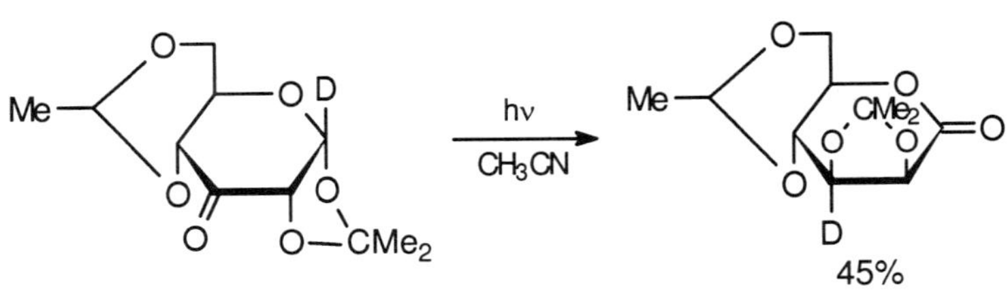

(2)

Equation 2

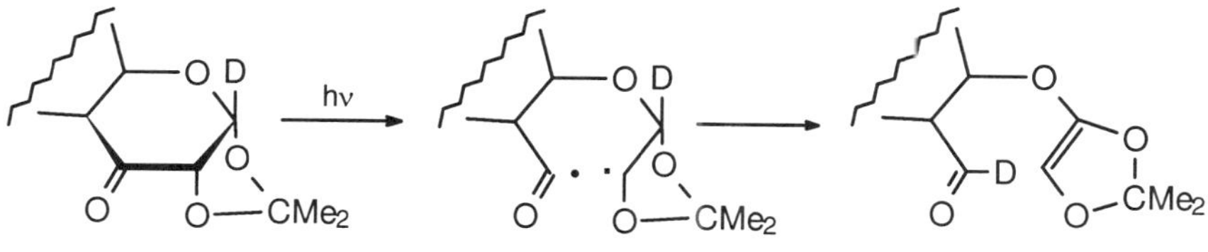

10

Scheme 2.14

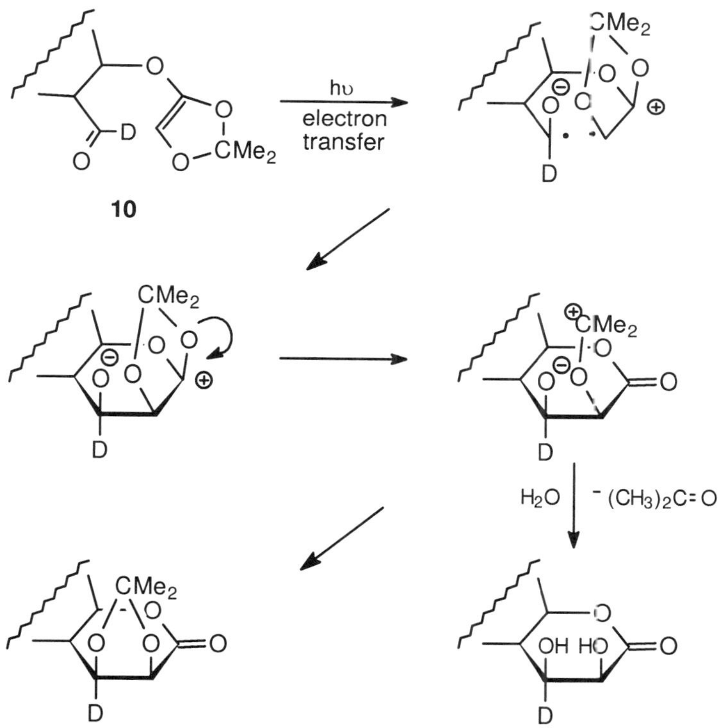

Scheme 2.15

observed product.[51,52] The details of this second process were left largely unspecified; however, if the suggested two-step mechanism is correct, it is reasonable to propose that the second reaction is an electron-transfer process (Scheme 2.15). This proposal draws support from the similar type of reaction shown in Scheme 2.16.[53]

Collins and Farnia investigated several additional 3-keto sugars in which, in each case, the keto group was part of a five-membered ring (i.e., a 3-pentulose).[25] The reaction products were lactones (Table 2.6), and the mechanism proposed for their formation began with an α-cleavage followed by an internal

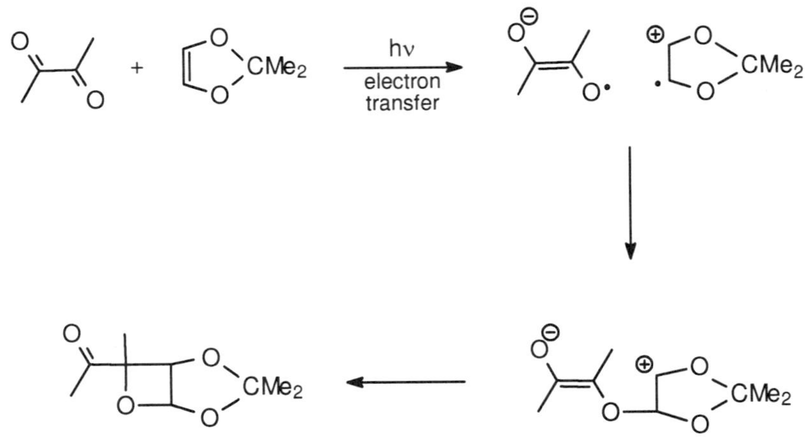

Scheme 2.16

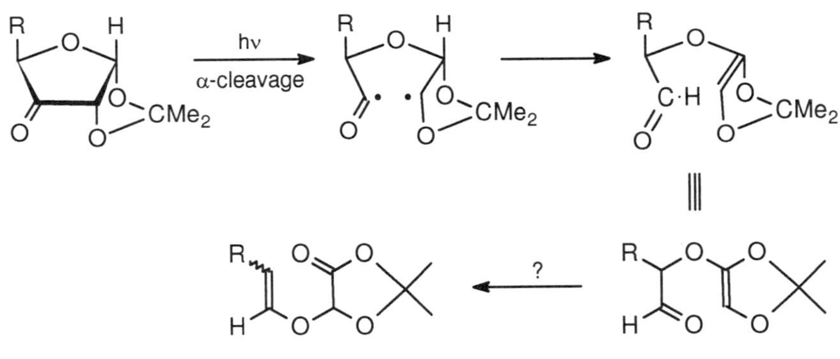

R = H, CH_3

Scheme 2.17

hydrogen atom transfer to give an intermediate, unsaturated aldehyde (Scheme 2.17). The mechanism for conversion of this aldehyde into the observed photoproducts remained uncertain.

Ring Opening of Levoglucosenone and Its Derivatives

Photolysis of levoglucosenone (**11**) causes it to undergo an α-cleavage reaction that forms the diradical **12**, an intermediate that ring-opens to give the ketene **13**.[54] Reaction of **13** with methanol then produces the unsaturated ester **14** (Scheme 2.18, Table 2.7). One aspect of this reaction that is different from the other reactions discussed in this chapter is that the carbonyl group in **11** is in conjugation with a carbon–carbon double bond. Many α,β-unsaturated ketones do not undergo Norrish Type I reaction because their reactive state, particu-

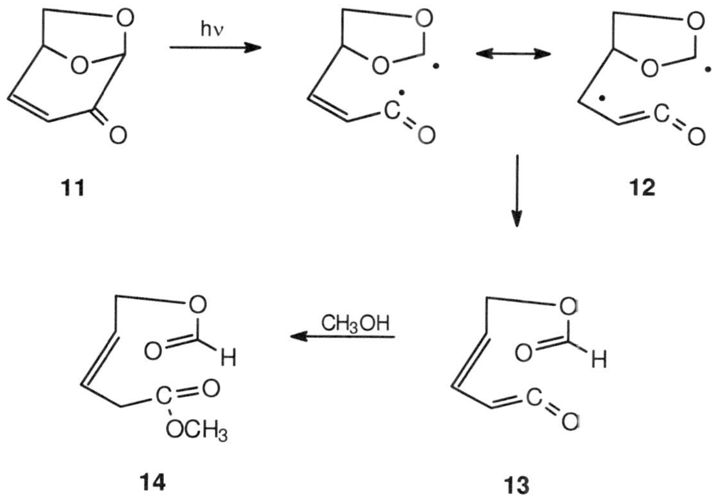

Scheme 2.18

larly if it is a triplet, is (π,π*) rather than (n,π*). Because (π,π*) states do not have an orbital on oxygen containing a single electron, the electronic structure associated with α-cleavage reactions (Scheme 2.1) is missing. Compounds with reactive (π,π*) states generally undergo other types of reaction, such as [2 + 2] addition or *cis–trans* isomerization (both discussed in Chapter 8).

Does reaction of levoglucosenone (**11**) occur from an (n,π*) or (π,π*) excited state? In studying levoglucosenone, Yamada and Matsumoto did not definitely identify the state responsible for reaction; however, they did show that this rearrangement can occur by way of a triplet state because sensitization with the triplet sensitizer acetophenone leads to the same product (in even better yield) that is formed by direct irradiation.[54] This experiment shows that reaction can occur by way of a triplet state; furthermore, because α-cleavage occurs during photolysis, it is likely that this triplet state is (n,π*).

In a continuation of their research on levoglucosenone (**11**), Yamada et al. found that α-cleavage reactions also occur upon photolysis of several derivatives of **11**.[55] These authors suggested that the initial bond fragmentation was followed by a hydrogen atom transfer to give an oxacarbene (Scheme 2.19). The oxacarbene then either reacted with the solvent (methanol) to give **15** or extruded carbon dioxide to produce the unsaturated aldehyde **16**.

Summary

Type I reactions are common from photolysis of carbonyl-containing carbohydrates. The basic photochemical reaction, fragmentation of a carbonyl to α-

Table 2.7. Reactions of levoglucosenone and its derivatives.

Reactant	*Product*	*Percent Yield*	*Ref.*
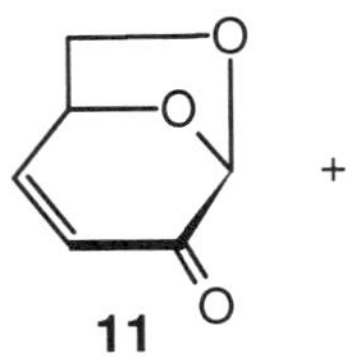 **11** + CH_3OH	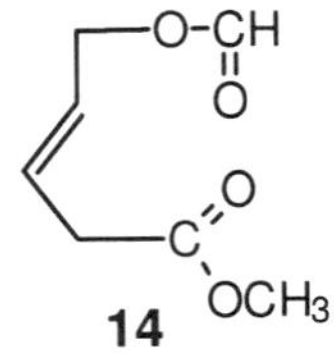 **14**	58	54
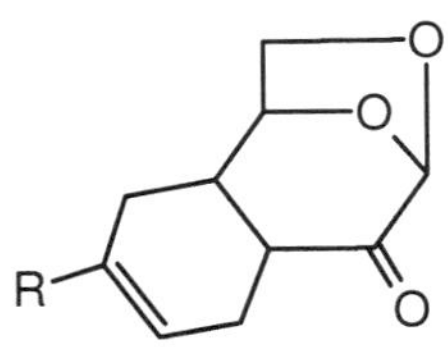	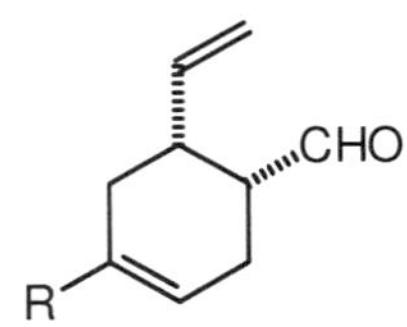 **16**		55
R = H		41	
R = Me		50	
R = C_6H_5		48	
R = $CH_2CH_2CH{=}CMe_2$		38	

carbon bond, forms the more stable radical pair (or diradical). The most widely observed process following photochemical α-cleavage is decarbonylation (Scheme 2.5, Tables 2.1–2.4). Other reactions that occur are disproportionation (Scheme 2.11) and combination (Scheme 2.5). Two less common but quite interesting reactions that also take place following α-cleavage are ring expansion (Scheme 2.13, Table 2.5) and rearrangement (Schemes 2.14 and 2.15, Table 2.6).

References

1. Zimmerman, H. E. In *Advances in Photochemistry;* Noyes, W. A., Jr.; Hammond, G. S.; Pitts, Jr., J. N. Eds.; Interscience: New York, 1963; Vol. 1, p 183.
2. Turro, N. J. *Modern Molecular Photochemistry;* Benjamin/Cummings: Menlo Park, CA, 1978; pp 364, 528.
3. Walling, C.; Gibrian, M. *J. Am. Chem. Soc.* **1964,** *86,* 3902.
4. Padwa, A. *Tetrahedron Lett.* **1964,** 3465.

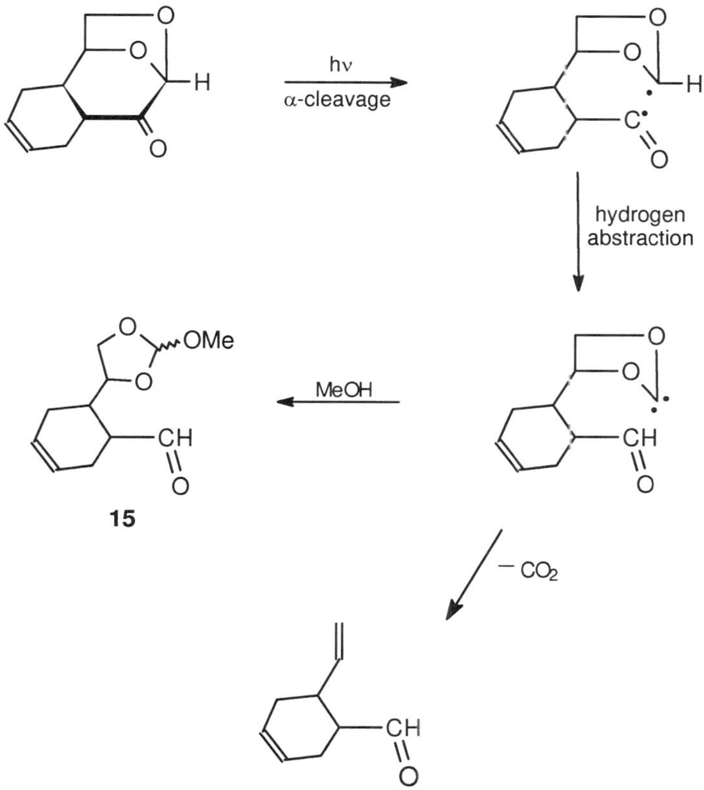

15

16

Scheme 2.19

5. Turro, N. J. *Modern Molecular Photochemistry;* Benjamin/Cummings: Menlo Park, CA, 1978; p 386.
6. Turro, N. J. *Modern Molecular Photochemistry;* Benjamin/Cummings: Menlo Park, CA, 1978; pp 530-532.
7. Dauben, W. G.; Salem, L.; Turro, N. J. *Acc. Chem. Res.* **1975,** *8,* 41.
8. Goodman, B. K.; Greenberg, M. M. *J. Org. Chem.* **1966,** *61,* 2.
9. Turro, N. J. *Modern Molecular Photochemistry;* Benjamin/Cummings: Menlo Park, CA, 1978; p 528.
10. Collins, P. M.; Gupta, P.; Iyer, R. *J. Chem. Soc., Chem. Commun.* **1970**, 1261.
11. Collins, P. M.; Iyer, R.; Travis, A. S. *J. Chem. Res. (S)* **1978**, 446.
12. Kuwabara, Lion, M. Y.; Riesz, P. *Int. J. Radiat. Bio.* **1981,** *39,* 451.
13. Seifert, K.-G. *Tetrahedron Lett.* **1974,** 4513.
14. Coyle, J. D. *Introduction to Organic Photochemistry;* John Wiley & Sons: New York, 1986; p 108.
15. Turro, N. J. *Molecular Photochemistry;* W. A. Benjamin: New York, 1965; p 224.
16. Collins P. M.; Gupta, P. *J. Chem. Soc., Chem. Commun.* **1969,** 1288.
17. Collins, P. M.; Gupta, P.; Travis, A. S. *J. Chem. Soc., Perkin Trans. I* **1980,** 277.

18. Heyns, K.; Neste, H.-R.; Thiem, J. *Chem. Ber.* **1981**, *114,* 891.
19. Collins, P. M. *J. Chem. Soc., Chem. Commun.* **1968**, 403.
20. Collins, P. M. *J. Chem. Soc. (C)* **1971**, 1960.
21. Collins, P. M.; Gupta, P. *J. Chem. Soc. (C)* **1971**, 1965.
22. Collins, P. M.; Farnia, F.; Travis, A. S. *J. Chem. Res. (S)* **1979**, 266.
23. Collins, P. M.; Travis, A. S. *J. Chem. Soc., Perkin Trans. I* **1980**, 779.
24. Heyns, K.; Neste, R.; Paal, M. *Tetrahedron Lett.* **1978**, 4011.
25. Collins, P. M.; Farnia, F. *J. Chem. Soc., Perkin Trans. I* **1985**, 575.
26. Triantaphylides, C.; Schuchmann, H.-P.; von Sonntag, C. *Carbohydr. Res.* **1982**, *100,* 131.
27. Araki, K.; Sakuma, M.; Shiraishi, S. *Chem. Lett.* **1983**, 665.
28. Araki, K.; Shiraishi, S. *Chem. Lett.* **1984**, 267.
29. Araki, K.; Sakuma, M.; Shiraishi, S. *Bull. Chem. Soc. Jpn.* **1984**, *57,* 997.
30. Araki, K.; Shiraishi, S. *Bull. Chem. Soc. Jpn.* **1986**, *59,* 229
31. Steenken, S.; Jaenicke-Zauner, W.; Schulte-Frohlinde, D. *Photochem. Photobiol.* **1975**, *21,* 21.
32. Schreiber, S. L.; Satake, K. *Tetrahedron Lett.* **1986**, *27,* 2575.
33. Triantaphylides, C.; Halmann, M. *J. Chem. Soc., Perkin Trans. II* **1975**, 34.
34. Triantaphylides, C.; Gerster, R. *J. Chem. Soc., Perkin Trans. II* **1977**, 1719.
35. Whistler, R. L.; Ong, K.-S. *J. Org. Chem.* **1971**, *36,* 2575.
36. Funcke, W.; von Sonntag, C.; Triantaphylides, C. *Carbohydr. Res.* **1979**, *75,* 305.
37. Hayward, L. D.; Angyal, S. J. *Carbohydr. Res.* **1977**, *53,* 13.
38. von Sonntag, C. *Adv. Carbohydr. Biochem.* **1980**, *37,* 7.
39. Ichikawa, S.; Tomita, I.; Hosada, A.; Sato, T. *Bull. Chem. Soc. Jpn.* **1988**, *61,* 513.
40. Giese, B.; Imwinkelried, P.; Petretta, M. *Synlett* **1994**, 1003.
41. Giese, B.; Dussy, A.; Elie, C.; Erdmann, P.; Schwitter, U. *Angew. Chem. Int. Ed. Engl.* **1994**, *33,* 1861.
42. Peukert, S.; Giese, B. *Tetrahedron Lett.* **1996**, *37,* 4365.
43. Yates P.; Kilmurry, L. *Tetrahedron Lett.* **1964**, 1739.
44. Yates P.; Loutfy, R. O. *Acc. Chem. Res.* **1975**, *8,* 209.
45. Yates, P. *J. Photochem.* **1976**, *5,* 91
46. Yates, P.; Verma, A. K.; Tam, J. C. L. *J. Chem. Soc., Chem. Commun.* **1976**, 933.
47. Collins, P. M.; Oparaeche, N. N.; Whitton, B. R. *J. Chem. Soc., Chem. Commun.* **1974**, 292.
48. Lee-Ruff, E.; Jiang, J.-L.; Wan, W.-Q. *Tetrahedron Lett.* **1993**, *34,* 261.
49. Lee-Ruff, E.; Wan, W.-Q.; Jiang, J.-L. *J. Org. Chem.* **1994**, *59,* 2114.
50. Lee-Ruff, E.; Xi, F.; Qie, J. H. *J. Org. Chem.* **1996**, *61,* 1547.
51. Collins, P. M.; Farnia, F.; Munasinghe, V. R. N.; Oparaeche, N. N.; Siavoshy, F. *J. Chem. Soc., Chem. Commun.* **1985**, 1038.
52. Collins, P. M.; Munasinghe, V. R. N.; Oparaeche, N. N. *J. Chem. Soc., Perkin Trans. I* **1977**, 2423.
53. Mattay, J.; Gersdorf, J.; Buchkremer, K. *Chem. Ber.* **1987**, *120,* 307.
54. Yamada Y.; Matsumoto, M. *Chem. Lett.* **1992**, 2273.
55. Yamada, S. H.; Ishikawa, S.; Matsumoto, M. *J. Chem. Soc., Chem. Commun.* **1994**, 2155.

3

Norrish Type II and Related Reactions

Hydrogen Abstraction from a γ-Carbon Atom: Norrish Type II Process

Reaction Mechanism

Basic Mechanistic Sequence

Hydrogen atom abstraction by an excited carbonyl group can be an intra- or intermolecular process. If the reaction is intramolecular, the most readily abstracted hydrogen atom is one attached to a γ-carbon atom. Two related reasons that γ-hydrogen atom abstraction is favored are that in an unconstrained system this hydrogen atom is most frequently approached by the excited carbonyl group, and γ-hydrogen atom transfer involves a nearly strain-free transition state. The result of γ-hydrogen abstraction is the formation of a 1,4-diradical that then undergoes either cyclization to form a cyclobutanol, cleavage to give an alkene and an enol, or internal hydrogen atom transfer to regenerate the starting carbonyl compound (Scheme 3.1). In carbohydrate systems the 1,4-diradical often contains an oxygen atom and, consequently, its cleavage produces a new carbonyl compound (rather than an alkene) and an enol (Scheme 3.2). Photochemical formation and reaction of a 1,4-diradical produced by hydrogen abstraction is termed a Type II process* (or Type II reaction) or, in honor of its discoverer, a Norrish Type II process.[3,4] In noncarbo-

*A distinction sometimes is made between the formation of the 1,4-diradical (a Type II process) and its further reaction (a Type II reaction),[1] but often this distinction is lost and the entire transformation is referred to as either a Type II reaction or a Type II process or some variation of this terminology. Wagner and Park have proposed that the cyclization part of this reaction (cyclobutanol formation) be called the Yang reaction after its discoverer.[2]

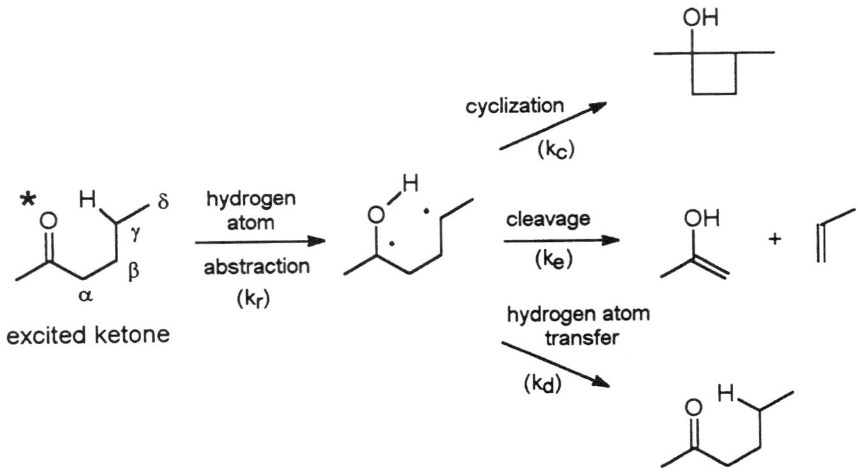

Scheme 3.1

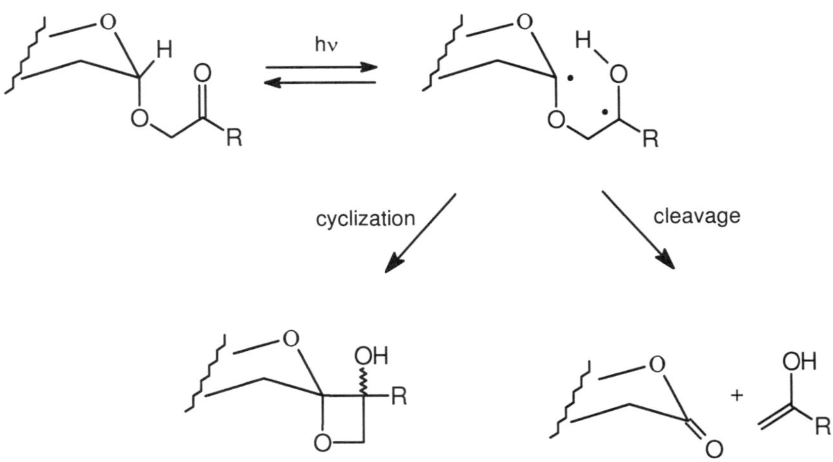

Scheme 3.2

hydrate and carbohydrate systems, these reactions are widely observed and, consequently, are among the most extensively studied of all photochemical processes.[1,5,6] Except for one group of reactions that will be discussed later in this chapter, in the section "Pyruvate Ester Photolysis", Type II reactions that involve carbohydrates are listed in Tables 3.1 and 3.2.

Competition between Cyclization and Cleavage Reactions

The ratio of cyclization to cleavage in Type II reactions varies widely. Sometimes both reactions take place (eq 1),[7] but in other instances only one of these

Table 3.1. Type II reactions of carbohydrates.

Reactant	*Product*	*Percent Yield*	*Ref.*
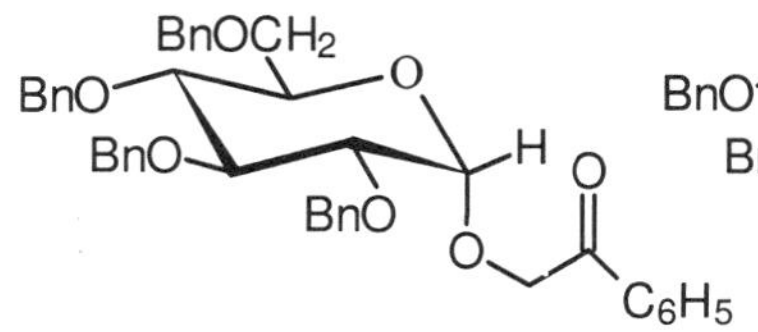	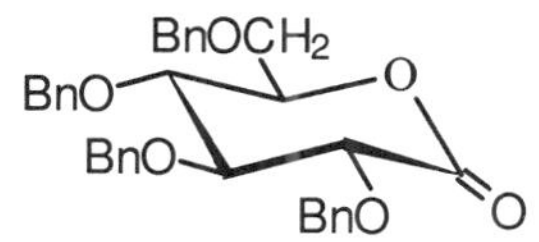 **A** + 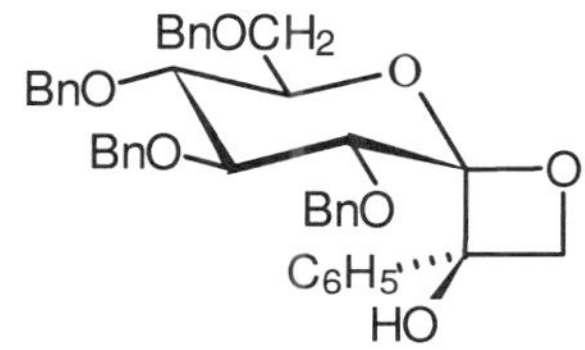 **B**	57 24	7
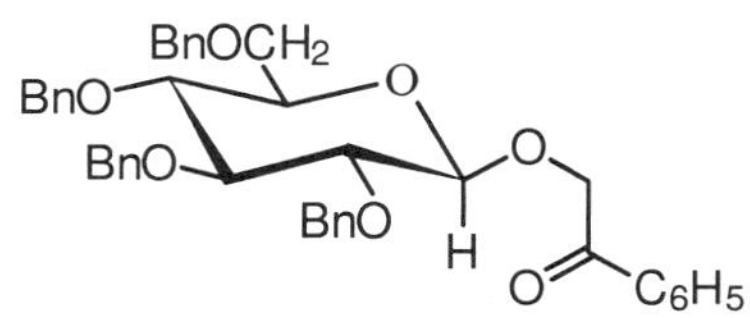	**A** + **B**	56 10	7
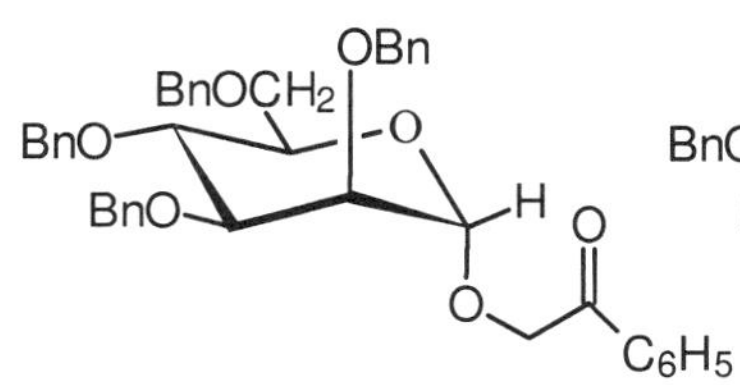	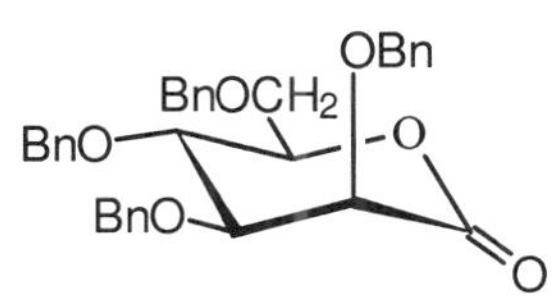	90	7
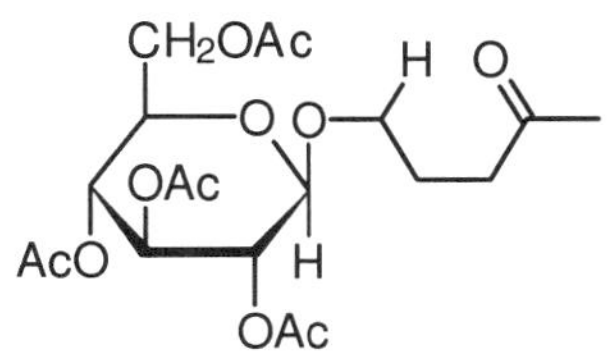	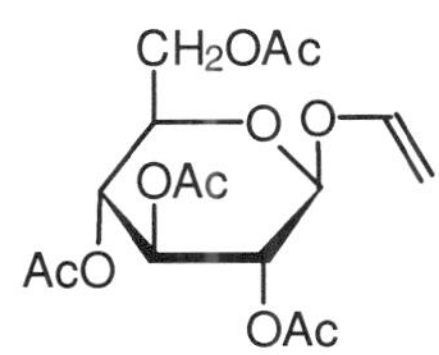	50	8
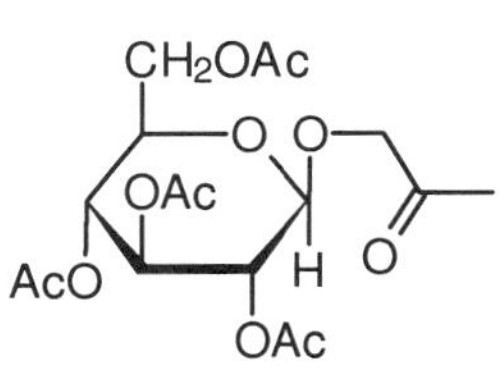	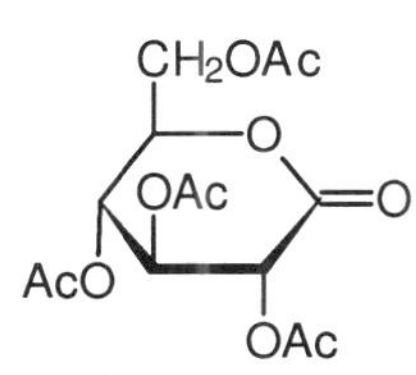	100	9

Continued on next page

Table 3.1. Type II reactions of carbohydrates—*Continued*

Reactant	*Product*	*Percent Yield*	*Ref.*
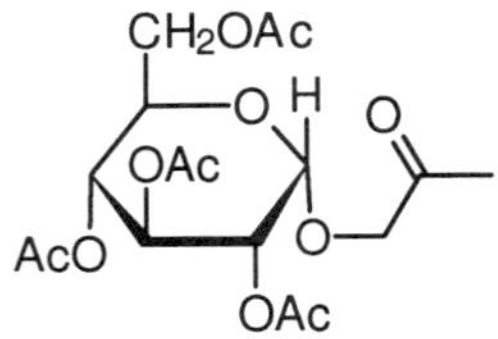	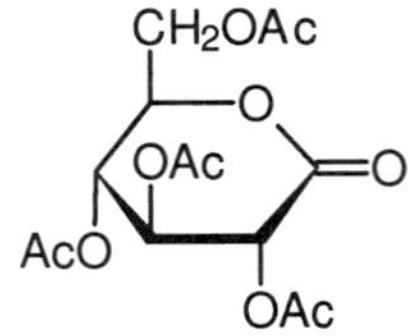	a	9
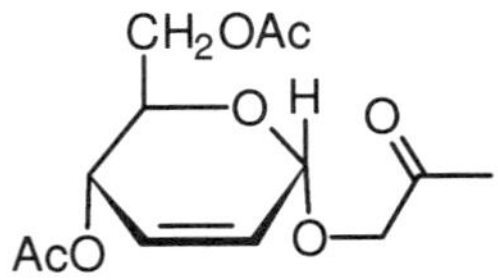	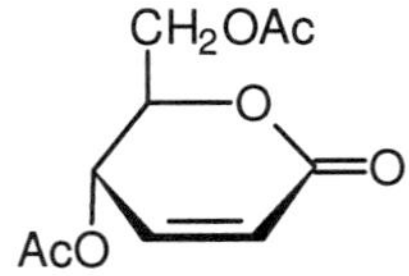	39	9
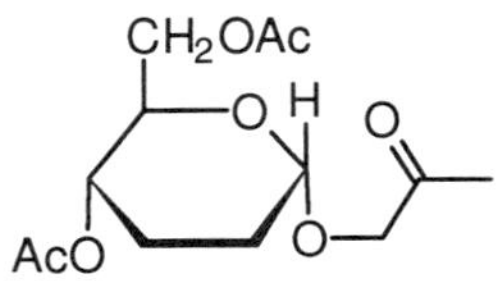	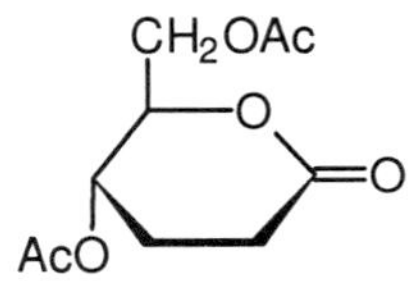	51	9
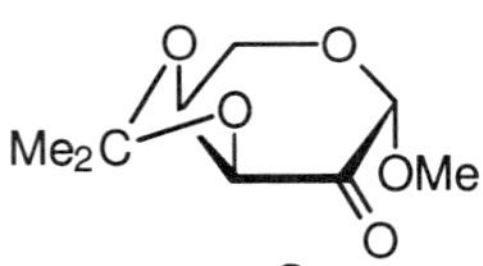 **C**	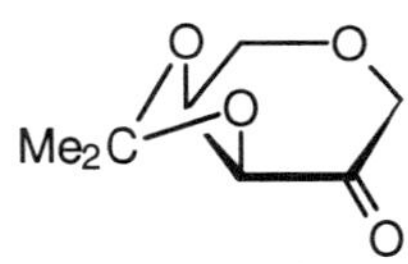 **D**	40	10, 11
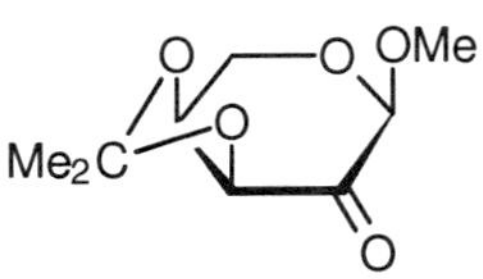	**C** **D**	10 a	10
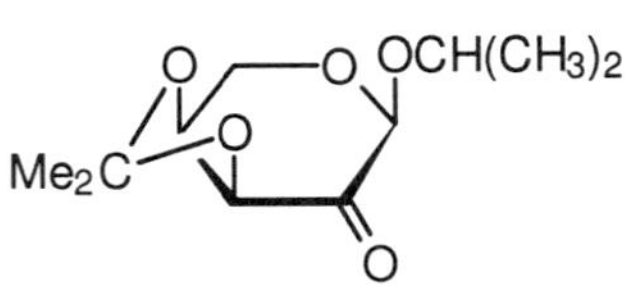	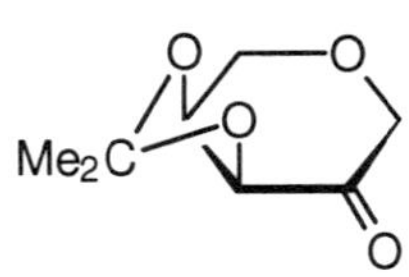	50	10
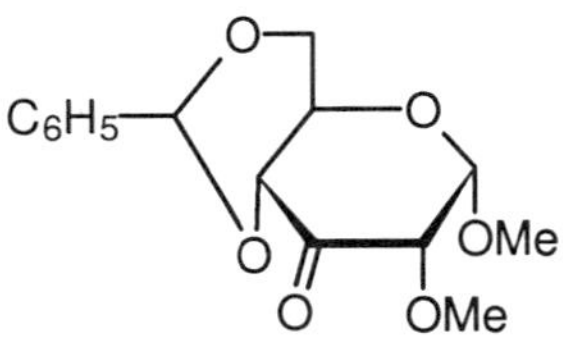	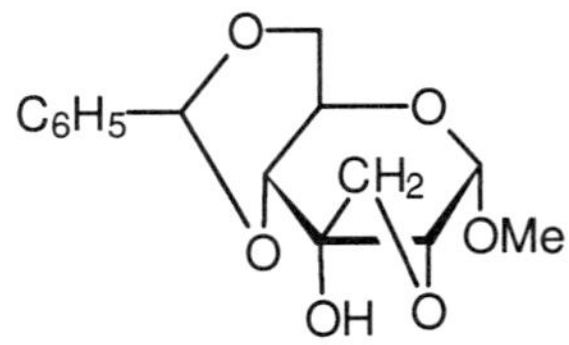	65[b,c]	10, 11

Continued on next page

Table 3.1. Type II reactions of carbohydrates—*Continued*

Reactant	*Product*	*Percent Yield*	*Ref.*
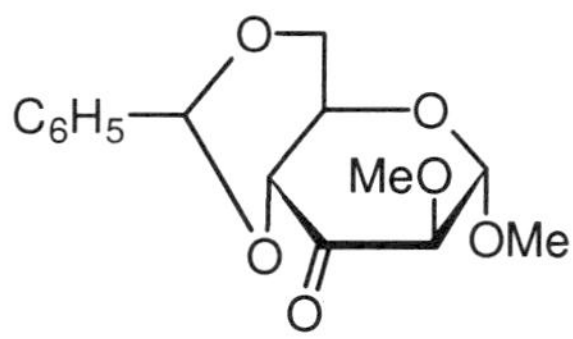	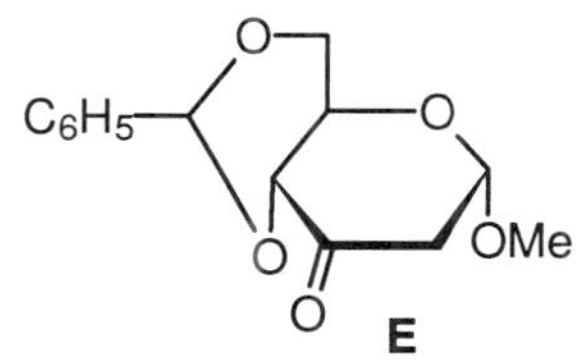 **E**	5[e]	10
	+ 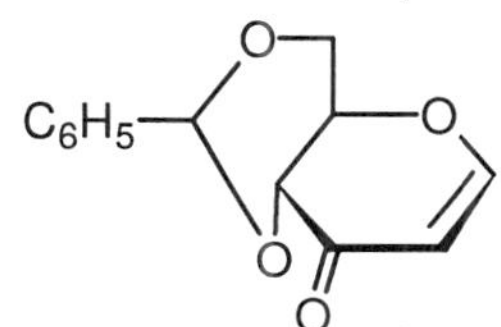 **F**	46	
	+ 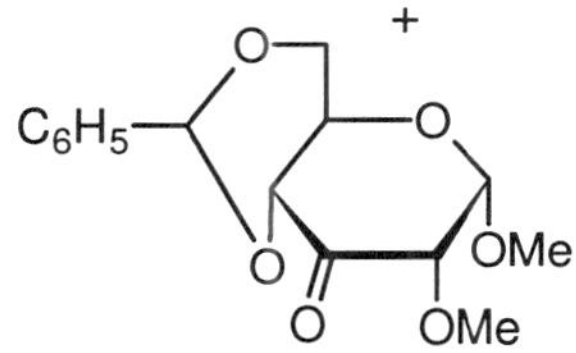	3	
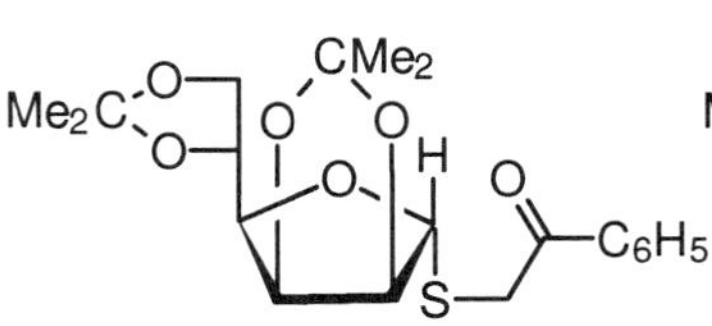	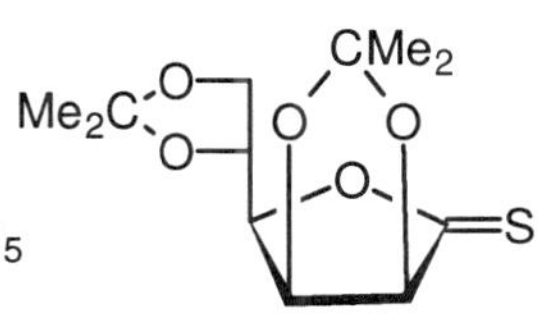	78-89	12
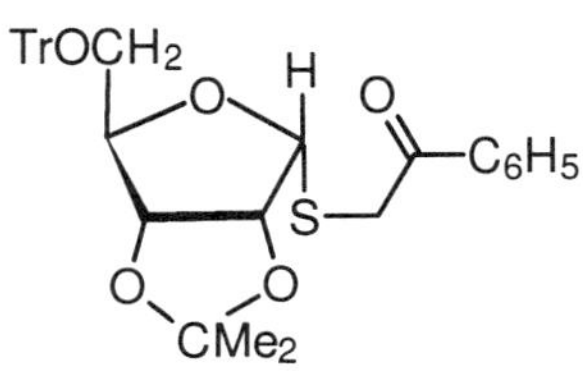	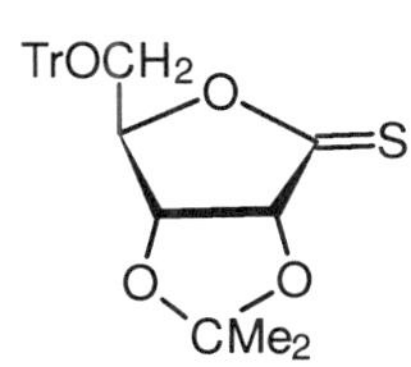	95	12
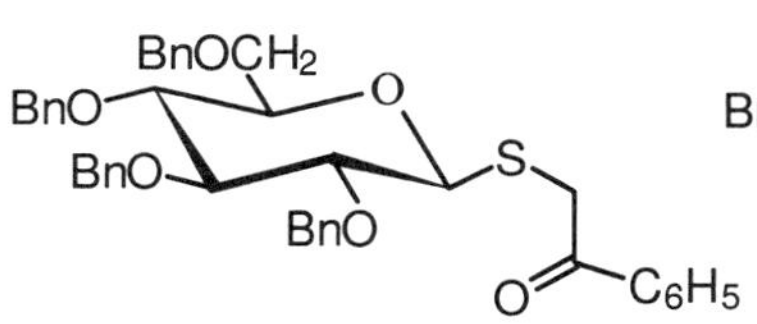	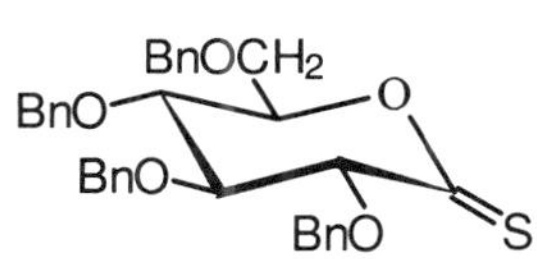	48[d]	12

Continued on next page

Table 3.1. Type II reactions of carbohydrates—*Continued*

Reactant	*Product*	*Percent Yield*	*Ref.*
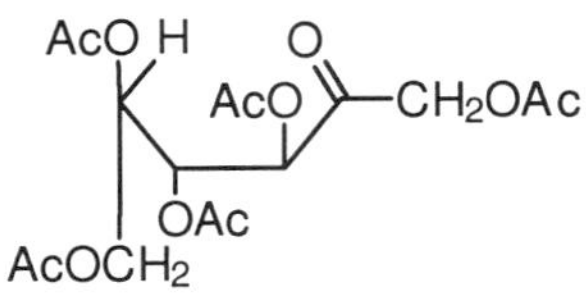	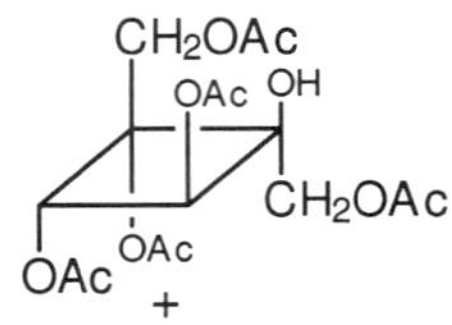 +	12	13
	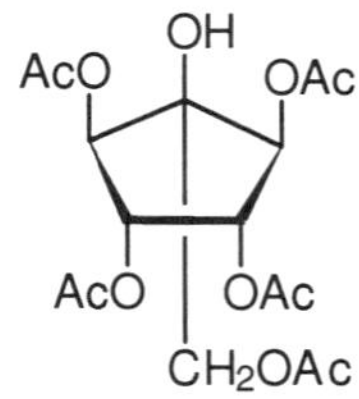	1	13
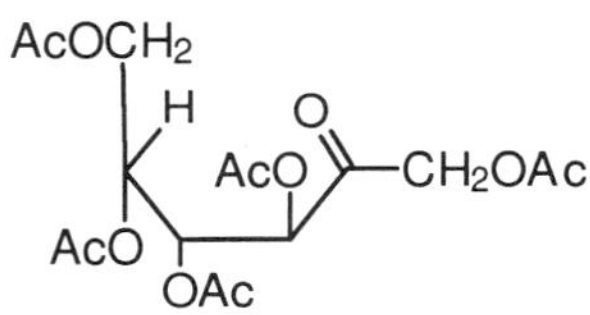	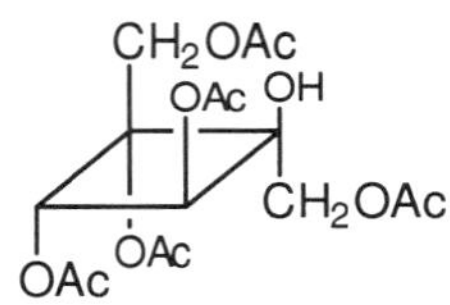	26	13
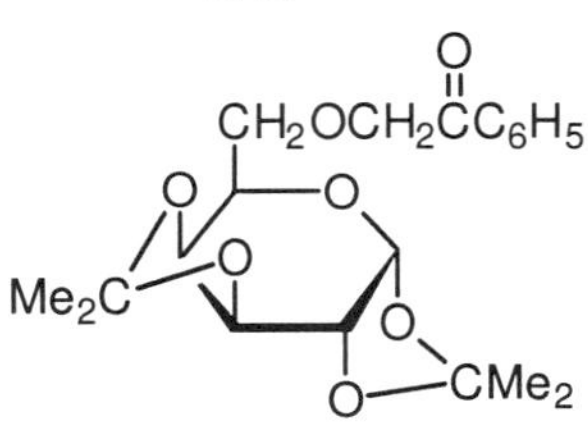	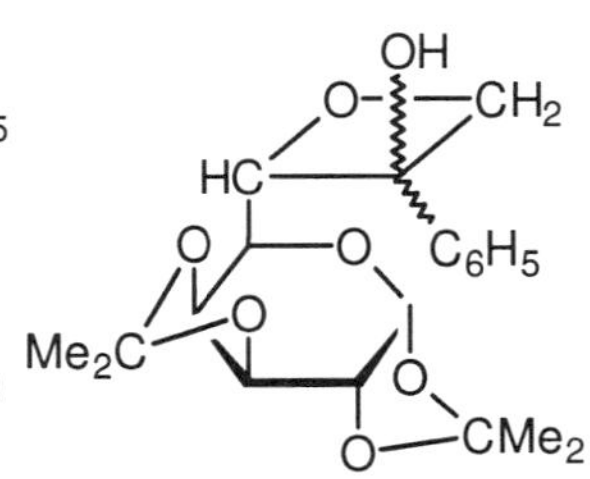	54	14
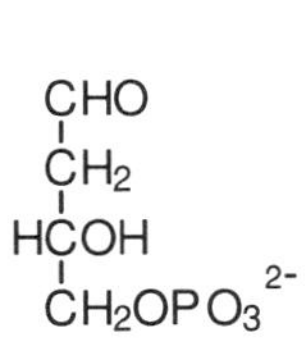	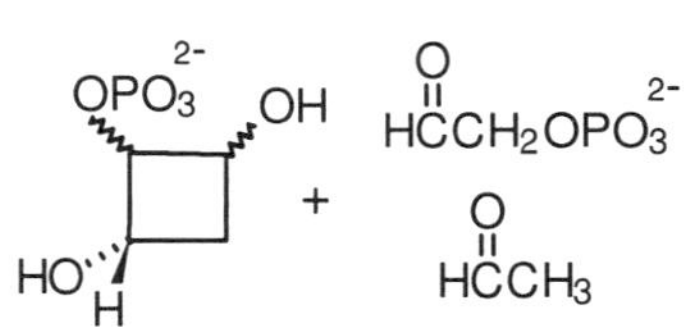	a	15

[a]No yield was reported.

[b]This reaction produced a 1% yield of E and 3% of F. [c]In ref. 11 compound F was identified as the major product from this reaction.

[d]The reaction product is photochemically unstable. Further photolysis causes 2-*O*-benzyl group loss.

[e]The corresponding ethylene derivatives were reported to undergo the same reactions. No yields were given.

Table 3.2. Type II reactions of model compounds.

Reactant	*Product and Percent Yield*	*Ref.*
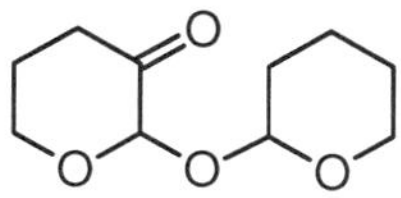	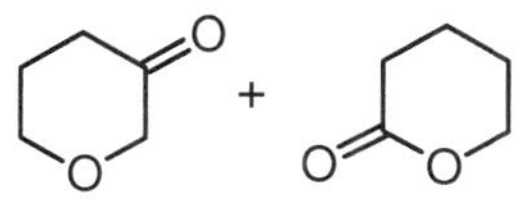 trace 27% (25 °C) 9% 45% (80 °C)	16
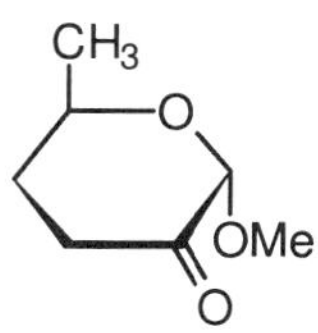	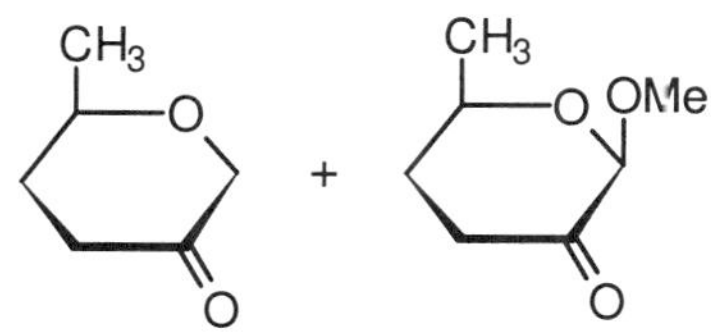 3% 17%	17
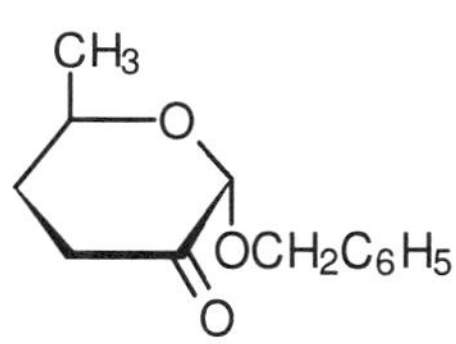	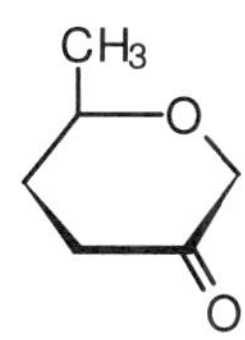 + C_6H_5CHO 64% 13%	17
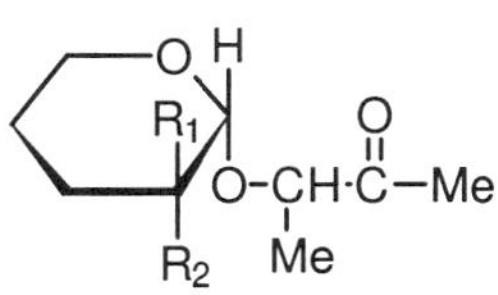 $R_1 = R_2 = H$ $R_1 = R_2 = CH_3$ $R_1 = H$, $R_2 = OMe$	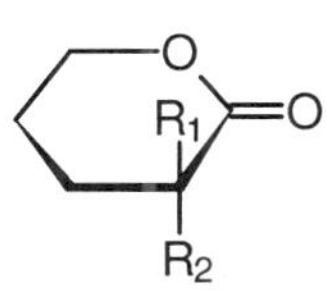 100% 100% 100%	18
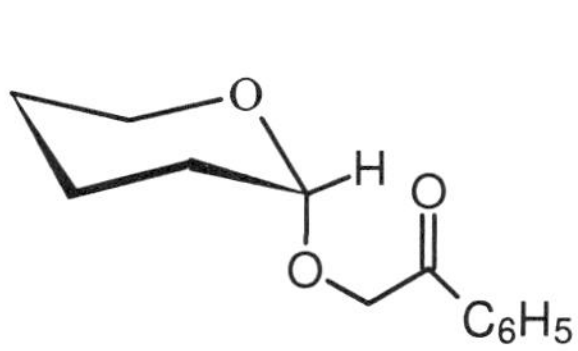	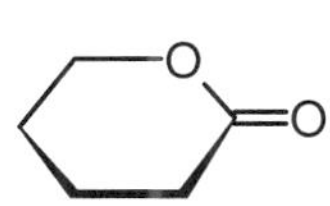 100%	7

Continued on next page

Table 3.2. Type II reactions of model compounds—*Continued*

Reactant	*Product*	*Percent Yield*	*Ref.*
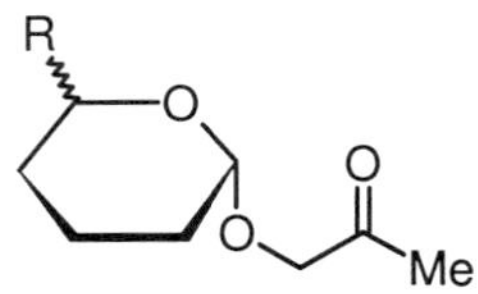	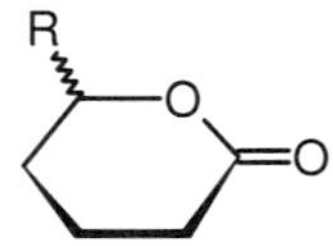		19
R = H		72	
R = OMe		59	
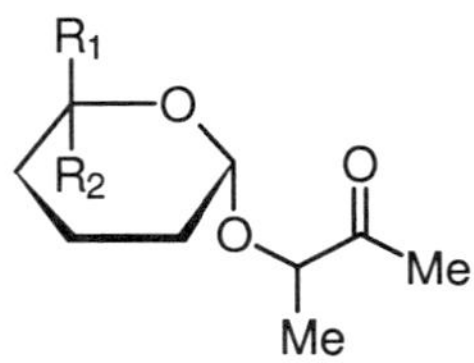	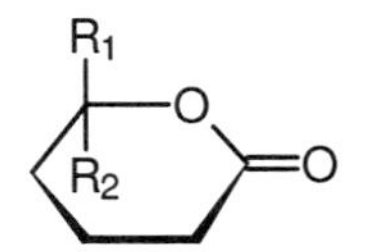		19
$R_1 = R_2 = H$		58	
$R_1 = H,\ R_2 = OMe$		69	
$R_1 = R_2 = Me$		63	
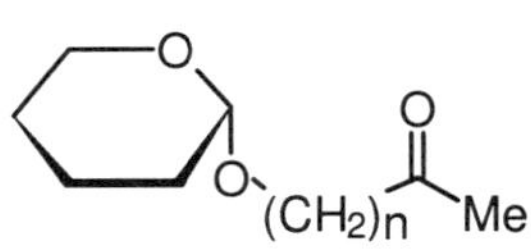	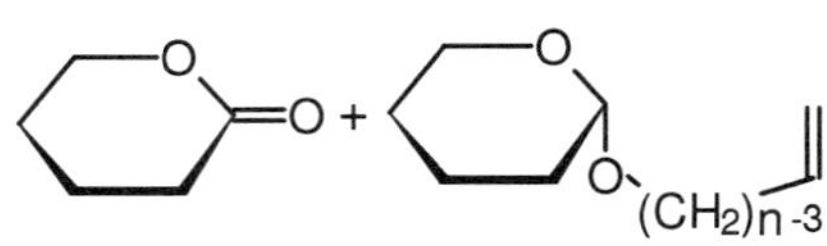		19
n = 3		60%	
n = 4	26%	42%	
n = 5	5%	36%	
n = 6		46%	

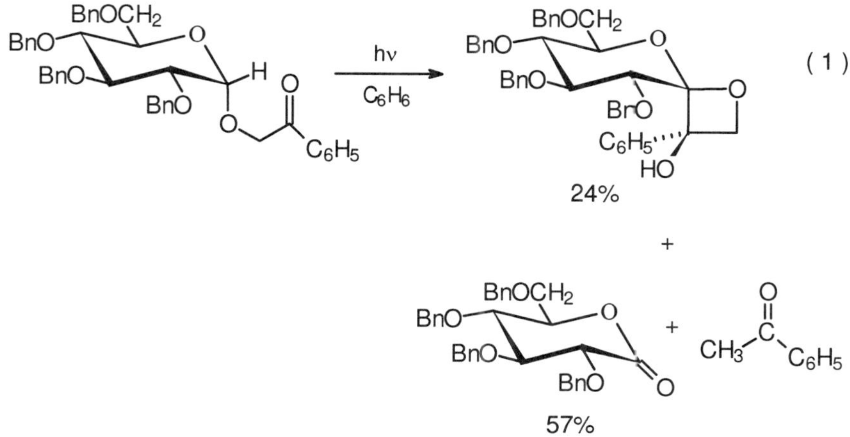

Equation 1

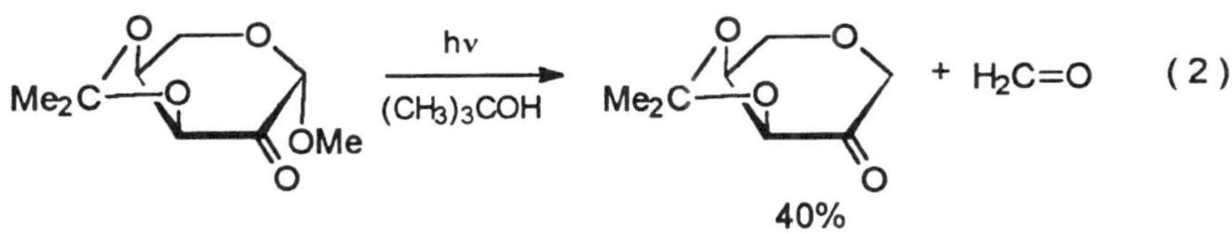

Equation 2

processes is observed (eq 2).[10,11] (Because there are many cases in which considerable starting material is unaccounted for (Tables 3.1 and 3.2), products from an unobserved process may be formed but remain undetected because they are too unstable or too volatile to be isolated by the procedures used.) It is reasonable to expect that the relative amounts of cyclization and cleavage products should vary considerably among carbohydrates because the ring systems that usually exist in these molecules before photolysis can place constraints on product formation. Cyclobutanol production can be limited by the strain introduced into a molecule by forming another ring system. Cleavage, on the other hand, requires an alignment of orbitals in the 1,4-diradical (Scheme 3.3) that may be difficult to attain in molecules for which access to the required conformation is hindered by an existing ring system.

The Nature of the Reactive Excited States

One of the characteristics of the Type II process is that it can occur from both the lowest singlet (S_1) and triplet (T_1) excited states in aliphatic carbonyl compounds.[20–22] For aromatic compounds, however, intersystem crossing is so rapid that essentially all reaction occurs from the triplet state.[23] Establishing the nature of the reactive excited state in Type II reactions of carbohydrates is a rare event. Whistler and Doner[13] did determine that the reaction of com-

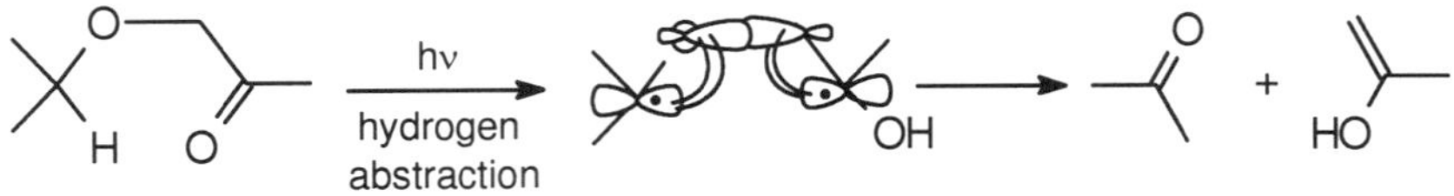

Scheme 3.3

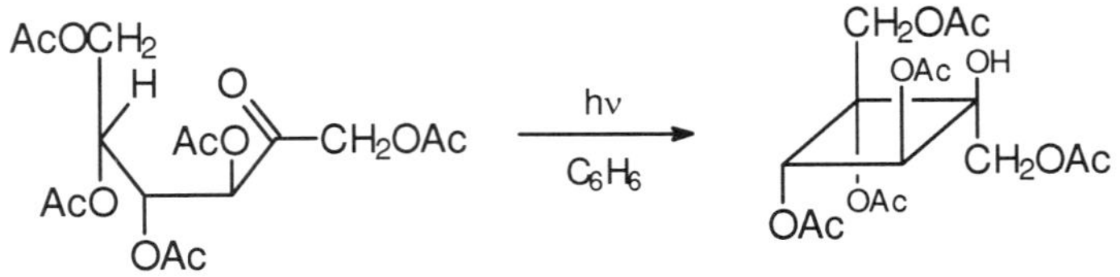

1

26%

(3)

Equation 3

pound **1** (eq 3) was quenched by piperylene (no Stern–Vollmer relationship was established), suggesting a reactive triplet excited state; however, identifying reactive excited states has not been a high-priority task for carbohydrate photochemists because their interests generally have been focused on synthetic applications rather than mechanistic studies. Because investigation of several aliphatic ketones has shown that reaction from excited singlet and triplet states is qualitatively similar,[1] not identifying the reactive state in excited carbohydrates probably has not significantly compromised the use of these reactions in synthesis. It is worth noting, however, that examples exist in which reaction from an excited singlet state is stereospecific while that from the triplet state is not;[24,25] therefore, selecting the reactive excited state (by addition of appropriate sensitizers or quenchers) merits consideration in the future in instances where stereochemistry of the product is an important aspect in the use of the reaction.

Solvent Effects

Wagner has noted that quantum yields for Type II reaction of several ketones approach unity when *t*-butyl alcohol is the reaction solvent.[5] He has explained this observation by proposing that hydrogen bonding between the solvent and the intermediate 1,4-diradical increases the quantum yield by effectively preventing return of the 1,4-diradical to starting material.[26,27] Transfer of a hydrogen atom back to the γ-carbon atom is made difficult because the stabilization provided by hydrogen bonding must be overcome in order for this reaction to take place (Scheme 3.4).[28] In addition to being able to participate in hydrogen bonding, *t*-butyl alcohol is the solvent of choice because it is a poor hydrogen atom donor and, therefore, does not participate readily in intermolecular reaction with excited carbonyl compounds. Although the effect of *t*-butyl alcohol on the quantum yield for Type II reaction of carbohydrates has

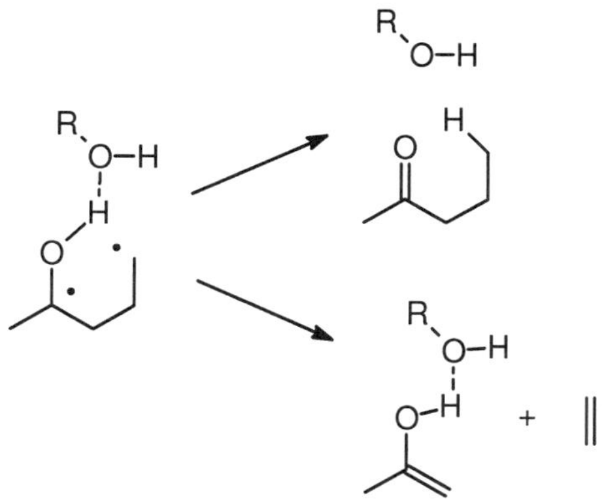

Scheme 3.4

not been studied, its advantage as a poorly hydrogen-donating solvent has been recognized.[10,11]

Reactivity of Anomeric Hydrogen Atoms

Studies of intra- and intermolecular hydrogen abstraction reactions have shown that differences can exist between the reactivity of α and β anomeric hydrogen atoms. (See "Hydrogen Abstraction from a δ-Carbon Atom" in this chapter and "Reaction of Anomeric Hydrogen Atoms" in Chapter 5.) Brunckova and Crich were interested in determining whether such a difference existed for hydrogen abstraction from anomeric carbon atoms during Type II reaction.[7] Their findings, based upon careful comparison of the reactivity of compounds **2** and **3** (Chart 3.1), showed that no difference existed. They concluded that intramolecular hydrogen abstraction in these systems was sufficiently facile that "the difference in activation barriers for removal of an axial or equatorial hydrogen is insignificant".[7] This situation stands in contrast to intermolecular, anomeric hydrogen abstraction by excited carbonyl compounds, in which axial hydrogen atoms are abstracted more easily than equatorial ones. This topic is discussed in detail in Chapter 5.

Range of Type II Reactions

Several Type II reactions have interesting aspects that merit further comment. For example, Cottier et al. found that Type II reaction can be used to introduce unsaturation into the aglycon portion of alkyl glycosides (eq 4).[8] Also, Collins and co-workers discovered that photolysis of keto sugars, in some cases, is a method for removing alkoxy substituents (eq 5).[10,11] Finally, Hürzeler et al. demonstrated that thiocarbonyl-containing carbohydrates could be synthesized by Type II reaction (eq 6).[12] As described in Chapter 6, Type II reactions also can occur from photolysis of other carbonyl compounds.

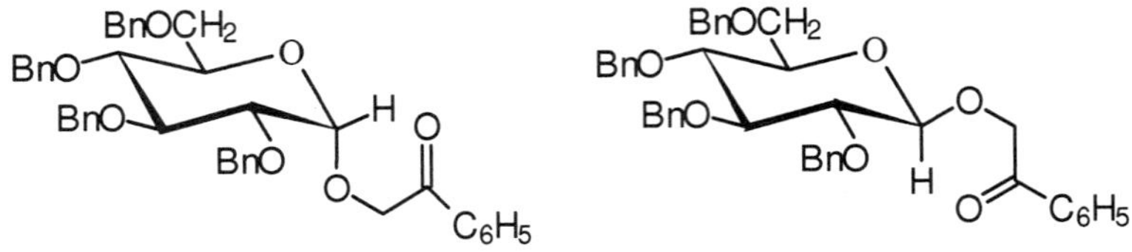

2 3

Chart 3.1. Compounds for testing anomeric hydrogen atom abstraction.

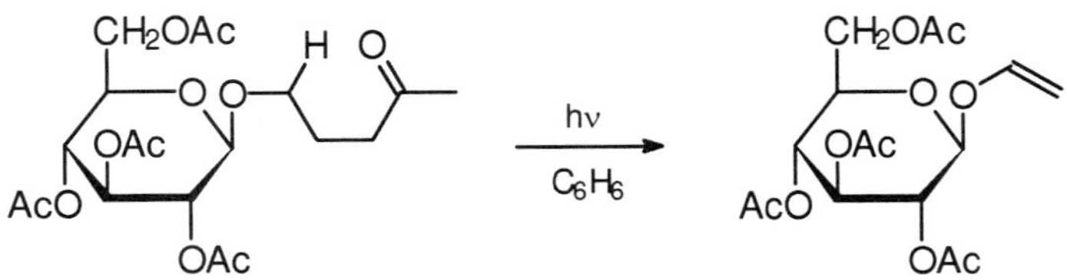

(4)

50%

Equation 4

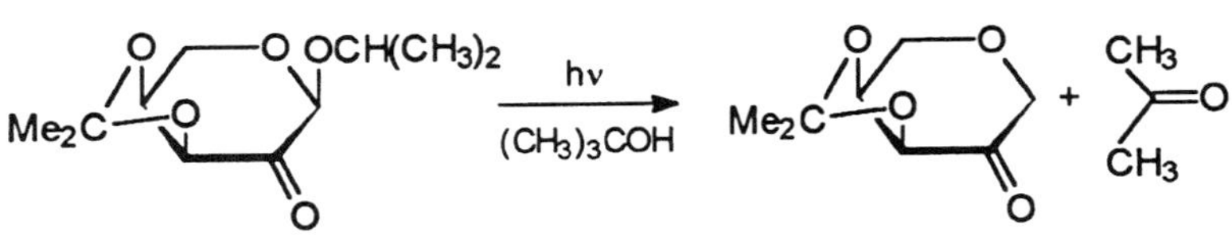

(5)

50%

Equation 5

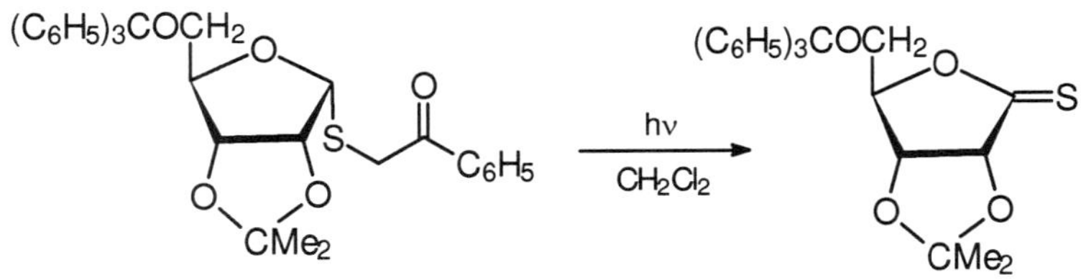

(6)

95%

Equation 6

Hydrogen Abstraction from a δ-Carbon Atom

Criteria for δ-Hydrogen Abstraction

Intramolecular hydrogen abstraction can occur from sites other than the γ-position if no γ-hydrogen atom is available or if structural features favor close approach of the excited carbonyl group to other hydrogen atoms. For equivalent C–H bonds, the δ-hydrogen abstraction rate is estimated to be about 5% of that for the γ-hydrogen.[29,30] Consequently, δ-hydrogen abstraction generally is of importance only in molecules that lack γ-hydrogen atoms. Because the 1,5-diradicals formed by δ-hydrogen abstraction do not undergo cleavage to give stable products, such as those generated by Type II reaction (Scheme 3.1), cyclization to produce five-membered rings becomes the favored reaction pathway (Scheme 3.5). (Even though reactions involving δ-hydrogen atom abstraction are not actually Type II processes, their similarity to Type II reactions makes it logical to discuss the two reactions in close proximity.) As with Type II reactions (Scheme 3.1), intramolecular disproportionation competes with cyclization to reduce the quantum yield for product formation by regenerating the starting carbonyl compound (Scheme 3.5).[29,30]

Formation of Spiro Compounds

If a δ-hydrogen atom is attached to a ring carbon atom, abstraction and then closure of the resulting 1,5-diradical will produce a spiro compound. Descotes and co-workers have used this reaction to synthesize a variety of "spiroketals" that contain pyranoid ring systems. Various examples of this type of reaction

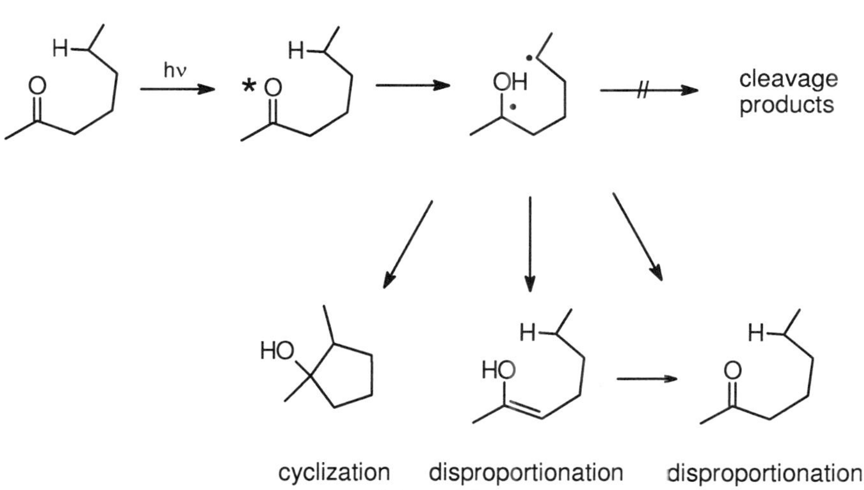

Scheme 3.5

are found in Tables 3.3–3.6. In nearly all instances, the carbonyl group is positioned in the aglycon portion of the molecule and the δ-hydrogen atom is attached to C1. The yields from these reactions are generally low; in fact, often less than half of the reacted starting material can be accounted for. Caution, therefore, needs to be exercised in drawing conclusions based upon product yields and ratios; nevertheless, interesting observations about the reactivity of anomeric hydrogen atoms have been made.

Reactivity of Anomeric Hydrogen Atoms

Descotes and co-workers have observed that if an anomeric hydrogen atom is attached to a carbon atom δ-related to the excited carbonyl group (and no γ-related hydrogen atom exists), intramolecular abstraction occurs more readily from β-glycosides than α-glycosides.[31,32] [This observation stands in contrast to the nearly equal reactivity of anomeric hydrogen atoms when they are attached to a carbon atom γ-related to the excited carbonyl group (see the section "Reactivity of Anomeric Hydrogen Atoms" in this chapter).[7]] A striking example of this difference in reactivity is seen in eqs 7 and 8, where the β-anomer forms a spiro compound but the α-anomer is unreactive. Descotes has proposed that a difference in reactivity of this type can be attributed to a difference in the strength of the two anomeric C–H bonds.[45–47] Supporting this general idea is the work of Wagner and Kemppainen, who determined from study of the photolysis of various γ-substituted ketones that weaker C–H bonds undergo hydrogen abstraction more easily that stronger ones.[5,48] If, therefore, bond strength is "at the heart" of the difference in reactivity illustrated by eqs 7 and 8, the anomeric C–H bond in the α-glycoside must be stronger than that in the β-glycoside.

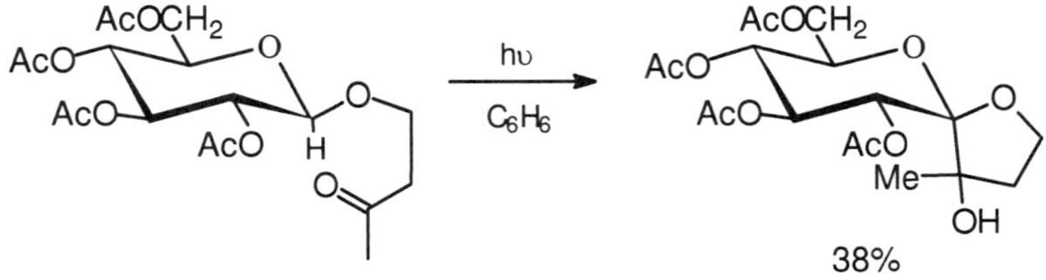

(7)

Equation 7

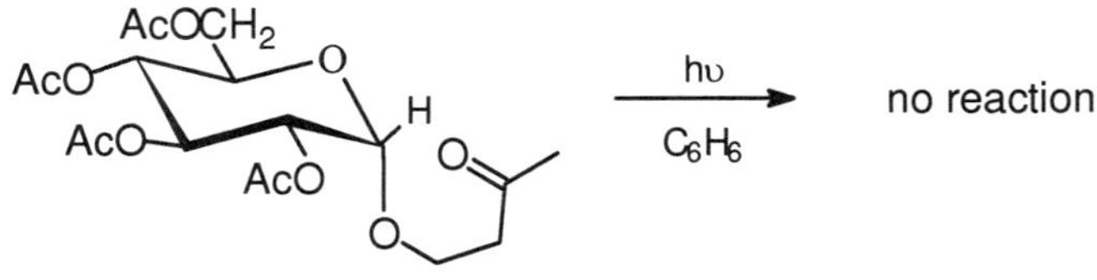

(8)

Equation 8

Table 3.3. Spiro compound formation from irradiation of D-gluco- and D-mannopyranosides.

Reactant	*Product*	*Percent Yield*	*Ref.*
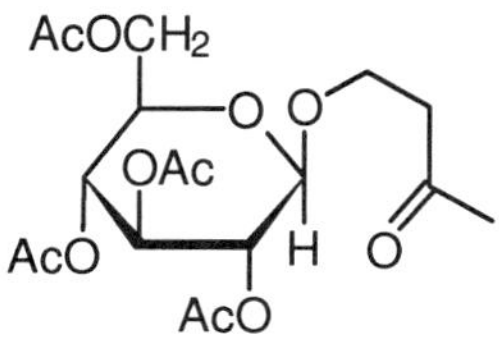	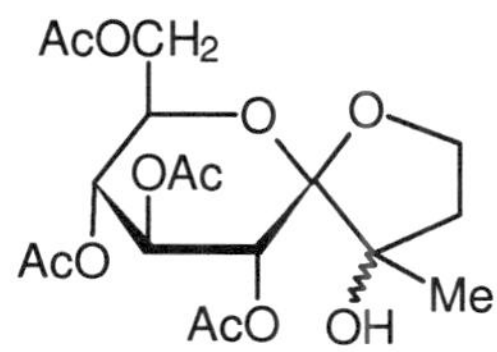	44	31, 32
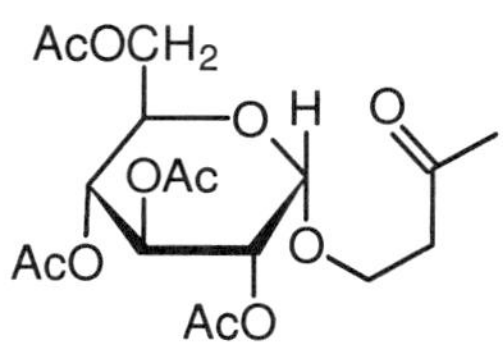	no reaction	----	31, 32
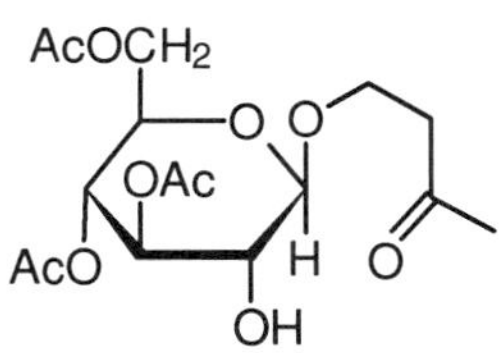	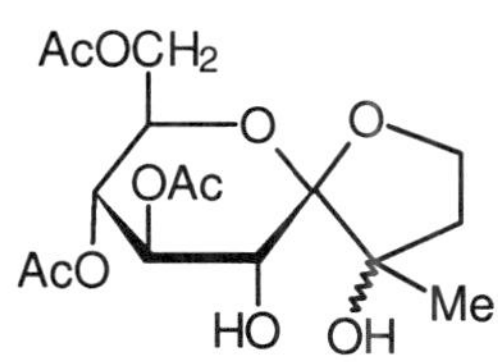	38	32
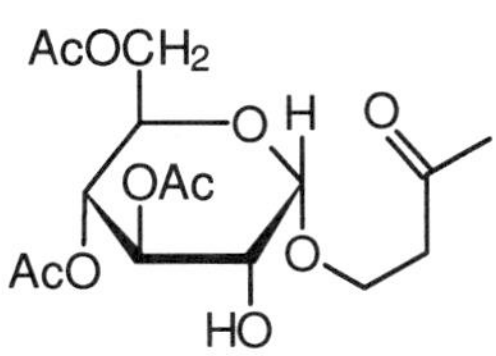	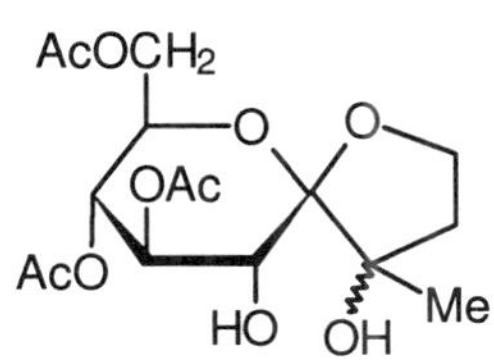	16	32
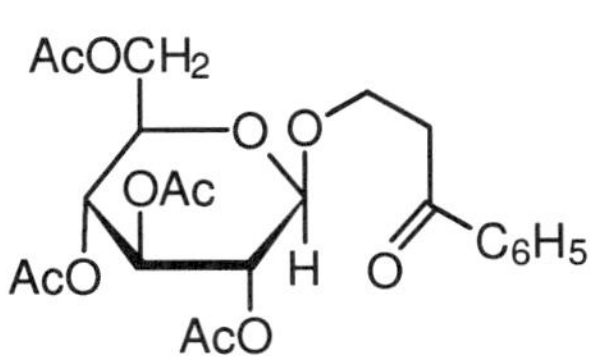	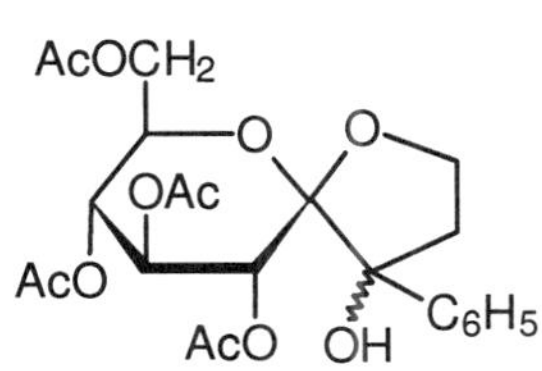	29	33
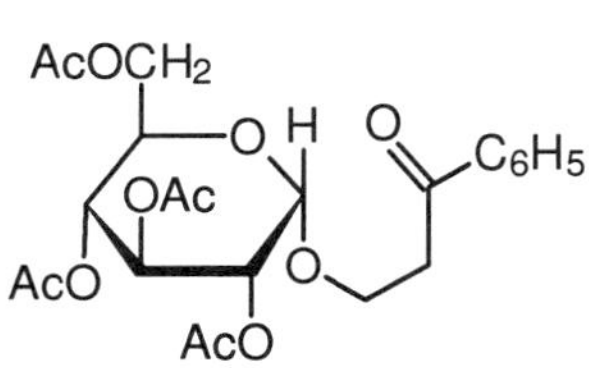	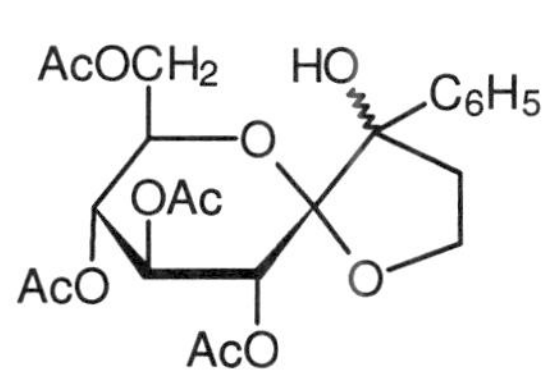	not reported	33

Continued on next page

Table 3.3. Spiro compound formation from irradiation of D-gluco- and D-mannopyranosides—*Continued*

Reactant	*Product*	*Percent Yield*	*Ref.*
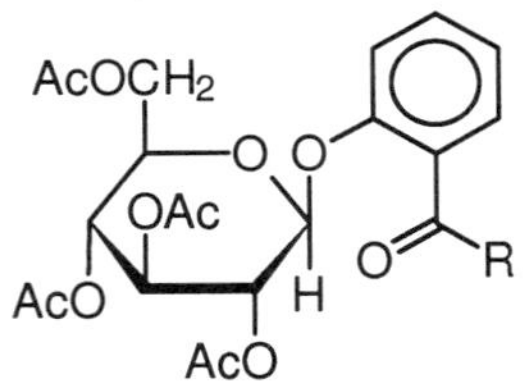	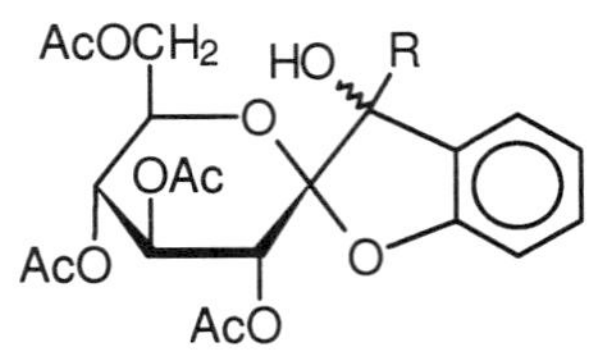		34, 35
R = Me		72[a]	
R = H		52[a]	
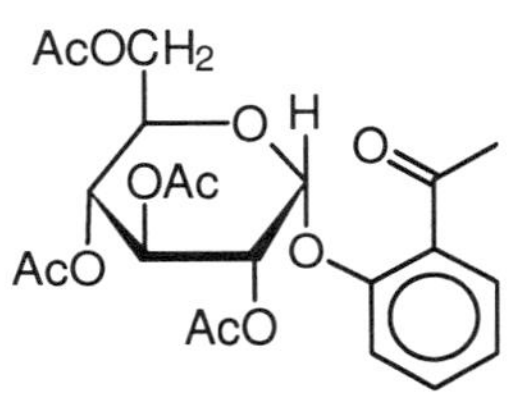	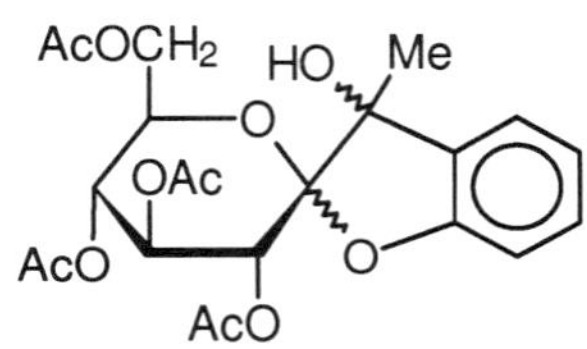	29-50	34, 35
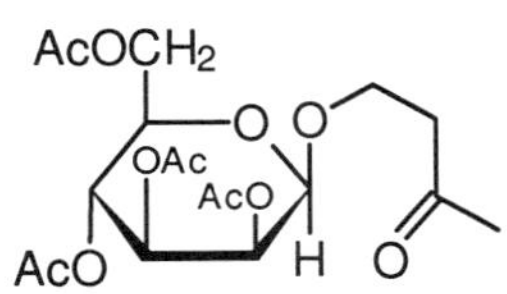	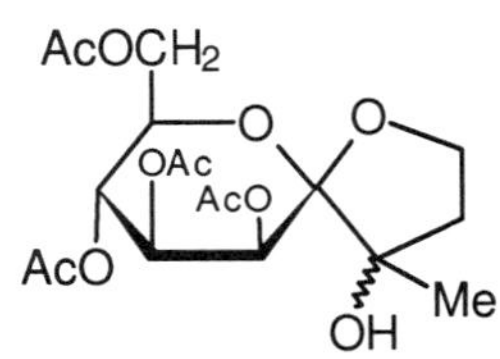	33	36
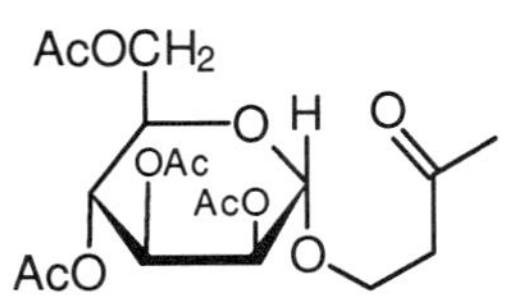	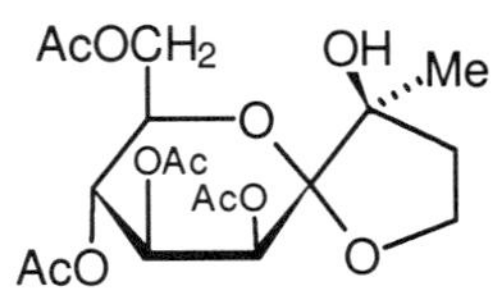	5	36
	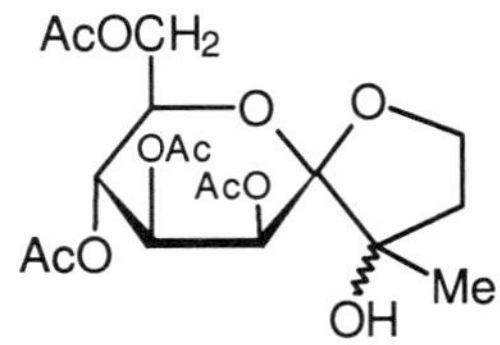	28	

[a]Photolysis mixtures were subject to acid-catalyzed isomerization.

Table 3.4. Spiro compound formation from glycoside irradiation.

Reactant	*Product*	*Percent Yield*	*Ref.*
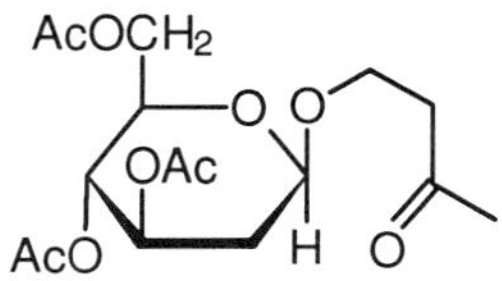	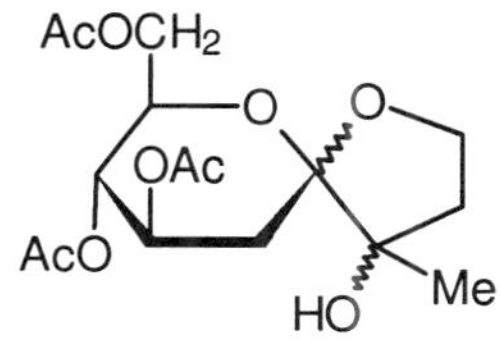	10	37
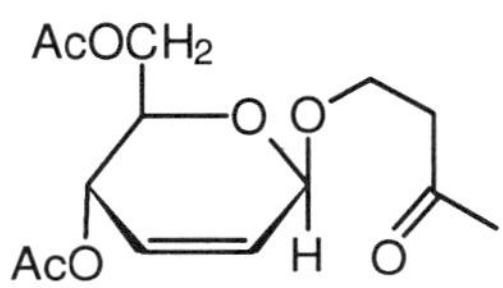	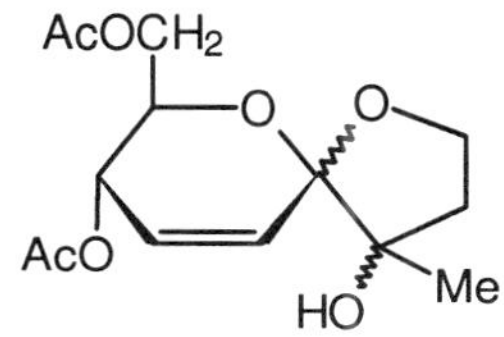	43	37
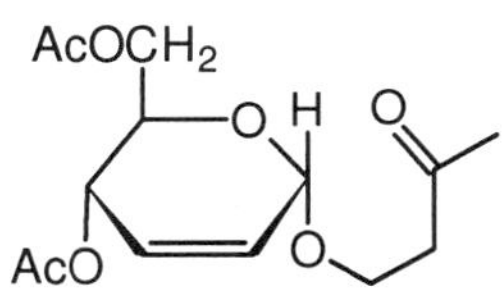	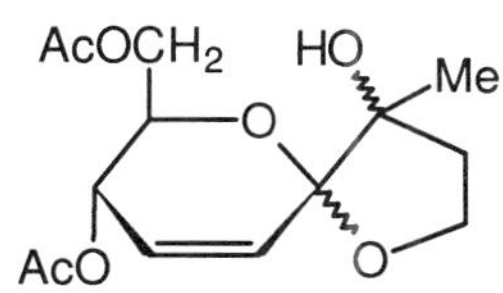	26	31, 32
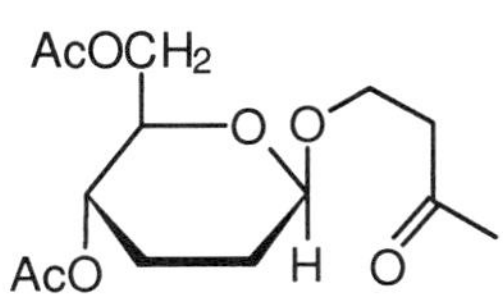	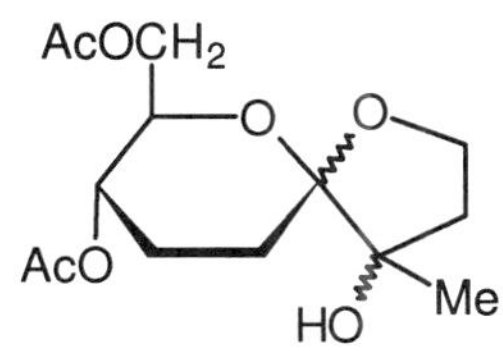	40	31, 32
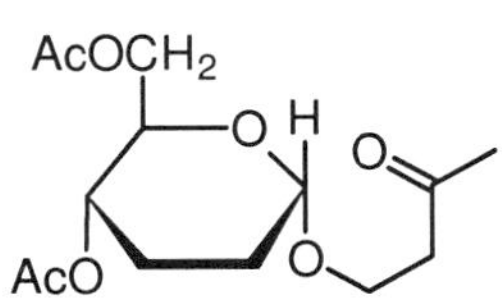	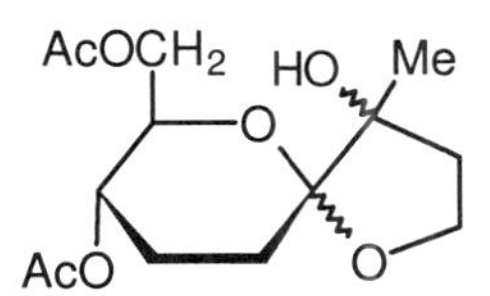	33	31
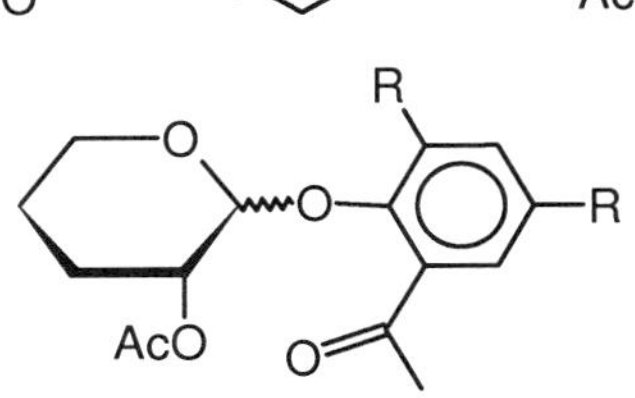	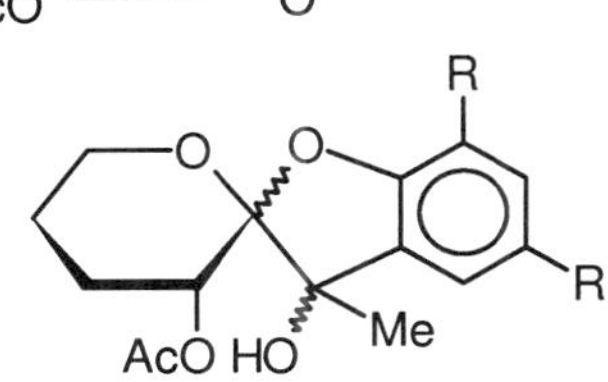		38
R = H		87	
R = OAc		20	
R = OMe		trace	

Continued on next page

Table 3.4. Spiro compound formation from glycoside irradiation—*Continued*

Reactant	*Product*	*Percent Yield*	*Ref.*
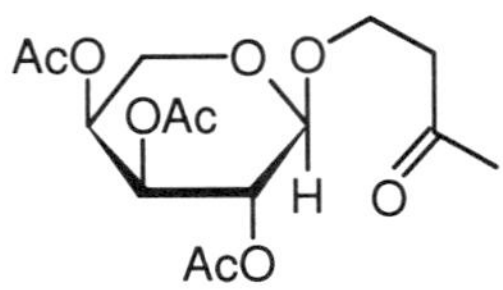	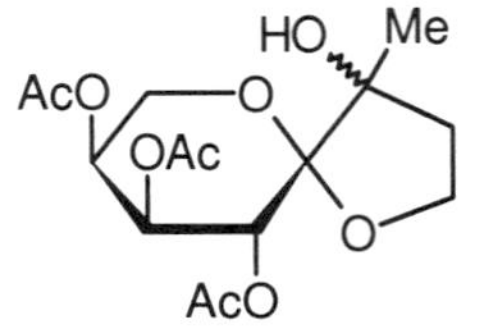 **A**	56	39
	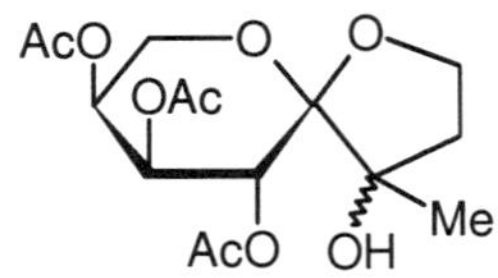 **B**	40	
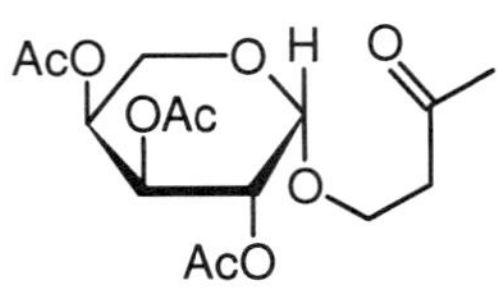	**A**	12	39
	B	13	

Table 3.5. Photocyclization of D-hexofuranos-5-uloses.

Reactant	Product and Percent Yield		Ref.
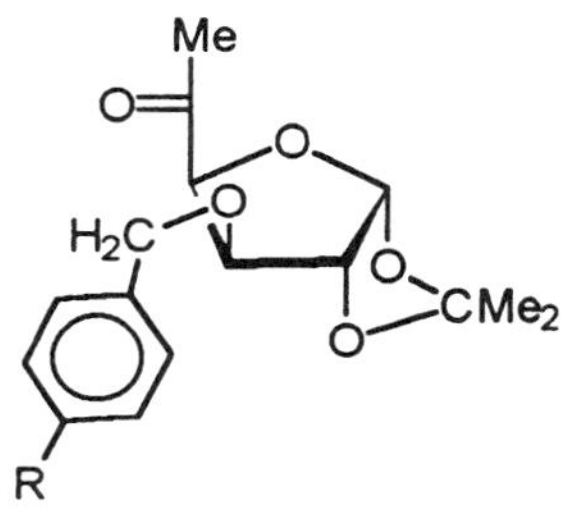	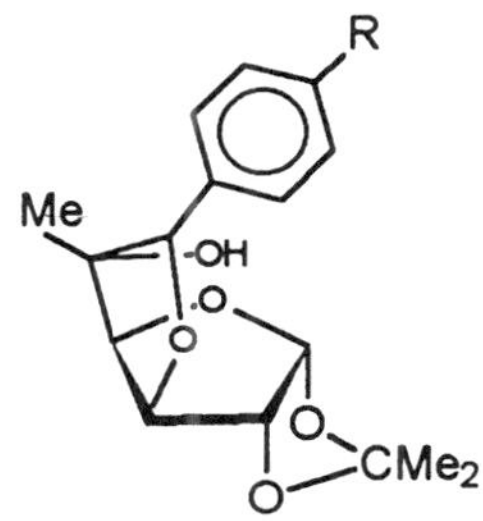 A	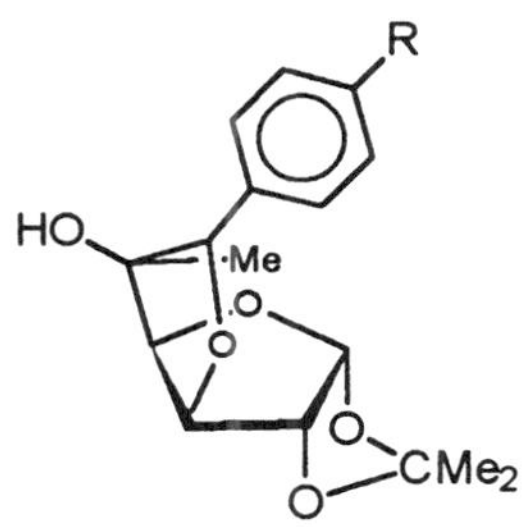 B	40
	A	B	
R = H	54%	11%	
R = OMe	56%	10%	
R = CN	29%	17%	

Table 3.6. Spiro compound formation from irradiation of model compounds.

Reactant	*Product*	*Percent Yield*	*Ref.*
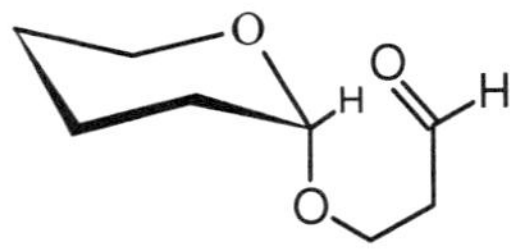	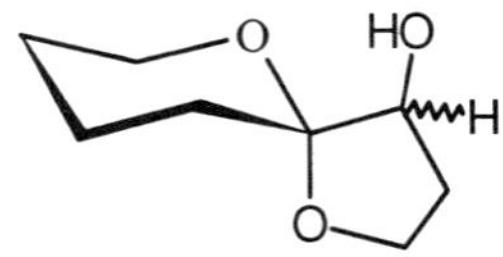	43	41
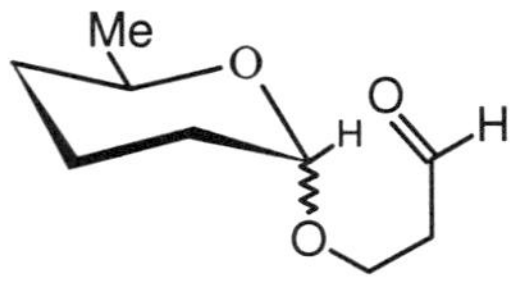	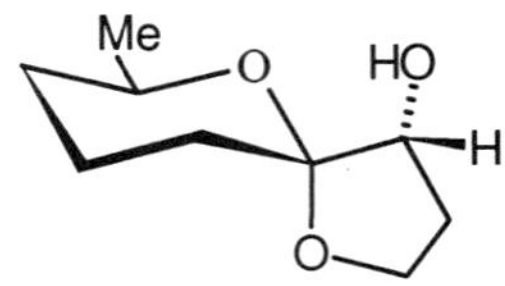	20(α) 7(β)	41
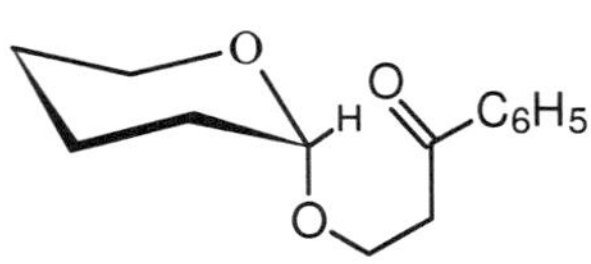	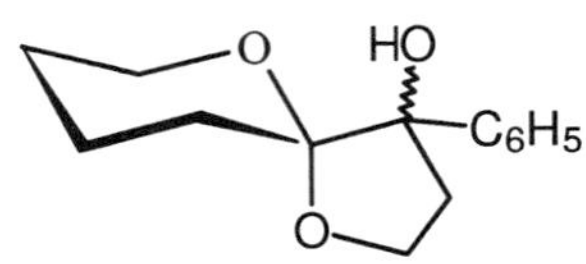	48	33
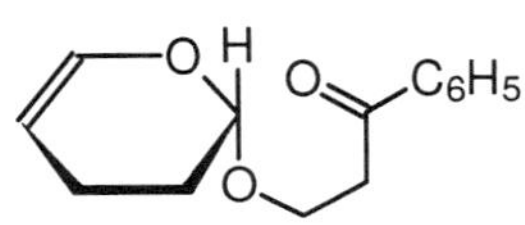	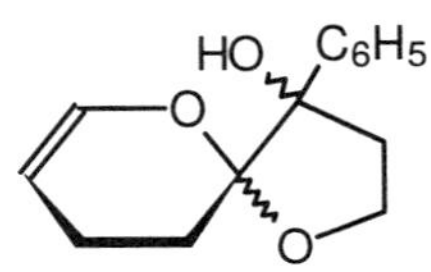	28[a]	33
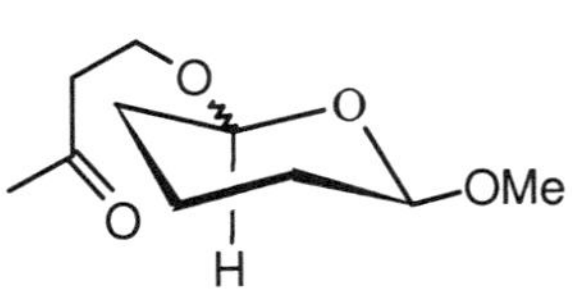	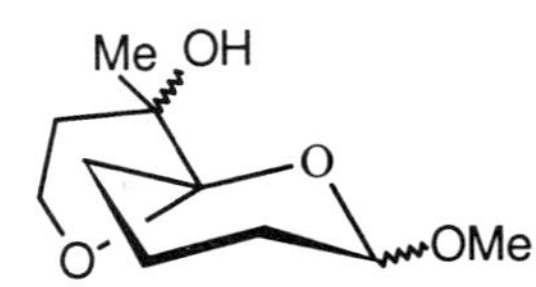	26-58	42
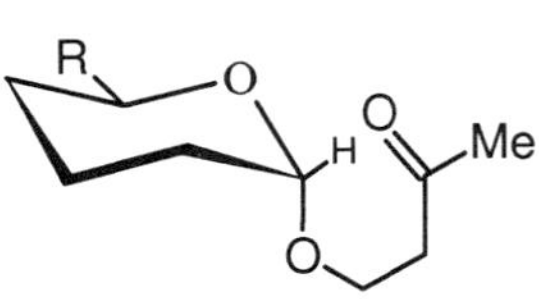	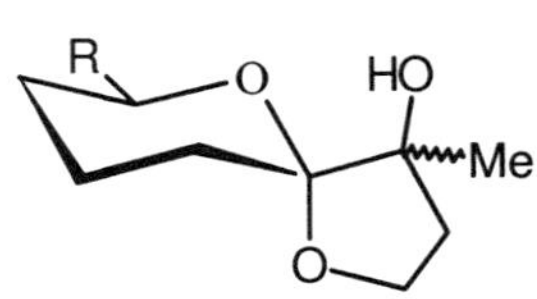		
R = H		77	18, 19
R = OMe		77[b]	18, 19

Continued on next page

Table 3.6. Spiro compound formation from irradiation of model compounds—*Continued*

Reactant	*Product*	*Percent Yield*	*Ref.*
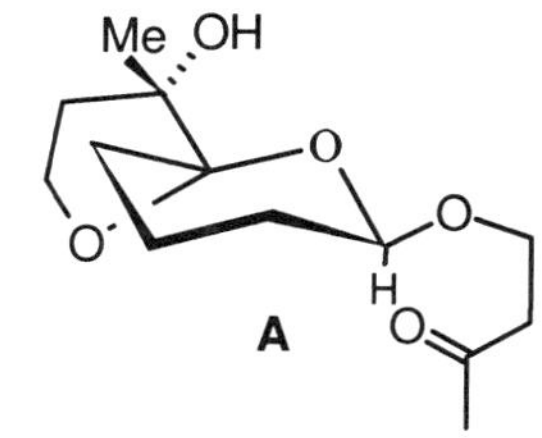 **A**	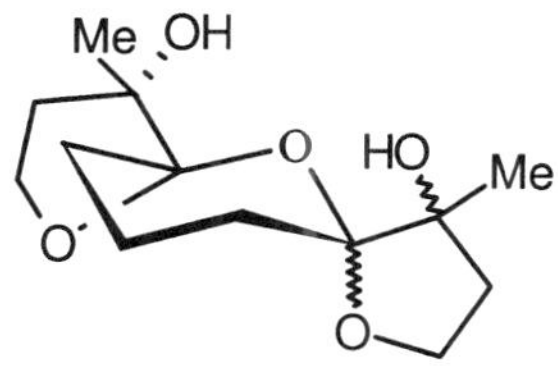	32-57[c]	42, 43
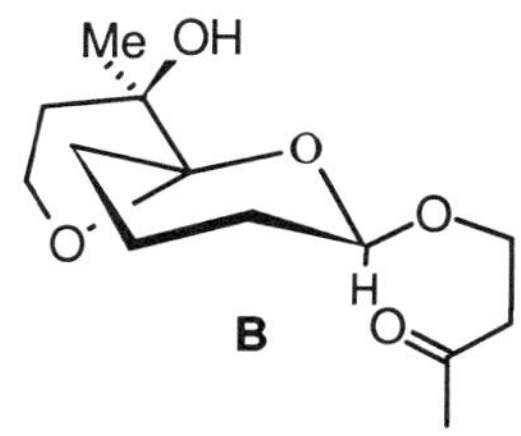 **B**	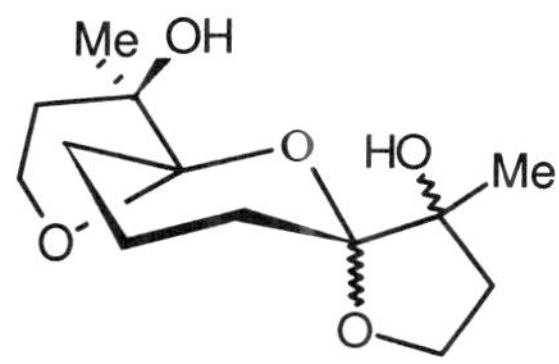	19-63[c]	
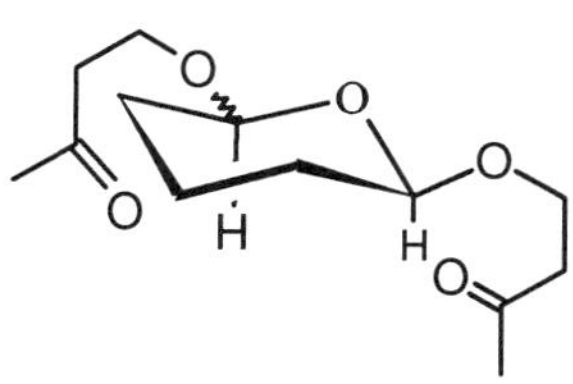	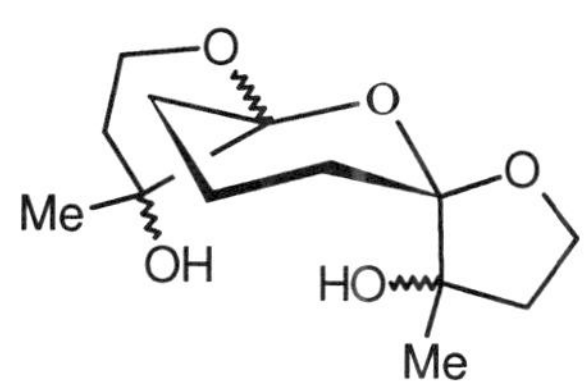	52	43
	A	5	
	B	17	
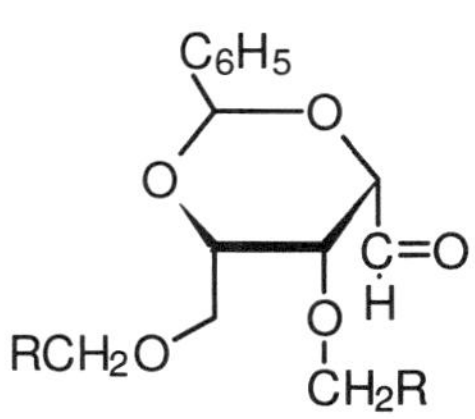	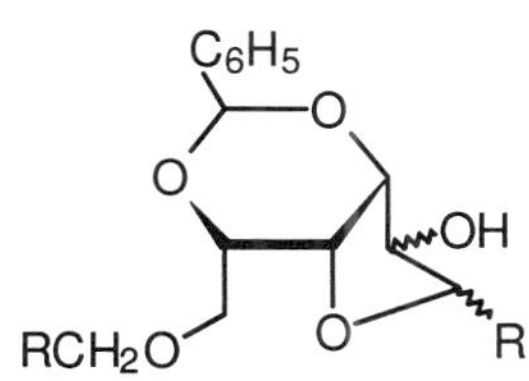		44[b]
R = C_6H_5–		57[b]	
R = CH_2=CH–		37[b]	
R = CH_3O—		44	

[a]The [2 + 2] cycloaddition product (5.5% yield) also was formed.
[b]Reaction was conducted under an inert atmosphere.
[c]Different yields occur under different reaction conditions.

Giese and Dupuis have argued that such a difference in bond strength should exist. These investigators have pointed out that an axially oriented, unshared electron pair on a ring oxygen atom will stabilize the developing anomeric radical, if the orbitals involved overlap effectively.[49–50] Because effective overlap (between an oxygen orbital and the back lobe of the orbital containing the developing radical) can occur for the α-anomeric radical (**4**) but not the β (**5**) (Figure 3.1), the prediction is that the α-anomeric C–H bond should break more easily because of the stabilization of the developing anomeric radical at the transition state.

The rate constant for anomeric hydrogen abstraction also will be affected by substituents attached to C2 if these substituents stabilize (or destabilize) the developing anomeric radical. Giese, Sustmann, and co-workers have shown that if there is an axially oriented substituent at C2, a stabilizing interaction should exist between the α-anomeric radical and the C2 bond.[49–52]

These workers found that this stabilization results from interaction between the SOMO (singly occupied molecular orbital) at C1 and the σ* orbital of the bond to the C2 substituent (Figure 3.2). If the substituent bond is equatorial, stabilizing interaction is reduced, and if C2 has no substituents (other than hydrogen atoms), interaction is even less important. (Presumably the σ* orbital for the axially oriented C–H bond at C2 is sufficiently high in energy that the interaction with the SOMO is considerably reduced.) Descotes

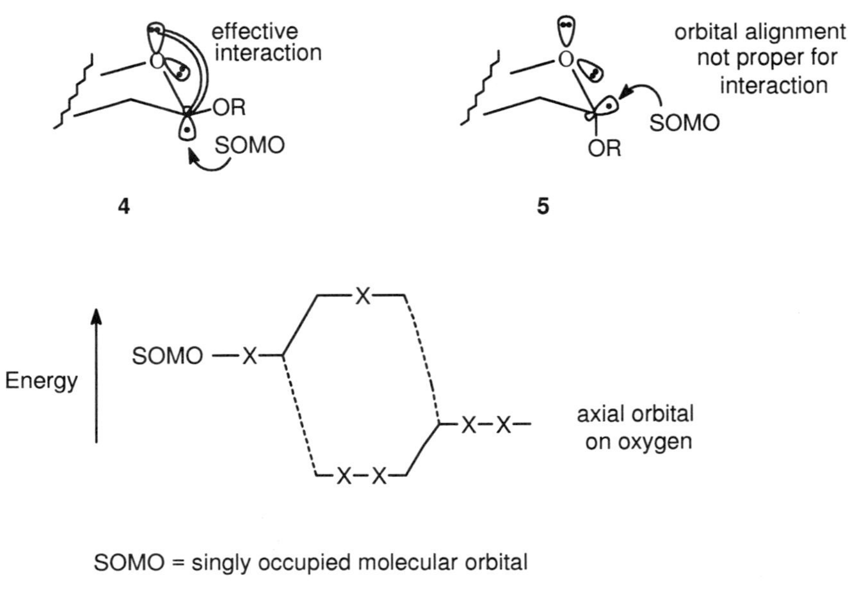

SOMO = singly occupied molecular orbital

Figure 3.1. Interaction between a nonbonding orbital on oxygen and a SOMO on carbon.

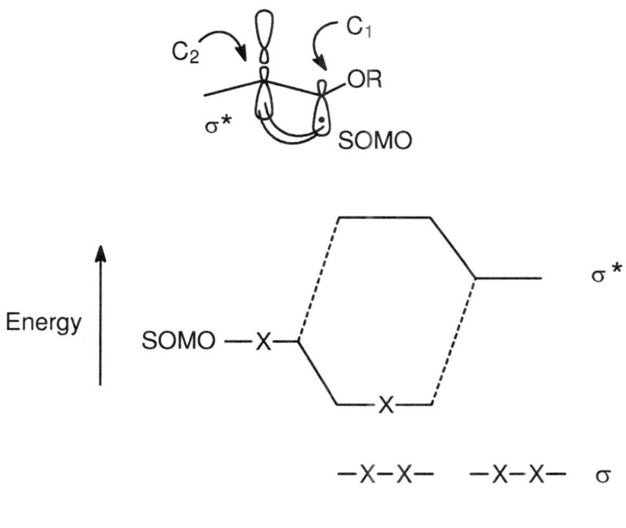

Figure 3.2. Interaction between a σ orbital and a SOMO on the neighboring carbon atom.*

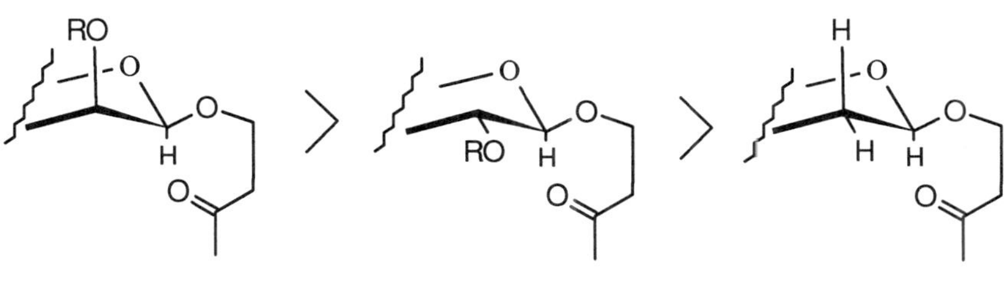

Figure 3.3. Rank order for producing spiro compounds.

has invoked this phenomenon to explain the differences in photochemical reactivity between compounds that have different substituents and stereochemistry at C2 (Figure 3.3).[45,46]

Although the Descotes explanation is a reasonable one, caution is necessary in assigning meaning to differences in the amount of product formed when comparing photochemical reactions. The reason that caution needs to be exercised is that comparisons should be made between rate constants for reaction rather than product yields. Because δ-hydrogen abstraction by an excited carbonyl group is the basic photochemical process in this reaction, the true measure of reactivity is the rate constant for hydrogen abstraction. The study of Type II reactions by Wagner and Kemppainen has shown that differences in rate constants for reaction of different compounds are not necessarily reflected in product yields.[48]

Part of the difficulty involved in relying on product yields has to do with the fact that compounds differ in their ability to absorb light. This particular problem can be overcome by comparing quantum yields rather than product yields. Wagner and Kemppainen have shown, however, that even differences in quantum yields do not correlate with differences in rate constants for Type II reactions.[48] These authors reasoned that because the 1,4-diradical produced by hydrogen abstraction appeared to be formed with 100% efficiency, quantum yields for reaction actually were determined by a competition between the cyclization, cleavage, and hydrogen transfer reactions that are possible from the 1,4-diradical (Scheme 3.1). Because any or all of the rate constants associated with these processes (k_c, k_e, and k_d, Scheme 3.1) may change when a γ-substituent is altered, the effect of such a change on the quantum yield is not necessarily predictable. This possible lack of correlation between the quantum yield and the γ-substituent is the reason for "sounding a note of caution" in interpreting differences in product or quantum yields as indicative of differences in reactivity.* Because the rate constant for reaction (k_r, Scheme 3.1) is a function of the ability of the excited carbonyl group to abstract a hydrogen atom, a lower C–H bond dissociation energy produces a larger k_r; therefore, a definite relationship exists between the rate constant for reaction and change in C–H bond energy brought about by a new γ-substituent.

Isomerization of Anomeric Radicals

One final factor relating to the stereoselectivity of the photochemical formation of spiro compounds deserves comment. Once the anomeric hydrogen atom has been abstracted from an α-glycoside, a competition exists for the 1,5-diradical between isomerization to the more stable α-anomeric radical and ring closure (Scheme 3.6).[32] In some instances ring closure occurs prior to isomerization.[33,41] In other cases the isomerization is complete before closure (eq 9),[32] and sometimes a mixture of products forms, probably reflecting incomplete isomerization (eq 10).[36,39]

Pyruvate Ester Photolysis

Reaction Mechanism

In 1961 Hammond et al. reported that ethyl pyruvate (**6**) reacted from its triplet excited state to give the products shown in eq 11.[54] This initial investigation

*Coyle provides a brief but excellent discussion of the relationship between quantum yields and reaction rate constants.[53]

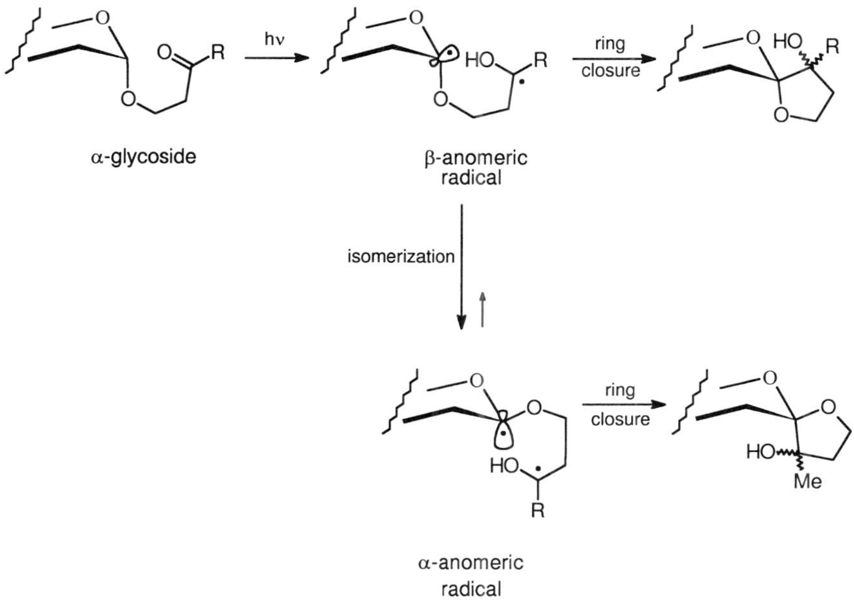

Scheme 3.6

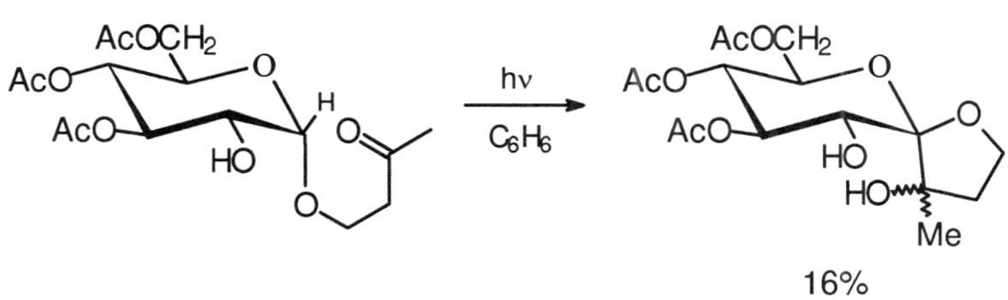

(9)

Equation 9

was followed by two papers by Leermakers and co-workers on further photochemical (and thermal) study of simple esters of pyruvic acid.[55,56] From this latter work, the authors concluded that loss of carbon monoxide was the primary process and that the reaction involved radical intermediates. Although they did not propose a specific reaction mechanism, one consistent with their findings is shown in Scheme 3.7. This mechanism is an example of a Norrish Type I process.

Because esters of pyruvic acid (and other α-keto acids) can, in principle, undergo Norrish Type II reaction (Scheme 3.8), one question that remained after these initial studies[54–56] was whether any esters of α-keto acids would react along a Type II pathway. Huyser and Neckers answered this question when they found that photolysis of esters of phenylglyoxalic acid took place by

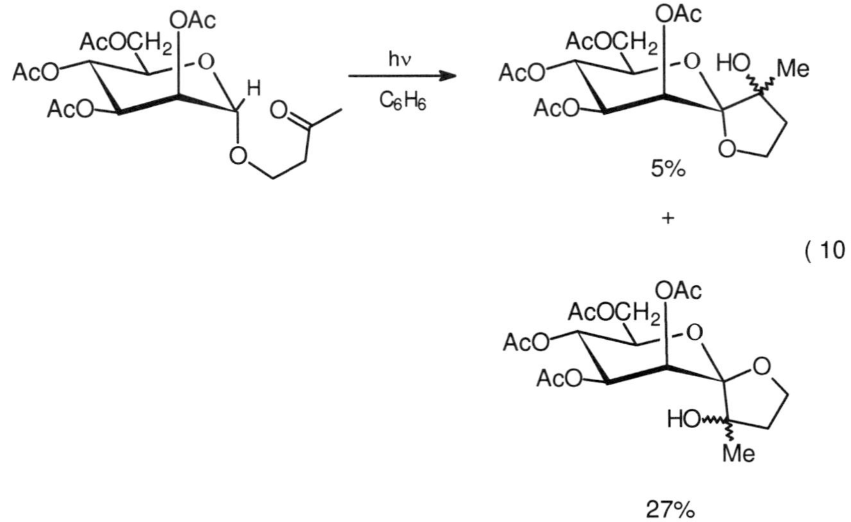

Equation 10

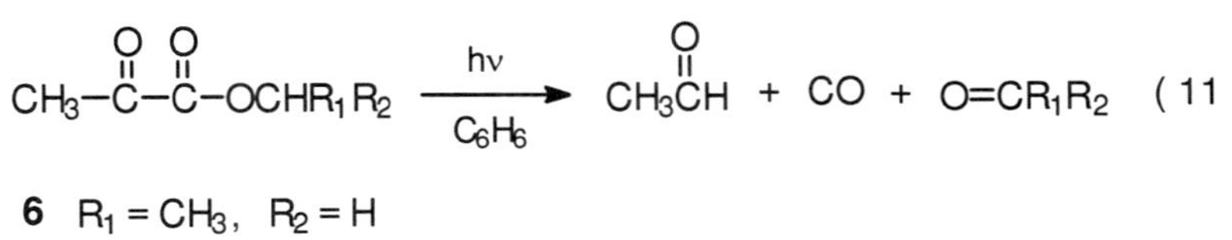

Equation 11

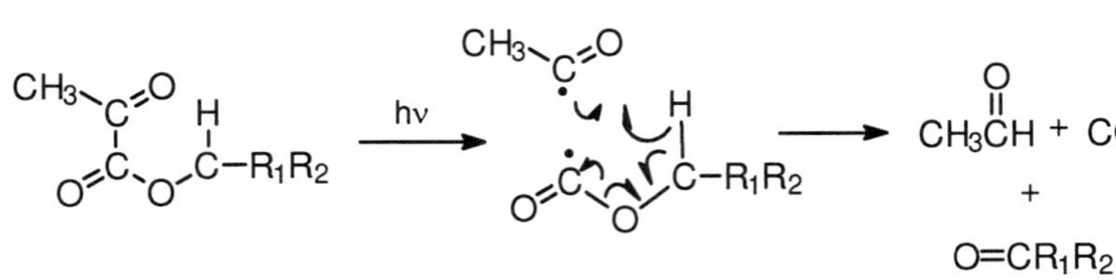

Scheme 3.7

a Type II process (Scheme 3.9).[57] Several years later, Nakadaira et al. discovered a pyruvate ester that also appeared to react by a Type II pathway.[58] This latter work, in particular, left the general mechanism for pyruvate photoreaction uncertain.

More than two decades after the original photolyses of alkyl pyruvates, Davidson and Goodwin reexamined the mechanism for alkyl pyruvate photoreaction and concluded that this reaction was a Type II process.[59] Their conclusion was based primarily on the observation that *t*-butyl pyruvate (no hydrogens attached to the γ-carbon atom) was photostable, unlike methyl, ethyl,

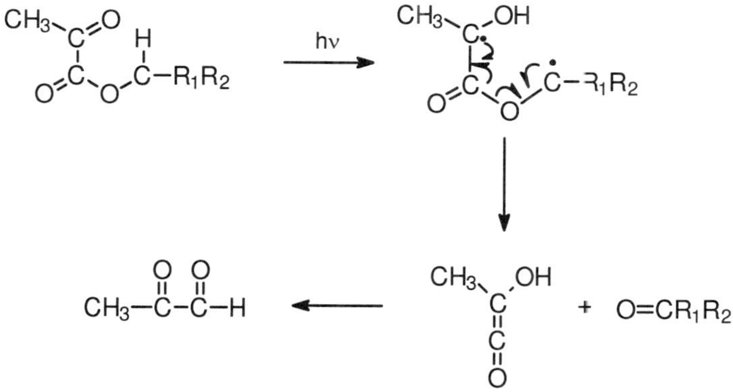

Scheme 3.8

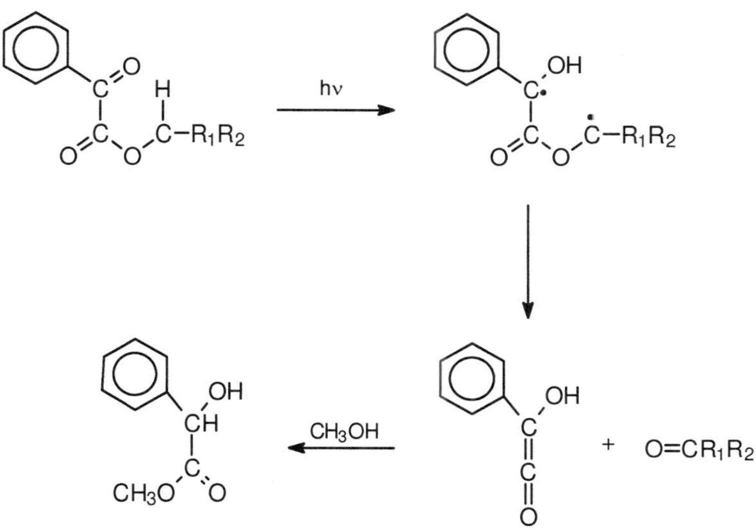

Scheme 3.9

isopropyl, and benzyl pyruvates (all having γ-hydrogen atoms), each of which underwent normal reaction (eq 11). Excited *t*-butyl pyruvate also had a much longer lifetime than any of the other pyruvates, indicating that a pathway for reaction of the other pyruvates had been eliminated in the *t*-butyl derivative. These results were consistent with a Type II mechanism but seemed incompatible with a Type I process. Scaiano and co-workers, using flash photolysis techniques, concurred with the proposal that pyruvate ester photoreaction was a Type II process.[60,61] More recent work by Hu and Neckers also supports a Type II reaction mechanism.[62]

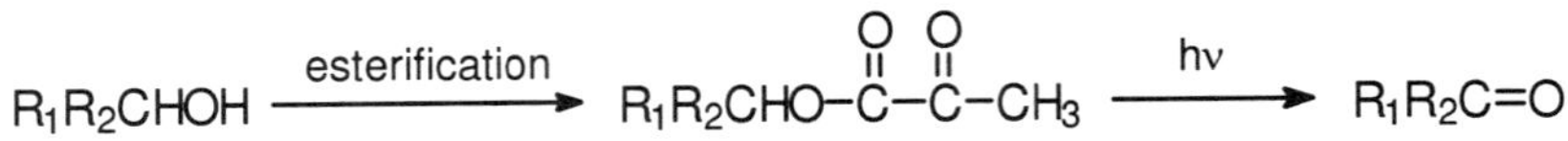

Scheme 3.10

(12)

Equation 12

Photochemically Based Alcohol Oxidation

When considered from a synthetic point of view, the photolysis of pyruvate esters described by the early investigators (eq 11) had the potential to become the "key" element in a photochemically based reaction sequence for the oxidation of alcohols (Scheme 3.10). The first step in this sequence would be the esterification of the alcohol to be oxidized with pyruvic acid (or an appropriate derivative), and the second step would be photochemical reaction. All that was needed to complete the sequence, therefore, was an effective, general method for synthesizing pyruvate esters. This requirement turned out to be a difficult one to meet because normal esterification procedures could not be used with pyruvic acid.* The problem was solved by Ottenheijm and de Man, [63] who discovered a mild procedure for synthesis of pyruvoyl chloride (eq 12) and, in so doing, found the missing element in a general procedure for photochemical oxidation of alcohols.

With pyruvoyl chloride as the esterifying agent, Binkley found that pyruvate ester formation and photolysis could be used to oxidize a variety of alcohols to the corresponding carbonyl compounds.[64,65] This technique was easily adapted to the oxidation of partially protected carbohydrates and nucleosides (Tables 3.7 and 3.8). Primary alcohols were oxidized to aldehydes, without further oxidation, and secondary alcohols were converted with equal ease into ketones. Reaction took place at room temperature, in neutral solution, and with no added inorganic reagents. Even though the yields of carbonyl compounds from pyruvate photolyses were good (Tables 3.7 and 3.8), this procedure offered no special advantage for the oxidation of most carbohydrates; however, in situations where the oxidation products themselves were unstable, photochemical oxidation became a technique of choice because it was conducted under such mild conditions. Experience showed pyruvate formation and photolysis to be successful where other oxidation reactions failed. One example of the "feather touch" nature of photochemical oxidation is seen in

* Pyruvoyl chloride is too unstable to be synthesized by normal procedures. Acid-catalyzed synthesis of pyruvate esters suffers from an equilibrium constant that is quite unfavorable for ester formation.[55]

Table 3.7. Type II reactions of pyruvate esters of sugars.

Reactant	*Product*	*Percent Yield*	*Ref.*
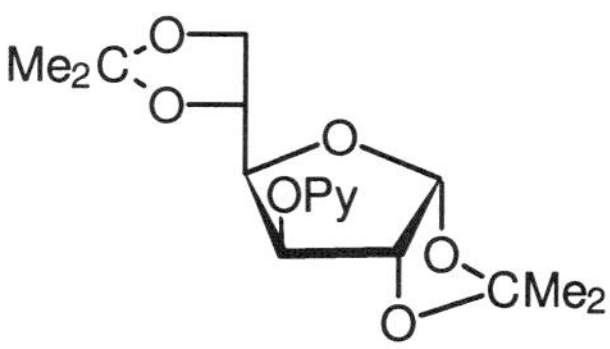	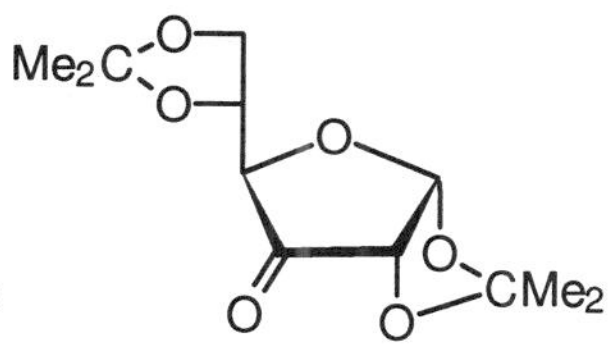	71	66, 67
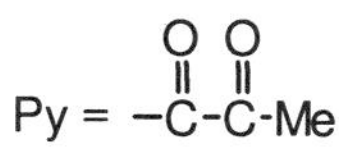			
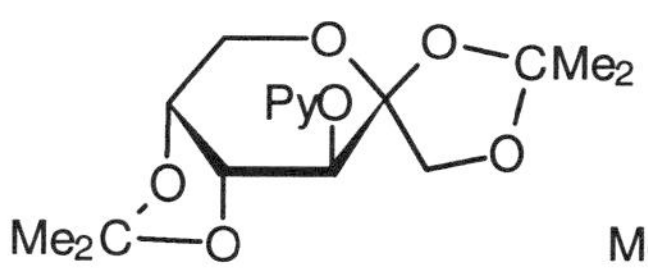	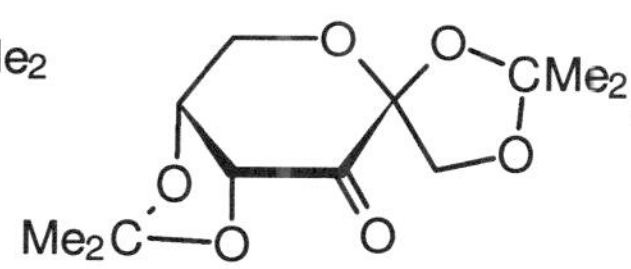	53	66, 67
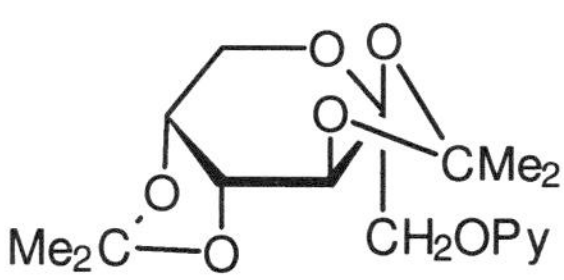	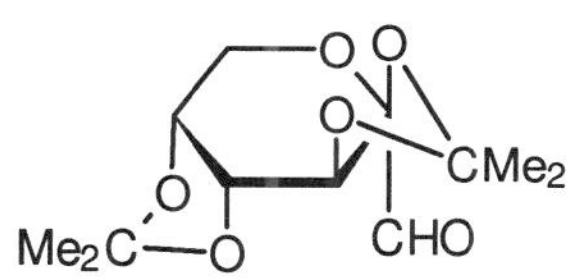	62	66, 67
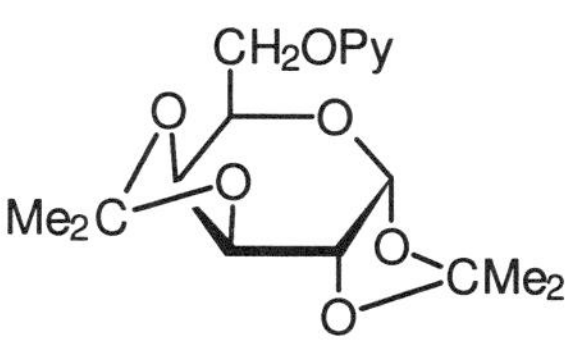	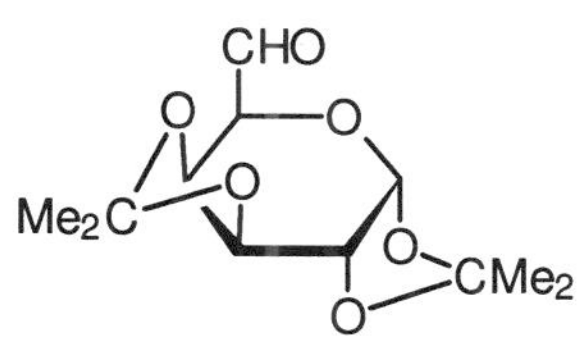	70	66, 67 68
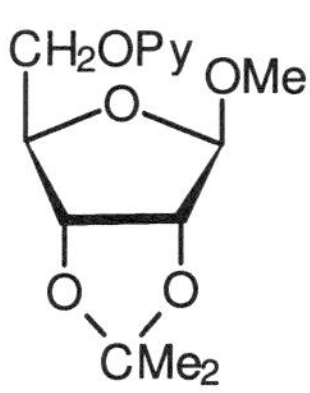	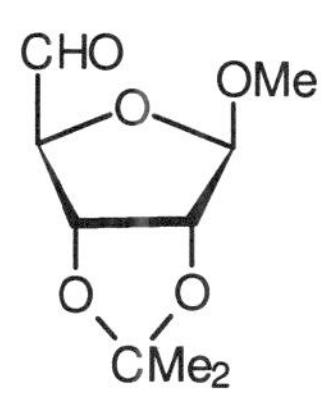	75	66, 67
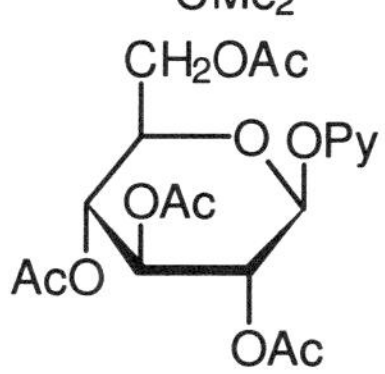	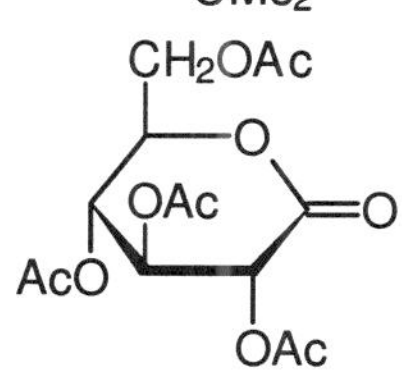	90	67

Continued on next page

Table 3.7. Type II reactions of pyruvate esters of sugars—*Continued*

Reactant	*Product*	*Percent Yield*	*Ref.*
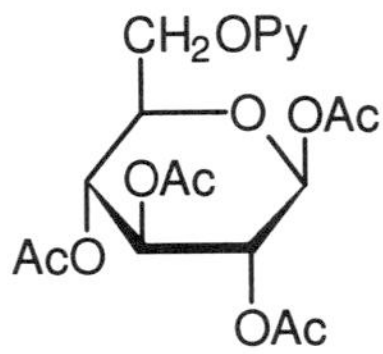	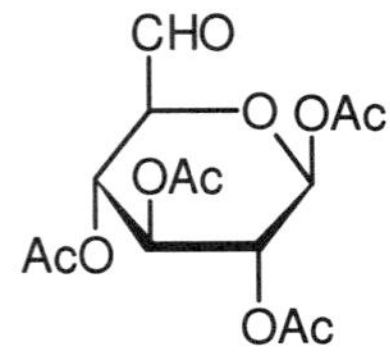	85	67
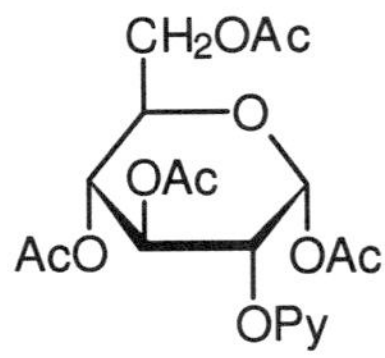	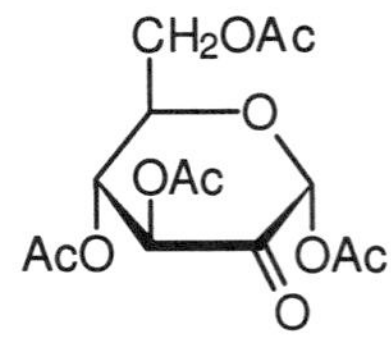	70	67
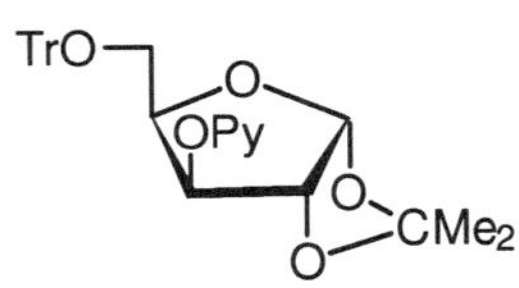	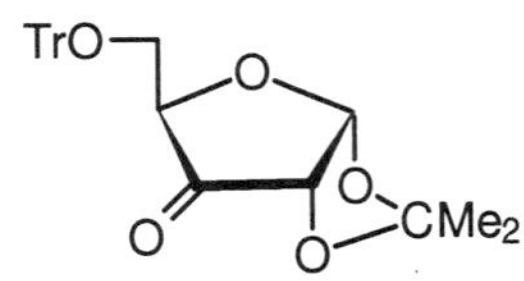	59[c]	67
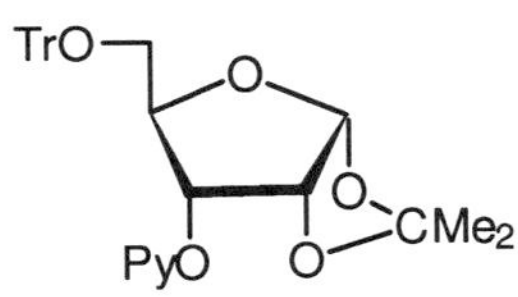 Tr = $(C_6H_5)_3C$	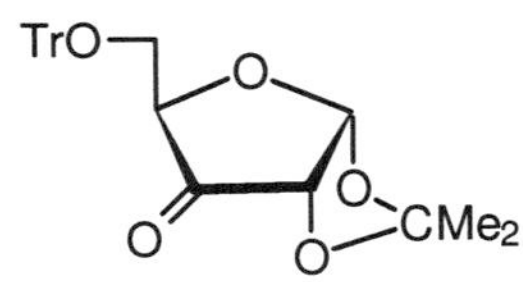	38[d]	69
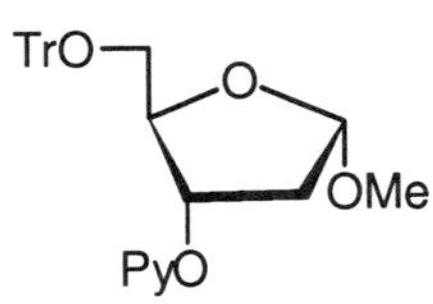	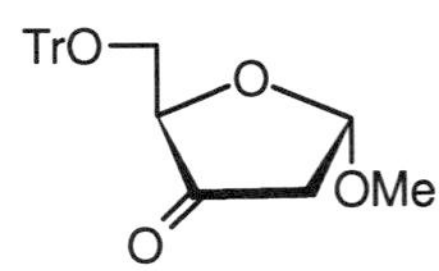	80	69
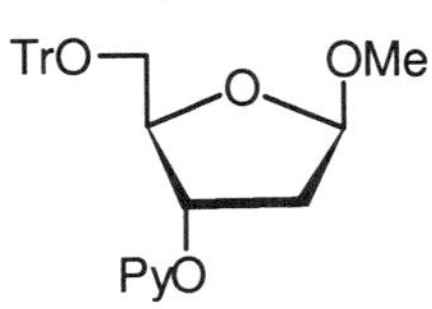	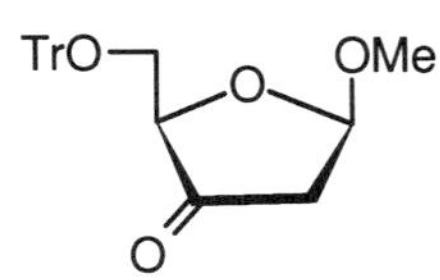	79	69
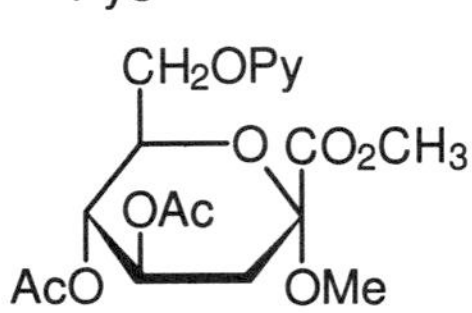	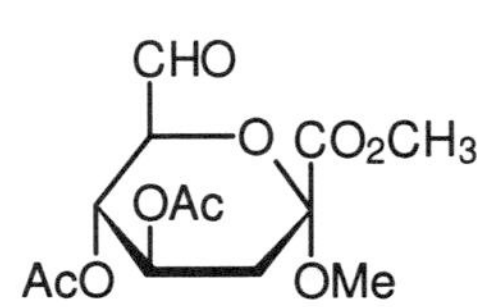	60	70

Continued on next page

Table 3.7. Type II reactions of pyruvate esters of sugars—*Continued*

Reactant	*Product*	*Percent Yield*	*Ref.*
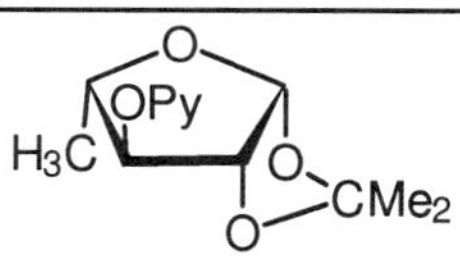	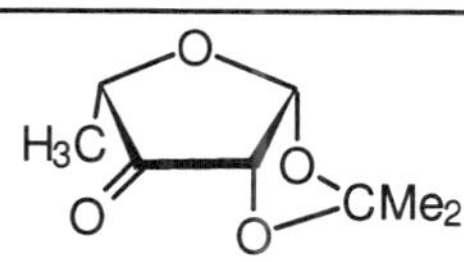	50	71
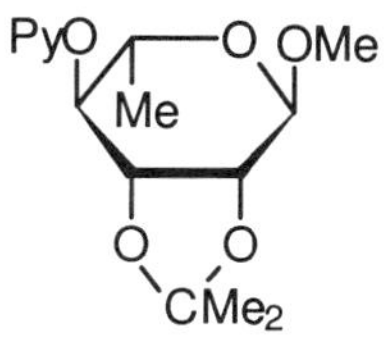	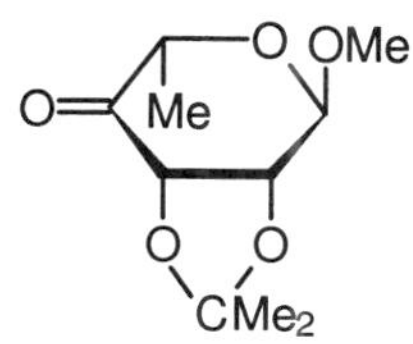	63	72
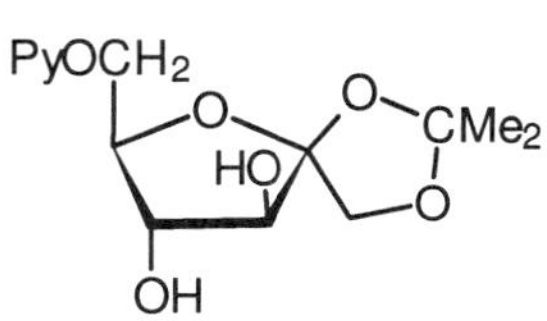	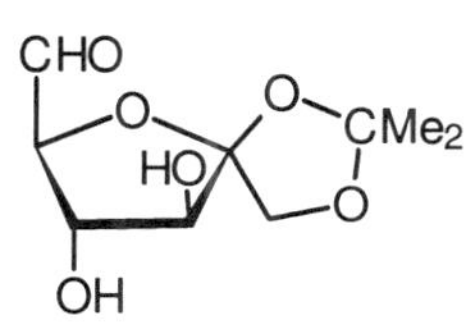	81	73
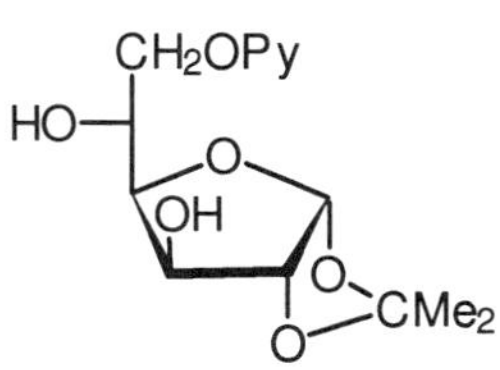	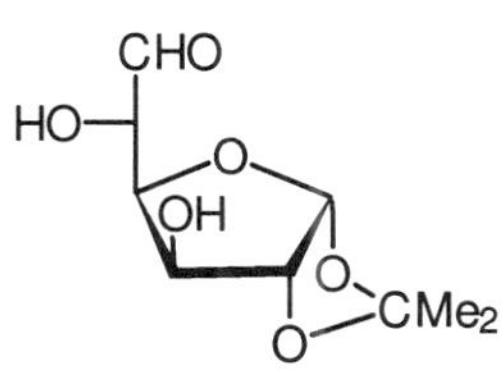	82	73

[a]Py represents a pyruvoyl group.
[b]Reaction was conducted under an inert atmosphere.
[c]The deesterified alcohol was isolated in 14% yield.
[d]The deesterified alcohol was isolated in 37% yield.

Table 3.8. Type II reactions of pyruvate esters of nucleosides.

Reactant	*Product*	*Percent Yield*	*Ref.*
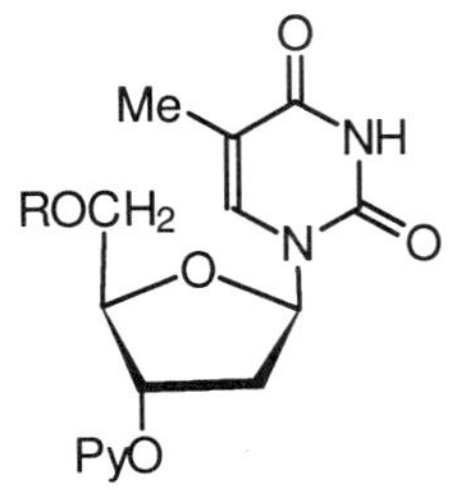	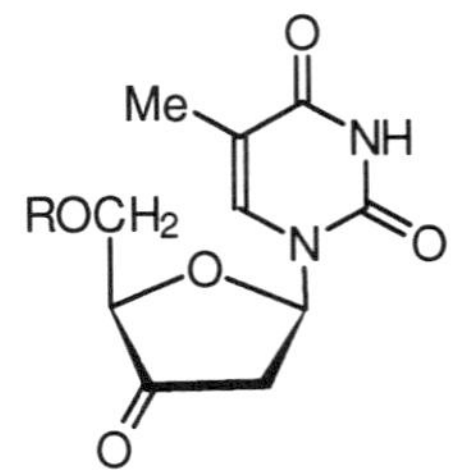		
$R = (C_6H_5)_3C-$		61	69, 74
$R = C_6H_5C(=O)-$		68	69, 74
$Py = CH_3C(=O)-C(=O)-$			
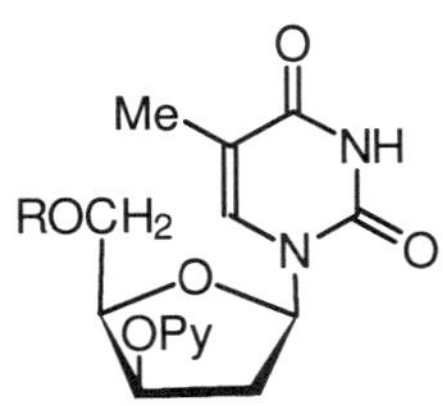	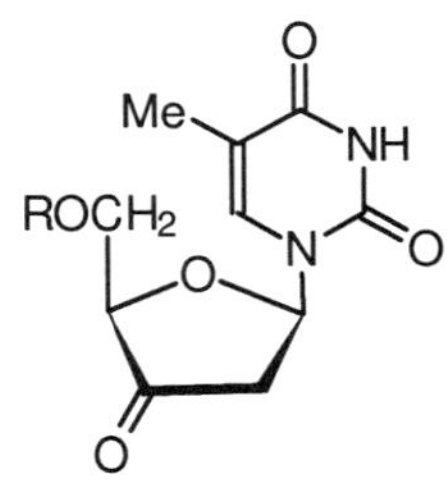		69, 74
$R = (C_6H_5)_3C-$		57	
$R = C_6H_5C(=O)-$		57	
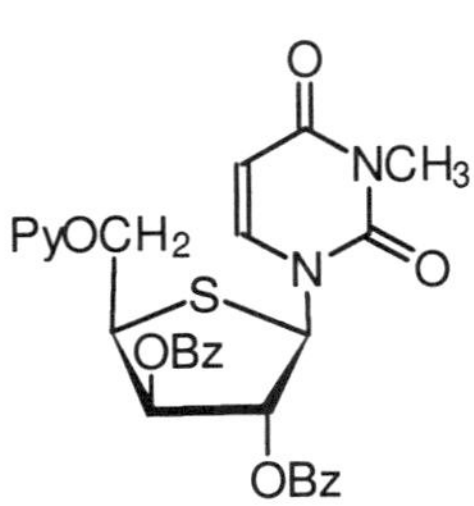	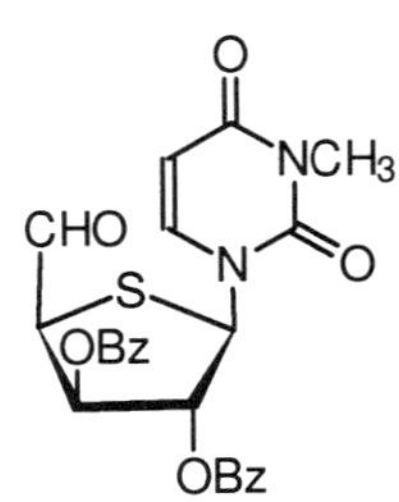	69	75

[a]Reactions were conducted under an inert atmosphere.

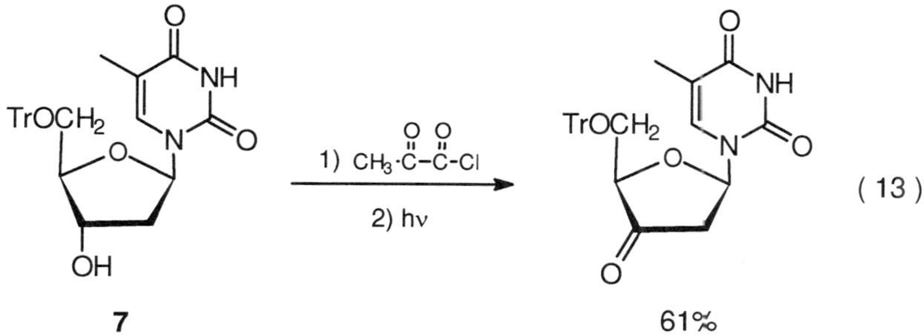

Equation 13

the synthesis of the 3′-ketonucleoside **7**.[69,74] This compound decomposes so easily that it cannot be chromatographed or even exposed to weak bases such as pyridine; however, it is synthesized easily by photochemical oxidation (eq 13). Pyruvate ester synthesis and photolysis now occupies a useful position among oxidation procedures for carbohydrates because it can be called upon in synthesizing carbonyl compounds that are too reactive to be produced by other oxidation reactions.

Summary

Norrish Type II reaction of carbohydrates produces the same type of cyclization and cleavage products that are formed from photolysis of aliphatic carbonyl compounds (Scheme 3.1, Tables 3.1 and 3.2). Some examples of the application of this reaction to carbohydrates result in introduction of unsaturation (eq 4), removal of alkoxy groups (eq 5), and generation of thiocarbonyl compounds (eq 6). Photolysis of pyruvate esters of carbohydrates is of interest because it is a Type II reaction that is part of a mild sequence for the oxidation of alcohols (Scheme 3.10, Tables 3.7 and 3.8).

When Type II reaction cannot occur due to lack of a hydrogen atom attached to a γ-carbon atom, abstraction will take place from a δ-carbon atom and result in formation of a five-membered ring (Scheme 3.5). This reaction has been used extensively to generate spiro compounds (Tables 3.3 and 3.4).

References

1. Turro, N. J. *Modern Molecular Photochemistry;* Benjamin/Cummings: Menlo Park, CA, 1978; pp 386-92.
2. Wagner, P.; Park, B.-S. *Org. Photochem.* **1991**, *11*, 227.
3. Norrish, R. G. W. *Trans. Faraday Soc.* **1937**, *33*, 1521.
4. Norrish, R. G. W.; Appleyard, M. *J. Chem. Soc.* **1934**, 874.
5. Wagner, P. J. *Acc. Chem. Res.* **1971**, *4*, 168.

6. Gilbert, A.; Baggott, J. *Essentials of Molecular Photochemistry;* CRC: Boca Raton, Fl, 1991; pp 310-28.
7. Brunckova, J.; Crich, D. *Tetrahedron* **1995**, *51*, 11945.
8. Cottier, L.; Remy, G.; Descotes, G. *Synthesis* **1979**, 711.
9. Bernasconi, C.; Cottier, L.; Descotes, G.; Rémy, G. *Bull. Soc. Chim. France* **1979**, 332.
10. Collins, P. M.; Gupta, P.; Iyer, R. *J. Chem. Soc., Perkin Trans. I* **1972**, 1670.
11. Collins, P. M.; Gupta, P. *J. Chem. Soc., Chem. Commun.* **1969**, 90.
12. Hürzeler, M.; Bernet, B.; Vasella, A. *Helv. Chim. Acta* **1993**, *76,* 995.
13. Whistler, R. L.; Doner, L. W. *J. Org. Chem.* **1973**, *38,* 2900.
14. Binkley, R. W.; Jarrell, H. F. *J. Carbohydr., Nucleosides, Nucleotides* **1980**, *7,* 347.
15. Triantoaphylides, C.; Gerster, R. *J. Chem. Soc., Perkin Trans. II.* **1977**, 1719.
16. Bernasconi, C.; Cottier, L.; Descotes, G.; Grenier, M. F.; Metras, F. *J. Heterocyclic Chem.* **1980**, *17,* 45.
17. Bernasconi, C.; Cottier, L.; Descotes, G.; Grenier, M. F.; Metras, F. *Nouv. J. Chim.* **1978**, *2,* 79.
18. Bernasconi. C.; Descotes, G. *C. R. Hebd. Seances Acad. Sci., Ser. C* **1975**, *280,* 469.
19. Bernasconi, C.; Cottier, L.; Descotes, G. *Bull. Soc. Chim. Fr.* **1977**, *1-2,* 107.
20. Coulson D. R.; Yang, N. C. *J. Am. Chem. Soc.* **1966**, *88,* 4511.
21. Wagner, P. J.; Hamond, G. S. *J. Am. Chem. Soc.* **1965**, *87,* 4009.
22. Dougherty, T. J. *J. Am. Chem. Soc.* **1965**, *87,* 4012.
23. Wagner P. J.; Hammond, G. S. *J. Am. Chem. Soc.* **1966**, *88,* 1245.
24. Stephenson, L. M.; Cavigli, P. R.; Parlett, J. L. *J. Am. Chem. Soc.* **1971**, *93,* 1984.
25. Casey, C. P.; Boggs, R. A. *J. Am. Chem. Soc.* **1972**, *94,* 6457.
26. Wagner, P. J. *J. Am. Chem. Soc.* **1967**, *89,* 5898.
27. Wagner, P. J. *Tetrahedron Lett.* **1967**, 1753.
28. Wagner, P. J.; Kelso, P. A.; Kemppainen, A. E.; McGrath, J. M.; Schott, H. N.; Zepp, R. G. *J. Am. Chem. Soc.* **1972**, 7506.
29. Wagner, P. J. *Acc. Chem. Res.* **1989**, *22,* 83.
30. Wagner, P. J.; Kelso, P. A.; Kemppainen, A. E.; Zepp, R. G. *J. Am. Chem. Soc.* **1972**, *94,* 7500.
31. Remy, G.; Cottier, L.; Descotes, G. *Tetrahedron Lett.* **1979**, 1847.
32. Remy, G.; Cottier, L.; Descotes, G. *Can. J. Chem.* **1980**, *58,* 2660.
33. Bron, P.; Cottier, L.; Descotes, G. *J. Heterocyclic Chem.* **1984**, *21,* 21.
34. Bernasconi, C.; Cottier, L.; Descotes, G.; Praly, J.-P.; Remy, G.; Grenier-Loustalot, M.-F.; Metras, F. *Carbohydr. Res.* **1983**, *115,* 105.
35. Praly, J.-P.; Descotes, G. *Carbohydr. Res.* **1981**, *95,* C1.
36. Remy, G.; Cottier, L.; Descotes, G. *J. Carbohydr. Chem.* **1982**, *1,* 37.
37. Remy, G.; Cottier, L.; Descotes, G. *Can. J. Chem.* **1983**, *61,* 434.
38. Cottet, F.; Cottier, L.; Descotes, G. *Can. J. Chem.* **1990**, *68,* 1251.
39. Remy G.; Descotes, G. *J. Carbohydr. Chem.* **1983**, *2,* 159.
40. Araki, Y.; Arai, Y.; Endo, T.; Ishido, Y. *Chem. Lett.* **1989**, 1.
41. Kozluk, T.; Cottier, L.; Descotes, G. *Tetrahedron* **1981**, *37,* 1875.
42. Cottier L.; Descotes, G. *Tetrahedron* **1985**, *41,* 409.
43. Cottier, L.; Descotes, G.; Grenier, M. F.; Metras, F. *Tetrahedron* **1981**, *37,* 2515.
44. Kraus, G. A.; Schwinden, M. D. *J. Photochem. Photobiol. A: Chem.* **1991**, *62,* 241.
45. Descotes, G. *Bull. Soc. Chim. Belg.* **1982**, *91,* 973.
46. Descotes, G. *J. Carbohydr. Chem.* **1988**, *7,* 1.
47. Descotes, G. *Top. Curr. Chem.* **1990**, *154,* 39.
48. Wagner, P. J.; Kemppainen, A. E. *J. Am. Chem. Soc.* **1968**, *90,* 5896.
49. Giese B.; Dupuis, J. *Tetrahedron Lett.* **1984**, *25,* 1349.
50. Giese, B. *Angew. Chem. Int. Ed. Engl.* **1989**, *28,* 969.
51. Dupuis, J.; Giese, B.; Rüegge, D.; Fischer, H.; Korth, H.-G.; Sustmann, R. *Angew. Chem. Int. Ed. Engl.* **1984**, *23,* 896.

52. Korth, H. G.; Sustmann, R.; Dupuis, J.; Giese, B. *J. Chem. Soc., Perkin Trans II* **1986**, 1453.
53. Coyle, J. D. *Introduction to Organic Photochemistry;* John Wiley & Scns: New York, 1986; pp 33-34.
54. Hammond, G. S.; Leermakers, P. A.; Turro, N. J. *J. Am. Chem. Soc.* **1961**, *83*, 2395.
55. Leermakers, P. A.; Warren, P. C.; Vesley, G. F. *J. Am. Chem. Soc.* **1964**, *86*, 1768.
56. Leermakers, P. A.; Ross, M. E.; Vesley, G. F.; Warren, P. C. *J. Org. Chem.* **1965**, *30*, 914.
57. Huyser E. S.; Neckers, D. C. *J. Org. Chem.* **1964**, *29*, 276.
58. Nakadaira, Y.; Hirota, Y.; Nakanishi, K. *J. Chem. Soc., Chem. Commun.* **1969**, 1469.
59. Davidson, R. S.; Goodwin, D. *J. Chem. Soc., Perkin Trans. II* **1982**, 993.
60. Scaiano, J. C.; Encinas, M. V.; Lissi, E. A.; Zanocco, A.; Das, P. K. *J. Photochem.* **1986**, *33*, 229.
61. Zanocco, A. L.; Soto, E. A.; Lissi, E. A.; Scaiano, J. C. *J. Photochem. Photobiol., A* **1990**, *53*, 77.
62. Hu, S.; Neckers, D. C. *J. Org. Chem.* **1996**, *61*, 6407.
63. Ottenheijm, H. C. J.; de Man, J. H. M. *Synthesis* **1975**, 163.
64. Binkley, R. W. *J. Org. Chem.* **1976**, *41*, 3030.
65. Binkley, R. W. *Synthetic Commun.* **1976**, *6*, 281.
66. Binkley, R. W. *Carbohydr. Res.* **1976**, *48*, C1.
67. Binkley, R. W. *J. Org. Chem.* **1977**, *42*, 1216.
68. Binkley, R. W. *Methods Carbohydr. Chem.* **1980**, *8*, 321.
69. Binkley, R. W.; Hehemann, D. G.; Binkley, W. W. *J. Org. Chem.* **1978**, *43*, 2573.
70. Myrvold, S.; Reimer, L. M.; Pompliano, D. L.; Frost, J. W. *J. Am. Chem. Soc.* **1989**, *111*, 1861.
71. Binkley, R. W.; Fan, J. C. *J. Carbohydr. Chem.* **1982**, *1*, 213.
72. Yamashita, M.; Long, P. T.; Shibata, M.; Inokawa, S. *Carbohydr. Res.* **1980**, *84*, 35.
73. den Drijver, L.; Holzapfel, C. W.; van Dyk, M. S.; Kruger, G. J. *Carbohydr. Res.* **1987**, *161*, 65.
74. Binkley, R. W.; Hehemann, D. G.; Binkley, W. W. *Carbohydr. Res.* **1977**, *58*, C10.
75. Bredenkamp, M. W.; Holzapfel, C. W.; Swanepoel, A. D. *S. Afr. J. Chem.* **1991**, *44*, 31.

4

The Paternò–Büchi Reaction

Introduction

In 1909 Paternò and Chieffi discovered that irradiating a solution of benzaldehyde in 2-methyl-2-butene produced an oxetane (equation 1).[1] Although this was one of the first organic photochemical reactions reported, its discovery remained largely unnoticed until 1954, when Büchi et al. confirmed the structures of the products, explored the scope of the reaction, and proposed a reaction mechanism.[2] Once the findings of Büchi and co-workers had been published, the stage appeared to be set for exploration of this reaction but, as Arnold pointed out in an excellent early review, this exploration did not happen right away because "the reactions Büchi tried gave very poor yields of the oxetanes, and thus the synthetic utility of the method was not generally recognized."[3] By 1968, however, this situation had changed; the mechanism of the reaction had been investigated more completely, and its synthetic potential had become evident. Photochemical cycloaddition between a carbonyl compound and an alkene to give an oxetane, which had become known as the Paternò–Büchi reaction, was firmly established as a valuable synthetic tool in organic chemistry.[3–7]

Reaction Mechanism

1,4-Diradical Intermediates

The Paternò–Büchi reaction is initiated in nearly every instance by $(n \rightarrow \pi^*)$ excitation of the carbonyl reactant.[8] Büchi et al. proposed that the excited carbonyl compound added to the alkene to give a 1,4-diradical; furthermore,

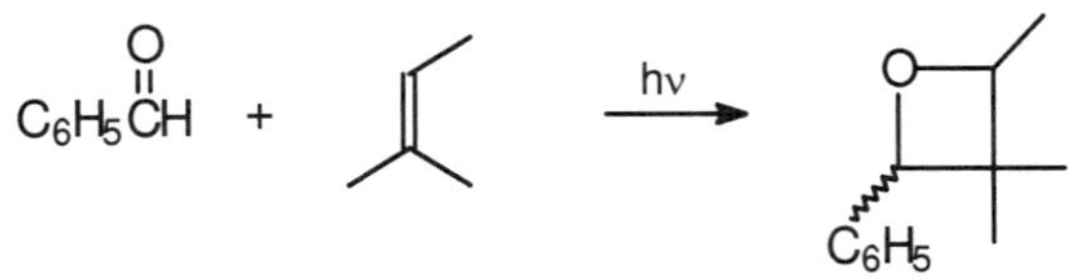

Equation 1

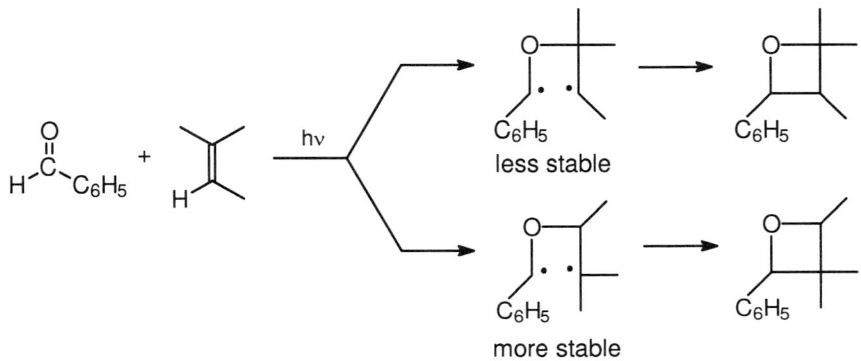

Scheme 4.1

when an asymmetrical alkene was involved, the more stable of the two possible 1,4-diradicals was formed (Scheme 4.1).[2] The major (sometimes exclusive) product, therefore, came from ring closure of the more stable 1,4-diradical. An example of this regioselectivity is found in Scheme 4.2, which describes a reaction used by Araki et al. as part of a branched-chain sugar synthesis.[9]

Evidence supporting the intermediacy of 1,4-diradicals in Paternò–Büchi reactions ranges from spectroscopic observation of these diradicals[10] to reactions, such as that shown in Scheme 4.3, in which diradicals are trapped by oxygen.[11] Other indirect evidence supporting the existence of 1,4-diradicals in Paternò–Büchi reactions is that products from a 1,5-hydrogen shift (i.e., internal diradical disproportionation) sometimes accompany oxetane formation (Scheme 4.4).[9]

Exciplex Formation

Once excitation of the carbonyl compound has occurred, the initial interaction between the Paternò–Büchi reactants is formation of an exciplex.[12–16] This exciplex, which is a precursor to the 1,4-diradical, possesses substantial charge-transfer character with the alkene acting as the electron donor and the excited carbonyl compound as the acceptor.[12,15] The importance of the exciplex as an intermediate in the Paternò–Büchi reaction is underscored in the statement by Yang and Eisenhardt that "oxetane formation is controlled by the ease of formation of the exciplex."[17]

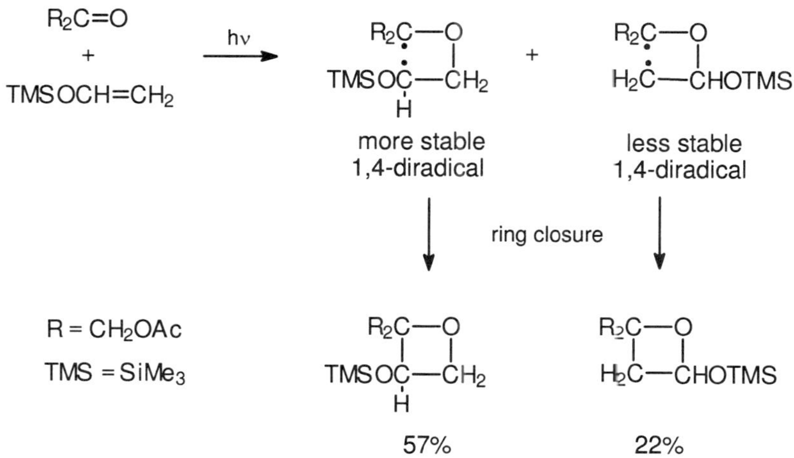

Scheme 4.2

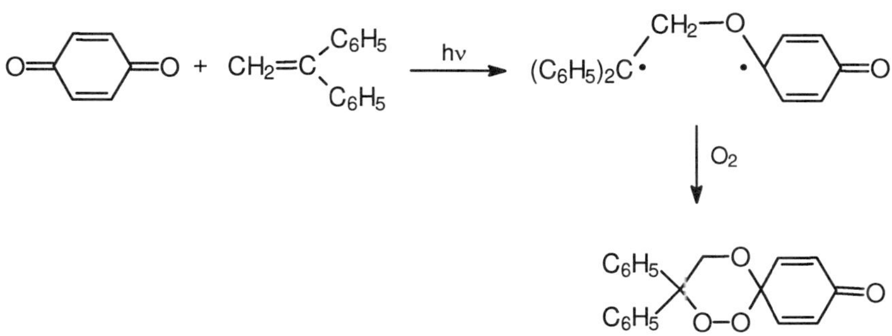

Scheme 4.3

Reactive Excited State

None of the reported Paternò–Büchi reactions of carbohydrates have included investigation of the excited state involved in the reaction; however, as mentioned in the section "1,4-Diradical Intermediates", studies of noncarbohydrate systems have shown that the (n,π*) excited state of the carbonyl-containing reactant nearly always initiates reaction.[8] For aromatic and α,β-unsaturated carbonyl compounds, intersystem crossing to the triplet state is sufficiently rapid that Paternò–Büchi reaction is almost exclusively a triplet state process.[18] The situation with respect to aliphatic aldehydes and ketones—the carbonyl compounds of particular interest in carbohydrate chemistry—is less "clear cut" because aliphatic carbonyl compounds can react in Paternò–Büchi reactions from either (n,π*) singlet or triplet states.

The identity of the reactive excited state is significant in determining stereoselectivity in Paternò–Büchi reactions. Cycloaddition taking place from an excited singlet state generally is highly stereoselective because it forms a sin-

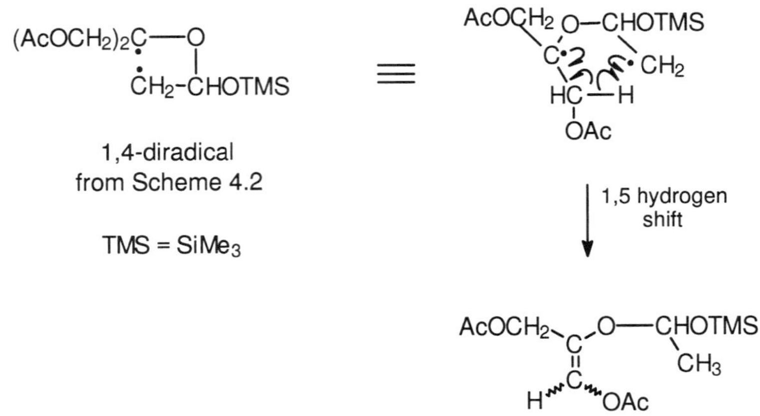

Scheme 4.4

glet 1,4-diradical, an intermediate that requires no spin change prior to ring closure. For highly stereoselective reactions, ring closure must compete effectively with bond rotation (Scheme 4.5).[17] For a triplet diradical, spin inversion is necessary before cyclization can occur; consequently, opportunity exists for bond rotation and loss of stereochemistry (Scheme 4.6). Carless demonstrated this point in his study of the addition of excited acetone to *cis*- and *trans*-2-butenes, reactions that proceed with little stereoselectivity.[19] He determined that these reactions were occurring nearly exclusively from acetone in its triplet state by showing that the presence of a 0.1 M concentration of the triplet state quencher 2,3-dimethyl-1,3-butadiene in the photolysis mixture quenched 98% of the reaction.

Electron-Transfer Reaction

In instances where an electron-rich alkene participates in the Paternò–Büchi reaction, the 1,4-diradical mechanism may be replaced by or compete with an electron-transfer process.[20–23] Mattay et al.[24] have found that excitation of biacetyl (**1**) in the presence of the electron-rich alkene **2** leads to an exciplex that then forms the radical ions **3** and **4** by complete electron transfer. These ions (**3** and **4**) couple to produce the zwitterion **5**, which then cyclizes to give an oxetane (Scheme 4.7).

Paternò–Büchi Reactions of Carbohydrates

Scope of Reported Reactions

The relatively few Paternò–Büchi reactions of carbohydrates that have been reported are listed in Table 4.1. Most of these reactions involve either the addition of an excited, carbonyl-containing carbohydrate to a furan (Scheme 4.8) or

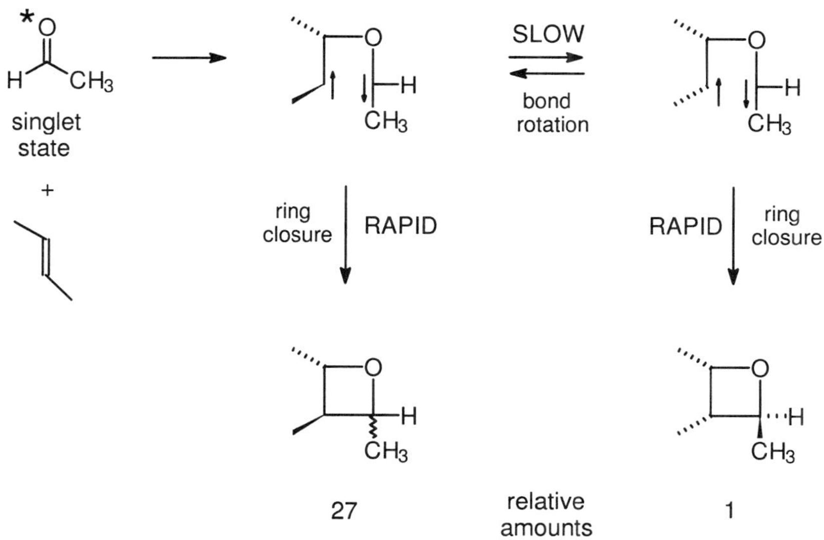

Scheme 4.5

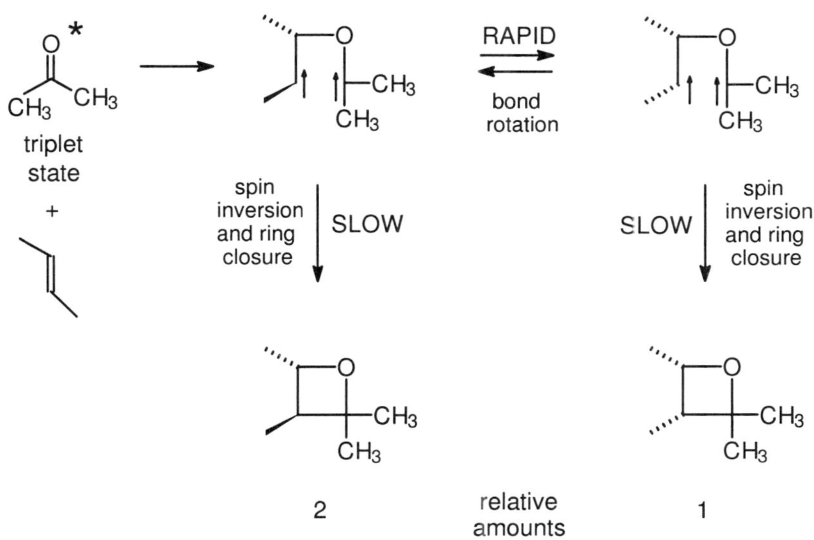

Scheme 4.6

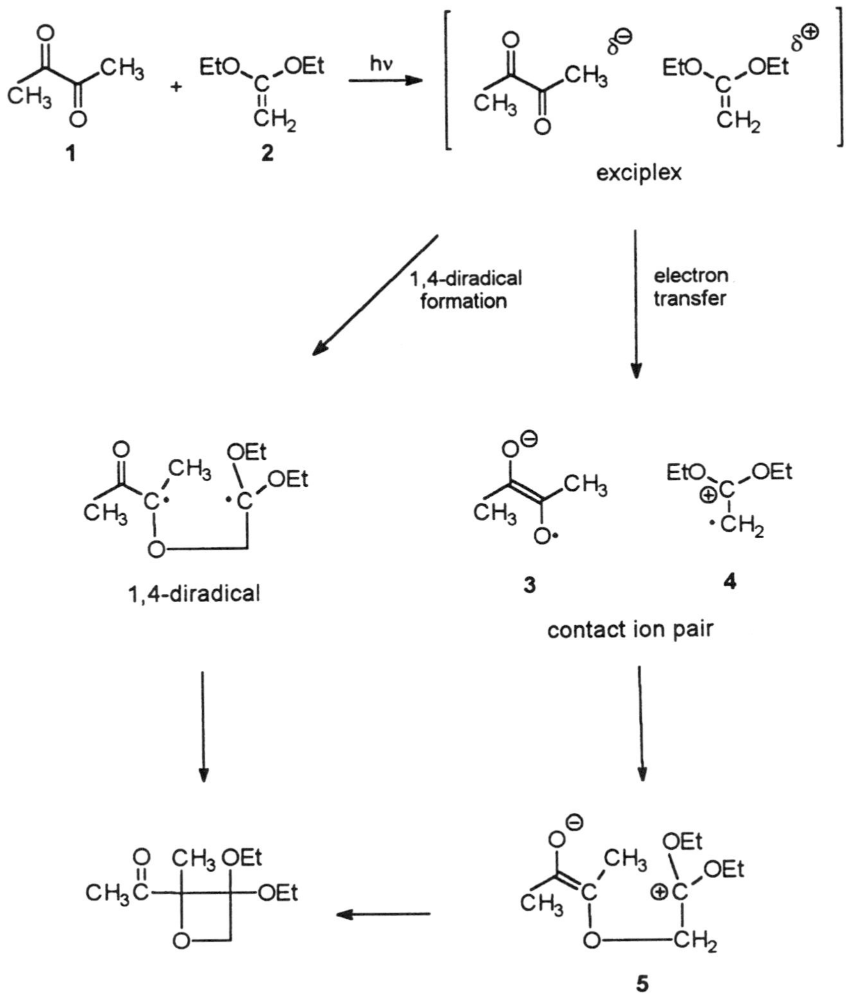
hν
exciplex
1,4-diradical formation
electron transfer
1,4-diradical
contact ion pair
1
2
3
4
5

Scheme 4.7

Table 4.1. Paternò–Büchi reactions of carbohydrates.

Reactant	*Product*	*Percent Yield*	*Ref.*
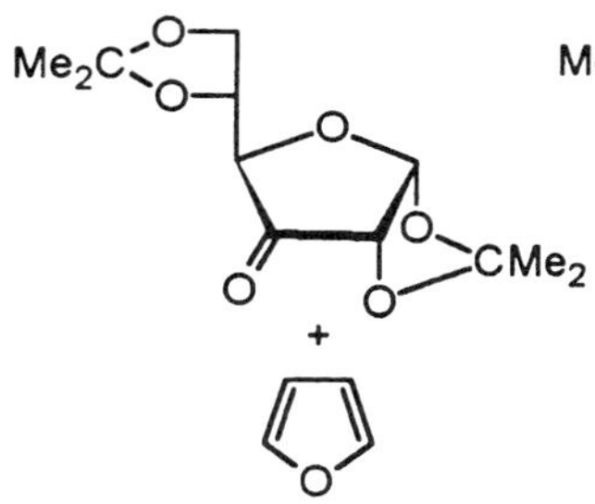	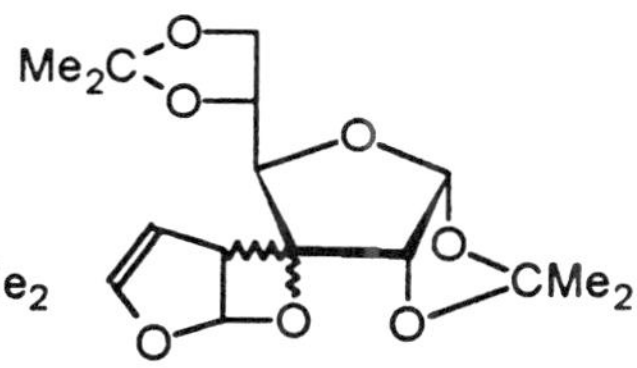 4 products	60 14	25 26
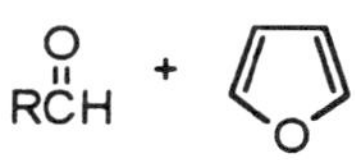	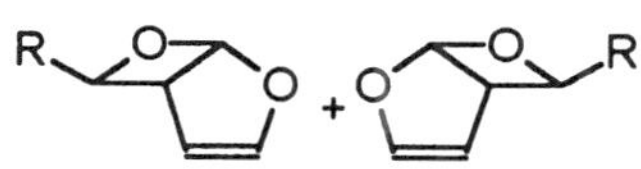		26
R = 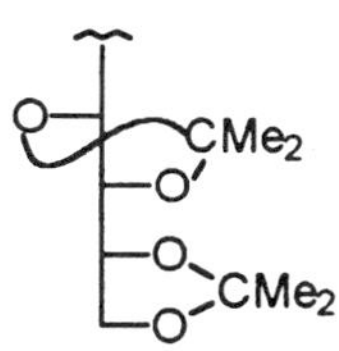 19%	R = 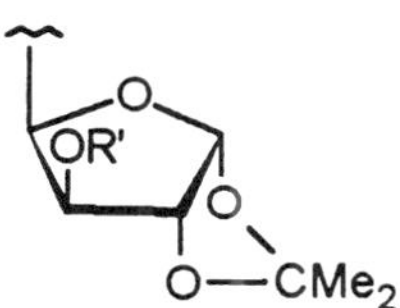 R' = $CH_2C_6H_5$ 24% R' = Me 19%		
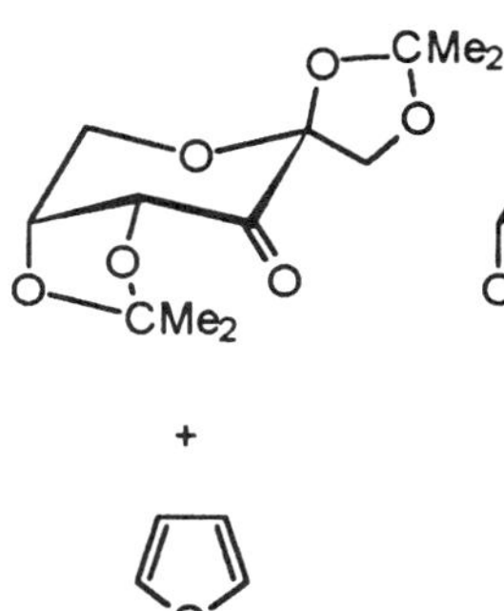	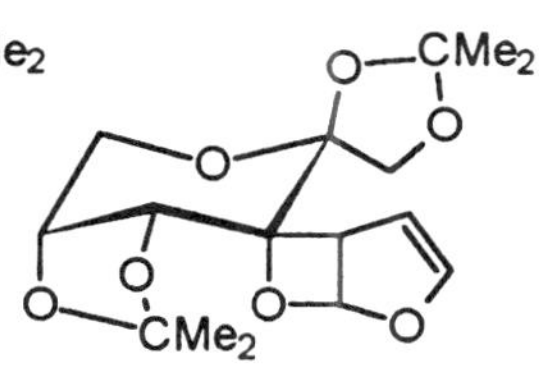 **A**	36	27
	+ 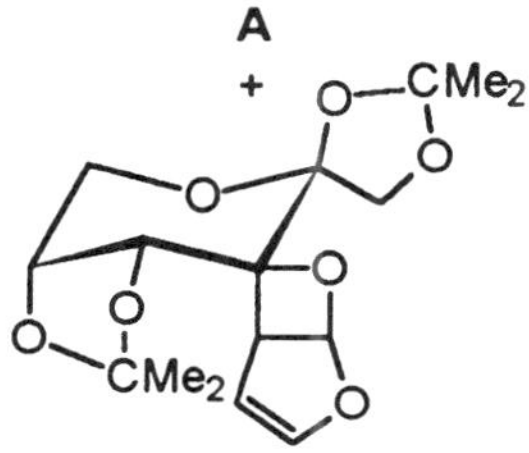 **B**	9	

Continued on next page

Table 4.1. Paternò–Büchi reactions of carbohydrates—*Continued*

Reactant	*Product*	*Percent Yield*	*Ref.*
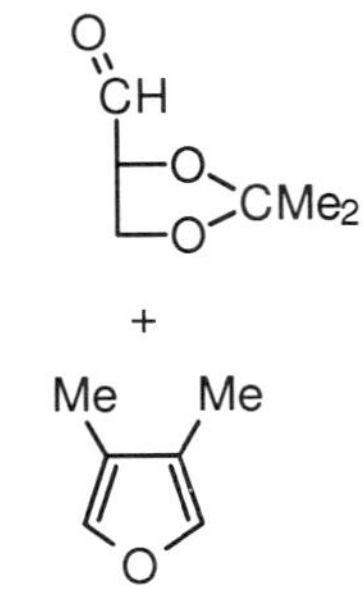	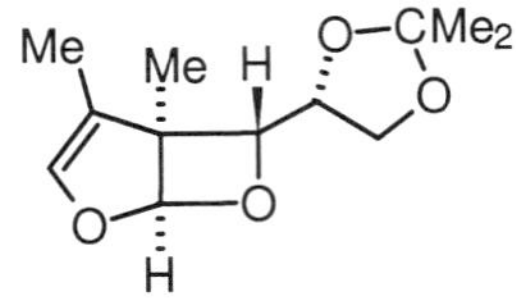	19[a]	28
	+ 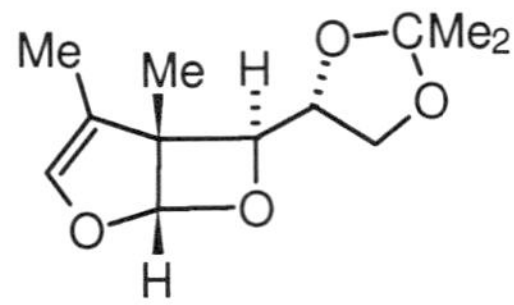	16[a]	
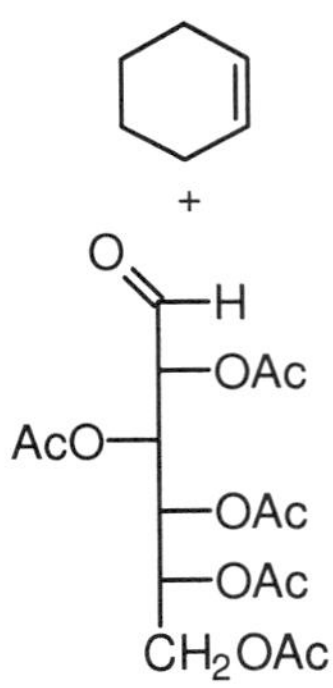	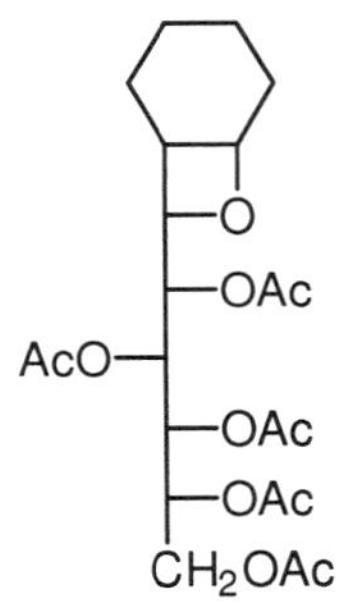	65	29
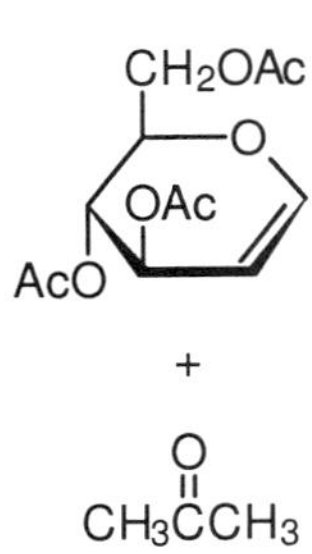	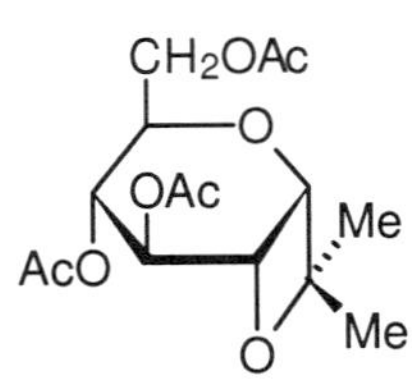	86[b] 33[c] 99[d] 27[e] 77[f]	30 31 32 33 30
	+ 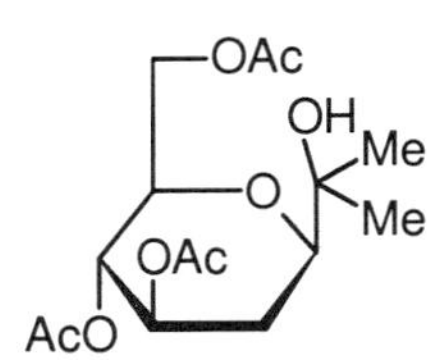	14[b]	30

Continued on next page

Table 4.1. Paternò–Büchi reactions of carbohydrates—*Continued*

Reactant	*Product*	*Percent Yield*	*Ref.*
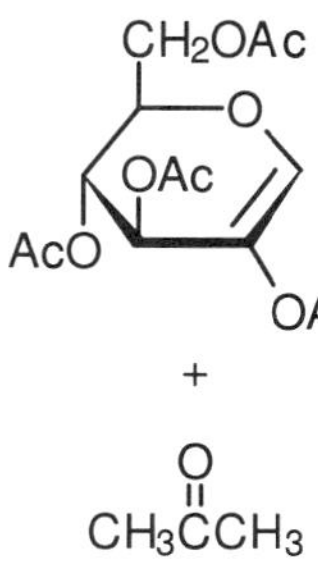	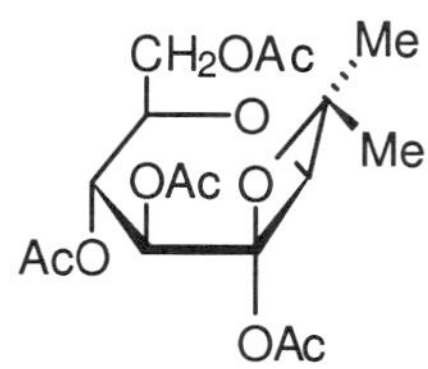	14	34
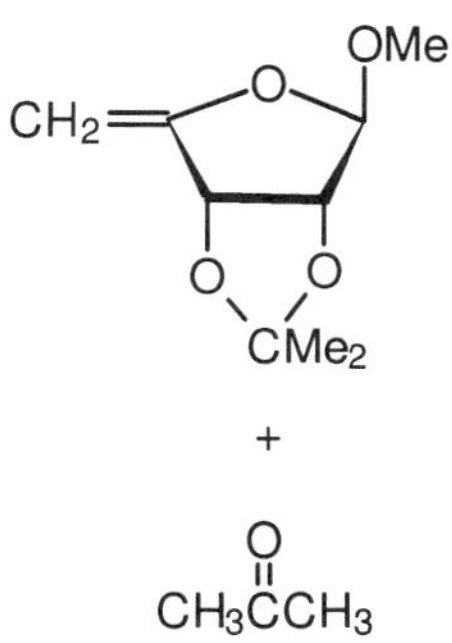	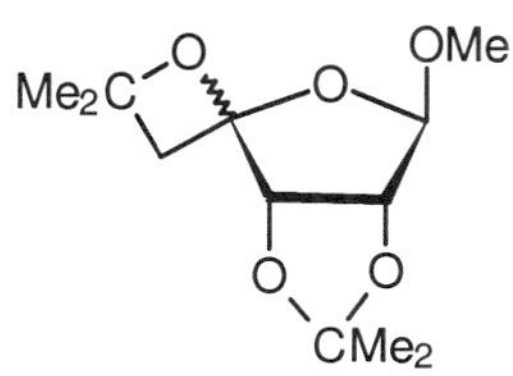 +	3	35
	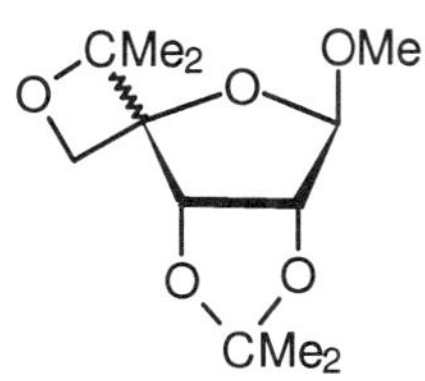	44	

[a]The reaction solvent was ethyl ether.
[b]Reaction was conducted in 2-propanol-acetone (1:9).
[c]Reaction was conducted in pure acetone.
[d]Reaction was conducted in 2-propanol-acetone (1:100).
[e]Reaction was conducted in 2-propanol-acetone (1:9). Several other products were reported.
[f]Reaction was conducted in acetonitrile.

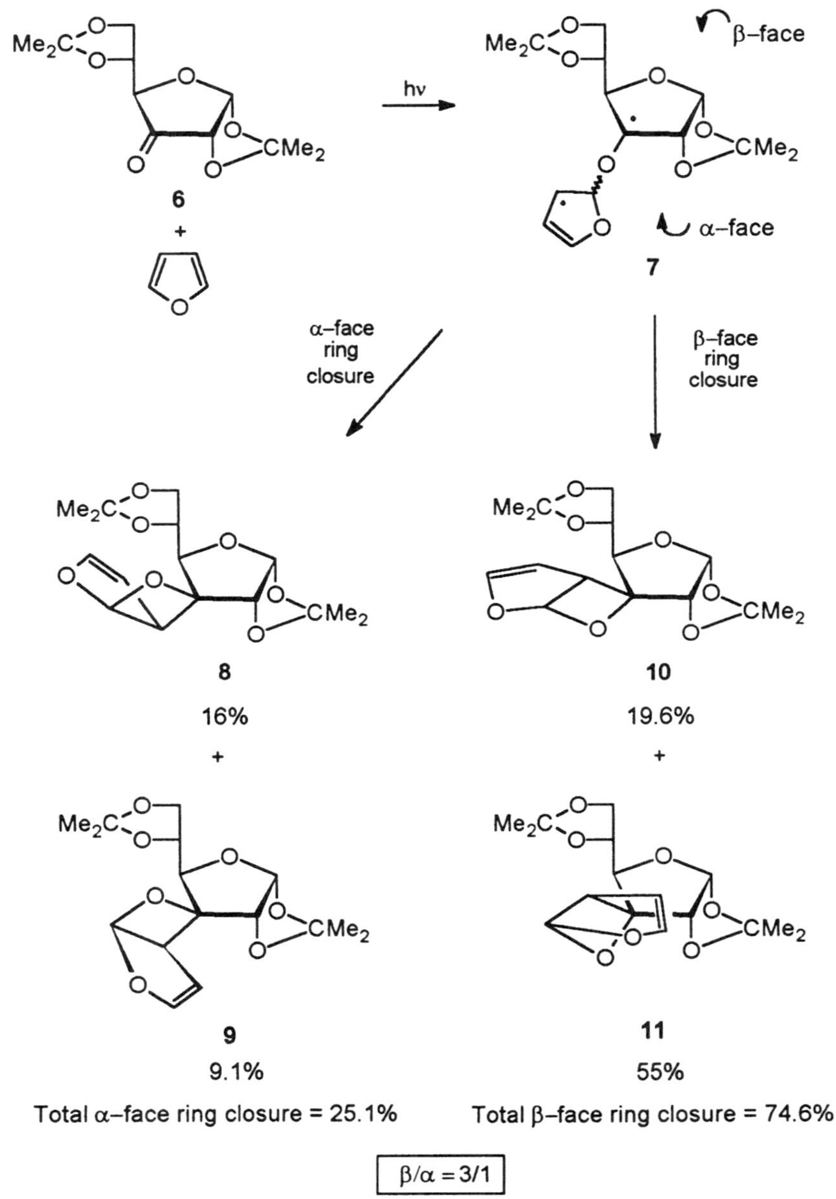
Me2C
CMe2
6
+
hν
β–face
α–face
7
α–face
ring
closure
β–face
ring
closure
8
16%
+
10
19.6%
+
9
9.1%
11
55%
Total α–face ring closure = 25.1%
Total β–face ring closure = 74.6%
β/α = 3/1

Scheme 4.8

photocyclization between acetone and an unsaturated sugar **12** (Scheme 4.9). The Paternò–Büchi reaction also is a critical step in the synthesis of several branched-chain sugars (Table 4.2).

Stereoselectivity in Product Formation

Stereoselectivity in Paternò–Büchi reactions involving carbohydrates has been a topic of some interest. In a particularly thorough study, Jarosz and Zamojski showed that the 3-ulose **6** reacts with furan to produce the four oxetanes (**8–11**) expected from this cycloaddition reaction (Scheme 4.8).[25] Ring closure of the 1,4-diradical **7** preferentially forms compounds **10** and **11**, structures for which the favored placement of the dihydrofuran ring (by a 3:1 ratio) is on the β-face of each molecule. Jarosz and Zamojski pointed out that while this stereoselectivity in ring closure is in the direction expected because the α-face of the 1,4-diradical **7** is more hindered, it was surprising that preference for β-face placement of the dihydrofuran ring was not greater. (They noted that in the hydride reduction of **6** (in the ground state), the exclusive product was that formed by β-face attack.)

Schreiber and Satake also observed a low sensitivity of the Paternò–Büchi reaction to steric effects when they found that stereoselectivity in ring closure of the 1,4-diradical **15**, formed by irradiation of compounds **13** and **14**, was nearly nonexistent (Scheme 4.10).[4,28] These authors made the further observation that the aldehyde **14** partially racemized during photolysis. This loss of stereochemistry in **14** suggested that an α-cleavage, followed by

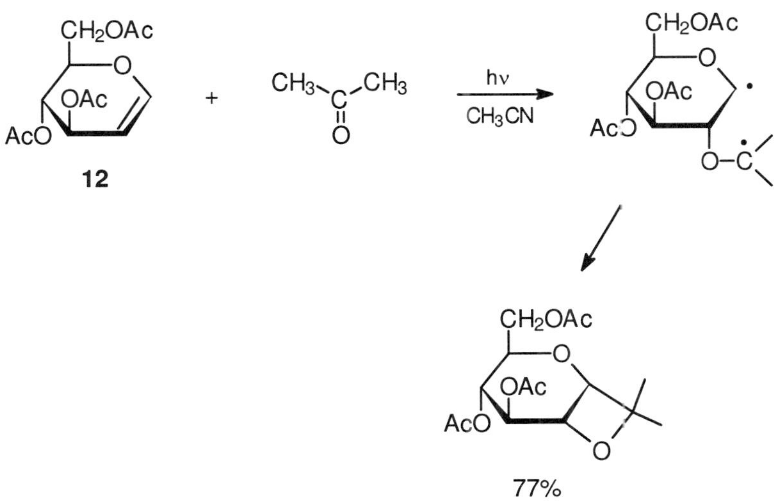

Scheme 4.9

Table 4.2. Synthesis of branched-chain sugars by Paternò–Büchi reaction.

Reactant	*Product and Percent Yield*	*Ref.*
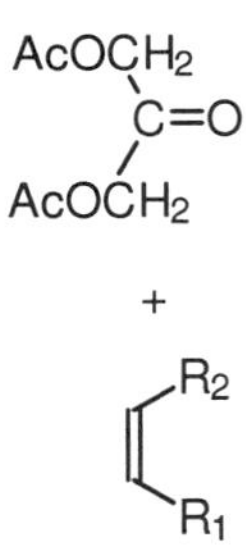	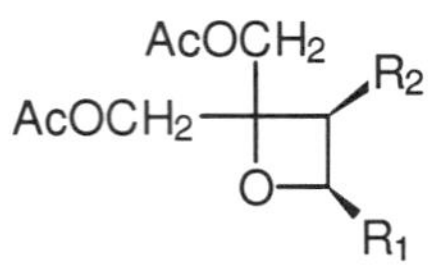 + 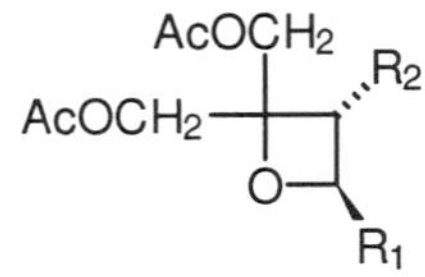	36
	A **B**	
$R_1 = R_2 = OAc$	A 37%, B 34%	
$R_1,R_2 = -OCH_2CH_2O-$	A 25%	
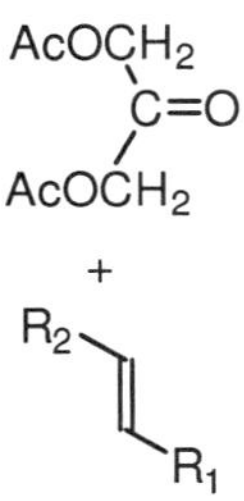	**A** + **B**	36
	46% 34%	
$R_1 = R_2 = OAc$		
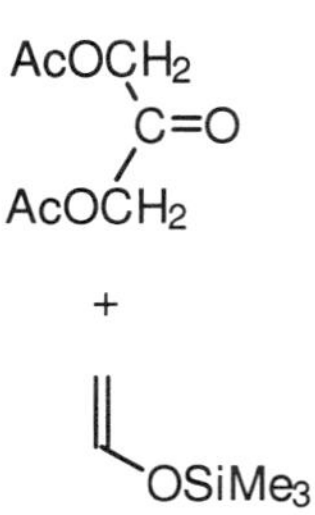	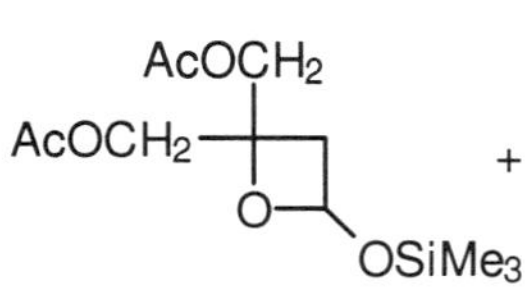 + 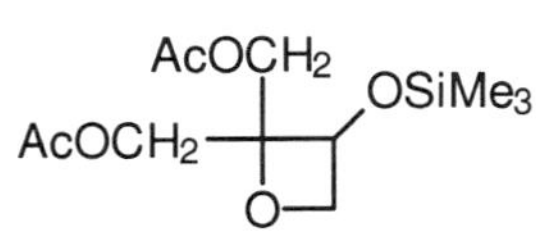	9
	57% 22%	

Continued on next page

Table 4.2. Synthesis of branched-chain sugars by Paternò–Büchi reaction—*Continued*

Reactant	*Product and Percent Yield*		*Ref.*
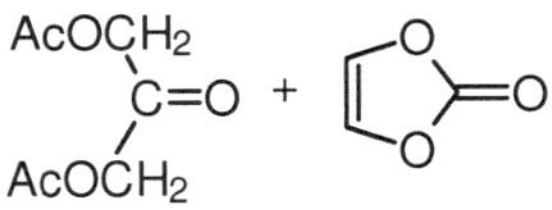	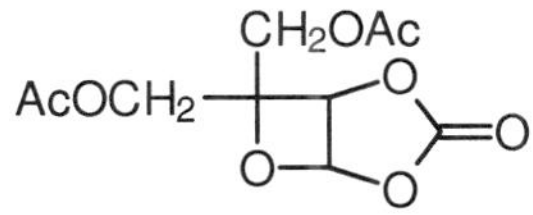 55%		36
	+		
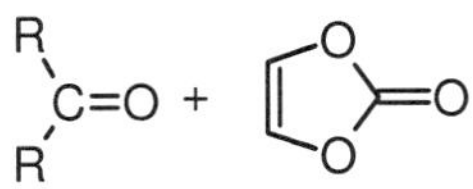	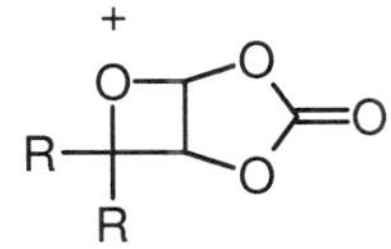		37
R = CH_2OAc	57%		
R = CO_2Et	23%		
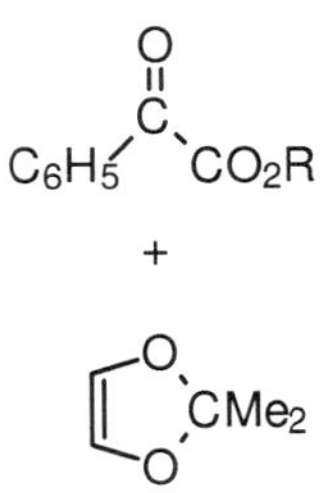	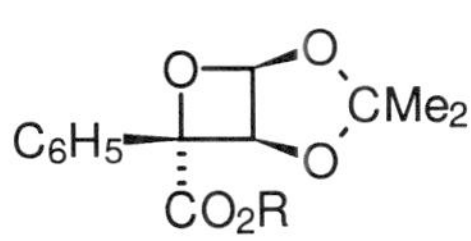 **C** + 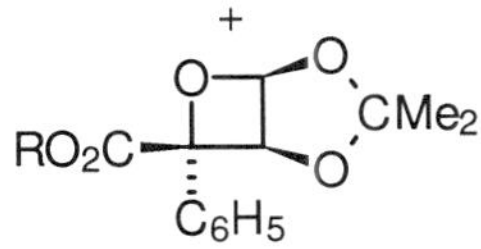 **D**		38
	C	**D**	
R = 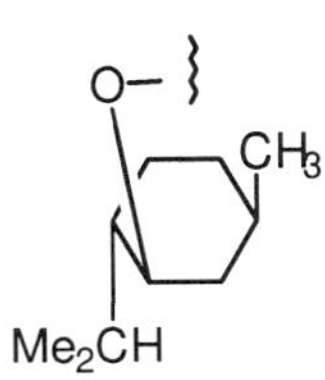	57%	22%	
R = 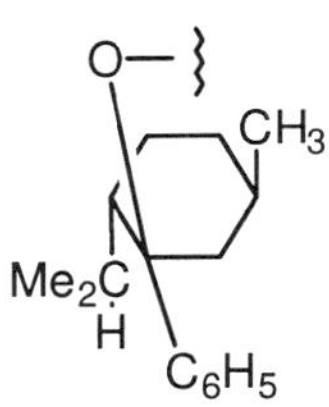	96%	3%	

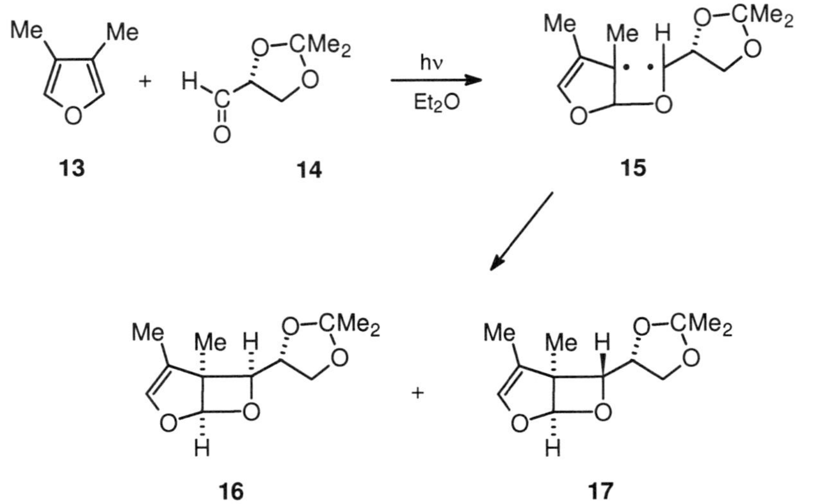

16 + **17** = 35% **16**/**17** = 1.2

Scheme 4.10

recombination (Scheme 4.11), was competing with the Paternò–Büchi reaction. (α-Cleavage reactions are discussed in Chapter 2.)

Competing Reactions

Because the Paternò–Büchi reaction is only one of the options open to an (n,π*) excited carbonyl compound, it is reasonable to expect that competing reactions also will take place. As mentioned in the previous paragraph and pictured in Scheme 4.11, α-cleavage is one possible competing reaction. Hydrogen abstraction is another competing process that occurs in the presence of effective hydrogen atom donors. (A brief discussion of the hydrogen-abstracting ability of excited carbonyl compounds is found in Chapter 2.) Araki et al. found such a competition when they studied the reaction between 3,4,6-tri-*O*-acetyl-1,5-anhydro-2-deoxy-D-arabino-hex-1-enitol (**12**, "tri-*O*-acetyl-D-glucal") and acetone.[33] Paternò–Büchi reaction took place in high yield in the poorly hydrogen-donating solvent acetonitrile (Scheme 4.9), but in an effective hydrogen-donating solvent such as 2-propanol, Paternò–Büchi reaction was completely suppressed in favor of hydrogen abstraction (Scheme 4.12). Matsuura et al. studied this competition between the Paternò–Büchi reaction and hydrogen abstraction in a variety of solvents and found that the course of reaction consistently depended upon the hydrogen-donating ability of the solvent.[30]

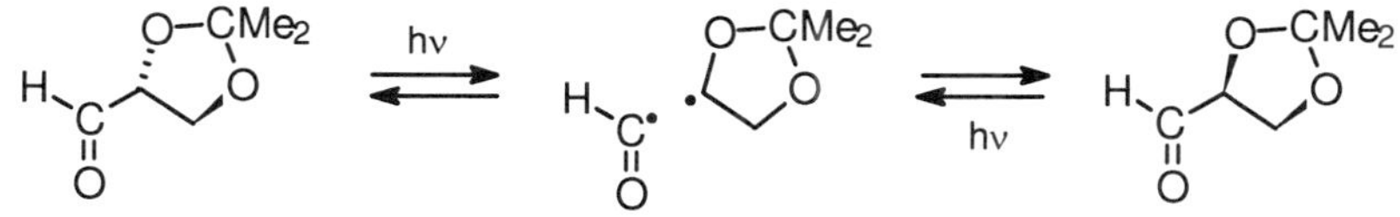

14

Scheme 4.11

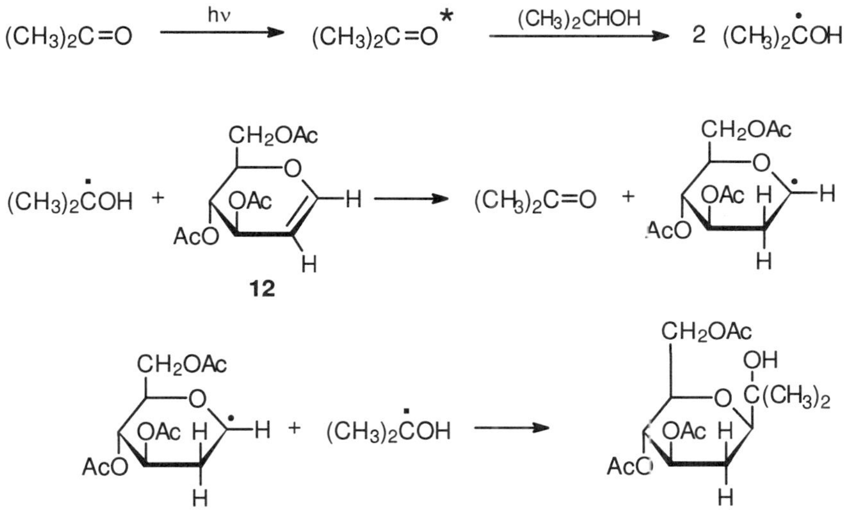

Scheme 4.12

Summary

The number of Paternò–Büchi reactions in which carbohydrates participate is not large. Most of those reported consist of the addition of a carbonyl-containing carbohydrate to furan (or a furan derivative) or addition of acetone to an unsaturated sugar (Table 4.1). In several instances Paternò–Büchi reactions are part of the synthesis of branched-chain sugars (Table 4.2).

There are no reported examples of intramolecular Paternò–Büchi reactions of carbohydrates, nor are there examples in which both reactants are carbohydrates. Although there is no report of an electron-transfer reaction leading to oxetane formation involving a carbohydrate, some unsaturated carbohydrates do contain electron-rich double bonds and, therefore, are candidates for Paterno–Büchi reaction by an electron-transfer pathway (Scheme 4.7).

References

1. Paternò, E.; Chieffi, G. *Gazz. Chim. Ital.* **1909**, *39*, 341.
2. Büchi, G.; Inman, C. G.; Lipinsky, E. S. *J. Am. Chem. Soc.* **1954**, *76*, 4327.
3. Arnold, D. R. *Adv. Photochem.* **1968**, *6*, 301.
4. Schreiber, S. L. *Science* **1985**, *227*, 857.
5. Jones, G., II *Org. Photochem.* **1981**, *5*, 1.
6. Carless, H. A. J. In *Synthetic Organic Photochemistry*; Horspool, W. M., Ed.; Plenum Press: New York, 1984; pp 425-487.
7. Gilbert, A.; Bagott, J. *Essentials of Molecular Photochemistry;* Blackwell Scientific Publications: London, 1991; pp 340-352.
8. Carless, H. A. J. In *Synthetic Organic Photochemistry*; Horspool, W. M., Ed.; Plenum Press: New York, 1984; p 427.
9. Araki, Y.; Nagasawa, J.-I.; Ishido, Y. *Carbohydr. Res.* **1981**, *91*, 77.
10. Frelich, S. C.; Peters, K. S. *J. Am. Chem. Soc.* **1981**, *103*, 6255.
11. Wilson, R. M.; Wunderly, S. W. *J. Chem. Soc., Chem. Commun.* **1974**, 461.
12. Caldwell, R. A.; Sovocool, G. W.; Cajewski, R. P. *J. Am. Chem. Soc.* **1973**, *95*, 2549.
13. Gupta, A.; Hammond, G. S. *J. Am. Chem. Soc.* **1976**, *98*, 1218.
14. Yang, N. C.; Hui, M. H.; Shold, D. M.; Hautala, R. R.; Dawes, K.; Dalton, J. C. *J. Am. Chem. Soc.* **1977**, *99*, 3023.
15. Mattay, J.; Gersdorf, J.; Buchkremer, K. *Chem. Ber.* **1987**, *120*, 307.
16. M. Demuth, M.; Mikhail, G. *Synthesis* **1989**, 145.
17. Yang, N. C.; Eisenhardt, W. *J. Am. Chem. Soc.* **1971**, *93*, 1277.
18. Gilbert, A.; Bagott, J. *Essentials of Molecular Photochemistry;* Blackwell Scientific Publications: London, 1991; p 341.
19. Carless, H. A. J. *Tetrahedron Lett.* **1973**, 3173.
20. Frelich, S. C.; Peters, K. S. *J. Am. Chem. Soc.* **1981**, *103*, 6255.
21. Mattay, J. *Angew. Chem. Int. Ed. Engl.* **1987**, *26*, 825.
22. Fox, M. A. *Advan. Photochem.* **1986**, *13*, 237.
23. Griesbeck, A. G.; Mauder, H.; Stadtmüller, S. *Acc. Chem. Res.* **1994**, *27*, 70.
24. Mattay, J.; Gersdorf, J.; Freudenberg, U. *Tetrahedron Lett.* **1984**, *25*, 817.
25. Jarosz, S.; Zamojski, A. *Tetrahedron* **1982**, *38*, 1453.
26. Tronchet, J. M. J.; Baehler, B. *J. Carbohydr., Nucleosides, Nucleotides* **1974**, *1*, 449.
27. Jarosz, S.; Zamojski, A. *Polish J. Chem.* **1983**, *37*, 455.
28. Schreiber, S. L.; Satake, K. *Tetrahedron Lett.* **1986**, *27*, 2575.
29. Kim, J. C.; Kim, M. S.; Whistler, R. L.; Anisuzzaman, A. K. M. *Org. Prep. Proced. Int.* **1983**, *15*, 161.
30. Matsuura, K.; Araki, Y.; Ishido, Y.; Murai, A.; Kushida, K. *Carbohydr. Res.* **1973**, *29*, 459.
31. Ong, K.-S.; Whistler, R. L. *J. Org. Chem.* **1972**, *37*, 573.
32. Matsuura, K.; Araki, Y.; Ishido, Y. *Bull. Chem. Soc. Jpn.* **1972**, *45*, 3496.
33. Araki, Y.; Senna, K.; Matsuura, K.; Ishido, Y. *Carbohydr. Res.* **1978**, *60*, 389.
34. Araki, Y.; Senna, K.; Ishido, Y. *Carbohydr. Res.* **1978**, *64*, 109.
35. Araki, Y.; Senna, K.; Matsuura, K.; Ishido, Y. *Carbohydr. Res.* **1978**, *65*, 159.
36. Araki, Y.; Nagasawa, J.-I.; Ishido, Y. *J. Chem. Soc., Perkin Trans. I* **1981**, 12.
37. Araki, Y.; Nagasawa, J.-I.; Ishido, Y. *Carbohydr. Res.* **1977**, *58*, C4
38. Nehrings, A.; Scharf, H.-D.; Runsink, J. *Angew. Chem., Int. Ed. Engl.* **1985**, *24*, 877

5

Intermolecular Hydrogen-Abstraction Reactions

Introduction

Hydrogen-abstraction reactions by excited carbonyl compounds can be divided into two groups: intramolecular and intermolecular reactions. If the abstraction is intramolecular, a Norrish Type II reaction typically occurs (Scheme 5.1). (Norrish Type II and related reactions are discussed in Chapter 3.) If, on the other hand, a reaction is intermolecular, abstraction produces a radical pair that generally undergoes one or more typical radical reactions (e.g., combination, disproportionation, and reaction with molecular oxygen). Hydrogen abstraction becomes mechanistically more complicated when effective electron donors are present in the reaction mixture. Amines, for example, are sufficiently good electron donors that they can combine with an excited carbonyl compound to form an exciplex prior to hydrogen abstraction; furthermore, in cases where a particularly good electron-donating amine is involved, exciplex formation can be followed by complete electron transfer. Finally, hydrogen abstraction is believed to be the first step in the synthesis of sugars from formaldehyde (the formose reaction) and cleavage of glycosidic linkages that are located in uronic acids and esters.

Hydrogen Abstraction by an Excited, Carbonyl-Containing Carbohydrate

Methanol as the Hydrogen-Atom Donor

The photolysis of compound **1** provides an example of a reaction of a carbohydrate that proceeds, in part, by an intermolecular hydrogen abstraction

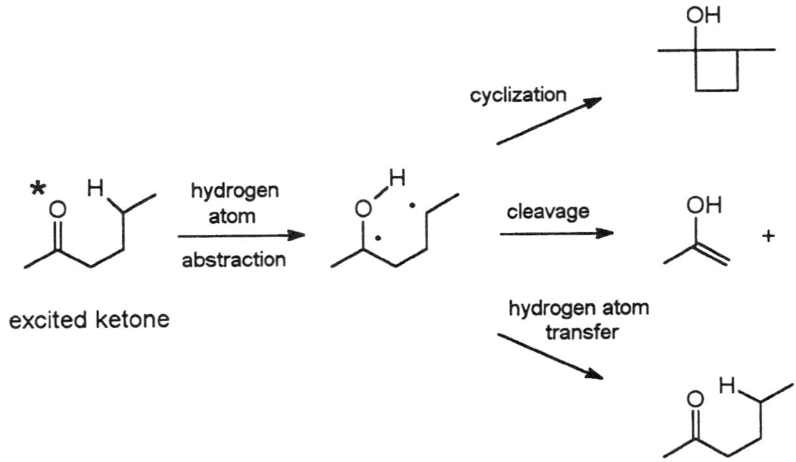

Scheme 5.1

(Scheme 5.2).[1] Combination of the radical pair **2**, formed by abstraction of a hydrogen atom from methanol by excited **1**, produces the pinacols **3** and **4** and the disproportionation product **5**. The reaction of **1** is rare insofar as few intermolecular, hydrogen-abstraction reactions are known in which the excited aldehydo or keto group is located in the carbohydrate reactant (Table 5.1). It is more common in intermolecular hydrogen-abstraction reactions for the carbohydrate to serve as the hydrogen-atom donor (Table 5.2).

A noteworthy aspect of the reaction shown in Scheme 5.2 is that the major product, the lactone **6**, arises from an unusual rearrangement. The mechanism for formation of **6** is not known with confidence, but the process appears to be a rather complicated one that begins with a Norrish Type I reaction producing the unsaturated aldehyde **7** (Scheme 5.3).[1] (The proposed mechanism for the entire reaction is discussed in Chapter 2.) Having Norrish Type I and hydrogen-abstraction reactions occur together is understandable because both are characteristic reactions of carbonyl compounds in (n,π*) states. (Chapter 2 contains a discussion of the tendency of carbonyl compounds in (n,π*) states to undergo both intermolecular hydrogen abstraction and Norrish Type I reaction.)

Triethylamine as an Electron and Hydrogen-Atom Donor

Perhaps the photochemical reaction that has been conducted by the largest number of individuals is the reaction of excited benzophenone with 2-propanol to produce benzpinacol. (This is a common experiment in undergraduate organic chemistry laboratories.) Pitts et al. proposed the mechanism outlined in Scheme 5.4 for this reaction.[8] It has been recognized, however, since the pioneering work of Cohen et al., that amines have a dramatic effect on this reaction (and reactions of other excited ketones).[9] For example,

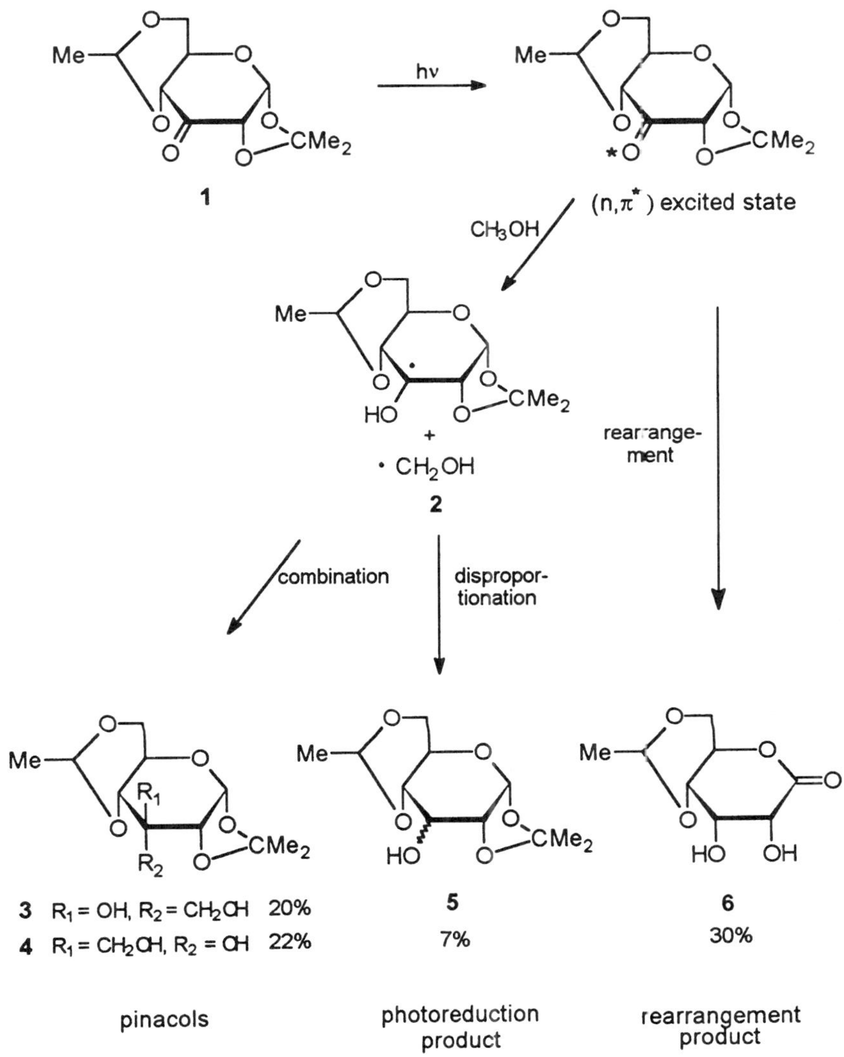

hν
1
(n,π*) excited state
CH3OH
rearrange-
ment
· CH2OH
2
combination
dispropor-
tionation
3 R1 = OH, R2 = CH2OH 20%
4 R1 = CH2OH, R2 = OH 22%
5
7%
6
30%
pinacols
photoreduction product
rearrangement product

Scheme 5.2

Table 5.1. Intermolecular hydrogen abstraction by excited carbohydrates.

Reactant	*Product*	*Percent Yield*	*Ref.*
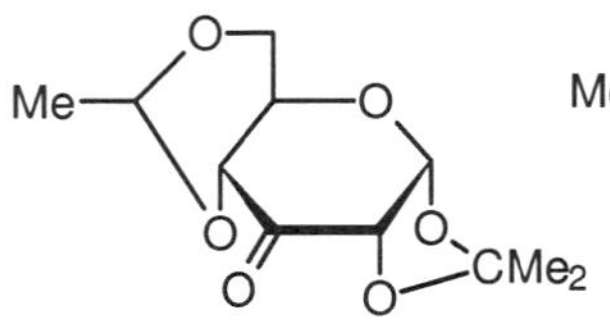 + CH_3OH	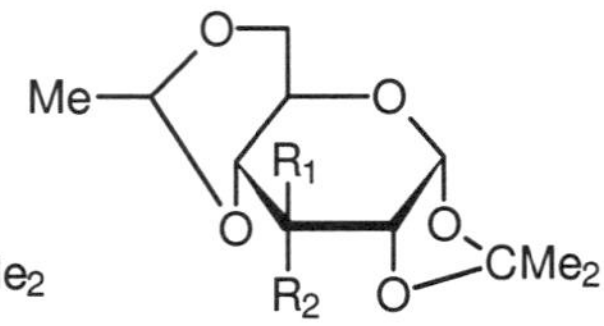		1
	$R_1 = OH$, $R_2 = CH_2OH$	20	
	$R_1 = CH_2OH$, $R_2 = OH$	22	
	+ 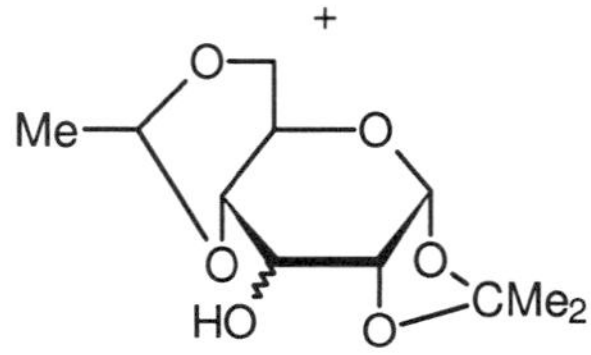	7[a]	
	+ 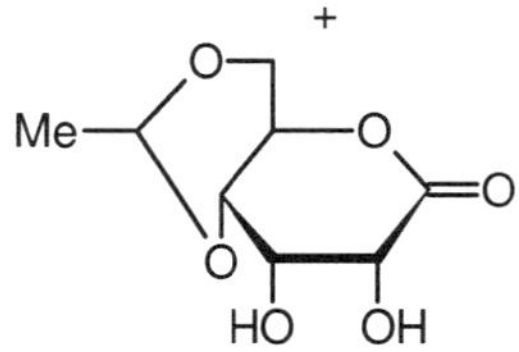	30[b]	
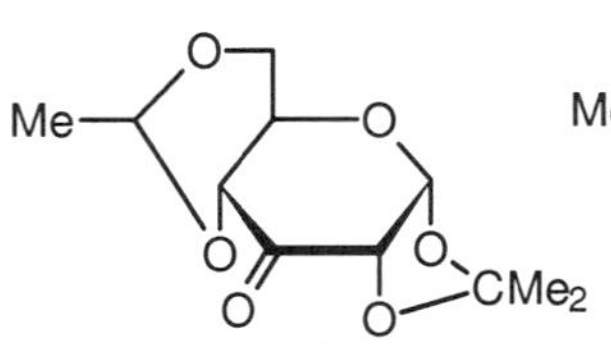 + CH_3OH	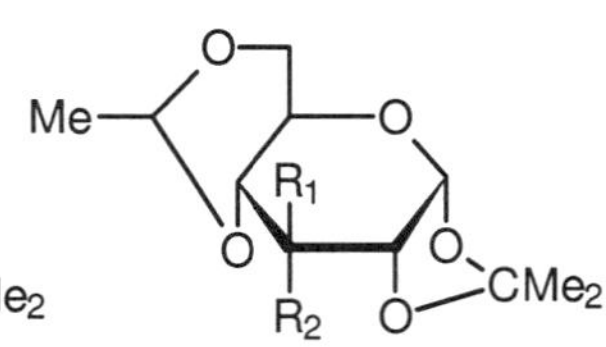		1
	$R_1 = OH$, $R_2 = CH_2OH$	25	
	$R_1 = CH_2OH$, $R_2 = OH$	30[c]	

[a]This is a combined yield of both compounds.
[b]The mechanism for formation of this compound is discussed in Chapter 2
[c]Other products were formed but not identified.

Table 5.2. Hydrogen abstraction from carbohydrates by excited ketones.

Reactant	*Product and Percent Yield*			*Ref.*
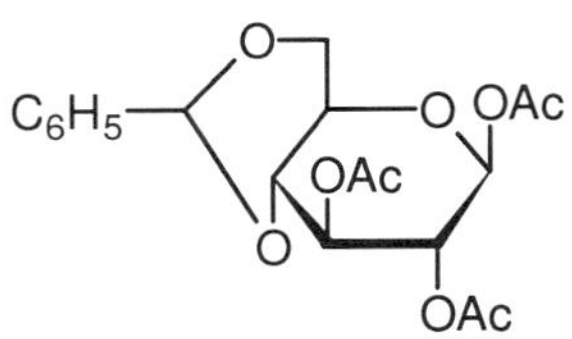 + 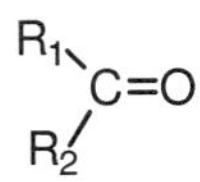	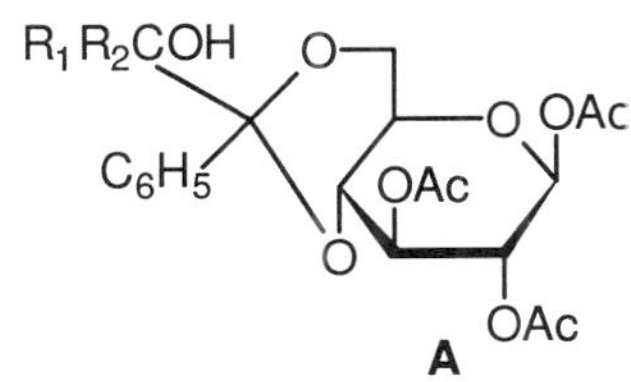 + 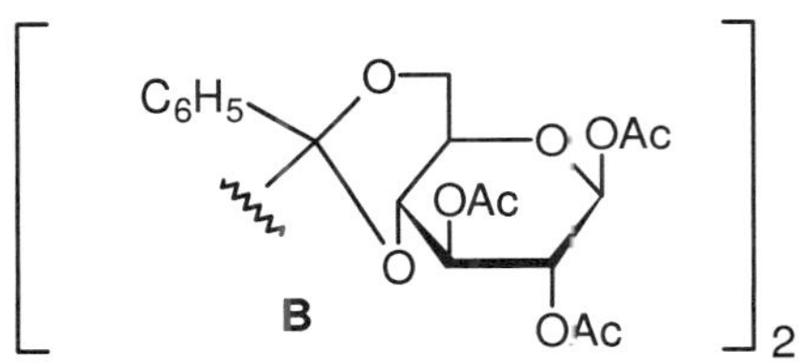			2a
		A	**B**	
	$R_1 = R_2 = C_6H_5$	38	trace	
	$R_1 = C_6H_5$, $R_2 = CH_3$	33	7	
	$R_1 = R_2 = CH_3$	23	9	
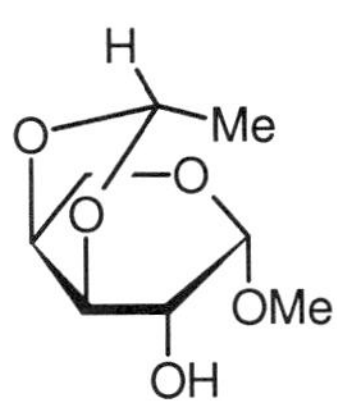 + $(CH_3)_2C{=}O$	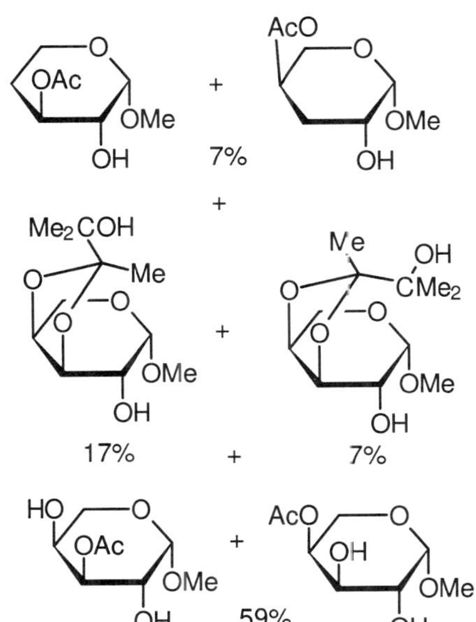			3b

Continued on next page

Table 5.2. Hydrogen abstraction from carbohydrates by excited ketones—*Continued*

Reactant	*Product and Percent Yield*	*Ref.*
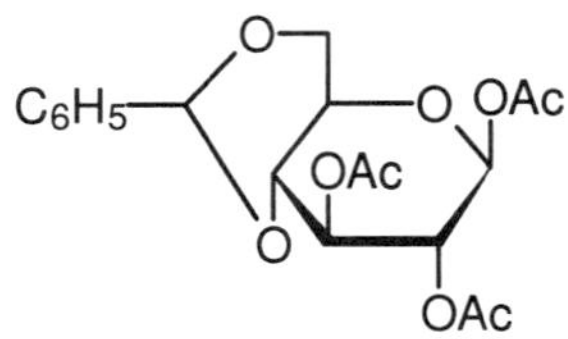 + O_2 + $R_2C{=}O$	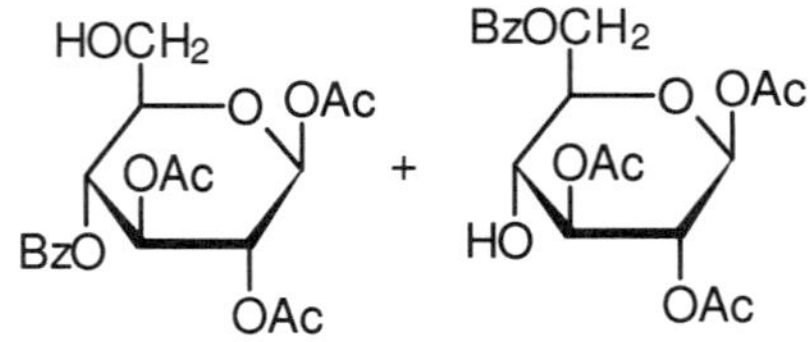 27% R = C_6H_5 19%; 41% R = CH_3 23%	2
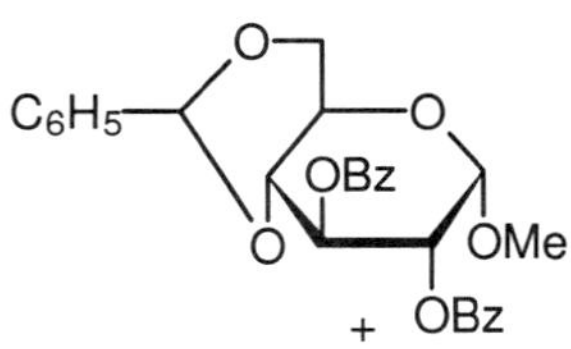 + O_2	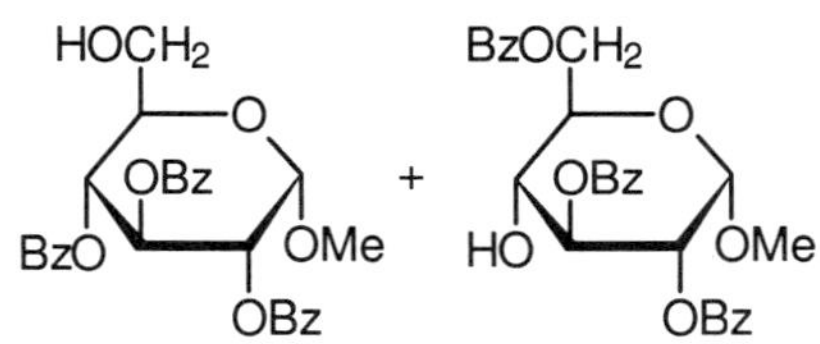 yield not reported	4
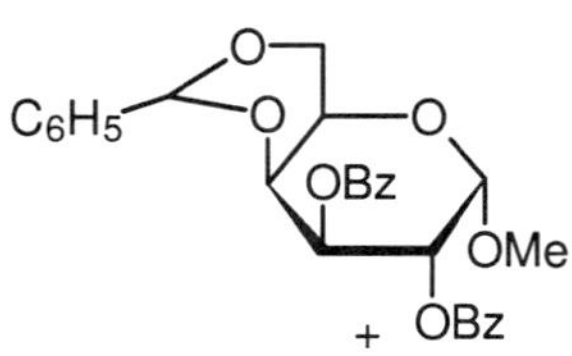 + O_2	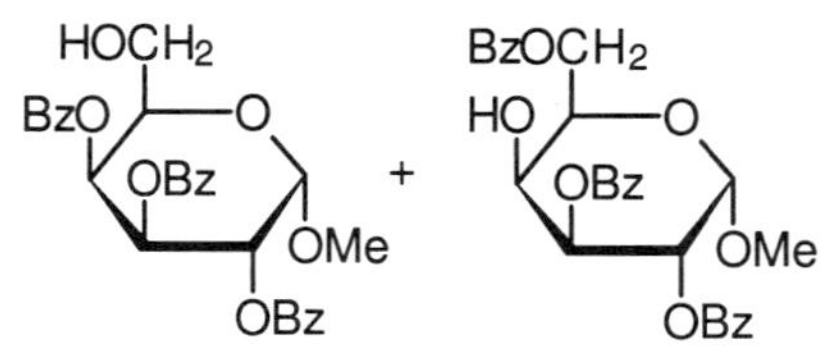 yield not reported	4
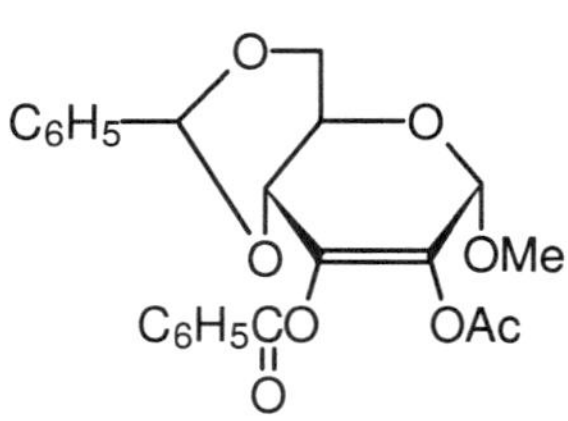	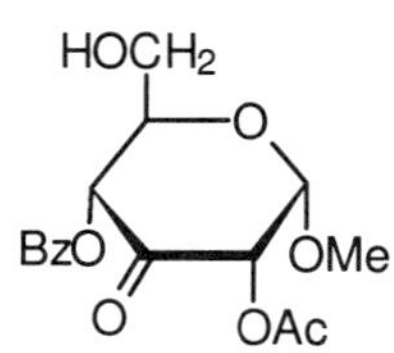 83%	5

Continued on next page

Table 5.2. Hydrogen abstraction from carbohydrates by excited ketones—*Continued*

Reactant	*Product*	*Percent Yield*	*Ref.*
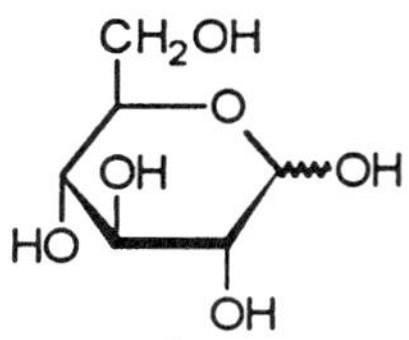 + $(CH_3)_2C{=}O$	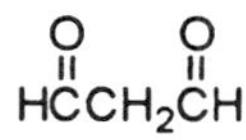	c	6
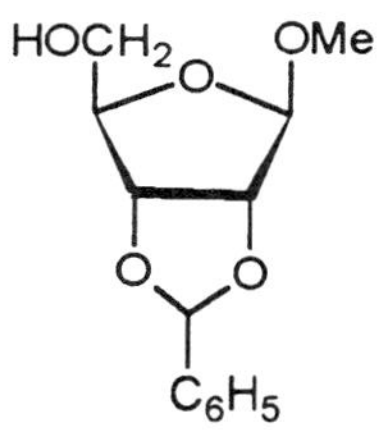	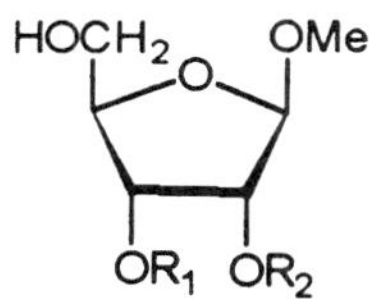	58	7
	R_1 = H, R_2 = C_6H_5CO R_1 = C_6H_5CO, R_2 = H	not reported	7
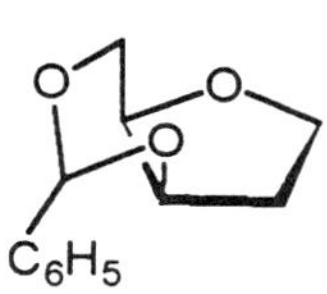	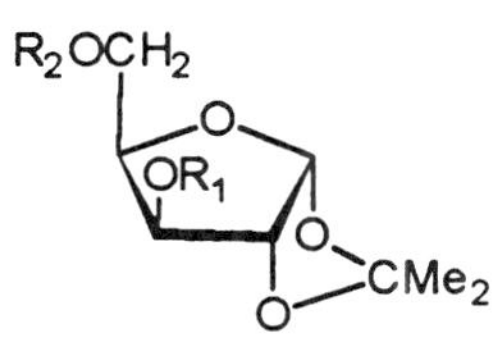 R_1 = H, R_2 = C_6H_5CO R_2 = C_6H_5CO, R_2 = H		7

[a]Pinacols not involving carbohydrates also were formed.
[b]Small amounts of oxygen were believed to be present.
[c]Sucrose also was reported to form malonaldehyde under similar conditions.

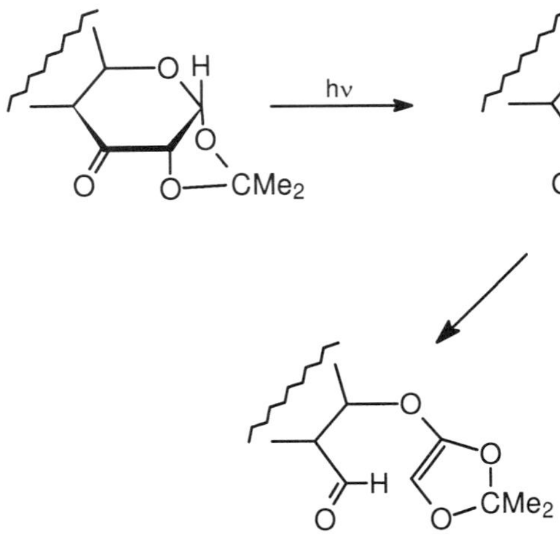

7

Scheme 5.3

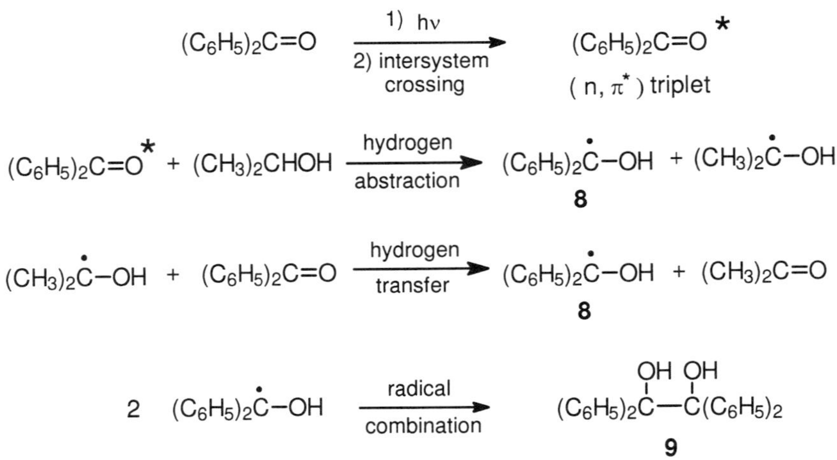

Scheme 5.4

although photolysis of benzophenone with either 2-propanol or triethylamine produces benzpinacol (**9**), reaction with triethylamine is 2000 times faster than that with 2-propanol. This difference in reactivity between 2-propanol and triethylamine is worth examining not only to determine its origin but also because it provides a useful background for understanding the β-cleavage reactions that occur when some carbohydrates are irradiated in the presence of triethylamine.

Cohen and co-workers have shown that excited benzophenone initially reacts with an amine to form an exciplex (**10**).[9,10] For triethylamine, and most aliphatic amines, rapid hydrogen transfer occurs within this exciplex to give the "benzhydrol" radical (**8**) and an α-amino radical (**11**) (Scheme 5.5).[9,11] Schaeffer and Peters actually have observed the exciplex **10** directly by flash-photolysis techniques and have followed its conversion into radicals **8** and **11**.[12] For an amine such as DABCO (1,4-diazabicyclo[2.2.2]octane) (**12**), a compound that is a better electron donor than most aliphatic amines, exciplex formation with excited benzophenone is followed by complete electron transfer to form radical ions (Scheme 5.6). These ions also have been observed in flash-photolysis experiments.[13] Even in reactions with complete electron transfer, the benzhydrol radical (**8**), which is formed by proton

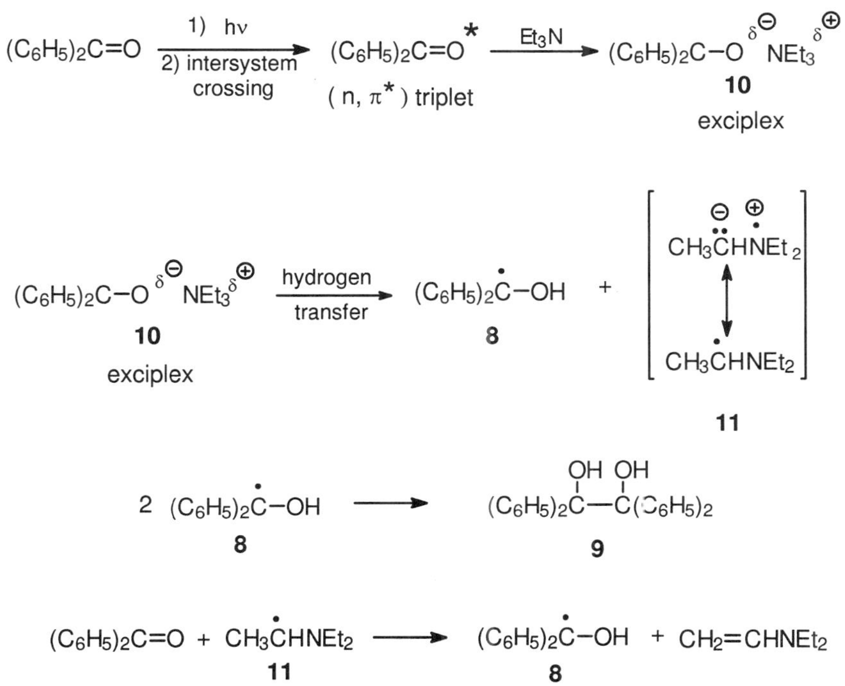

Scheme 5.5

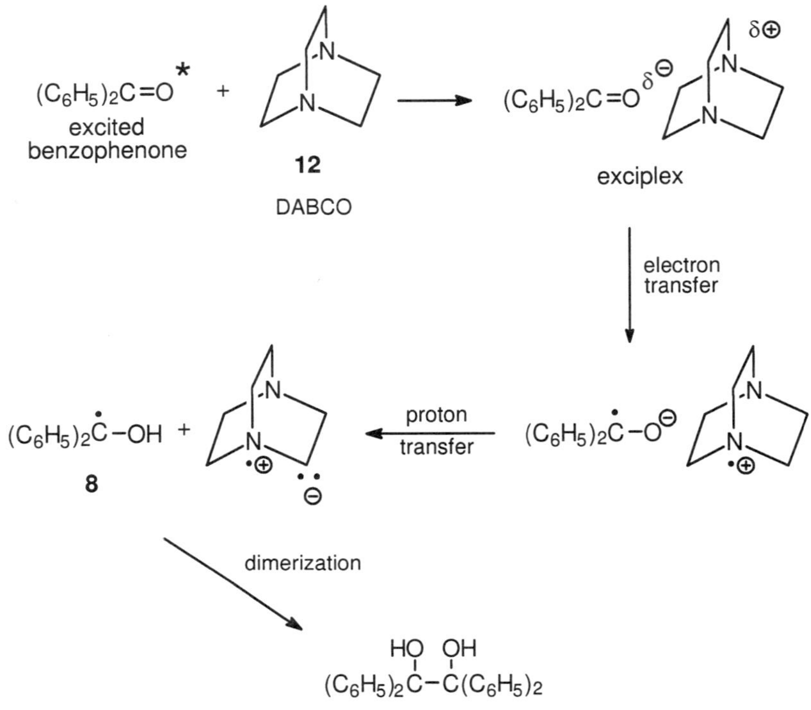

Scheme 5.6

transfer between radical ions, remains the precursor of benzpinacol (**9**) (Scheme 5.6).

When a carbonyl compound with an oxygen atom bonded to an α-carbon atom is irradiated in the presence of triethylamine, a β-cleavage reaction takes place (Scheme 5.7).[14] Electron transfer from triethylamine followed by β-cleavage has been used in the synthesis of deoxy sugars and in the cleavage of allyl glycosides (Table 5.3).[14–16] It is reasonable to expect that an exciplex is formed in these reactions and that complete electron transfer to generate radical ions may occur. Even though complete transfer is not observed between excited benzophenone and triethylamine, electron transfer is a possibility between triethylamine and other excited carbonyl compounds; in fact, those investigators that have offered mechanistic interpretation for such reactions have favored radical-ion intermediates.[16,19] More mechanistic investigation of this reaction would be valuable; in particular, it would be interesting to know whether radical ions are actually formed and, if so, whether β-cleavage can occur directly from the radical anion itself (Scheme 5.8).

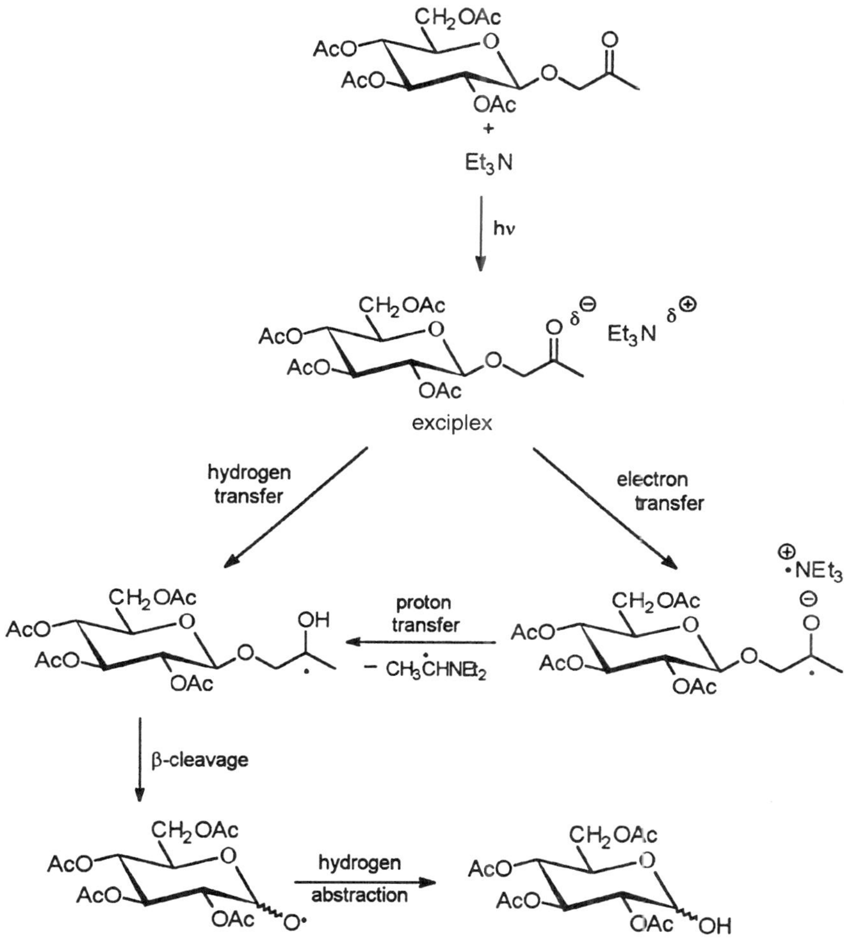

hν
exciplex
hydrogen transfer
electron transfer
proton transfer
hydrogen abstraction
β-cleavage

Scheme 5.7

Table 5.3. β-Cleavage reactions. Triethylamine was added to the reaction mixture.

Reactant	*Product and Percent Yield*	*Ref.*
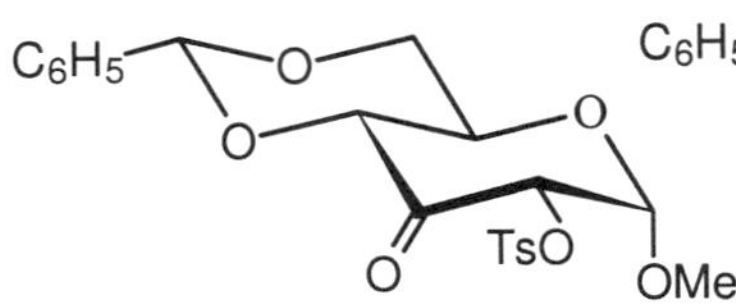	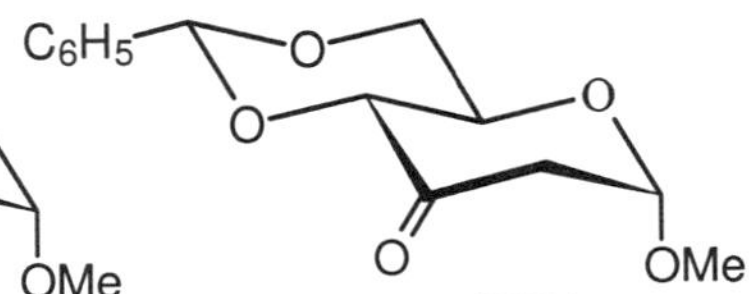 38%	15
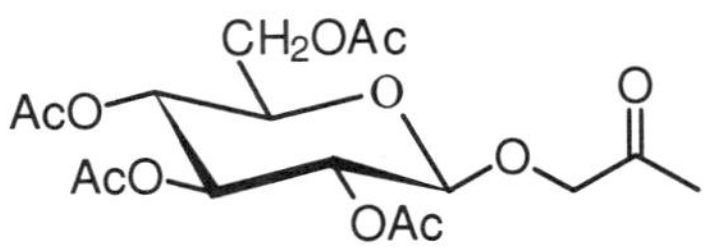	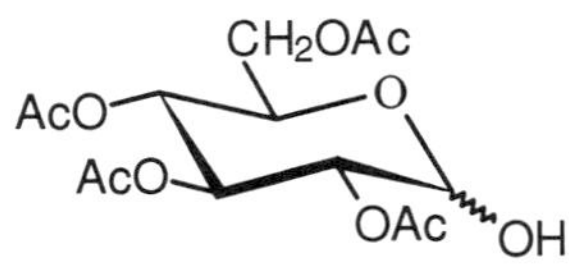 82%	14
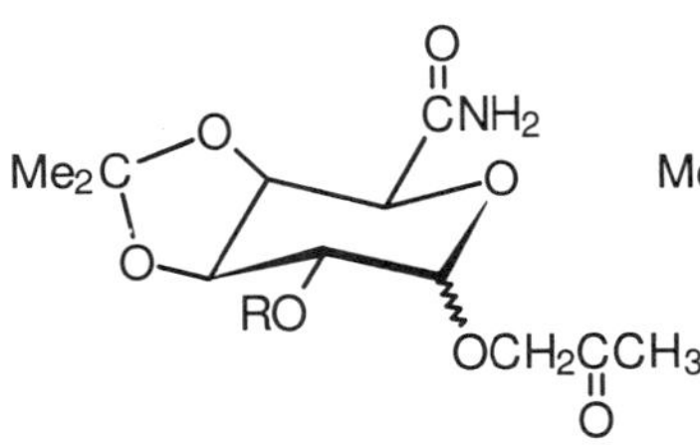 R = 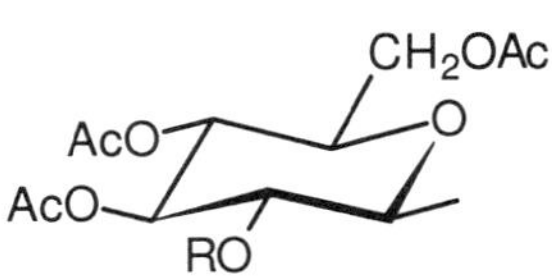	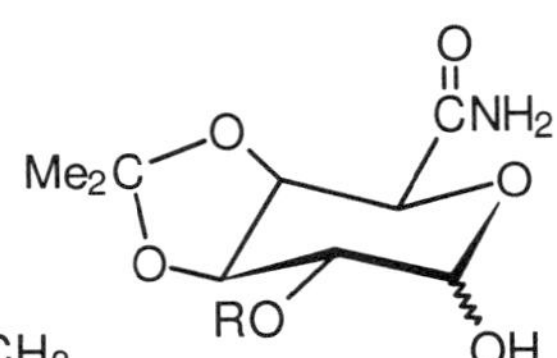 67%	14b
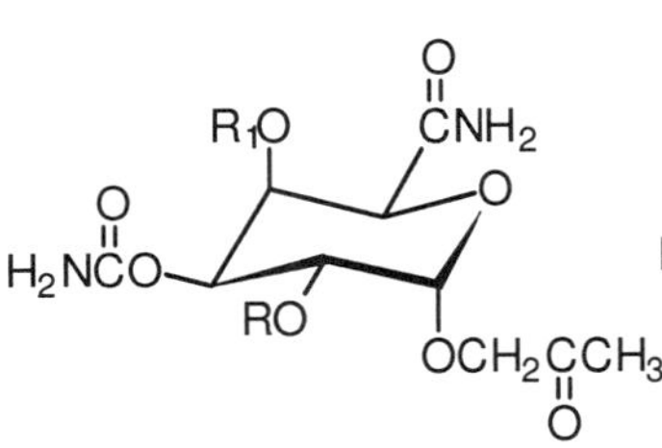 R_1 = Ac R_1 = $COOCH_2CCl_3$	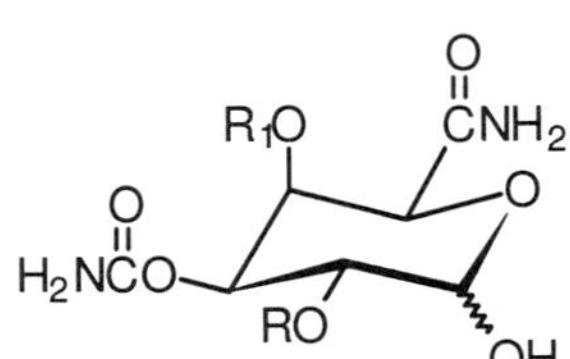 98% not reported	14
R is the same as in the previous reaction.		

Continued on next page

Table 5.3. β-Cleavage reactions. Triethylamine was added to the reaction mixture—*Continued*

Reactant	*Product*	*Percent Yield*	*Ref.*
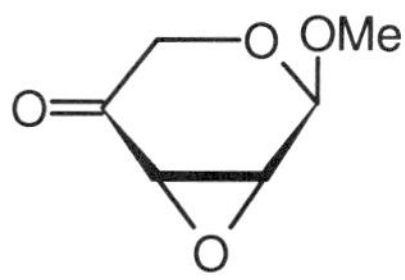	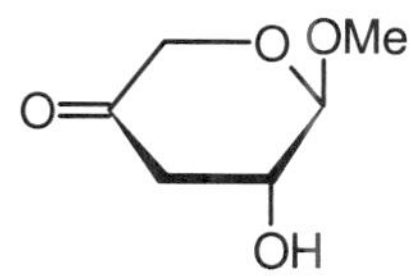	63	16
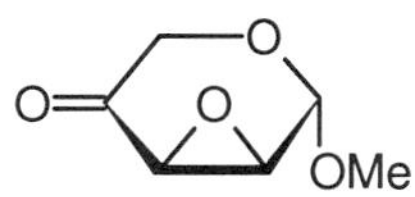	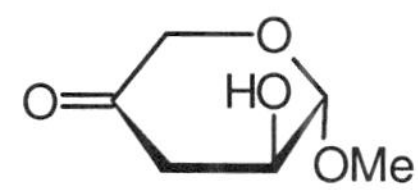	66	16
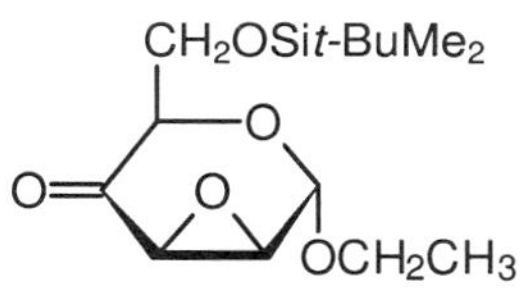	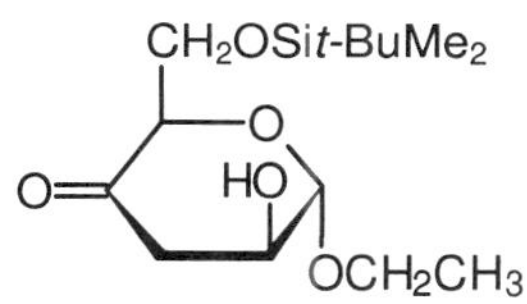	68	16
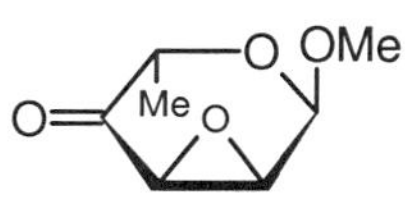	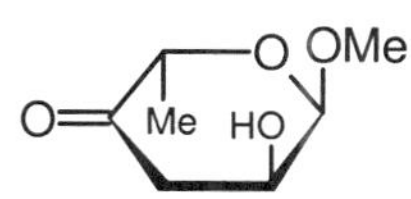	60	16
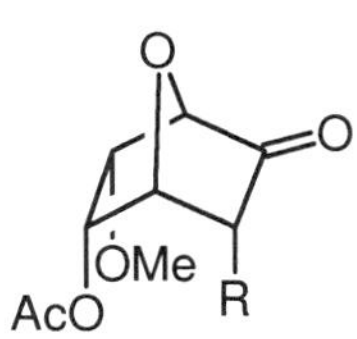	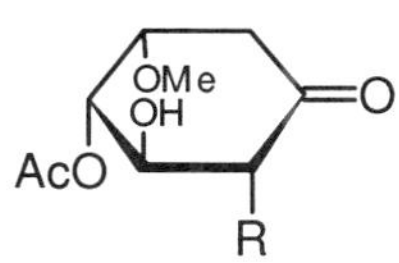	28 (R_1) 28 (R_2)	17 18
R_1 = 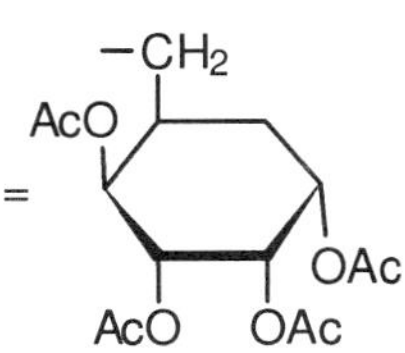	R_2 = 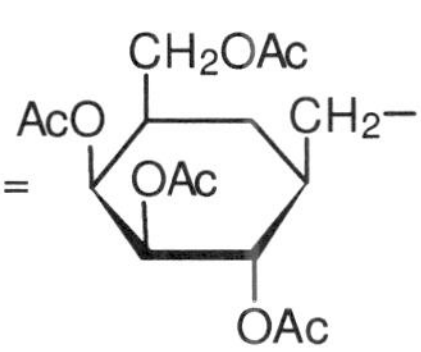		
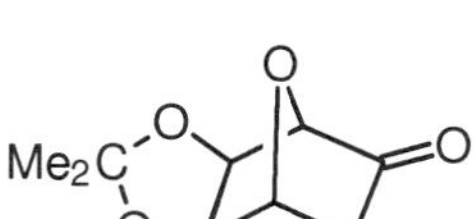	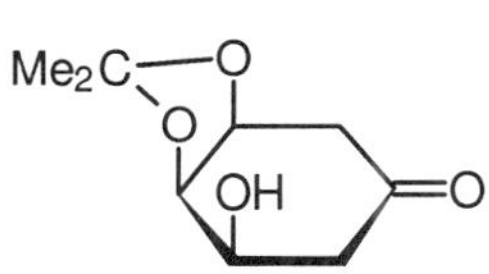	85	19, 20

[a]Triethylamine was added to the reaction mixture.
[b]Reaction also occurred with R = H, but no yield was reported.

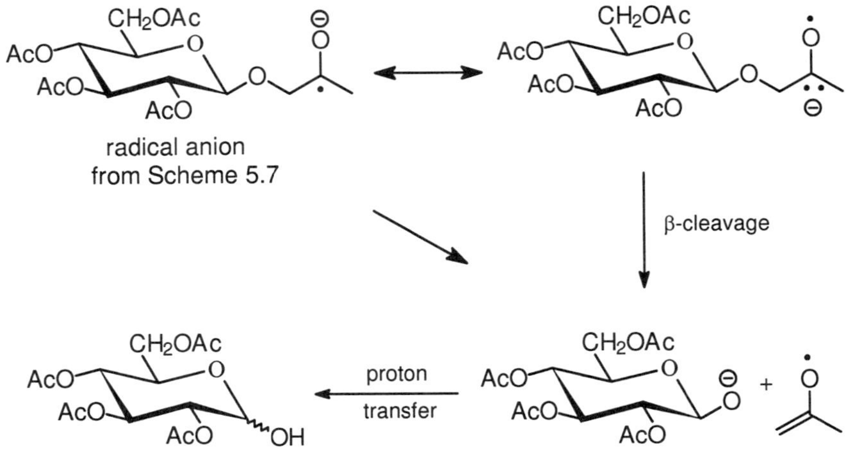

Scheme 5.8

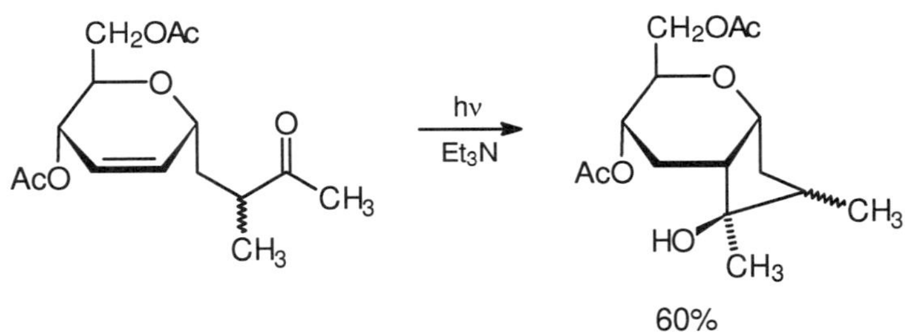

(1)

Equation 1

Electron transfer between triethylamine and an excited carbonyl compound also provides the basis for photoreductive cyclization of δ,ε-unsaturated aldehydes and ketones (eq 1). Cossy and Ibhi have used this reaction to synthesize a group of bicyclic cyclopentanols from unsaturated *C*-glycosides (Table 5.4).[21]

Hydrogen Abstraction from a Carbohydrate by an Excited Carbonyl Compound

Reaction in an Inert Atmosphere

An excited ketone, usually acetone or benzophenone, readily abstracts a hydrogen atom from a carbohydrate to produce a radical pair. This process generally is followed by radical combination reactions and, depending upon the reactants and the reaction conditions, all possible combination products can be formed (Scheme 5.9).[2] Reactions of this type are listed in Table 5.2.

Table 5.4. Photoreductive cyclization of δ,ϵ-unsaturated aldehydes and ketones.

Reactant	*Product*	*Percent Yield*	*Ref.*
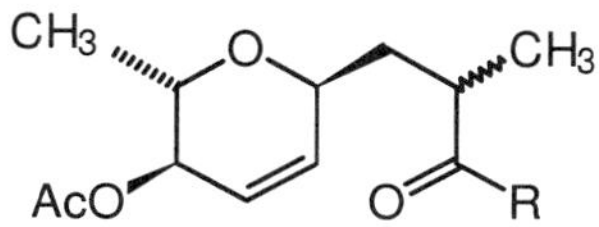	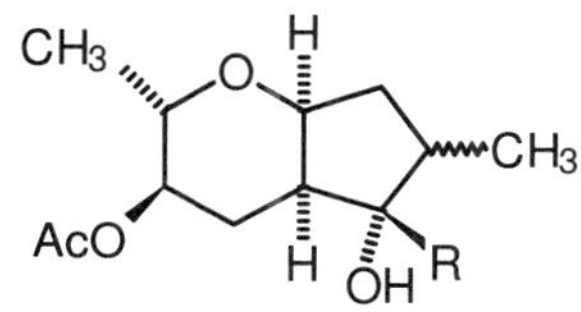		21
R = H		73	
R = CH_3		70	
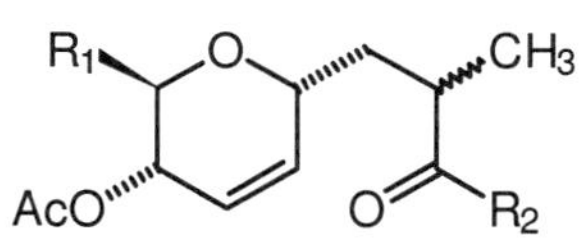	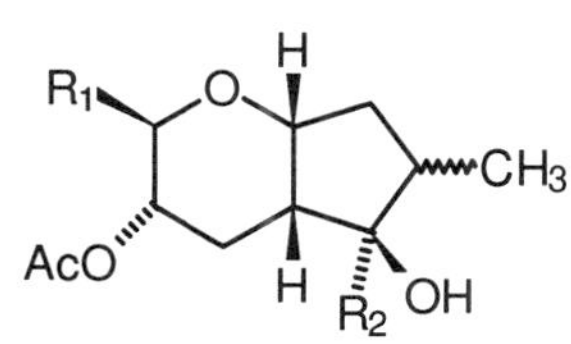		21
R_1 = H, R_2 = H		77	
R_1 = CH_2OAc, R_2 = H		68	
R_1 = CH_2OAc, R_2 = CH_3		60	
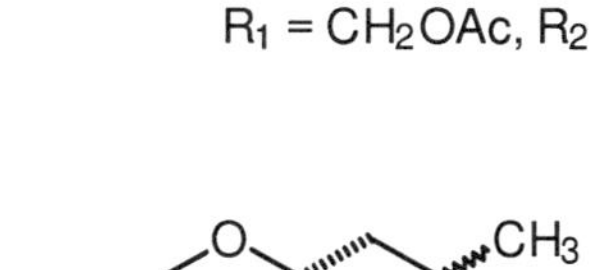	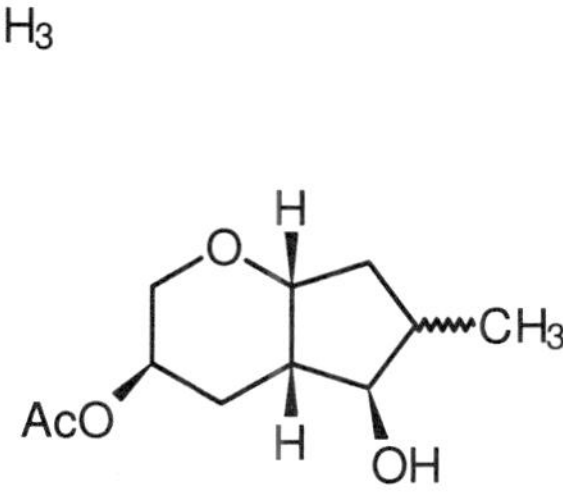	80	21

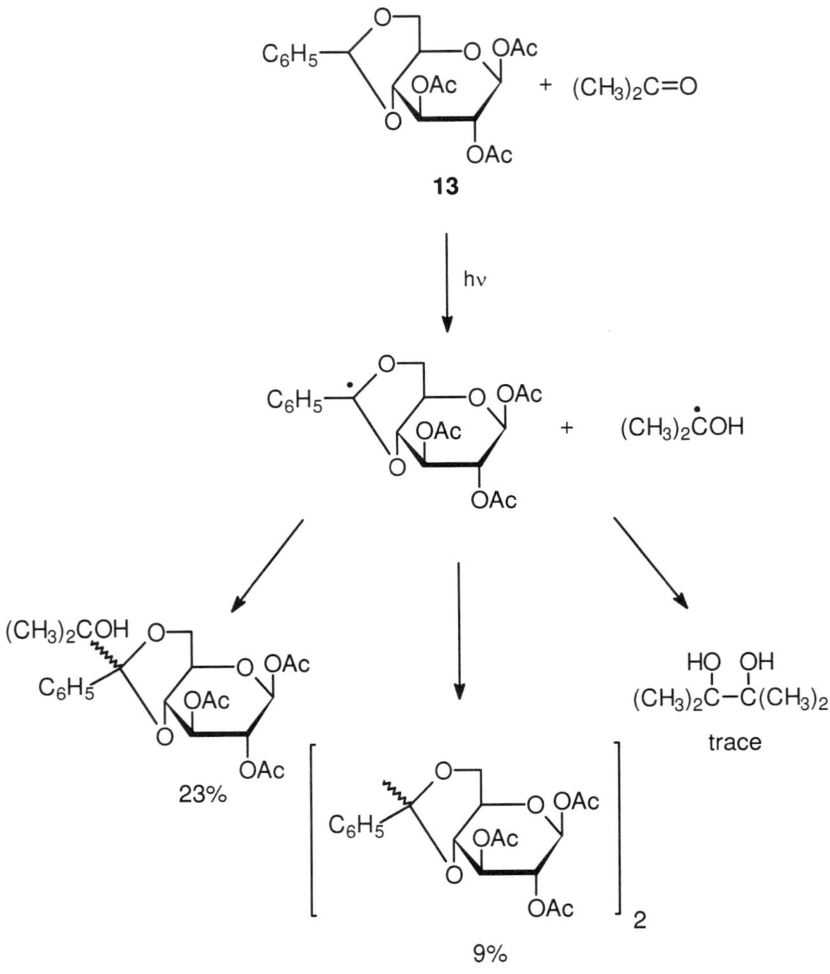

C6H5
OAc
OAc
OAc
13
(CH3)2C=O
hν
(CH3)2COH
(CH3)2COH
23%
HO OH
(CH3)2C−C(CH3)2
trace
2
9%

Scheme 5.9

Reaction in the Presence of Molecular Oxygen

If molecular oxygen (an excellent radical scavenger) is present in a photolysis mixture, even in small quantities, it has a dramatic effect on product formation. Suzuki et al. provided a striking example of the influence of molecular oxygen on a hydrogen-abstraction reaction when they found that during photolysis of the D-glucose derivative **13** and acetone, oxygen completely suppressed the formation of products produced by irradiation in an inert atmosphere (Scheme 5.9) in favor of the two hydroxybenzoates **17** and **18** (Scheme 5.10).[2] Formation of these compounds indicated that the radical **14** was reacting rapidly with oxygen to give the peroxy radical **15**. This rapid reaction is not unexpected because study of autoxidation reactions has shown that the addition of molecular oxygen to radicals is diffusion controlled in many instances.[22] Peroxy radicals, such as **15**, can enter into a variety of reactions that include hydrogen abstraction and self-addition.[22,23] These reactions, some of which are chain processes, provide several possible pathways leading from a peroxy radical (e.g., **15**) to an alkoxy radical (e.g., **16**). A radical such as **16** is one logical precursor for the hydroxybenzoate photoproducts **17** and **18** (Scheme 5.10). (Another possible mechanism for hydroxybenzoate formation that does not require an alkoxy radical intermediate is described in reference 24.)

Reaction of Anomeric Hydrogen Atoms

Considerable interest in hydrogen abstraction from carbohydrates has centered around reaction at the anomeric carbon atom. The motivation for investigation of the reactivity of anomeric hydrogen atoms is both theoretical and practical. The theoretical justification for such research comes from a desire to understand the difference in reactivity between axial and equatorial hydrogen atoms. The practical interest arises from observations that, under proper conditions, anomeric radicals are intermediates in the synthesis of β-glycosides, compounds that are often difficult to produce by classical methods of glycoside formation. (The synthesis of β-glycosides by free-radical reaction is discussed in Chapter 15.)

Because of the often-noted similarity in reactivity between (n,π*) excited carbonyl compounds and alkoxy radicals,[25–27] the reactivity of several compounds with the *t*-butoxy radical provides a beginning point for discussing the reactivity of anomeric carbon–hydrogen bonds in carbohydrates. Jenkins has shown that introduction of an oxygen atom into a ring system facilitates abstraction of a hydrogen atom on an α-carbon atom (Figure 5.1).[28] The activating effect of an oxygen atom is attributed to the nonbonding electrons on oxygen stabilizing the partial positive charge on the α-carbon atom in the transition state (Figure 5.2). This stabilization is greater than the inductive destabilization caused by electron withdrawal due to the electronegativity of oxygen (Figure 5.2). An oxygen atom attached to a β-carbon atom, however,

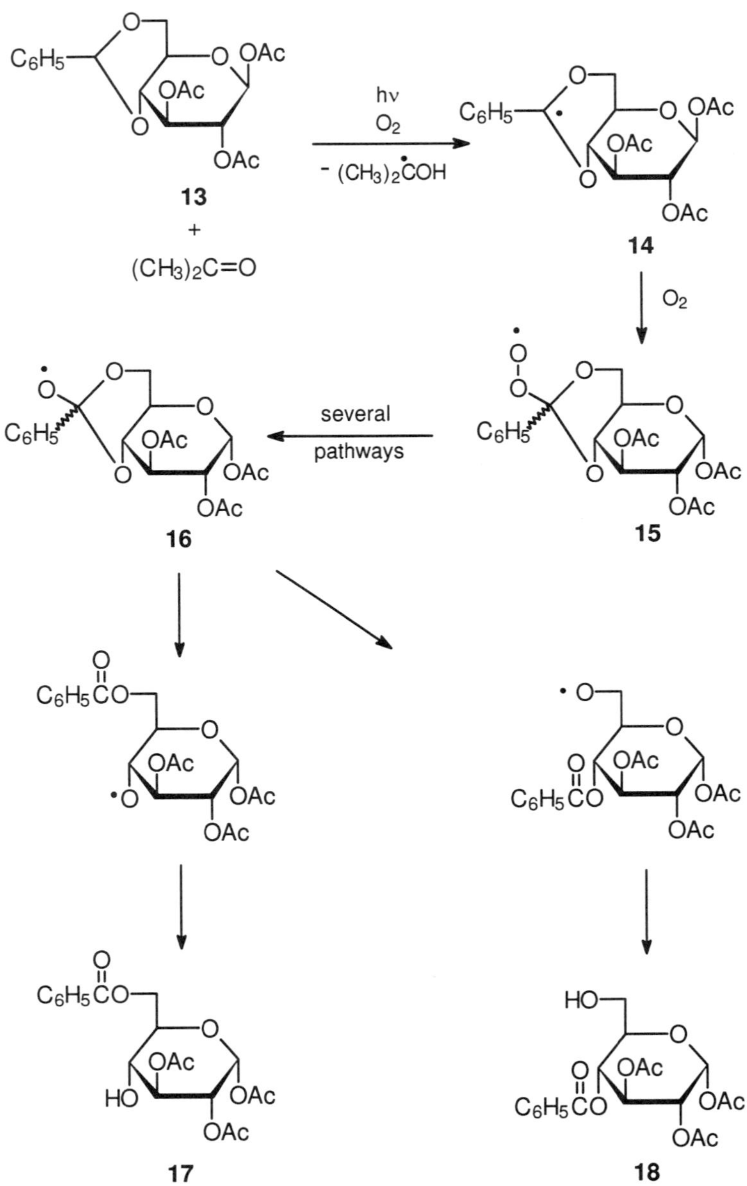

Scheme 5.10

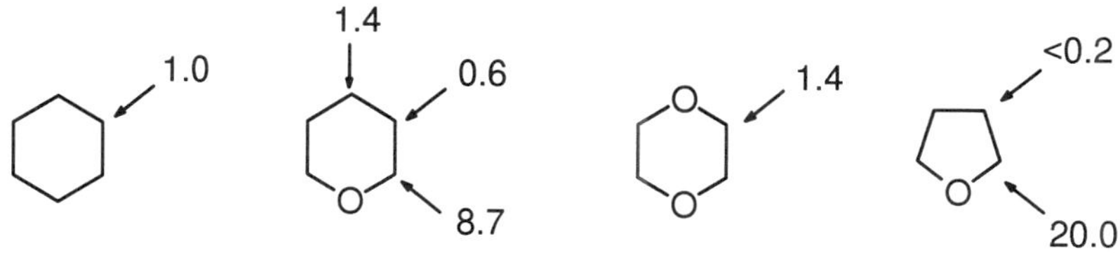

Figure 5.1. Relative rates of methylene hydrogen abstraction by t-butoxy radicals.

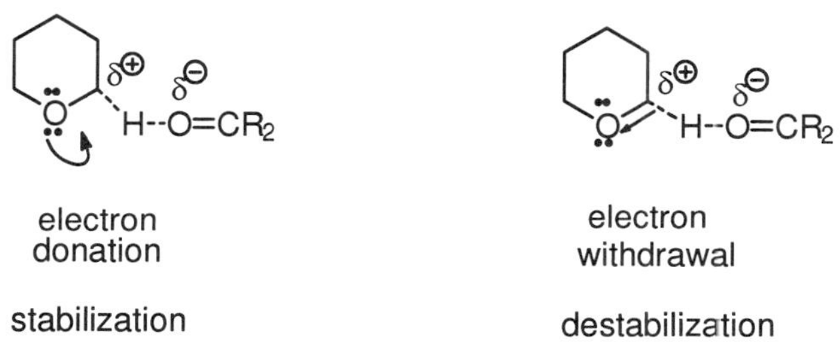

Figure 5.2. Hydrogen abstraction from an α-carbon atom by an (n,π) excited carbonyl compound.*

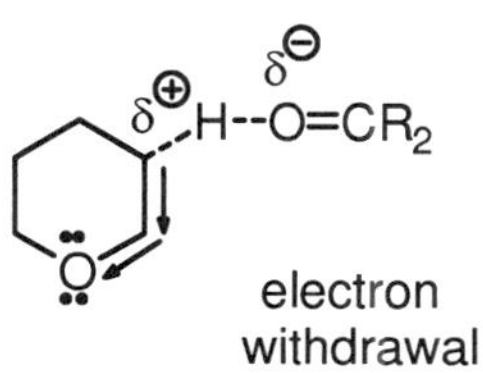

Figure 5.3. Hydrogen abstraction from a β-carbon atom by an (n,π) excited carbonyl compound.*

retards hydrogen abstraction because the inductive electron withdrawal has no stabilizing electron donation to "counterbalance" it (Figure 5.3).[28] The deactivating effect of oxygen atoms is important only for electron-deficient abstracting agents such as an alkoxy radical[28] or the oxygen atom of an (n,π*) excited carbonyl group[26,27] because these reactants create a partial positive charge on a carbon atom donating a hydrogen atom during the abstraction process (Figures 5.2 and 5.3).[28]

Essentially all of the studies of the relative reactivity of anomeric hydrogen atoms have been conducted on model compounds. Study of model compounds simplifies matters considerably because fully functionalized carbohydrates undergo hydrogen abstraction from all possible carbon atoms.[29,30] In contrast, for appropriate model compounds, hydrogen abstraction is exclusively from the atom corresponding to the anomeric carbon atom. For example, when McKelvey investigated the reaction of 2-methoxytetrahydropyran

(**19**) with excited benzophenone, he found that all the products could be explained by reaction beginning with hydrogen abstraction from the atom analogous to the anomeric carbon atom (Scheme 5.11).[31]

Hayday and McKelvey considered the question of whether an axial or equatorial, anomeric hydrogen atom was more easily abstracted.[32] To address this question, they investigated the *cis*- and *trans*-2-methoxy-4-methyltetrahydropyrans **20** and **21**, respectively, and determined from the relative reactivities of these two compounds that axial, anomeric hydrogen atoms were eight times more reactive toward abstraction by excited benzophenone than were equatorial ones. They explained this difference by noting that only for the *cis*-isomer is one of the orbitals on the ring oxygen atom properly positioned to stabilize the radical center developing on the anomeric carbon atom during hydrogen abstraction (Figure 5.4).

The conformational mobility of compounds **20** and **21** introduces uncertainty into the determination of the relative reactivity of the anomeric hydrogen atoms, owing to the possibility of reaction occurring through a conformer present in low concentration. This uncertainty prompted McKelvey and Iwamura to investigate conformationally less mobile compounds.[33] These addi-

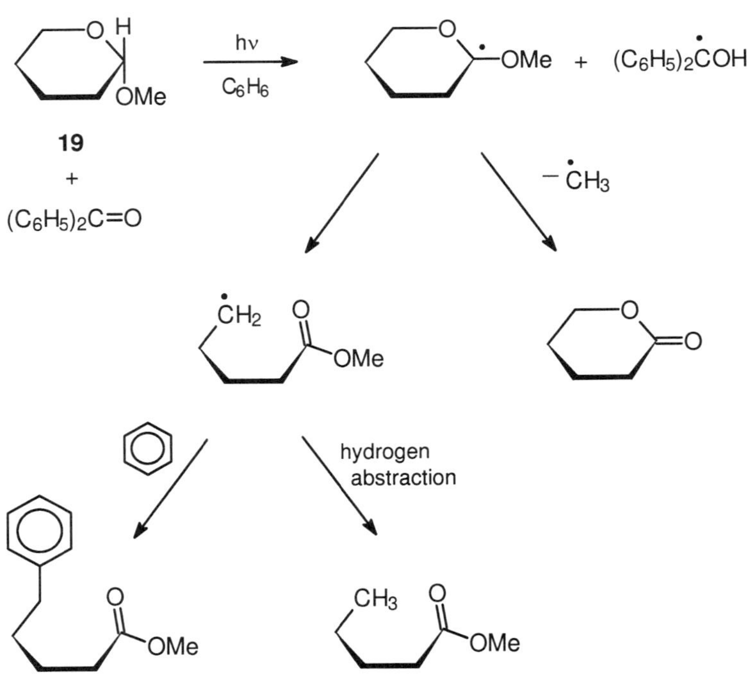

Scheme 5.11

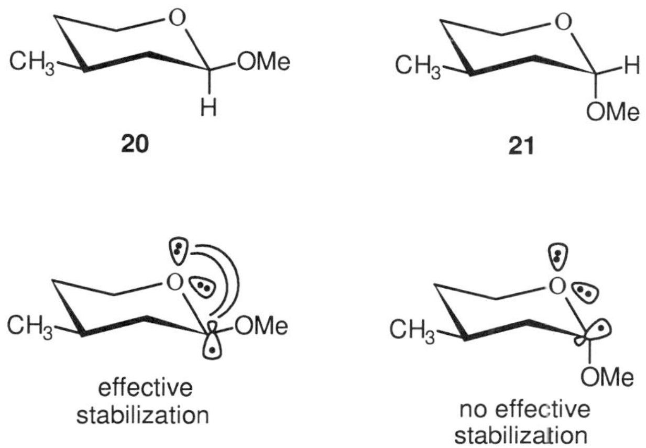

Figure 5.4. Radical stabilization during hydrogen abstraction.

tional compounds showed an even greater preference for abstraction of axial over equatorial anomeric hydrogen atoms.

Beckwith and Easton also have investigated the difference in reactivity between axial and equatorial hydrogen atoms in cyclic, oxygen-containing systems.[34] Malatesta and Ingold have conducted similar studies.[35] Each of these groups considered other possible explanations for the difference in reactivity between axial and equatorial hydrogen atoms and concluded that the stereoelectronic interpretation used by McKelvey was the best. Both groups did point out that the model used by McKelvey for explaining the stabilization provided by a ring oxygen atom (Figure 5.4) was not the best one. Experimental and theoretical findings indicated that the oxygen atom in a tetrahydropyranyl ring has one pair of nonbonding electrons in an orbital with considerable p-character and another pair in an orbital with considerable s-character; consequently, the stabilization provided by the ring oxygen atom should be seen as arising from electrons in a "p-like" orbital on oxygen (Figure 5.5). Such hybridization would permit stabilization of the developing anomeric radical similar to that proposed by McKelvey.

Photochemical Formose Reaction

Formose is a complex mixture of carbohydrates usually synthesized by heating aqueous formaldehyde in the presence of an inorganic base such as calcium hydroxide.[36] This reaction has been suggested to be the first documented synthesis of a sugar-like mixture in the chemical literature.[37] Formose-type products also can be produced by irradiation of aqueous formaldehyde in the presence of base.[38–45] Photochemical and nonphotochemical formose reactions

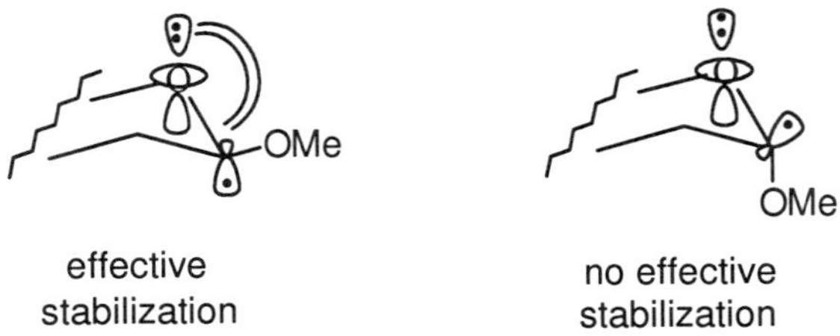

Figure 5.5. Radical stabilization by nonbonded electrons.

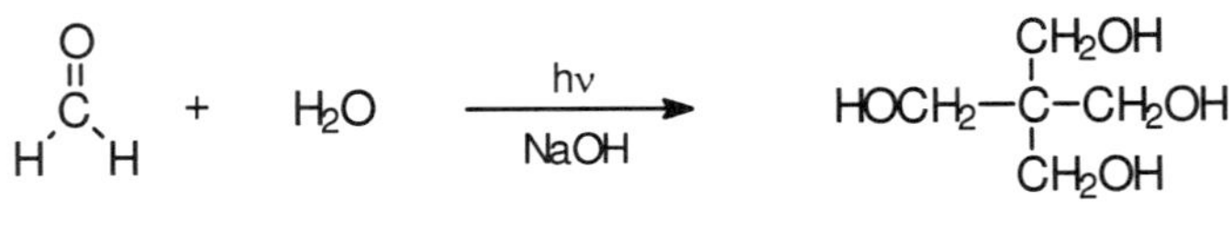

(2)

22

Equation 2

typically produce complex mixtures of sugars and related compounds.[36,41,42] (Seel et al.[41] identified 19 components from a photochemical formose reaction, while Shigemasa et al.[42] found evidence for at least 29 compounds in a product mixture.) The major photoproduct under most conditions is "pentaerythritol" (**22**); in fact, Irie reported this to be the only reaction product from photolysis of formaldehyde in aqueous sodium hydroxide (eq 2).[43]

Relatively little is known about the mechanism of the photochemical formose reaction, although many of the products appear to arise from radical reactions initiated by hydrogen abstraction. One significant mechanistic finding has been made by Schwartz and de Graaf, who determined that irradiation of aqueous formaldehyde produced acetaldehyde and that, upon standing in the dark, the amount of pentaerythritol (**22**) in the photolysis mixture increased over a two-week period.[44,45] Because base is required for the formose reaction to occur, the work of Schwartz and de Graaf suggests that the photochemically generated acetaldehyde undergoes a series of aldol condensations with formaldehyde, followed by a Cannizzaro reaction, to produce pentaerythritol (**22**). Beyond these findings, however, little is known about the mechanism of the photochemical formose reaction.

Hydrogen Abstraction by Excited Uronic Acids and Their Esters

Intermolecular hydrogen abstraction is involved in photochemical glycosidic cleavage of uronic acids and esters. This topic is discussed along with other reactions of carboxylic acids and esters in Chapter 6.

Summary

Intermolecular hydrogen-abstraction reactions produce radical pairs that then undergo typical radical reactions (Scheme 5.2). Carbohydrates can function as either the abstracting agents or the hydrogen-atom donors (Tables 5.1 and 5.2). Because radicals are generated by hydrogen abstraction, these reactions are quite sensitive to molecular oxygen. This sensitivity is illustrated by the reactions in Schemes 5.9 and 5.10, which show that molecular oxygen completely changes the course of reaction. Intermolecular hydrogen abstraction also is an integral part of the photoformose reaction (eq 2).

References

1. Collins, P. M.; Munasinghe, V. R. N.; Oparaeche, N. N. *J. Chem. Soc., Perkin Trans. 1* **1977**, 2423.
2. Suzuki, M.; Inai, T.; Matsushima, R. *Bull. Chem. Soc. Jpn.* **1976**, *49*, 1585.
3. Szarek, W. A.; Beveridge, R. J.; Kim, K. S. *J. Carbohydr., Nucleosides, Nucleotides* **1978**, *5*, 273.
4. Szeja, W.; Lapkowski, M. *Polish J. Chem.* **1978**, *52*, 673.
5. Box, V. G. S.; Box, L. L.; Roberts, E. V. E. *Carbohydr. Res.* **1982**, *102* 308.
6. Datta, R. K.; Rao, K. N. *Indian J. Chem.* **1976A**, *14A*, 122.
7. Matsuura, K.; Maeda, S.; Araki, Y.; Ishido, Y. *Bull. Chem. Soc. Jpn.* **1971**, *44*, 292.
8. Pitts, J. N., Jr.; Letsinger, R. L.; Taylor, R. P.; Patterson, J. M.; Recktenwald, G.; Martin, R. B. *J. Am. Chem. Soc.* **1969**, *81*, 1068.
9. Cohen, S. G.; Parola, A.; Parsons, G. H., Jr. *Chem. Rev.* **1978**, *73*, 141.
10. Guttenplan, J. B.; Cohen, S. G. *J. Am. Chem. Soc.* **1972**, *94*, 4040.
11. S. G. Cohen, S. G.; Stein, N. M. *J. Am. Chem. Soc.* **1971**, *93*, 6542.
12. Schaeffer, C. G.; Peters, K. S. *J. Am. Chem. Soc.* **1980**, *102*, 7566.
13. Peters, K. S.; Frelich, S. C.; and Schaeffer, C. G. *J. Am. Chem. Soc.* **1980**, *102*, 5701.
14. Lüning, J.; Möller, U.; Debski, N.; Welzel, P. *Tetrahedron Lett.* **1993**, *34*, 5871.
15. Szarek, W. A.; Dmytraczenko, A. *Synthesis* **1974**, 579.
16. Cossy, J.; Ibhi, S. *Carbohydr. Res.* **1994**, *259*, 141.
17. Ferritto, R.; Vogel, P. *Tetrahedron Lett.* **1995**, *36*, 3517.
18. Cossy, J.; Ranaivosata, J.-L.; Bellosta, V.; Ancerewicz, J.; Feritto, R.; Vogel, P. *J. Org. Chem.* **1995**, *60*, 8351.
19. Cossy, J.; Aclinou, P.; Bellosta, V.; Furet, N.; Baranne-Lafont, Sparfel, J. D.; Souchaud, C. *Tetrahedron Lett.* **1991**, *32*, 1315.
20. Cossy, J.; Bouzide, A.; Ibhi, S.; Aclinou, P. *Tetrahedron* **1991**, *47*, 7775.
21. Cossy, J.; Ibhi, S. *Carbohydr. Res.* **1996**, *291*, 189.
22. Ingold, K. U. *Accounts Chem. Res.* **1969**, *2*, 1.
23. Mayo, F. R. *Accounts Chem. Res.*, **1968**, *1*, 193.
24. Seto, H.; Yoshida, K.; Yoshida, S.; Shimizu, T.; Seki, H.; Hoshino, M. *Tetrahedron Lett.* **1996**, *37*, 4179.
25. Turro, N. J. *Modern Molecular Photochemistry;* Benjamin/Cummings: Menlo Park, CA, 1978; pp 364, 528.
26. Wagner, P. J. *Acc. Chem. Res.* **1971**, *4*, 168.
27. Wagner, P.; Park, B.-S. *Org. Photochem.* **1991**, *11*, 227.
28. Jenkins, I. D. *J. Chem. Soc., Chem. Commun.* **1994**, 1227.
29. Madden, K. P.; Fessenden, R. W. *J. Am. Chem. Soc.* **1982**, *104*, 2578.
30. Shkrob, I. A.; Depew, M. C.; Wan, J. K. S. *Chem. Phys. Lett.* **1993**, *202*, 133.

31. McKelvey, R. D. *Carbohdyr. Res.* **1975**, *42,* 187.
32. Hayday, K.; McKelvey, R. D. *J. Org. Chem.* **1976**, *41,* 2222.
33. McKelvey, R. D.; Iwamura, H. *J. Org. Chem.* **1985**, *50,* 402.
34. Beckwith, A. I. J.; Easton, C. J. *J. Am. Chem. Soc.* **1981**, *103,* 615.
35. Malatesta, V.; Ingold, K. U. *J. Am. Chem. Soc.* **1981**, *103,* 609.
36. Binkley, R. W. *Modern Carbohydrate Chemistry;* Marcel Dekker: New York, NY, 1988; p 106.
37. Hudlicky, T.; Entwistle, D. A.; Pitzer, K. K.; Thorpe, A. J. *Chem. Rev.* **1996**, *96,* 1195.
38. Moore, B.; Webster, T. A. *Proc. R. Soc. London, Ser. B.* **1918**, *90,* 168.
39. Baly, E. C. C.; Heilbron, I. M.; Barker, W. F. *J. Chem. Soc.* **1921**, 1025.
40. Baly, E. C. C. *Ind. Eng. Chem.* **1924**, *16,* 1016.
41. Seel, F.; Schaum, W.; Simon, G. *Z. Naturforsch.* **1981**, *36b,* 1451.
42. Shigemasa, Y.; Matsuda, Y.; Sakazawa, C.; Matsuura, T. *Bull. Chem. Soc. Jpn.* **1977**, *50,* 222.
43. Irie, S. *Chem. Lett.* **1984**, 2153.
44. Schwartz, A. W.; de Graaf, R. M. *Tetrahedron Lett.* **1993**, *34,* 2201.
45. Schwartz, A. W.; de Graaf, R. M. *J. Mol. Evol.* **1993**, *36,* 101.

6

Reactions of Carboxylic Acids and Esters

Introduction

One of the problems often faced by synthetic chemists working with carbohydrates is the need to select an appropriate method of replacing a hydroxyl group in a molecule with a hydrogen atom. The importance of this problem can be appreciated when one considers the variety of biologically active compounds that have deoxy sugars incorporated into their structures.[1] Many antibiotics and anticancer agents, for example, are glycosides of deoxy sugars. Several photochemically based reaction sequences for deoxygenation of alcohols have been applied to carbohydrates. These include synthesis and photochemical reaction of (1) 2-thiopyridones with hydrogen atom donors (discussed in Chapter 15), (2) halogenated compounds with alcohols (discussed in Chapter 19), (3) carboxylic acid esters with hexamethylphosphoramide (HMPA) (the primary topic of the current chapter), and (4) *m*-(trifluoromethyl)benzoates with *N*-methylcarbazole (an attractive alternative to irradiation in HMPA that also is discussed in this chapter).

Also considered in this chapter are several hydrogen-abstraction reactions. Hydrogen abstraction by excited esters is the beginning point for Norrish Type II and photodeconjugation reactions similar to those observed for aldehydes and ketones. (Norrish Type II and related reactions of aldehydes and ketones are discussed in Chapter 3). The final topic in this chapter is the hydrogen abstraction from uronic acids and esters that initiates a series of events leading to glycosidic cleavage.

Photolysis of Carboxylic Acid Esters in HMPA

Background

Beugelmanns et al. first raised the possibility that synthesis and photolysis of carboxylic acid esters in HMPA could be an effective procedure for replacement of hydroxyl groups with hydrogen atoms (eq 1).[2] It was Pète and coworkers, however, who transformed this reaction into a synthetically useful one when they discovered that the addition of water to the photolysis mixture dramatically increased product yields (eq 1).[3,4] Two years after their initial publications, Pète and Portella broadened the scope of the procedure by demonstrating that photolysis of acetylated carbohydrates in aqueous HMPA was a general method for producing deoxy sugars.[5] In the intervening years, other investigators have used this reaction to the extent that the number of compounds that have been investigated now is substantial (Table 6.1).

Reaction Mechanism

The discovery that deoxy sugars could be synthesized by photolysis of acetylated carbohydrates naturally raised questions about the scope and mechanism of this reaction. Why did water so dramatically affect product yields? What was the source of the hydrogen atoms that replaced the ester groups? What, if any, intermediates (exciplexes, radicals, and radical ions) were involved in the reaction? Which reactant was being excited, and what was the reactive excited state? Some of these questions were answered quickly, others waited many years for a satisfactory explanation, and for some the answers are still unknown.

In the first mechanistic study of photodeoxygenation in aqueous HMPA, Portella et al. established that for nonaromatic esters most of the incident light at 254 nm was absorbed by HMPA.[13] When these investigators combined this fact with the low ionization potential of HMPA, they proposed an electron-transfer mechanism for the reaction. This proposal, shown in Scheme 6.1, involved the formation of an exciplex prior to electron transfer.[13,14] It also

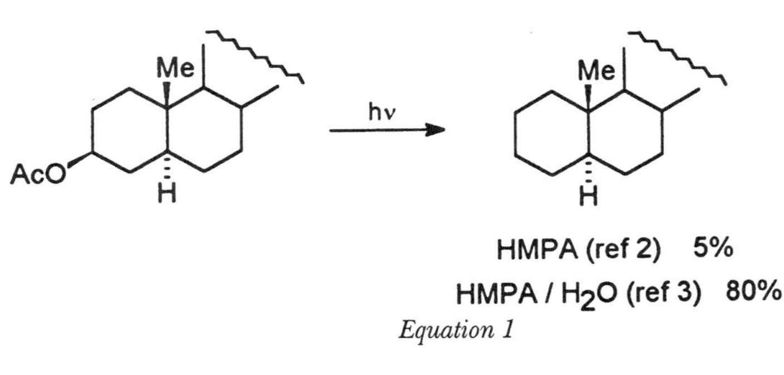

(1)

Equation 1

Table 6.1. Photochemical reactions of acetates in aqueous HMPA.

Reactant	*Product*	*Percent Yield*	*Ref.*
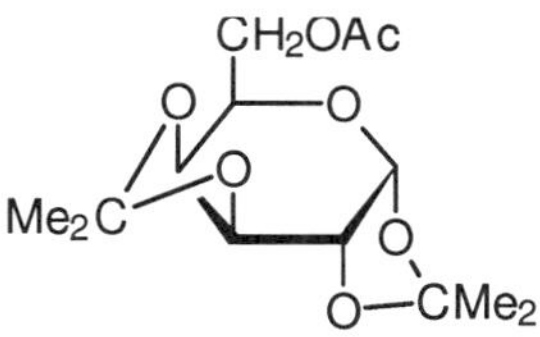	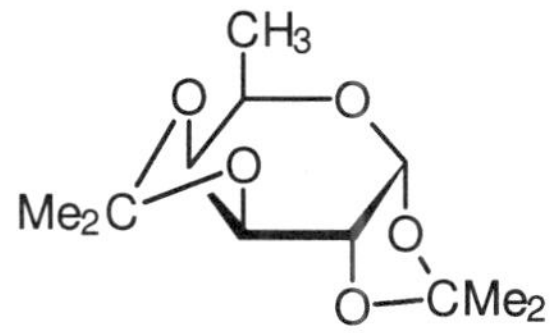	85 65	5, 7 6
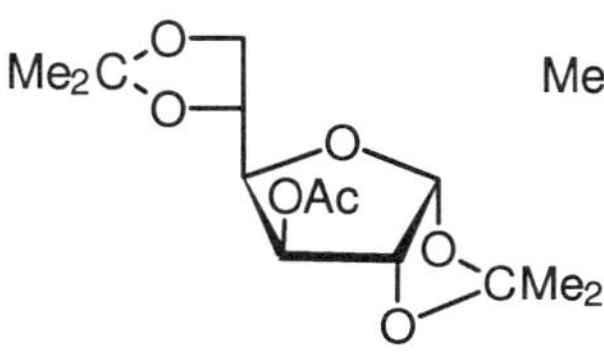	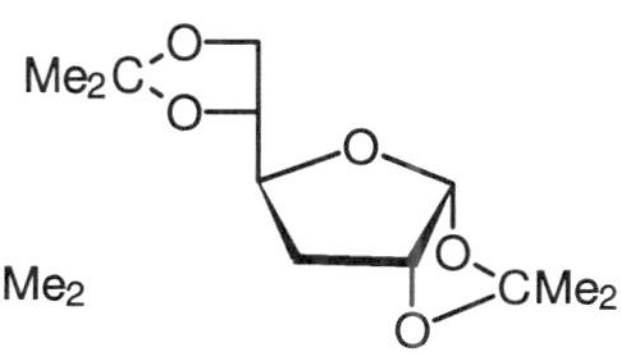 +	65	5, 6[a]
	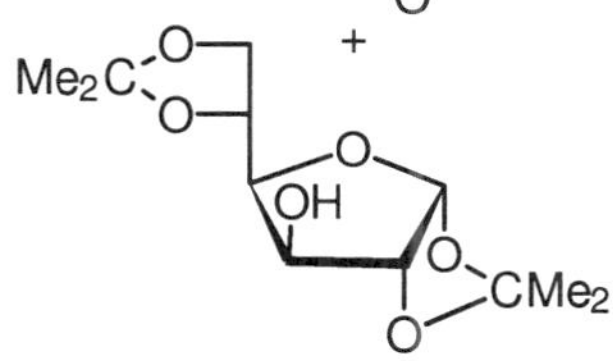	30	
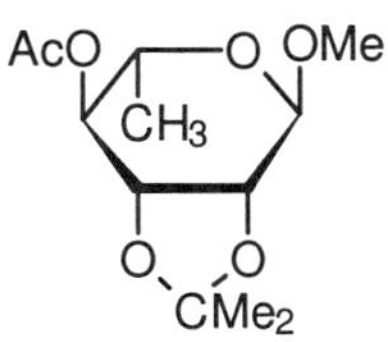	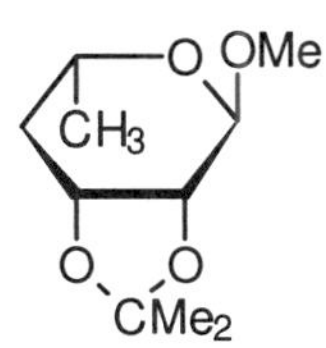	81 70	5 6
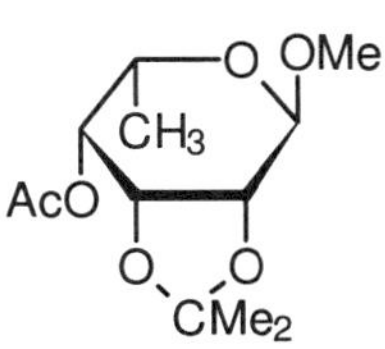	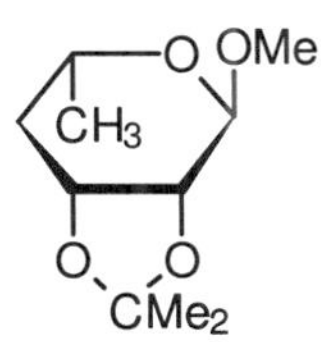	b	6
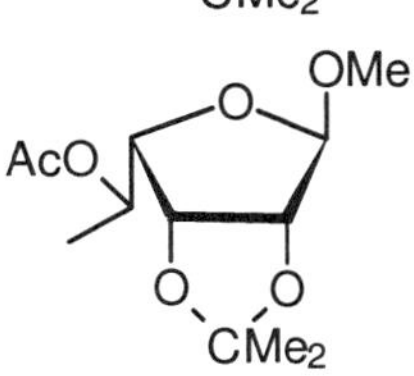	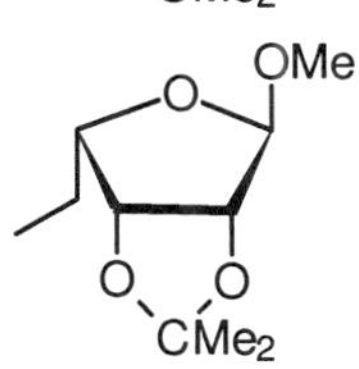	78	5
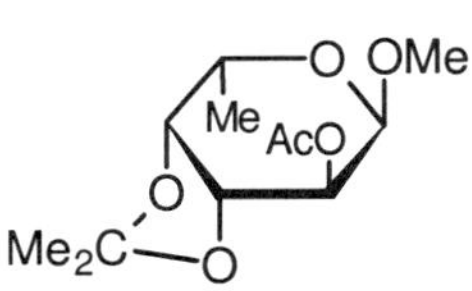	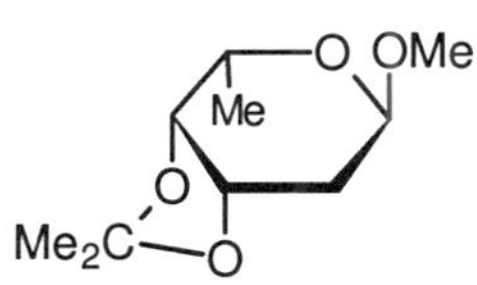	60	5

Continued on next page

Table 6.1. Photochemical reactions of acetates in aqueous HMPA—*Continued*

Reactant	*Product*	*Percent Yield*	*Ref.*
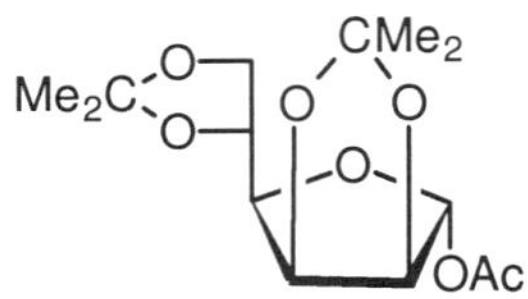	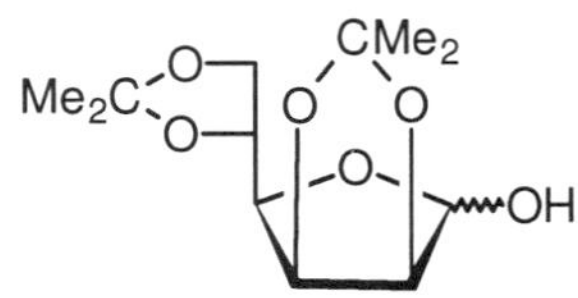	76	5
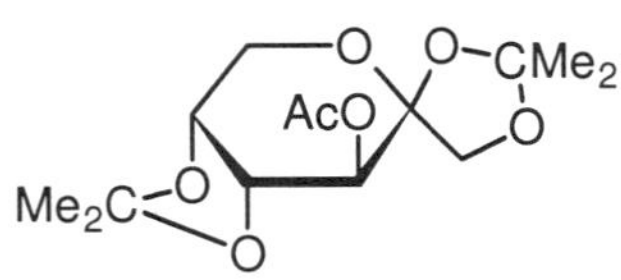	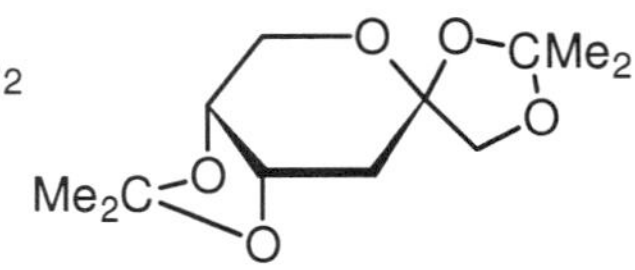	55	6
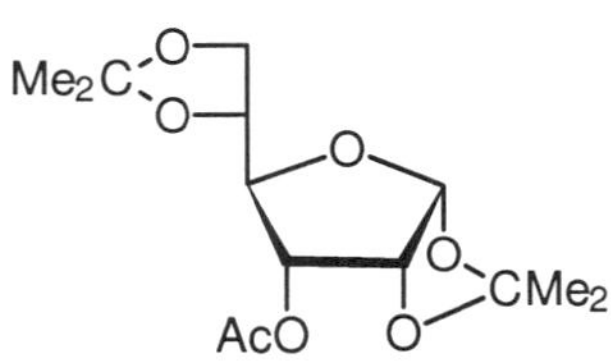	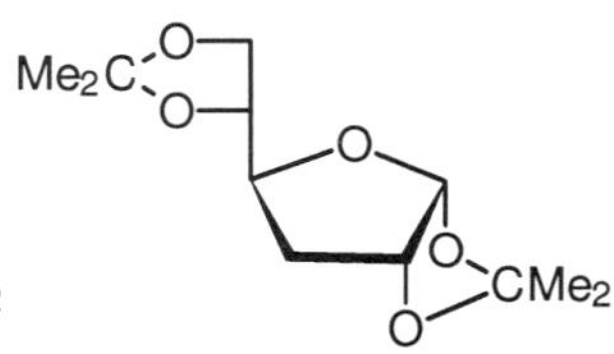	65	6
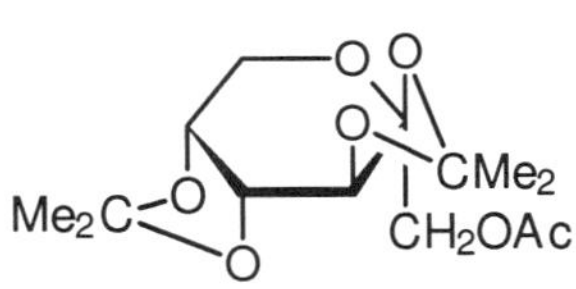	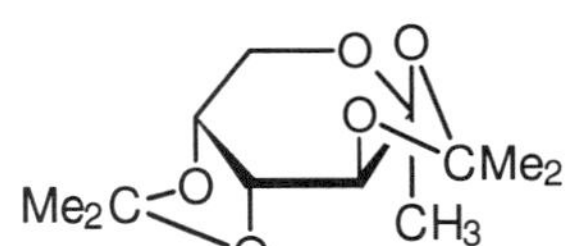	70	6
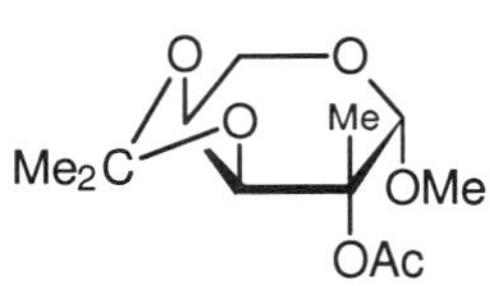	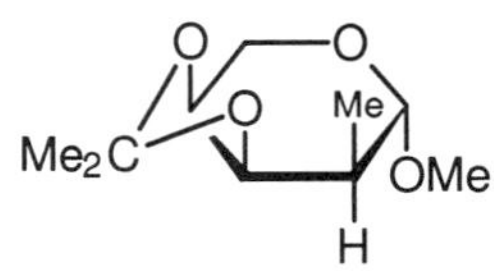 +	51	6
	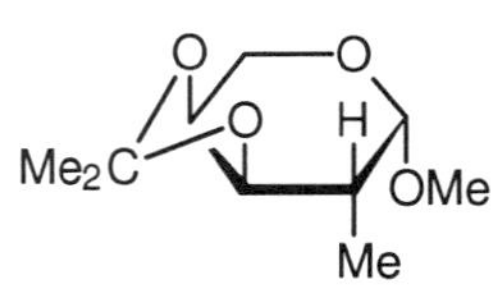	9	
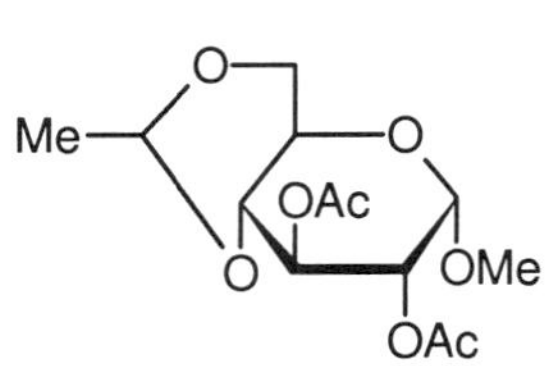	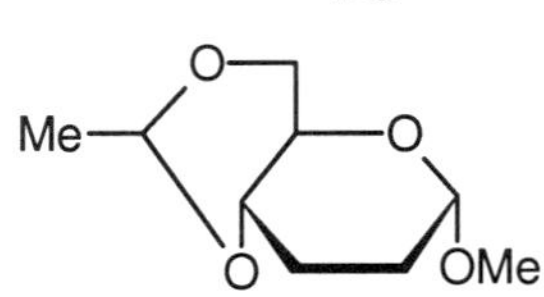	50	6

Continued on next page

Table 6.1. Photochemical reactions of acetates in aqueous HMPA—*Continued*

Reactant	*Product*	*Percent Yield*	*Ref.*
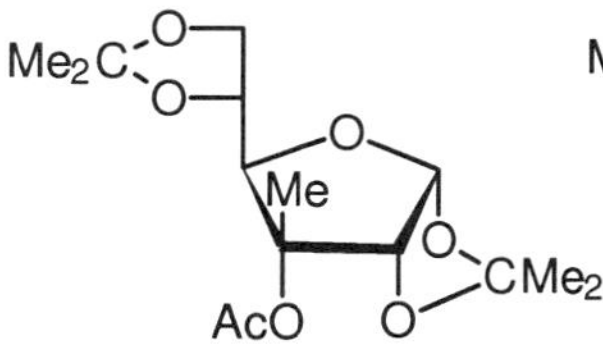	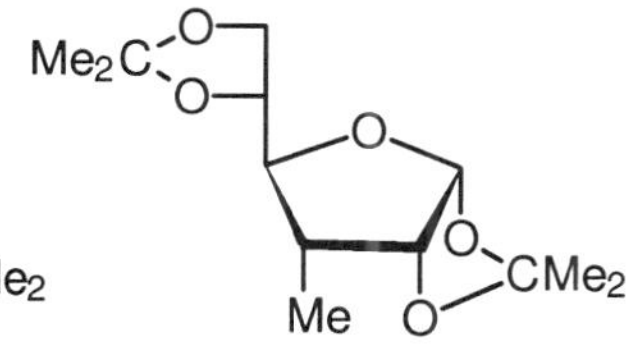	43	8
	+ 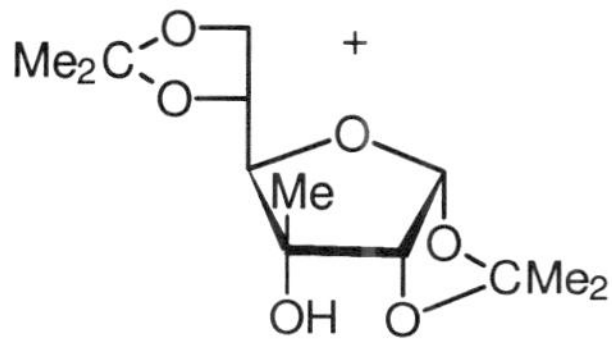	43	
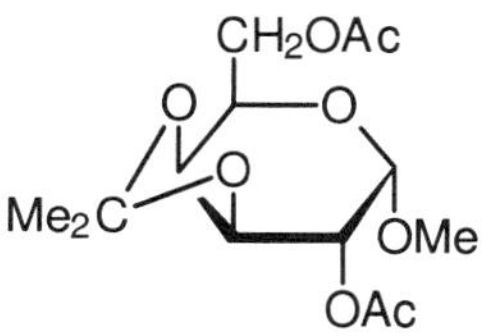	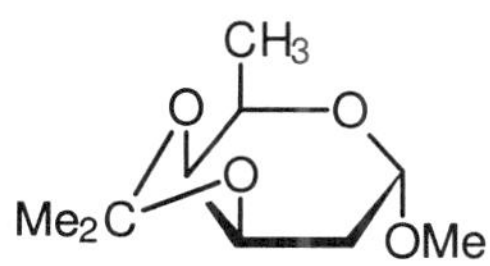	64	9
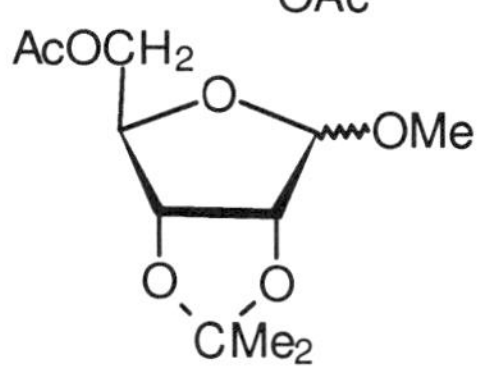	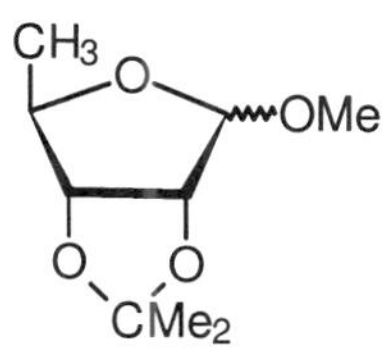	90	10
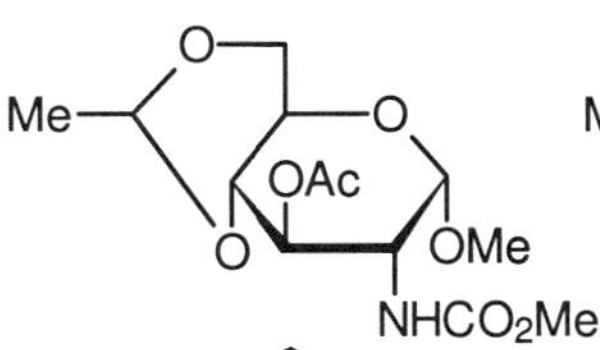	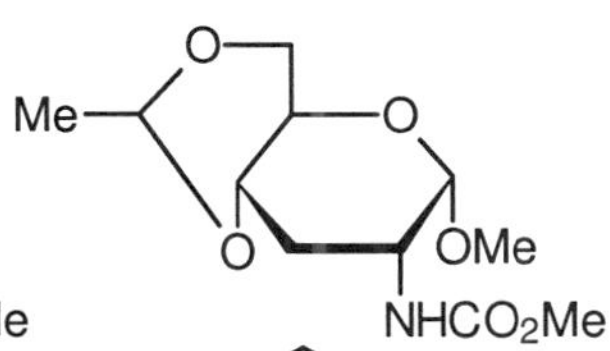	93	11
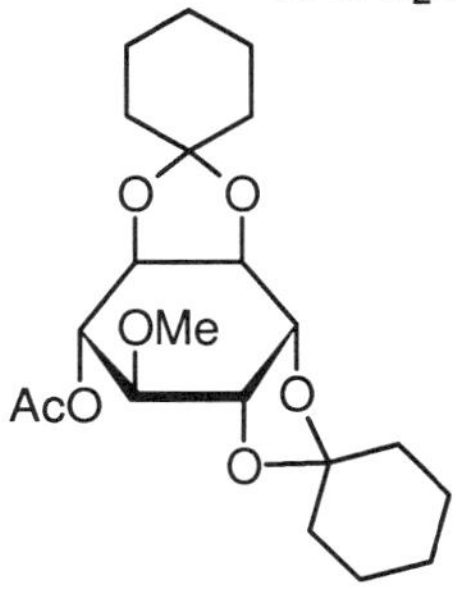	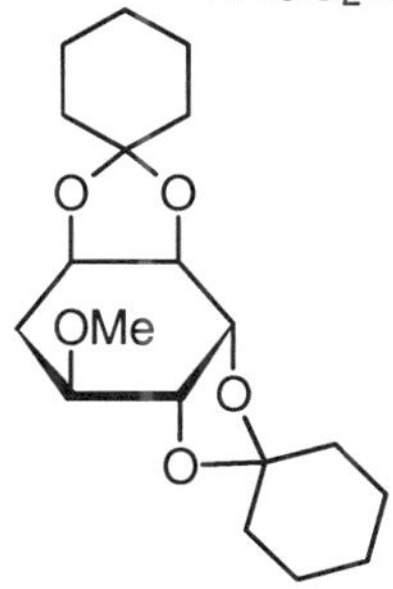	b	12

[a]Only the deoxy sugar was reported in reference 6.
[b]No yield was reported.

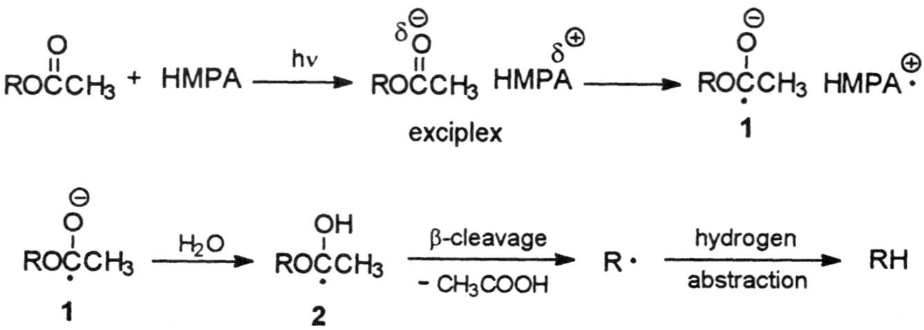

Scheme 6.1

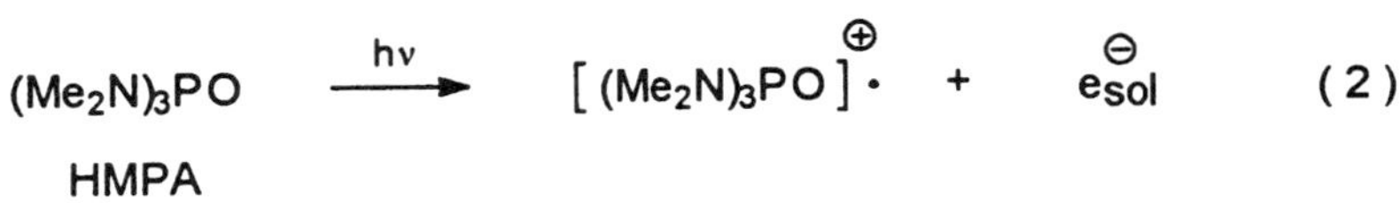

Equation 2

contained a suggestion for the role of water by indicating that water was the proton source for converting the radical anion **1** into the radical **2** immediately before β-cleavage (Scheme 6.1).

A particularly valuable investigation of ester photolysis in aqueous HMPA was reported in 1989 by Scharf and co-workers, who had been studying and using this photodeoxygenation process for several years.[15] When conducting flash-photolysis experiments on solutions of water in HMPA, these workers found that two transients were formed. They identified these two as the HMPA radical cation and the solvated electron and proposed that excited HMPA was ejecting an electron into solution (eq 2). Other experiments showed that adding water to HMPA increased the lifetime of the solvated electron. All of this investigation was done in the absence of any ester in the solution. When further flash-photolysis experiments were conducted with progressively greater amounts of an ester introduced into the aqueous HMPA, the solvated electron concentration decreased. (A plot of $1/e_{sol}$ vs. [ester] was found to be linear with a positive slope.) The conclusion reached from these experiments was that the solvated electron was reacting with the ester to give a radical anion (eq 3). Because water increased the lifetime of the solvated electron in the photolysis mixture, water also increased the probability that a solvated electron would exist long enough in solution to encounter an ester molecule.

One of the early observations about ester photolysis in aqueous HMPA was that photodeoxygenation products sometimes were accompanied by the alcohol from which the ester was synthesized (eq 4).[5] In their mechanistic study, Scharf and co-workers found that alcohol formation became more important as the concentration of the ester in the reaction mixture

Equation 3

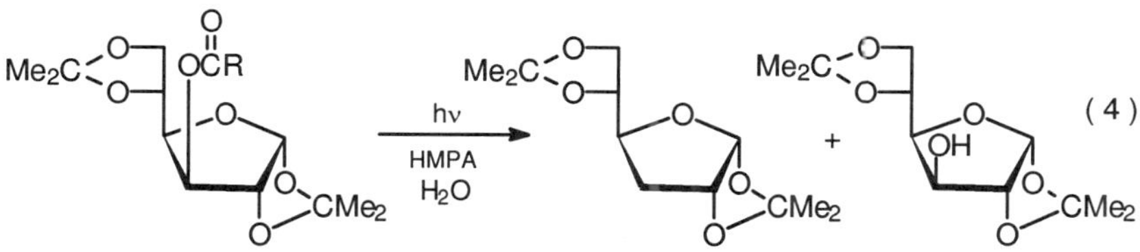

R = CH_3	65%	30%	(ref. 5)
R = $C(CH_3)_3$	75%	not reported	(ref. 6)

Equation 4

increased.[15] They explained this result by proposing that the radical anion generated by capture of the solvated electron had two pathways open to it. The first of these was β-cleavage to give a carboxylate anion and a sugar radical. This sugar radical became, after hydrogen abstraction, the deoxygenated photoproduct (Scheme 6.2). The second pathway involved a bimolecular reaction in which hydrogen transfer from another molecule of ester to either of the intermediates **3** or **4** produced the anion **5**. This anion (**5**) then protonated to continue along the pathway leading to the alcohol photoproduct (Scheme 6.2). Raising the concentration of the ester in the photolysis mixture increased the rate of the hydrogen transfer (bimolecular) reaction but not the β-cleavage process. This change in concentration, therefore, increased the amount of alcohol formed at the expense of the deoxygenation product.

One idea that follows from the mechanism shown in Scheme 6.2 is that the ease of β-cleavage of the radical anion (**3** or **4**) to give a carbon radical is likely to depend upon the stability of the radical being formed. β-Cleavage to generate a primary radical should be more difficult than fragmentation to give a secondary or tertiary radical. This idea is supported by photolysis of compounds that contain both primary and secondary acyloxy groups because the secondary groups appear to be replaced more rapidly, although the difference in reactivity is not large (Table 6.2).[6,18]

By conducting deuterium labeling experiments, Scharf and co-workers gathered information about the stereoselectivity of photodeoxygenation and also about the identity of the hydrogen-atom donor. They showed that hydrogen-atom transfer occurred to the less-hindered side of the intermediate radical (eq 5) and that HMPA was one, but not the exclusive, source of hydrogen atoms (eq 6).[15]

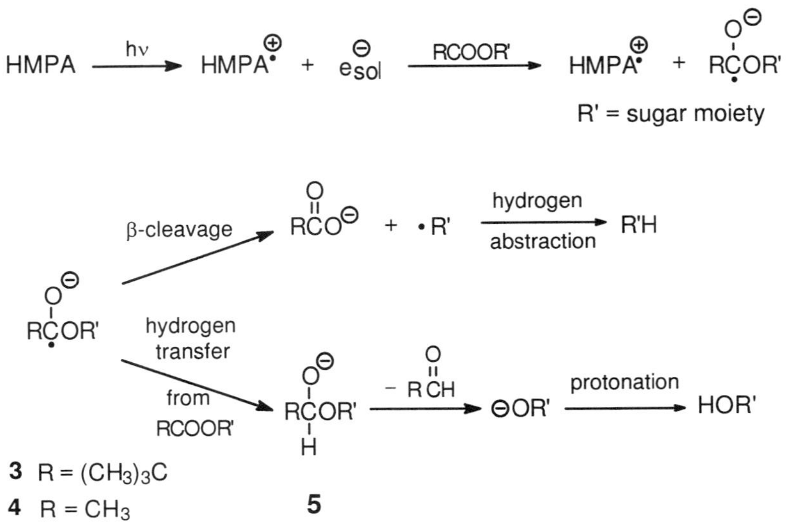

Scheme 6.2

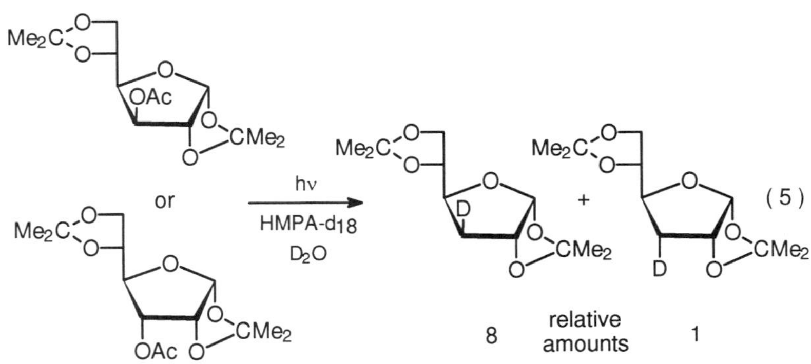

Equation 5

Reactivity of Different Esters

Although in most photodeoxygenation reactions acetylated compounds are the reactants (Table 6.1), a significant number of esters of pivalic acid also have been irradiated (Table 6.2). Photolysis of pivalates has been reported to give better yields of photodeoxygenation products than reaction of acetates (eq 4).[6,18,19] A generally better yield from pivalate irradiation is understandable if the reaction follows the mechanistic pathway shown in Scheme 6.2 because the transfer of a hydrogen atom to a radical anion should be more difficult to a carbon atom containing a *tert*-butyl group (**3**) than to one substituted by a methyl group (**4**). This difficulty in hydrogen-atom transfer should slow the

Table 6.2. Photochemical reactions of pivalates in aqueous HMPA.

Reactant	*Product and Percent Yield*	*Ref.*
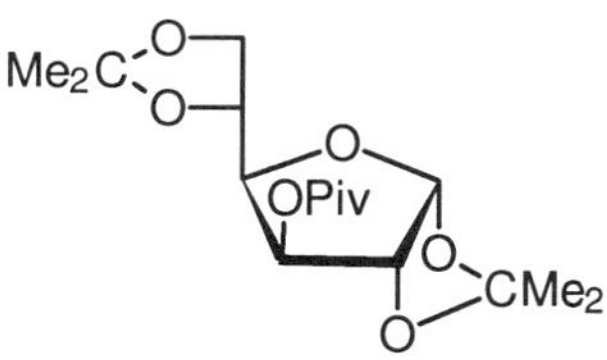 Piv = $(CH_3)_3CCO$	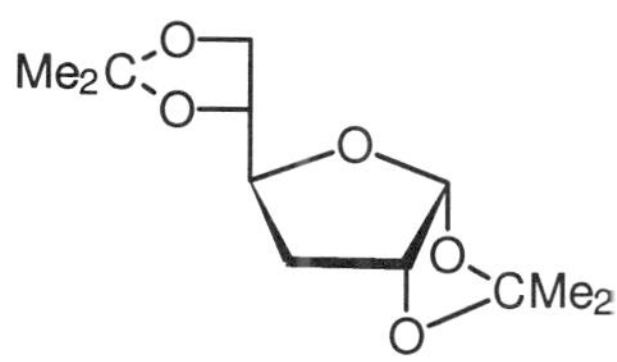 75%	6
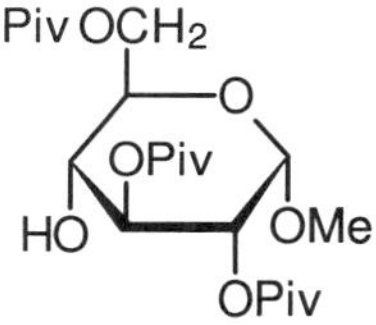	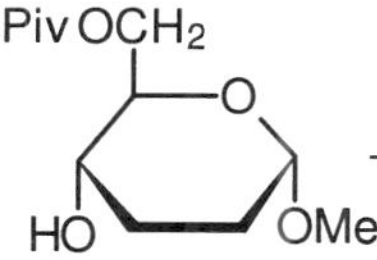 10% + 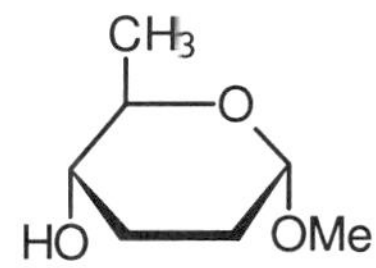 20%	6
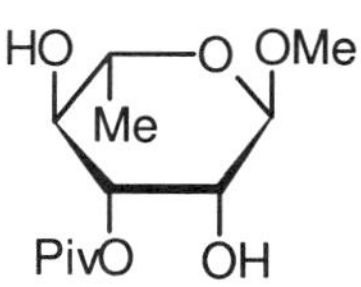	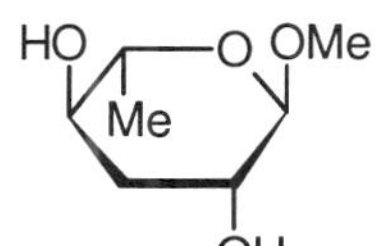 59%	16
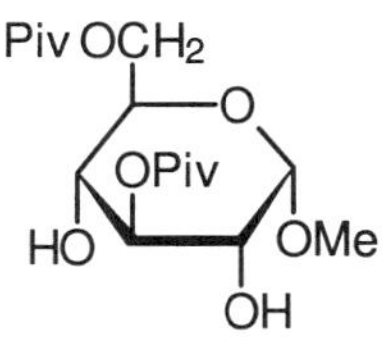	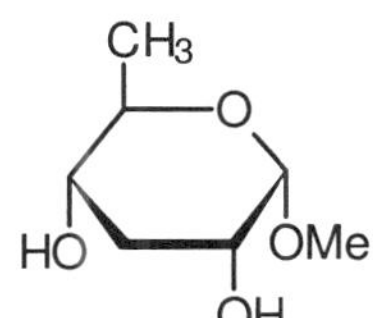 47%	17[a]
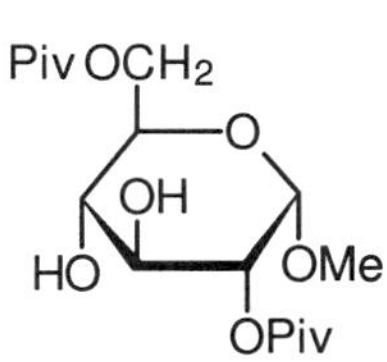	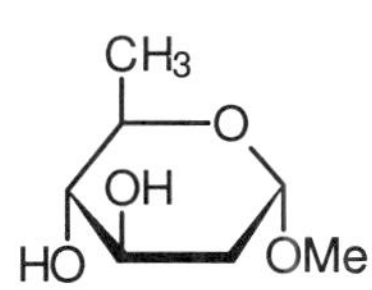 14% + 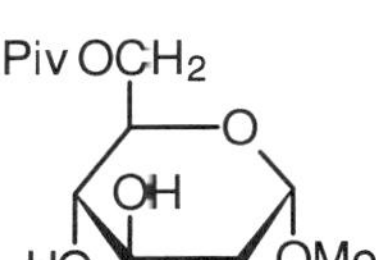 10%	18
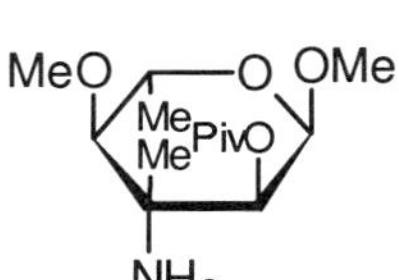	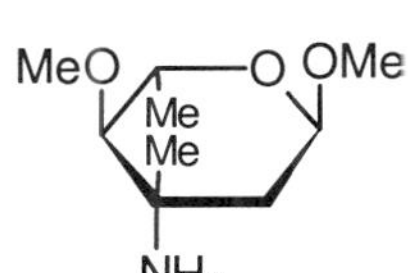 15%	19

Continued on next page

Table 6.2. Photochemical reactions of pivalates in aqueous HMPA—*Continued*

Reactant	*Product and Percent Yield*	*Ref.*
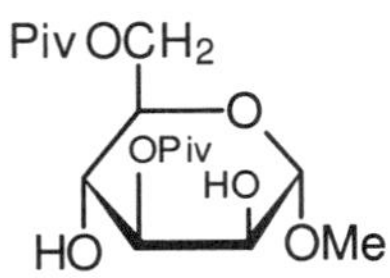	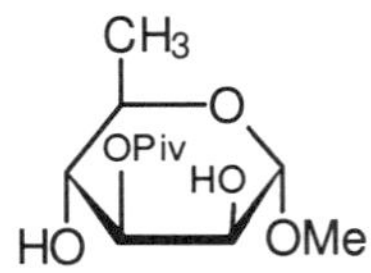 6% + 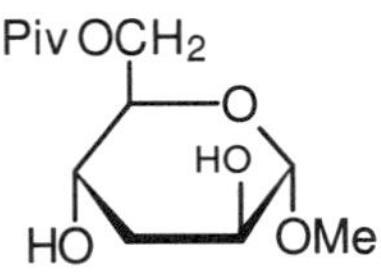 10% + 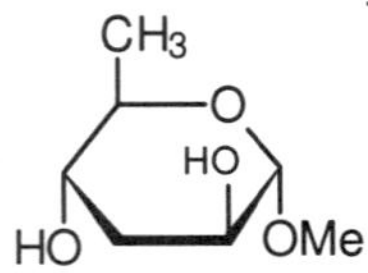 16%	18
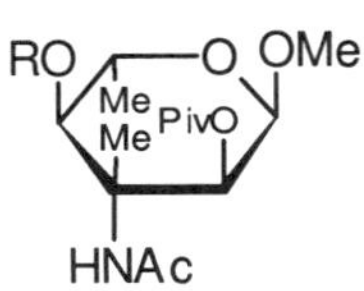 R = H R = CH_3	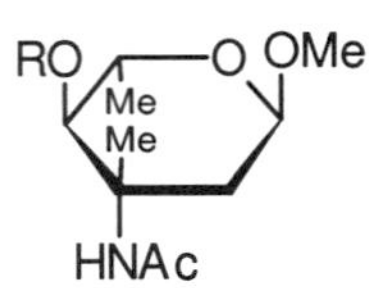 58% 79%	19
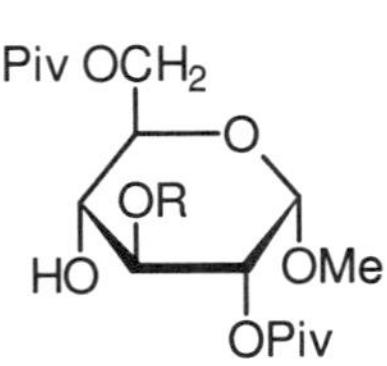 R = *t*-BuMe$_2$Si	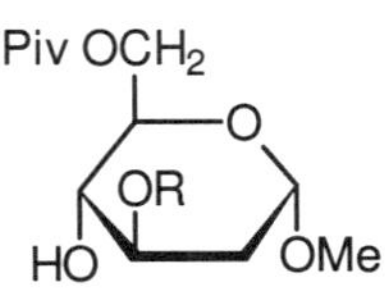 + 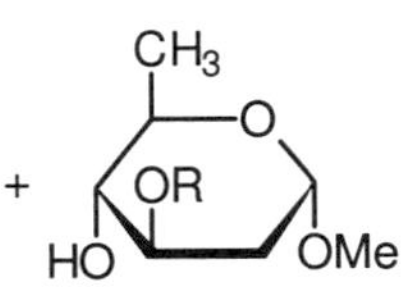 2.9% + 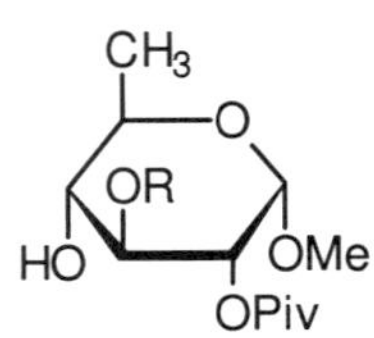 42%	20
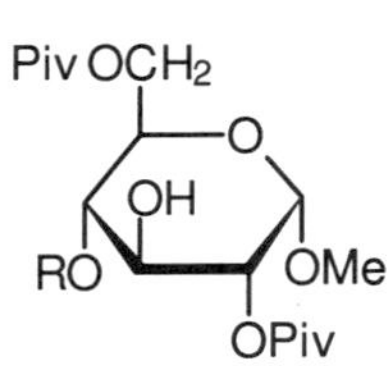 R = *t*-BuMe$_2$Si	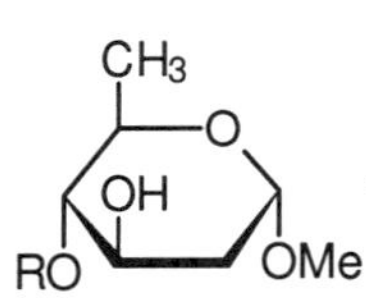 37% + 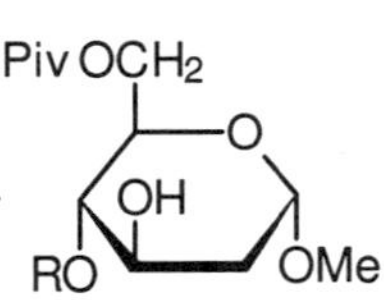 2%	20

Continued on next page

Table 6.2. Photochemical reactions of pivalates in aqueous HMPA—*Continued*

Reactant	*Product*	*Percent Yield*	*Ref.*
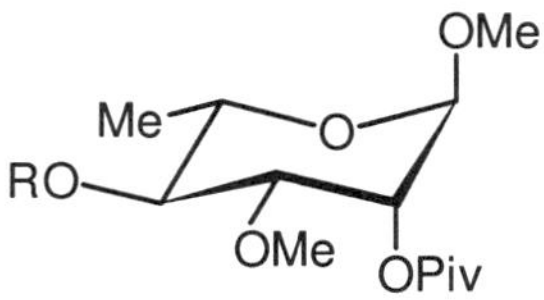 R = 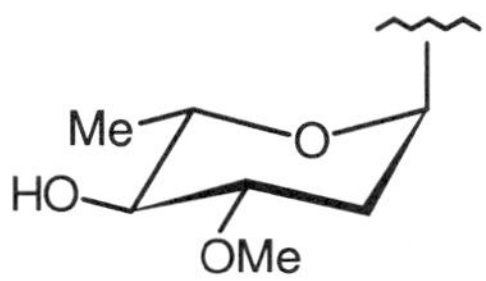	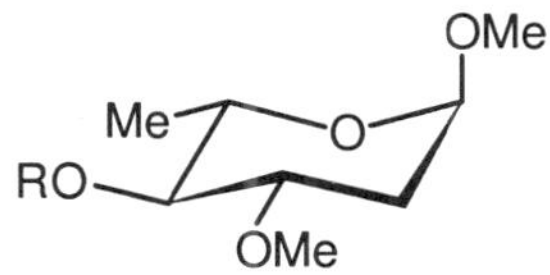	75	21
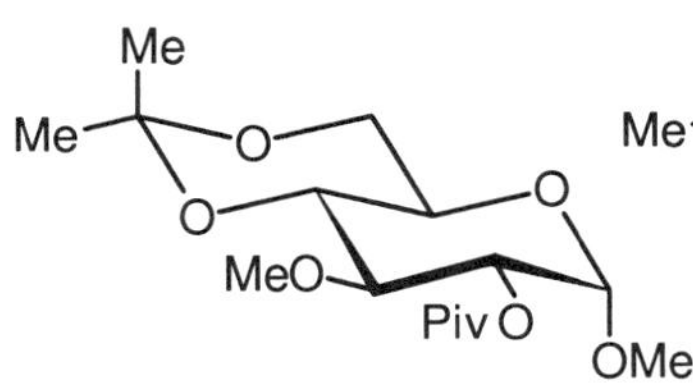	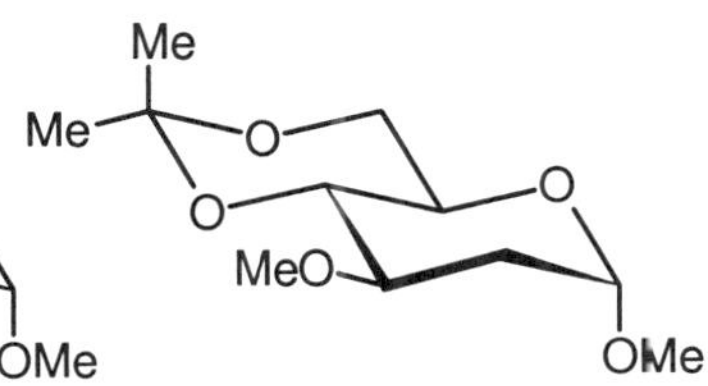	60	22
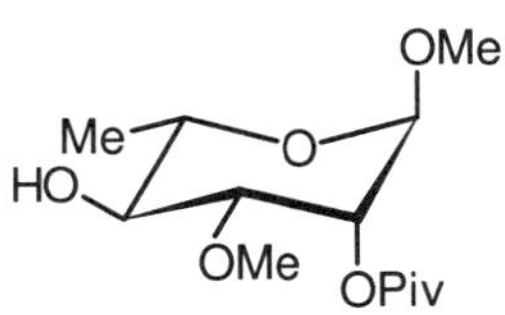	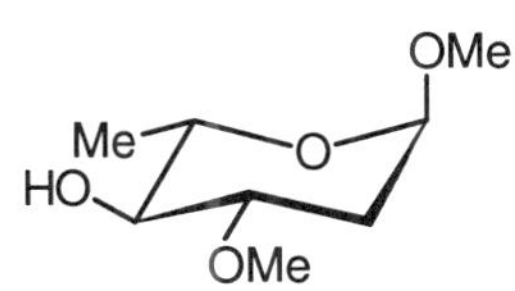	80	21

[a]The deuterio compound was investigated also.

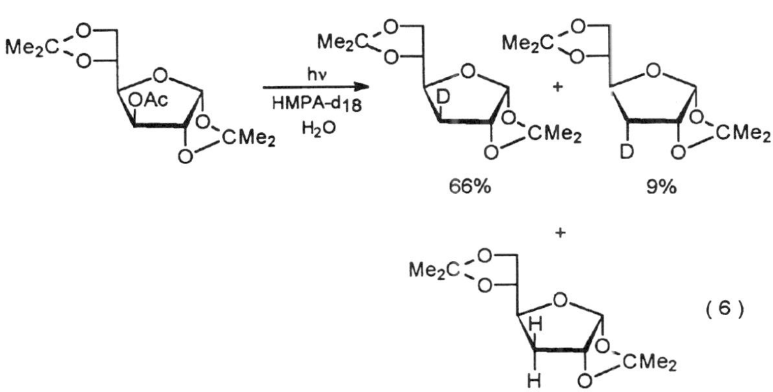

(6)

Equation 6

reaction leading to alcohol formation but should not greatly effect the β-cleavage reaction leading to deoxygenation.

Even though the limitations of photolysis in aqueous HMPA have not been thoroughly investigated, one disadvantage certainly is that any functional group that competes with the acyloxy group in electron capture can cause reactions other than photodeoxygenation to occur. Another drawback is that there are functional groups often present in carbohydrates (e.g., benzylidene and benzoyl groups) that are more effective at absorbing UV light than HMPA. Such groups are capable of stopping photodeoxygenation. One example of a potential photodeoxygenation reaction that appears to be thwarted by a light-absorbing substituent group is shown in eq 7, where loss of the benzylidene group occurs rather than the pivaloyloxy group.[22]

Photolysis of N-*Methylcarbazole and* m-*(Trifluoromethyl)benzoates*

In 1986 Saito et al. described a photodeoxygenation sequence in which an alcohol first was esterified to give the corresponding *m*-(trifluoromethyl)benzoate, an ester that then was converted into a deoxy compound by irradiation in the presence of *N*-methylcarbazole (Scheme 6.3).[23] Using this procedure, Saito et al. found that the 2′,3′-dideoxyadenosine derivative **7** could be synthesized from 2′,3′,5′-tri-*O*-[*m*-(trifluoromethyl)benzoyl]adenosine (**6**) (eq 8). This reaction sequence now has been used successfully by others for deoxygenation of several different types of compounds (Table 6.3). It is worth noting that reaction occurs at secondary carbon atoms but not primary ones. Presumably, the β-cleavage reaction shown in Scheme 6.3 does not occur when the radical to be formed is primary. Study of additional compounds would be valuable in determining whether the lack of reactivity of primary esters is a general phenomenon.

Photolysis of *m*-(trifluoromethyl)benzoates and *N*-methylcarbazole has some distinct advantages over irradiation of acetates or pivalates in aqueous

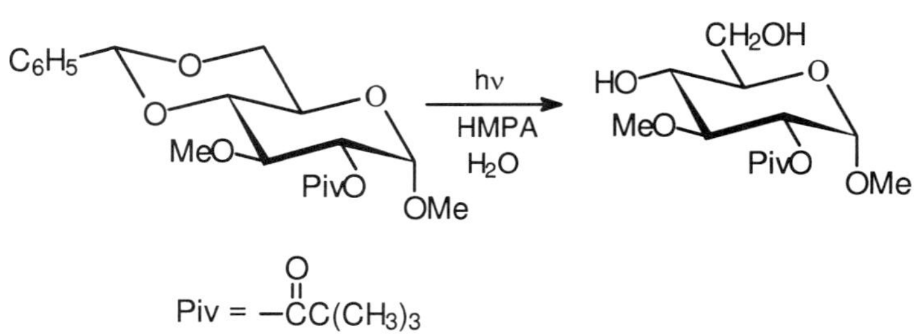

(7)

Equation 7

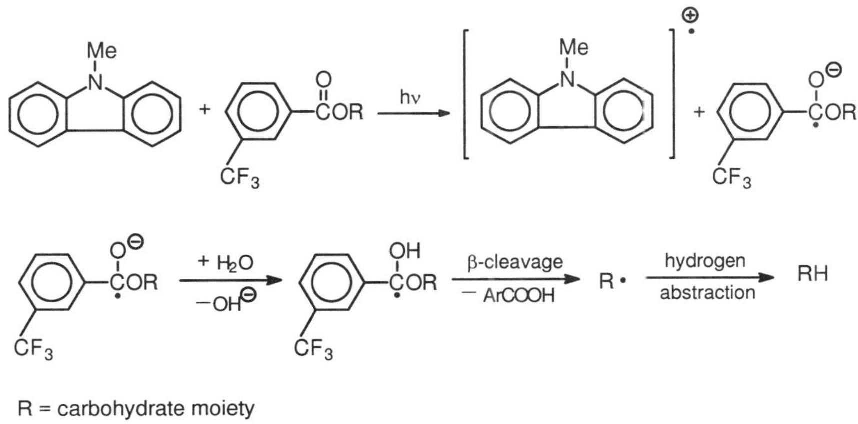

Scheme 6.3

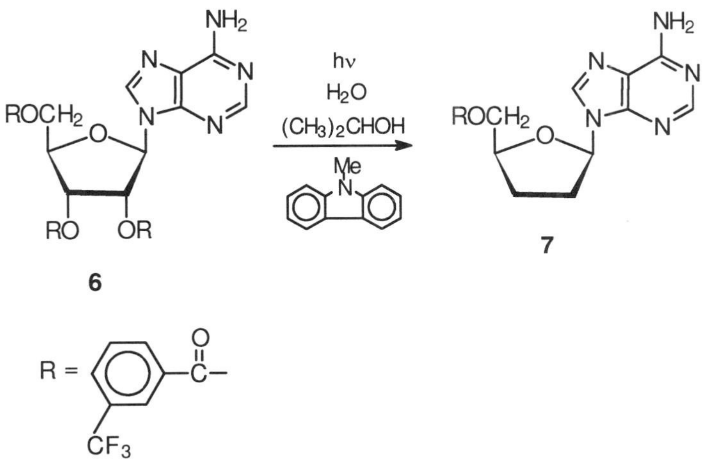

6 (8) 7

Equation 8

HMPA. From a safety point of view, circumventing HMPA avoids a highly toxic, cancer-suspect agent. From a synthetic point of view, the *m*-(trifluoromethyl)benzoyl group is a better electron acceptor than an acetyl or pivaloyl group; consequently, when the *m*-(trifluoromethyl)benzoyl group is used rather than the acetyl or pivaloyl groups, a greater variety of protecting groups can be present in the carbohydrate reactant without concern that these groups will compete successfully with the acyl group for the electron donated by an excited *N*-methylcarbazole (Table 6.3). Also, because *N*-methylcarbazole has longer wavelength absorption and greater molar absorptivity than HMPA, photodeoxygenation reactions that cannot take place with HMPA because the

Table 6.3. Photochemical reactions of *m*-(trifluoromethyl)benzoates with *N*-methylcarbazole.

Reactant	*Product*	*Percent Yield*	*Ref.*
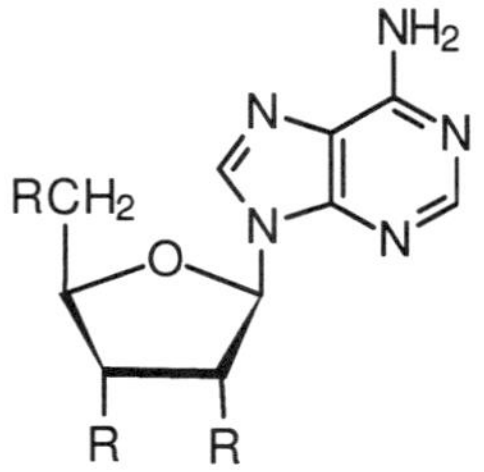 **6**	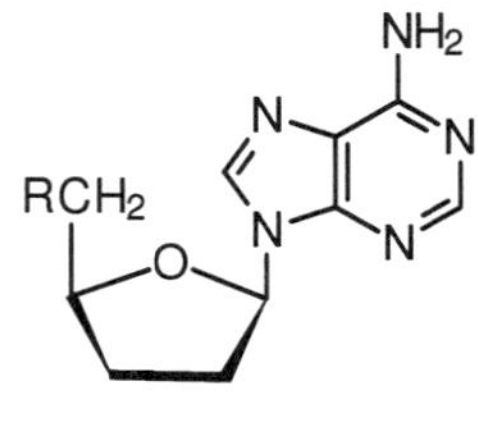 **7**	73	23
R = 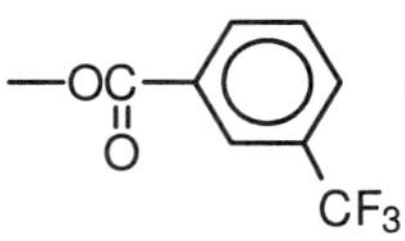			
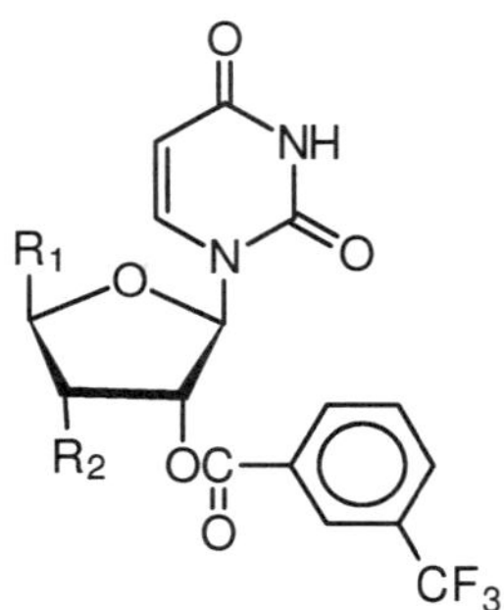	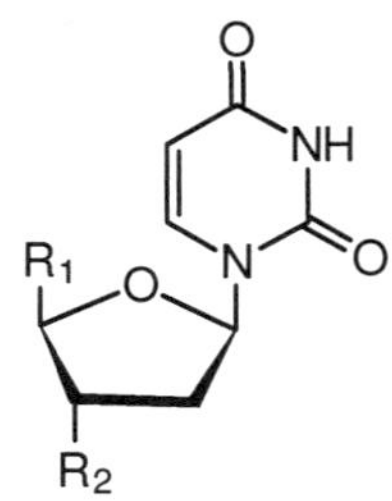	85	23
	R_1, R_2 = 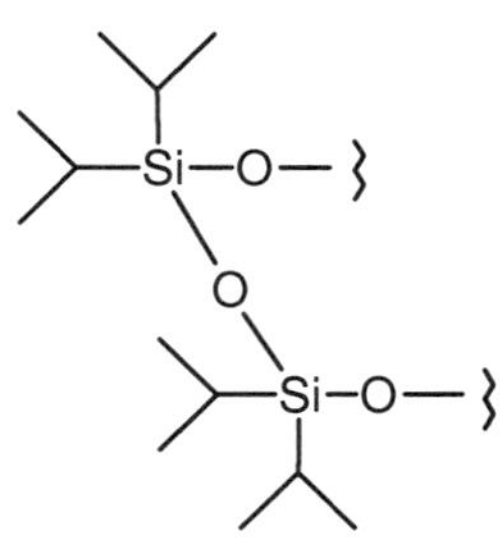		
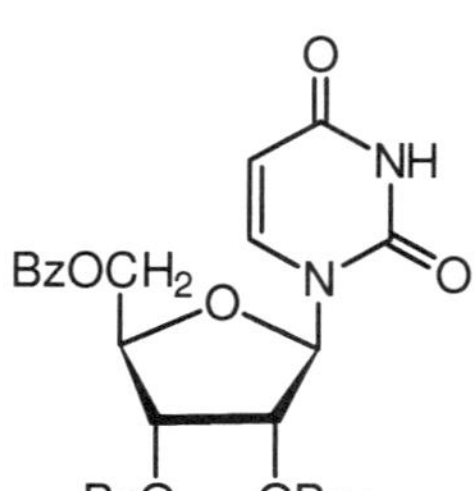	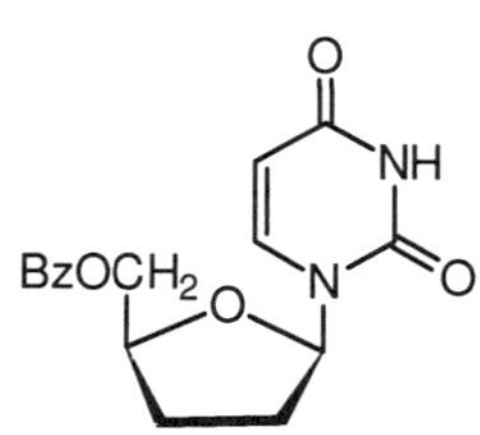	85[a]	23

Continued on next page

Table 6.3. Photochemical reactions of *m*-(trifluoromethyl)benzoates with *N*-methylcarbazole—*Continued*

Reactant	*Product*	*Percent Yield*	*Ref.*
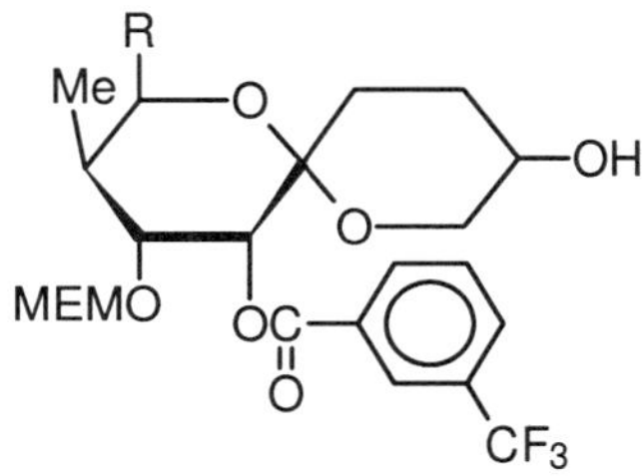	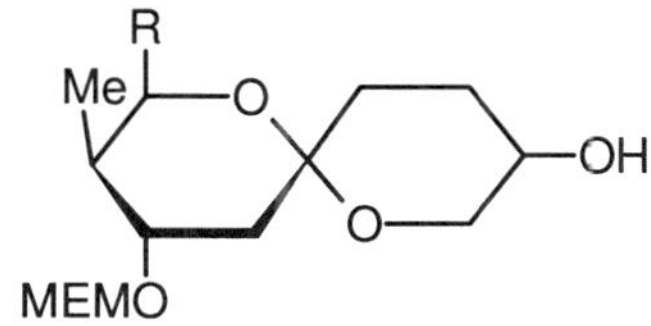	56	24
R = 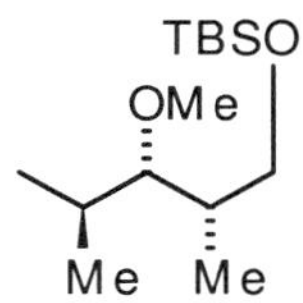 TBS = *t*-BuMe$_2$Si; MEM = $-CH_2OCH_2CH_2OCH_3$			
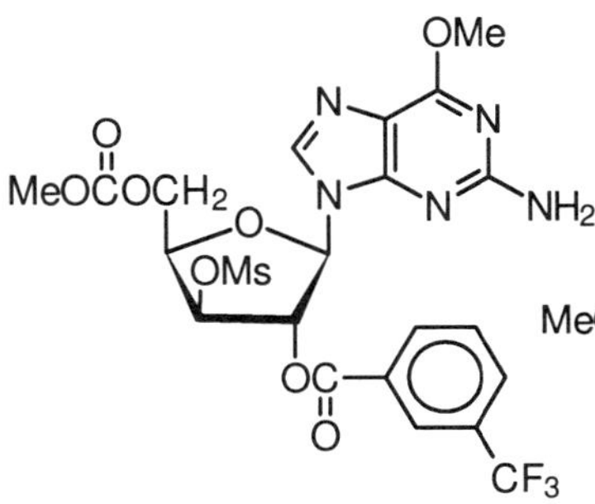	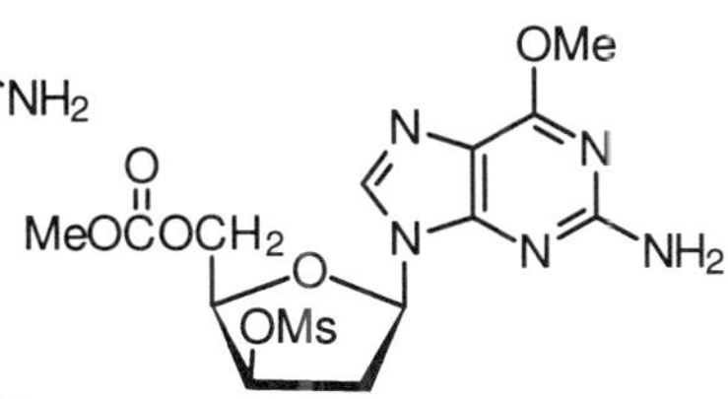	34	25
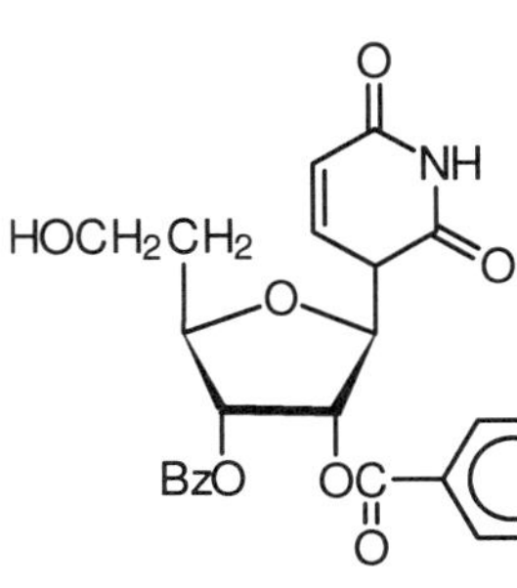	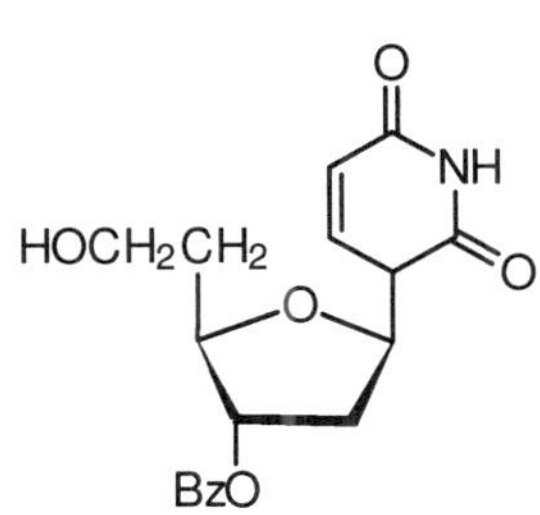	85[b]	26

Continued on next page

Table 6.3. Photochemical reactions of *m*-(trifluoromethyl)benzoates with *N*-methylcarbazole—*Continued*

Reactant	*Product*	*Percent Yield*	*Ref.*
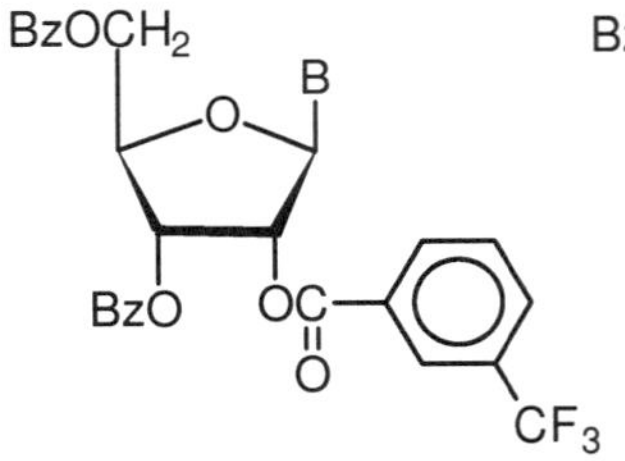	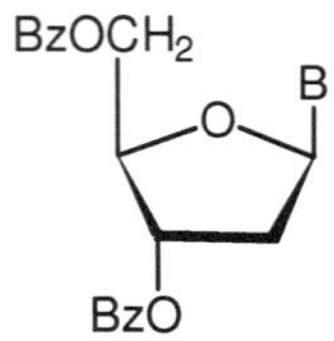		27
B = 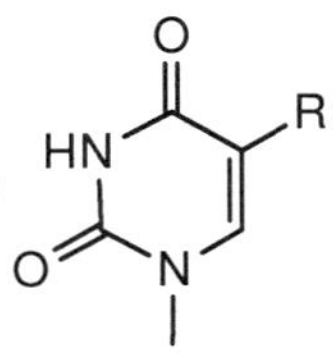	R = Me R = F R = CF_3	73 53 44	
B = 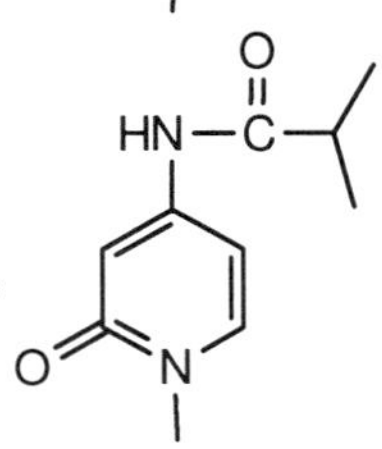		62	

[a]Magnesium perchlorate accelerated the reaction.

[b]A cytidine derivative was mentioned as undergoing photochemical deoxygenation in lower yield, but no further information was given.

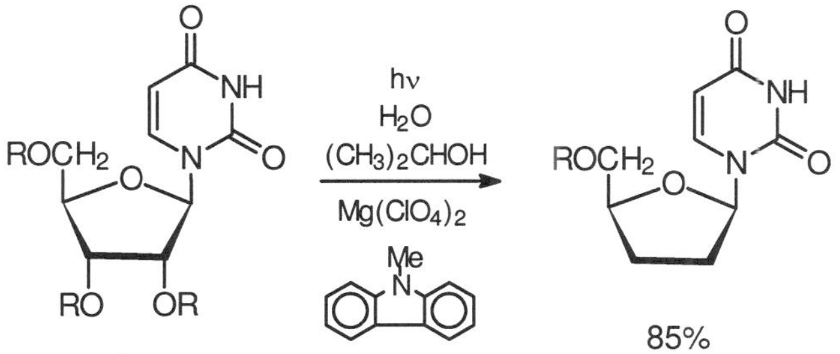

8

R = C_6H_5CO

(9)

Equation 9

reactant carbohydrate preferentially absorbs the incident light (e.g., eq 7) are possible when *N*-methylcarbazole is the light-absorbing reactant.

In a final comment about photochemical reactions of benzoates, Saito et al. mentioned the intriguing observation that the unsubstituted benzoic acid ester **8** underwent photodeoxygenation in high yield when irradiated in the presence of magnesium perchlorate and *N*-methylcarbazole (eq 9).[23] It would be interesting and potentially valuable to know whether this reaction of unsubstituted benzoates is a general one. This seems possible if the proposal of Abe and Oku is correct, that is, if magnesium perchlorate enhances the efficiency of electron transfer by stabilizing the ketyl radical formed and, in so doing, suppressing back-electron transfer.[28] This proposal is supported by the work of Mizuno et al., who found that if any one of a number of perchlorate salts or lithium tetrafluoroborate were present in the photolysis mixture, the quantum yield for photoisomerization of substituted cyclopropanes increased.[29]

Hydrogen-Abstraction Reactions

Three types of photochemical reaction of esters, all of which involve hydrogen abstraction, are discussed in this section. The first two are intramolecular in nature and, thus, could have been included in Chapter 3 along with the Norrish Type II and related reactions. The third is intermolecular and, consequently, might have been added to Chapter 5. However, because they all involve reactions of carboxylic acid esters, we decided to include them in the current chapter.

Norrish Type II Reaction of Esters

Although the Norrish Type II reaction of aldehydes and ketones is a frequently observed and extensively studied photochemical process, Type II reaction of esters and other carbonyl compounds is much less common. One reason for this reduced frequency of occurrence is related to the possible reactive excited states. For aldehydes and ketones, Type II reaction can occur from either the lowest excited singlet or triplet state because both of these are (n,π*) and, therefore, both have the partially filled, nonbonding orbital on oxygen that is associated with hydrogen abstraction (see Chapter 3).[30] Type II reaction of esters occurs only through the lowest excited singlet state because this state is (n,π*), while the lowest triplet state is (π,π*).[31,32] Hydrogen abstraction is not a characteristic process for a compound in a (π,π*) excited state, because of the absence of a nonbonding orbital on oxygen with an unpaired electron. (Wagner and Park have pointed out that even though (π,π*) triplet states are not totally unreactive toward hydrogen abstraction, their reactivity is much lower than that of (n,π*) triplet states.[33]) Effectively eliminating triplet states as participants in Type II reactions of esters naturally decreases the probability of such a reaction occurring.

One indication of the reduced ability of esters to participate in Type II reactions is that there are no unequivocal examples of this type of reaction reported for esterified carbohydrates. (Esters of pyruvic acid do undergo Type II reaction (Chapter 3); however, the carbonyl group participating in hydrogen abstraction is not the ester carbonyl.) Binkley and Meinzer reported that irradiation of the galactose derivatives **9** yielded the elimination product **10**, a compound that could have been formed by a Type II process (eq 10).[34]

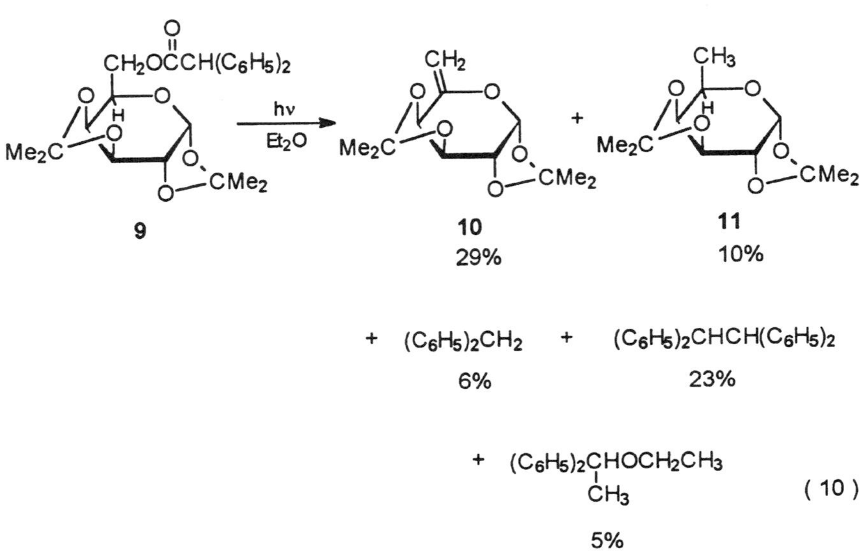

Equation 10

Because other products from this reaction [e.g., $(C_6H_5)_2CHCH(C_6H_5)_2$] indicated that free radicals were being generated, formation of **10** could have occurred by a reaction that was not a Type II process (Scheme 6.4).

Photodeconjugation of α,β-Unsaturated Esters

Internal hydrogen abstraction in an α,β-unsaturated ester can lead to photodeconjugation, that is, isomerization of the double bond to the β,γ position (Scheme 6.5).[35] An analogous reaction takes place for α,β-unsaturated ketones (Chapter 3). Photodeconjugation of esters occurs from the lowest (n,π*) singlet state.[37,38] (As mentioned in the previous section, the lowest triplet state of esters is much less reactive in hydrogen-abstraction reactions.) In one of the few examples of this reaction involving photolysis of a carbohydrate, Piva and Pète induced chirality into the carboxylic acid portion of an ester by using the 1,2:5,6-di-*O*-isopropylidene-α-D-glucofuranosyl group as a chiral auxiliary (Scheme 6.5 and Table 6.4).

Glycosidic Cleavage in Uronic Acids and Their Esters

Kitagawa and co-workers found that photochemical reaction can, in some situations, provide a useful alternative to acid-catalyzed hydrolysis of glycosides.[39–53] Situations in which this reaction is of importance are those where a uronic acid or its ester is attached by a glycosidic linkage to an acid-labile aglycon (equation 11).[39] Uronic acids and their esters are more resistant to acid-catalyzed hydrolysis than most other glycosides; consequently, during hydrolysis the aglycon moiety is at risk because of the vigorous reaction conditions typically required (in nonphotochemical hydrolysis). Photolysis provides a mild alternative to cleavage of the glycosidic linkage; consequently, the possibility for recovery of an "intact" aglycon is improved. Unfortunately, the sugars formed by photochemical reaction often have not been identified. In those cases where sugar identification was attempted, fragmentation of the sugar moiety occurred during photolysis. Bradbury and von Sonntag have conducted a careful mechanistic investigation of uronate photolyses and concluded that the major reaction pathway leads to ring opening and loss of the C2 substituent (Scheme 6.6).[54] This work explains the isolation of only carbohydrate fragments from photolysis of uronic acids and esters.

Carbohydrates as Chiral Auxiliaries

Carbohydrates have been used as chiral auxiliaries in several types of photochemical reactions. One example is the isomerization of the double bond shown in Scheme 6.5.[35] The typical situation is that the carbohydrate portion

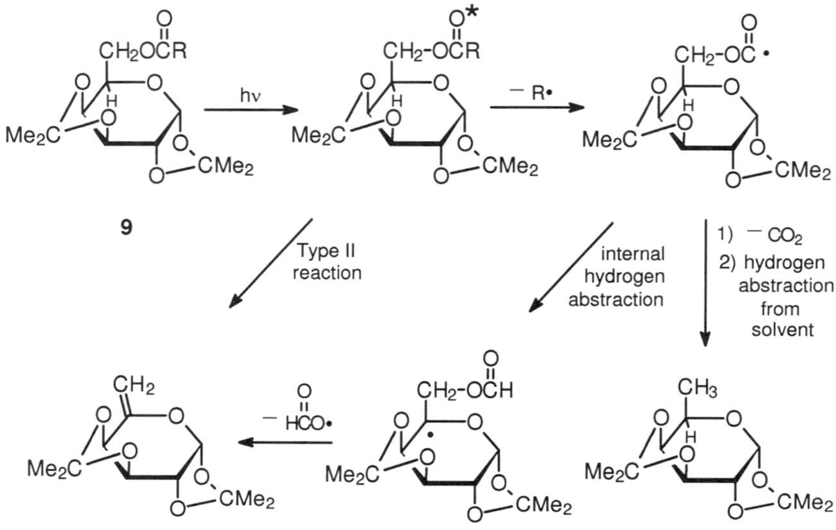

9 10 11

Scheme 6.4

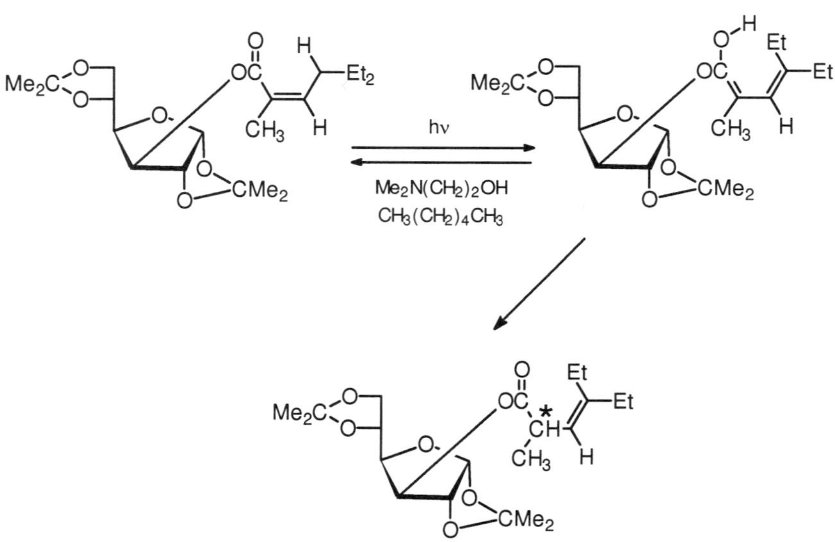

de = 88%

% conversion = >98%

Scheme 6.5

Table 6.4. Hydrogen-abstraction reactions.

Reactant	*Product and Percent Yield*		*Ref.*
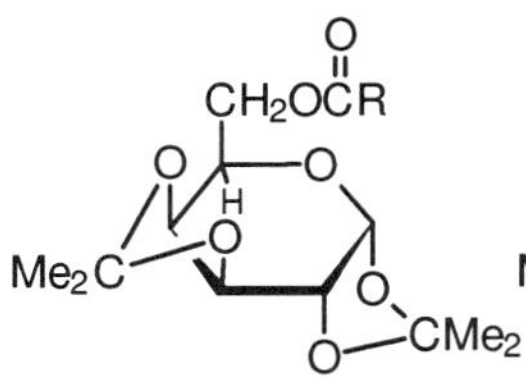	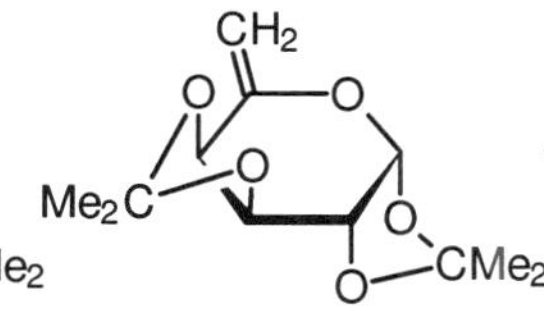 **10**	+ 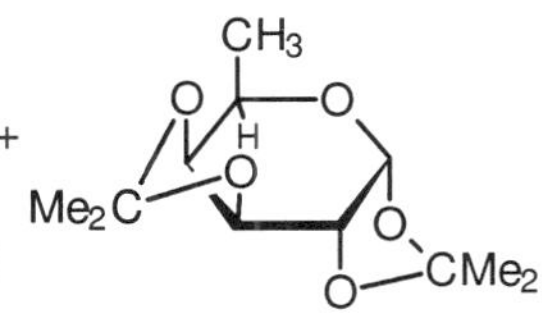 **11**	34
9a R = $-CH_2C_6H_5$	50%	26%	
9b R = $-CH_2CH_2C_6H_5$	8%	3%	
9c R = $-CH(C_6H_5)_2$	29%	10%	
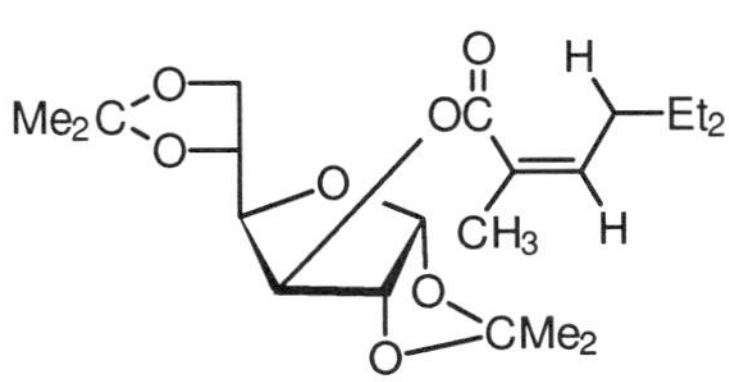	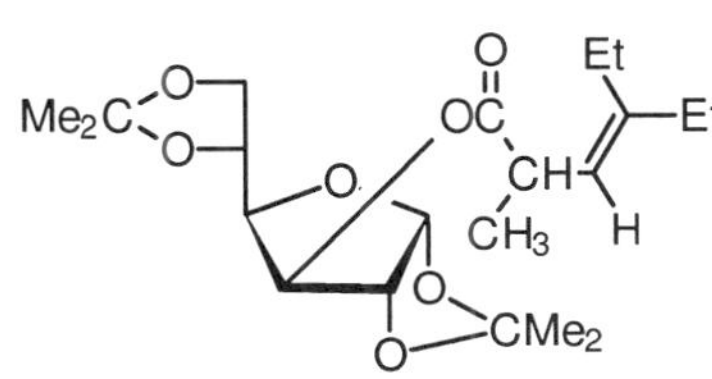		35[a]
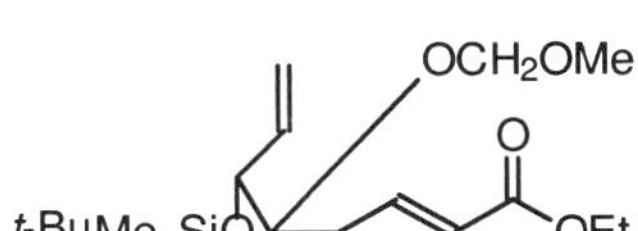	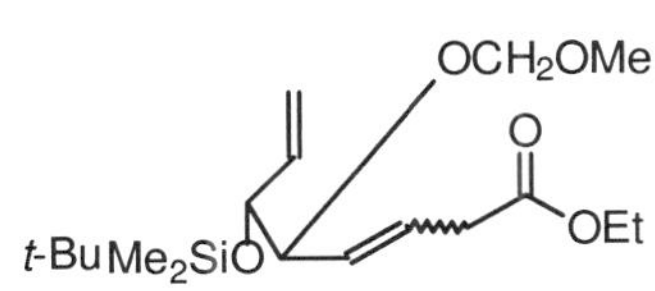 72%		36

[a]The de (diastereomeric excess) is 88% at a conversion greater than 98%. A large number of different solvents and different reaction conditions were used

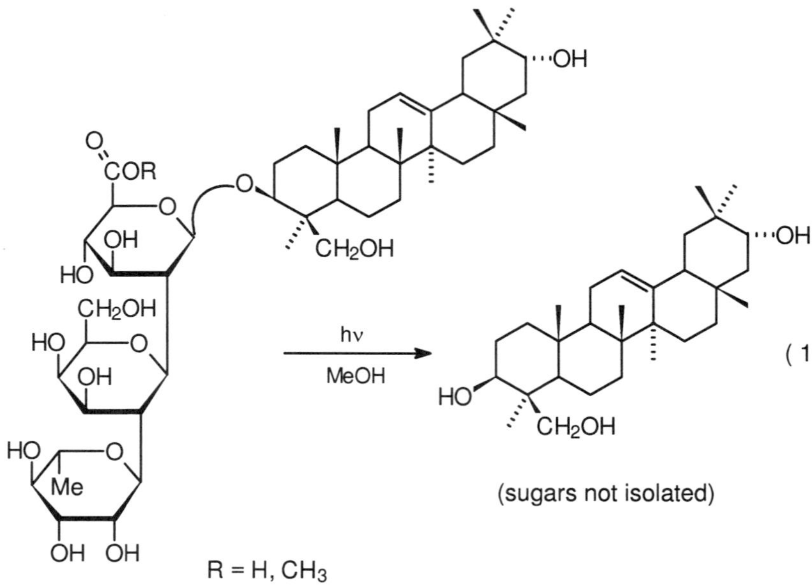

(11)

Reaction 11

of the molecule undergoes no change in bonding but influences structural changes occurring elsewhere in the molecule. Because in most of these reactions the carbohydrate moiety is connected by an ester linkage to the portion of the molecule undergoing change, these reactions have been collected together here in Table 6.5.

Summary

Photolysis of esterified carbohydrates under the proper conditions is a useful method for their deoxygenation. Two basic processes are effective. The more widely used of the two involves irradiating either carbohydrate acetates (Table 6.1) or pivalates (Table 6.2) in aqueous HMPA. The other calls for synthesis of *m*-(trifluoromethyl)benzoates and their subsequent photolysis in the presence of *N*-methylcarbazole (Table 6.3). Both of these are electron-transfer reactions that involve a β-cleavage step (Schemes 6.2 and 6.3).

Hydrogen-abstraction reactions occur from photolysis of esterified carbohydrates. If the abstraction is intramolecular, Type II and photodeconjugation reactions occur. Intermolecular hydrogen abstraction is the first step in cleavage of a glycosidic linkage in uronic acids and esters (Scheme 6.6). Photodeconjugation of α,β-unsaturated esters can be used to induce chirality into the β,γ-unsaturated esters formed (Scheme 6.5).

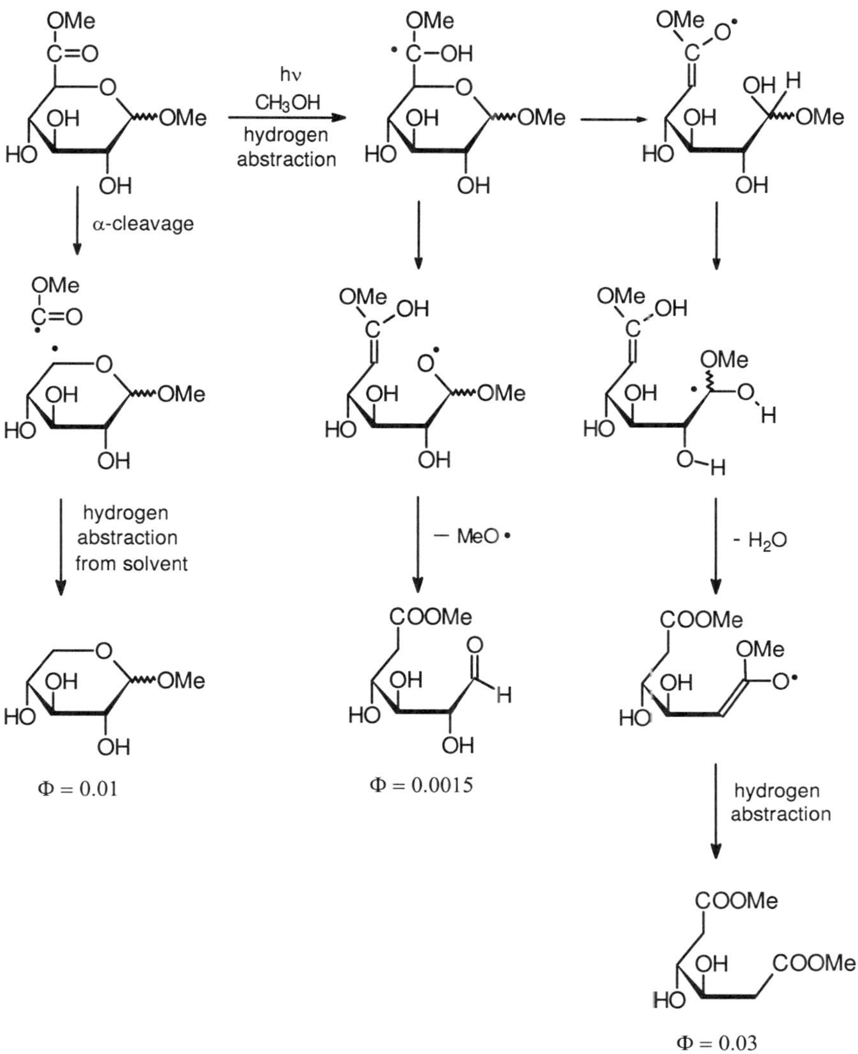

OMe
C=O
O
OH
OMe
HO
OH
hν
CH3OH
hydrogen
abstraction
C–OH
α-cleavage
hydrogen
abstraction
from solvent
− MeO•
- H2O
COOMe
H
Φ = 0.01
Φ = 0.0015
Φ = 0.03

Scheme 6.6

Table 6.5. Reactions with carbohydrates as chiral auxiliaries.

Reactant	*Product*	*Percent Yield*	*Ref.*
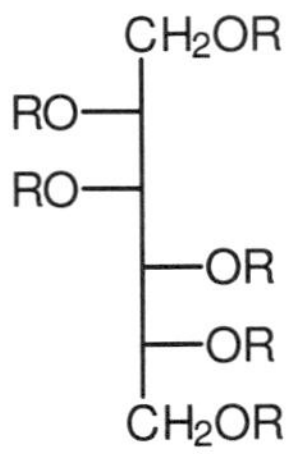	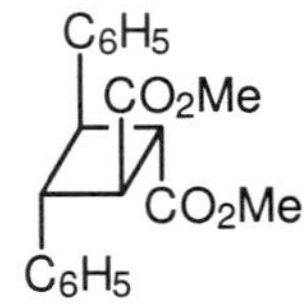	25-40 (ee = 38 - 46%)	55[a], 56[a,b] 57[a,c], 58[a]
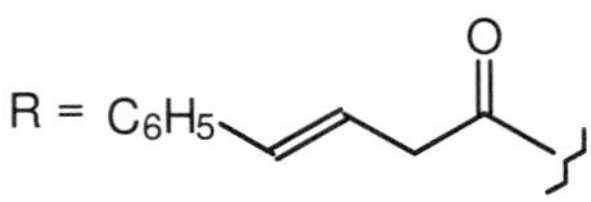	+ 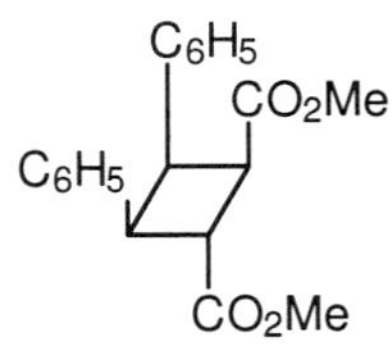	10	
	+ 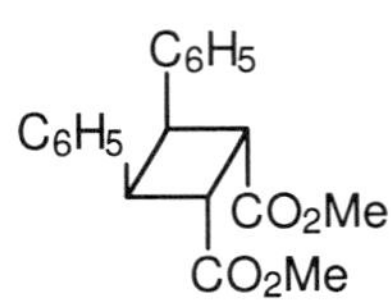	20	
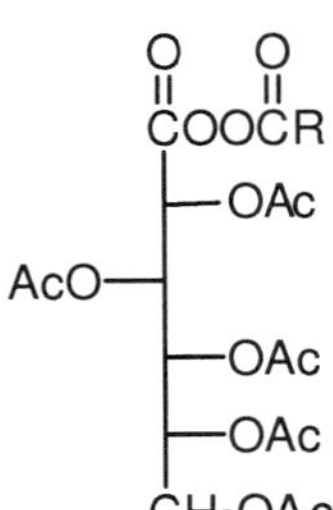	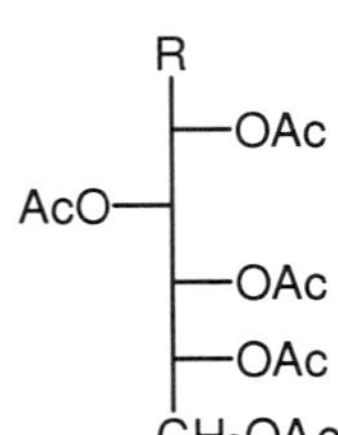	34-41 (ee = 82 - 87%)	59

Continued on next page

Table 6.5. Reactions with carbohydrates as chiral auxiliaries—*Continued*

Reactant	*Product*	*Percent Yield*	*Ref.*
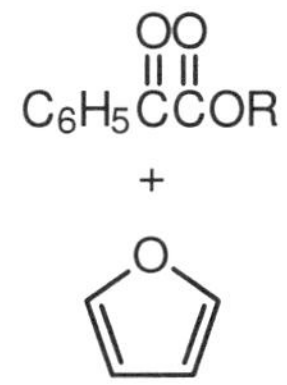 R = 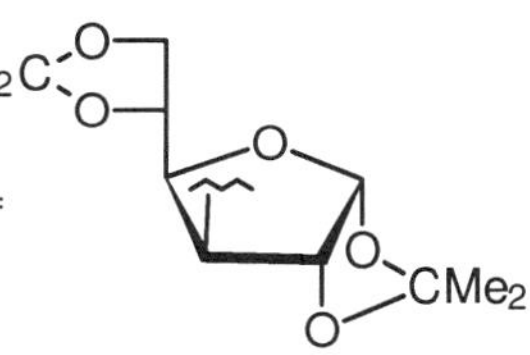	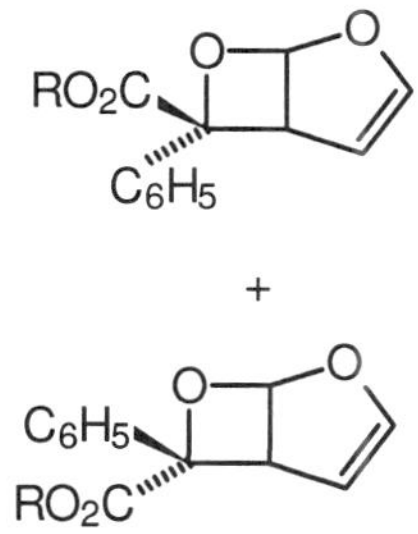	79[d] (de = 48%)	60
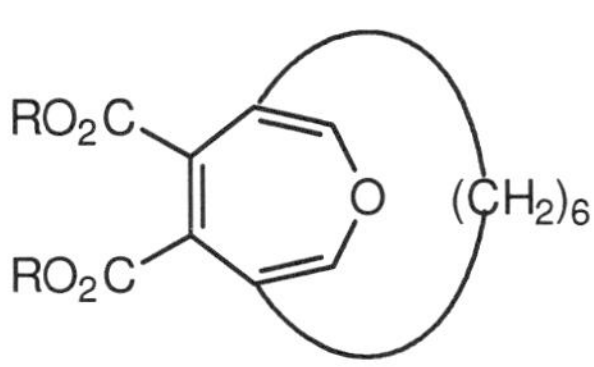 R = 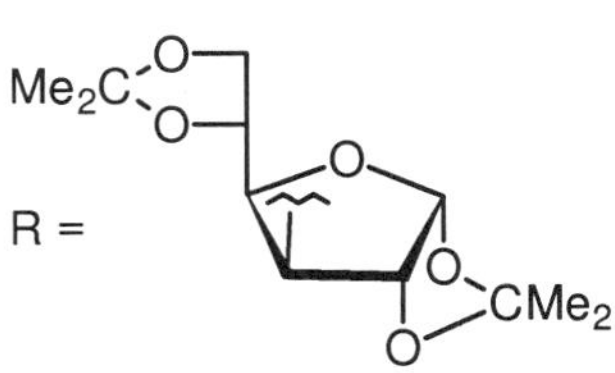	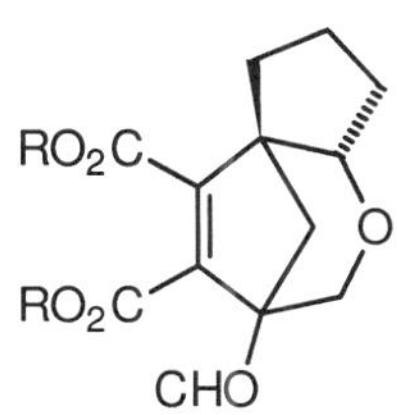	54[e] (de = 92%)	61

[a]The products were isolated after hydrolysis in acidic methanol.
[b]Five other derivatives of D-mannitol were irradiated.
[c]Similar products were formed from irradiation of an ester of L-erythritol. The ee is enantiomeric excess.
[d]Five other carbohydrate-containing R groups were used.
[e]The de (diastereomeric excess) is for reaction in aqueous suspension.

References

1. Williams, N. R.; Wander, J. D. In *The Carbohydrates;* Pigman, W.; Horton, H. Eds.; Academic Press: New York, 1972; Vol. IB, p 761.
2. Beugelmanns, R.; LeGoffand, M. T.; Compaignon de Marcheville, H. *C. R. Acad. Sci., Ser. C* **1969**, *269*, 1309.
3. Deshayes, H.; Pète, J.-P.; Portella, C.; Scholler, D. *J. Chem. Soc., Chem. Commun.* ***1975***, 439.
4. Deshayes, H.; Pète, J.-P.; Portella, C. *Tetrahedron Lett.* **1976**, 2019.
5. Pète, J.-P.; Portella, C. *Synthesis* **1977**, 774.
6. Collins, P. M.; Munasinghe, V. R. Z. *J. Chem. Soc., Chem. Commun.* **1977**, 927.
7. Pète, J.-P.; Portella, C.; Scholler, D. *J. Photochem.* **1984**, *27*, 128.
8. Bonjouklian, R.; Ganem, B. *Carbohydr. Res.* **1979**, *76*, 245.
9. Thiem, J.; Meyer, B. *Chem. Ber.* **1980**, *113*, 3058.
10. Kiss, J.; D'Souza, R. *J. Carbohydr., Nucleosides, Nucleotides* **1980**, *7*, 141
11. Miyashita, M.; Chida, N.; Yoshikohsi, A. *J. Chem. Soc., Chem. Commun.* **1982**, 1354.
12. Angyal, S. J.; Odier, L. *Carbohydr. Res.* **1982**, *101*, 209.
13. Portella, C.; Deshayes, H.; Pète, J.-P.; Scholler, D. *Tetrahedron* **1984**, *40*, 3635.
14. Hartwig, W. *Tetrahedron* **1983**, *39*, 2609.
15. Klausener, A.; Beyer, G.; Leismann, H.; Scharf, H.-D.; Müller, E.; Runsink, J.; Görner, H. *Tetrahedron* **1989**, *45*, 4989.
16. Jütten, P.; Scharf, H.-D. *J. Carbohydr. Chem.* **1990**, *9*, 675.
17. Klausener, A.; J. Runsink, J.; Scharf, H.-D. *Liebigs Ann. Chem.* **1984**, 783.
18. Klausener, A.; Müller, E.; Runsink, J.; Scharf, H.-D. *Carbohydr. Res.* **1983**, *116*, 295.
19. Jütten, J.; Scharf, H.-D. *Carbohydr. Res.* **1991**, *212*, 93.
20. Dornhagen, J.; Klausener, A.; Runsink, J.; Scharf, H.-D. *Liebigs Ann. Chem.* **1985**, 1838.
21. Rainer, H.; Scharf, H.-D.; Runsink, J. *Liebigs Ann. Chem.* **1992**, 103.
22. Dornhagen, J.; Scharf, H.-D. *J. Carbohydr. Chem.* **1986**, *5*, 115.
23. Saito, I.; Ikehira, H.; Kasatani, R.; Watanabe, M.; Matsuura, T. *J. Am. Chem. Soc.* **1986**, *108*, 3115.
24. Ireland, R. E.; Armstrong, J. D., III; Lebreton, J.; Meissner, R. S.; Rizzacasa, M. A. *J. Am. Chem. Soc.* **1993**, *11*, 7152.
25. Almond, M. R.; Collins, J. L.; Reitter, B. E.; Rideout, J. L.; Freeman, G. A.; St Clair, M. H. *Tetrahedron Lett.* **1991**, *32*, 5745.
26. Huang, Z.; Schneider, K. C.; Benner, S. A. *J. Org. Chem.* **1991**, *56*, 3869.
27. Park, M.; Risso, C. J. *J. Org. Chem.* **1996**, *61*, 6092.
28. Abe, M.; Oku, A. *J. Org. Chem.* **1995**, *60*, 3063.
29. Mizuno, K.; Ichinose, N.; Otsuji, Y. *Chem. Lett.* **1985**, 455.
30. Zimmerman, H. E. In *Advances in Photochemistry*; Noyes, W. A., Jr.; Hammond, G. S.; Pitts, J. N., Jr., Eds.; Interscience: New York, 1963; Vol 1, p 183.
31. Kopecký, J. *Organic Photochemistry: A Visual Approach;* VCH Publishers: New York, 1992; p 156.
32. Coyle, J. D. *Chem. Rev.* **1978**, *78*, 97.
33. Wagner, P.; Park, B.-S. *Org. Photochem.* **1991**, *11*, 227.
34. Binkley, R. W.; Meinzer, J. L. *J. Carbohydr., Nucleosides, Nucleotides* **1975**, *2*, 465.
35. Piva, O.; Péte, J.-P. *Tetrahedron: Asymmetry* **1992**, *3*, 759.
36. Ferrier, R. J.; Tyler, P. C. *Carbohydr. Res.* **1985**, *136*, 249.
37. Barltrop, J. A.; Wills, J. *Tetrahedron Lett.* **1969**, 4987.
38. Jorgenson, M. J.; Bundel, L. *Tetrahedron Lett.* **1968**, 4991.
39. Kitagawa, I.; Yoshikawa, M. *Heterocycles* **1977**, *7*, 783.
40. Kitagawa, I.; Yoshikawa, M.; Imakura, Y.; Yosioka, I. *Chem. Ind.* **1973**, 276.
41. Kitagawa, I.; Yoshikawa, M.; Imakura, Y.; Yosioka, Y. *Chem. Pharm. Bull.* **1974**, *22*, 1339.

42. Kitagawa, I.; Im, K. S.; Fujimoto, Y. *Chem. Pharm. Bull.* **1977**, *25*, 800.
43. Kitagawa, I.; Wang, H. K.; Saito, M.; Yoshikawa, M. *Chem. Pharm. Bull.* **1983**, *31*, 664.
44. Kitagawa, I.; Wang, H. K.; Saito, M.; Yoshikawa, M. *Chem. Pharm. Bull.* **1983**, *31*, 674.
45. Kitagawa, I.; Wang, H. K.; Saito, M.; Yoshikawa, M. *Chem. Pharm. Bull.* **1983**, *31*, 683.
46. Kitagawa, I.; Yoshikawa, M.; Yosioka, I. *Tetrahedron Lett.* **1973**, 3997.
47. Kitagawa, I.; Wang, H. K.; Yoshikawa, M. *Chem. Pharm. Bull.* **1983**, *31*, 716.
48. Tschesche, R.; Wagner, R.; Widera, W. *Liebigs Ann. Chem.* **1983**, 993.
49. Kitagawa, I.; Saito, M.; Taniyama, T.; Yoshikawa, M. *Chem. Pharm. Bull.* **1985**, *33*, 1069.
50. Kitagawa, I.; Saito, M.; Taniyama, T.; Yoshikawa, M. *Chem. Pharm. Bull.* **1985**, *33*, 598.
51. Yoshikawa, M.; Wang, H. K.; Kayakiri, H.; Taniyama, T.; Kitagawa, I. *Chem. Pharm. Bull.* **1985**, *33*, 4267.
52. Taniyama, T.; Nagahama, Y.; Yoshikawa, M.; Kitagawa, I. *Chem. Pharm. Bull.* **1988**, *36*, 2829.
53. Kitagawa, I.; Taniyama, T.; Nagahama, Y.; Okubo, K.; Yamauchi, T.; Yoshikawa, M. *Chem. Pharm. Bull.* **1988**, *36*, 2819.
54. Bradbury, A. G. W.; von Sonntag, C. *Carbohydr. Res.* **1978**, *60*, 183.
55. Green, B. S.; Rabinsohn, Y.; Rejtö, M. *J. Chem. Soc., Chem. Commun.* **1975**, 313.
56. Green, B. S.; Rabinsohn, Y.; Rejtö, M. *Carbohydr. Res.* **1975**, *45*, 115.
57. Green, B. S.; Hagler, A. T.; Rabinsohn, Y.; Rejtö, M. *Isr. J. Chem.* **1976/77**, *15*, 124.
58. Bernstein, J.; Green, B. S.; Rejtö, M. *J. Am. Chem. Soc.* **1980**, *102*, 323.
59. Lomölder, R.; Schäfer, H. J. *Angew. Chem. Int. Ed. Engl.* **1987**, *26*, 1253.
60. Pelzer, R.; Jütten, P.; Scharf, H.-D. *Chem. Ber.* **1989**, *122*, 487.
61. Tochtermann, W.; Schlösser, U.; Ott, F.; Popp, B.; Sdunnus, N.; Snatzke, G.; Peters, E.-M.; Peters, K.; von Schnering, H. G. *Chem. Ber.* **1993**, *126*, 1733.

7

Glycoside Photolysis

Introduction

Although photolysis of glycosides is common in carbohydrate photochemistry, actual photochemical reaction of the acetal group that forms the glycosidic linkage is rare. This is the situation because even though carbohydrates often exist as glycosides, the glycosidic linkage itself contains no chromophore capable of absorbing significant UV radiation. (Some investigators have argued that glycosides have a weakly absorbing, "acetal chromophore";[1] however, there is no general agreement that such a chromophore exists[2] or that an acetal is capable of light absorption beyond that which is associated with other carbon–oxygen single bonds.) Generating an excited glycoside, therefore, depends upon the presence of a light-absorbing chromophore. If excitation is to lead to reaction of the glycosidic linkage, an absorbing chromophore needs to be in close proximity. Reactions that have been observed in which the glycosidic linkage is directly involved in the photochemical process have been reactions of aryl glycosides, that is, compounds in which an aromatic ring is in direct conjugation with the exocyclic, glycosidic oxygen atom.

Direct Irradiation of Aryl Glycosides

Early studies on aryl glycosides served primarily to show that a photochemically promoted reaction was occurring, but these investigations were incomplete in that they did not include product identification [3,4,5] Later work[6,7] demonstrated that aryl glycoside photolysis gave "free" sugar and a phenol; for example, irradiation of phenyl β-D-glucopyranoside in aqueous solution produced D-glucose and phenol (eq 1, Table 7.1).[7] Filby et al. noted that the

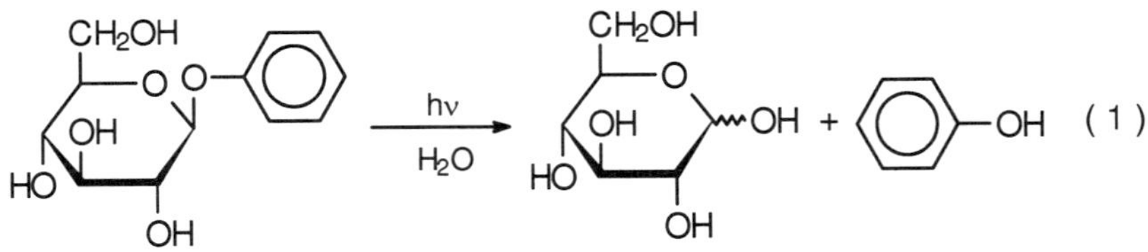

Equation 1

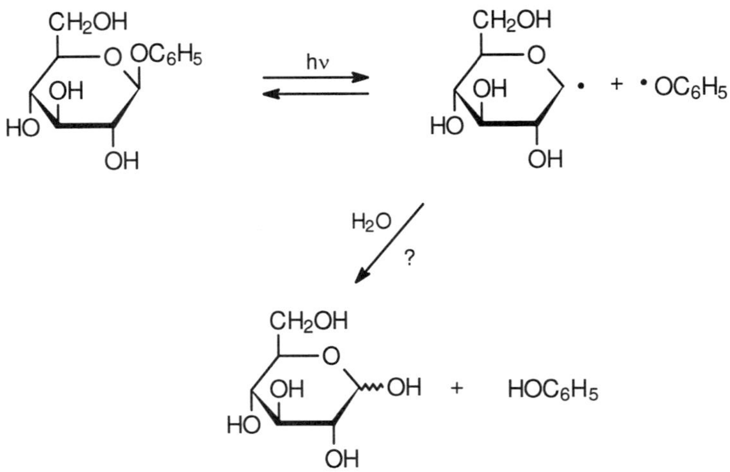

Scheme 7.1

presence of molecular oxygen in the reaction mixture did not reduce the amount of D-glucose formed during photolysis and concluded that reaction occurred from an excited singlet state.[7] (This conclusion may be correct, but the question of the reactive excited state is not settled by a lack of quenching by molecular oxygen. Other possibilities, such as a triplet state that reacts rapidly, could explain this lack of quenching.*) These investigators were unable, however, to decide upon a mechanism that satisfactorily explained product formation. They noted that flash photolysis experiments indicated a homolytic, carbon–oxygen bond cleavage, but they were unable to reconcile the radical intermediates produced by C–O bond homolysis with the formation of D-glucose and phenol (Scheme 7.1). Filby et al. pointed out that it is difficult to understand how a 1-glucosyl radical could produce D-glucose in aqueous solution when this radical, generated in other ways, was known to form keto and deoxy sugars.[7] This mechanistic puzzle was left unresolved at the

*Molecular oxygen can quench both singlet and triplet states, but it is more effective at quenching triplets due to the generally longer triplet lifetime that increases the opportunity for encounter with the quencher.[8]

Table 7.1. Reactions from Direct Irradiation of Aryl Glycosides.

Reactant	*Product and Percent Yield*[a]		*Ref.*
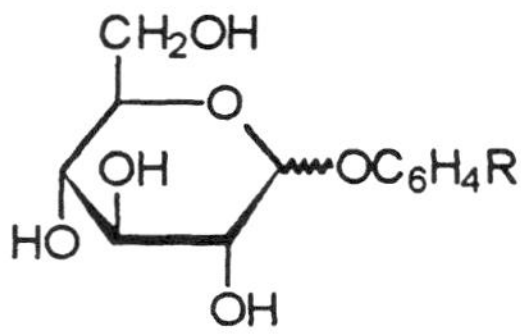 R = H, *o*-Me, *m*-Me *p*-Me, *p*-Cl, *o*-OMe	D-glucose (% yield not reported)		6
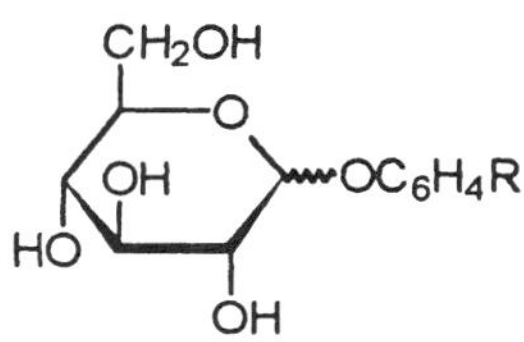	D-glucose +	HOC_6H_4R	7[b]
	Quantum Yields		
R = H	0.006	0.005	
R = *o*-Me,	0.008	0.004	
R = *m*-Me	0.010	0.008	
R = *p*-Me	0.008	0.008	
R = *p*-Cl	------	0.010	
R = *o*-OMe	0.006	0.003	

[a]Noncarbohydrate products were not identified.
[b]Reaction occurs in the presence of oxygen.

time; however, later work by other investigators provided a basis for its solution. Before considering the mechanism of this reaction further, however, it is useful to discuss some of the information about electron-transfer reactions from more recent investigations.

Electron-Transfer Reactions

Timpa et al. became interested in studying aryl glycosides because these compounds have a similarity in basic structure to lignocellulosic compounds found in biomass.[9,10] They reasoned that photolysis of aryl glycosides could serve as a model reaction for solar, biomass degradation, and they adopted an approach to glycosidic bond cleavage in which an aryl glycoside was as an electron donor in a photochemically initiated, electron-transfer reaction. Irradiation of phenyl α or β-D-glucopyranoside in the presence of the electron acceptor 1,4-dicyanonaphthalene (DCN) cleaved the glycosidic linkage to give D-glucose and phenol (Scheme 7.2).[9,10] These workers further suggested that D-glucose was formed by reaction of the 1-glucosyl cation (**1**) with water.[11] In support of this proposal they found that irradiation conducted in methanol produced the corresponding methyl glucosides. Irradiation of other aryl glycosides gave similar results (Table 7.2). An important element in this process was the formation of the resonance-stabilized phenoxy radical. Timpa and Griffin found that if the structure of the glycoside was such that a phenoxy radical could not be produced, no glycosidic cleavage products formed (eq 2).[10]

Hashimoto et al. also chose a photochemical, electron-transfer reaction as a method for cleaving aryl glycosides.[12] This reaction depended upon a chain of events, outlined in Scheme 7.3, that started with absorption of UV light by phenanthrene and proceeded to formation of the 1-glycosyl cation **2**. This cation (**2**) then combined with a nucleophile to produce either an anhydro sugar by intramolecular reaction (Scheme 7.3) or a new glycoside by an intermolecular process (eq 3).

Reaction Mechanism for Direct Irradiation of Aryl Glycosides

After considering the reactions of aryl glycosides under electron-transfer conditions, it is useful to return to the discussion of the reaction of these compounds upon direct photolysis. As mentioned in the section "Direct Irradiation of Aryl Glycosides", Phillips et al. noted that the photoproducts formed were difficult to explain if homolytic cleavage were the only reaction taking place (Scheme 7.1).[7] If, however, one considers the products formed from electron-transfer reactions and assumes that electron transfer could occur

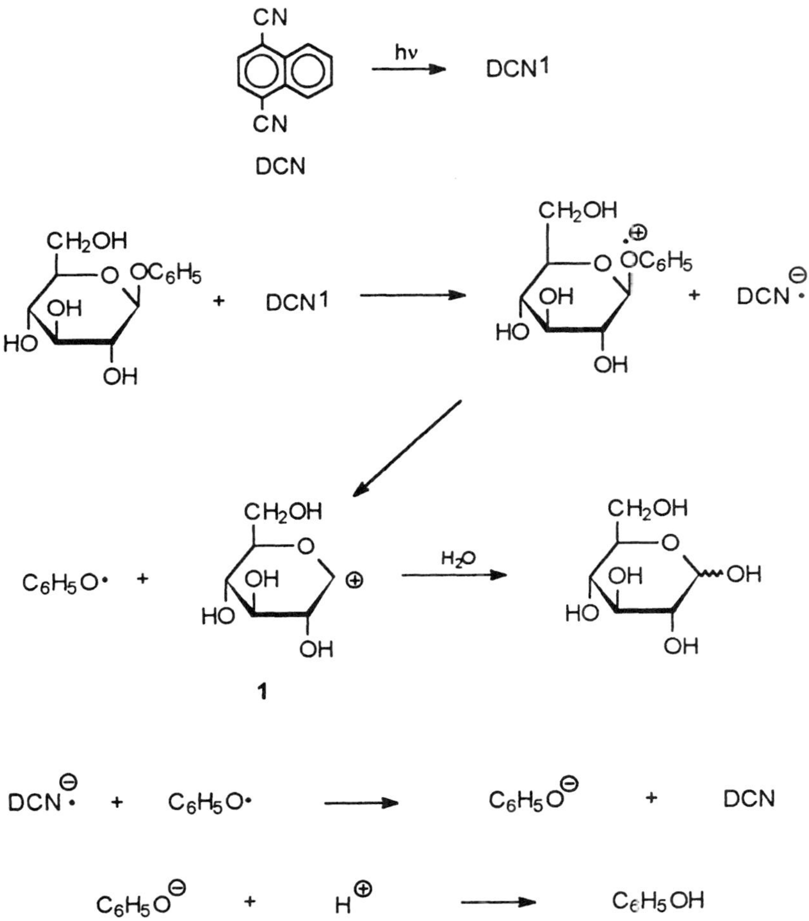

Scheme 7.2

Table 7.2. Electron-Transfer Reactions of Aryl Glycosides[a].

Reactant	*Product*	*Percent Yield*	*Ref.*
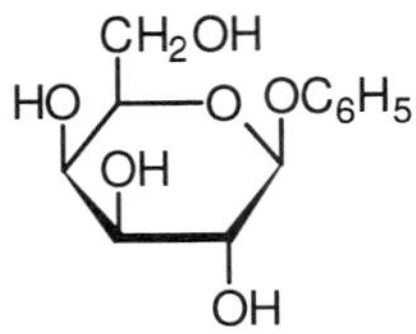	D-galactose	6-60[b]	9,10,11
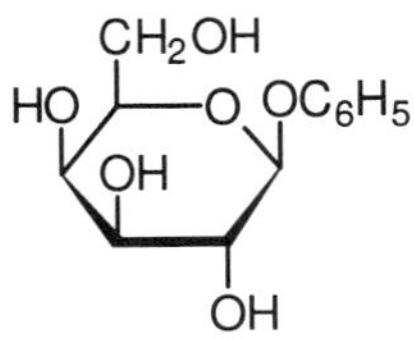 + MeOH	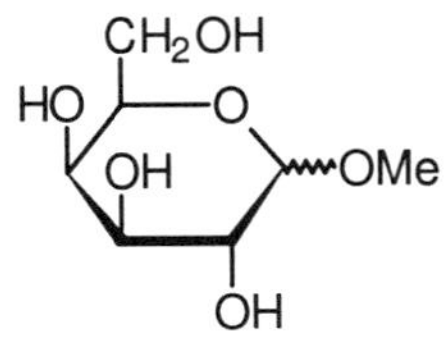	85-100[b]	9,10
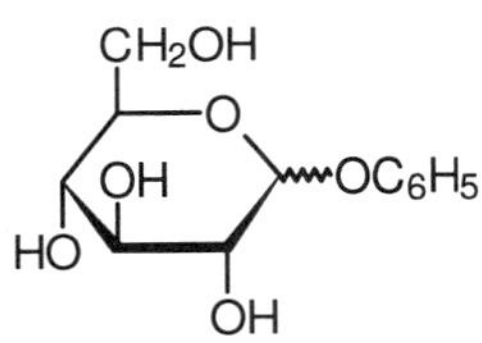			9,10,11
α-anomer	D-glucose	4-12[b]	
β-anomer	D-glucose	17-40b,c	
β-anomer + MeOH	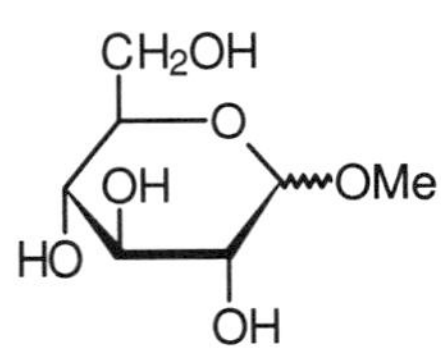	α (33-41)[b] β (27-39)[b]	
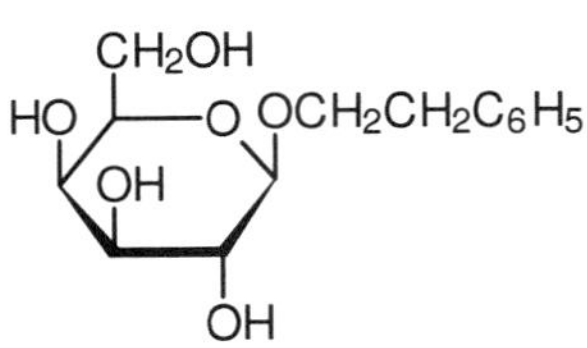	no cleavage products reported	d	9,10

Continued on next page

Table 7.2. Electron-Transfer Reactions of Aryl Glycosides[a]—*Continued*

Reactant	*Product*	*Percent Yield*	*Ref.*
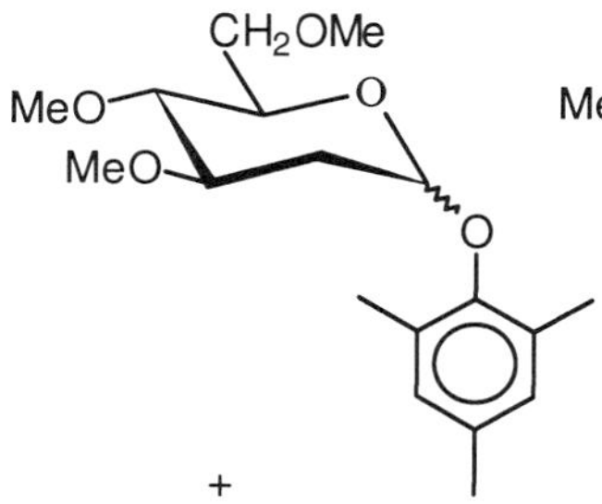 + $CH_3(CH_2)_7OH$	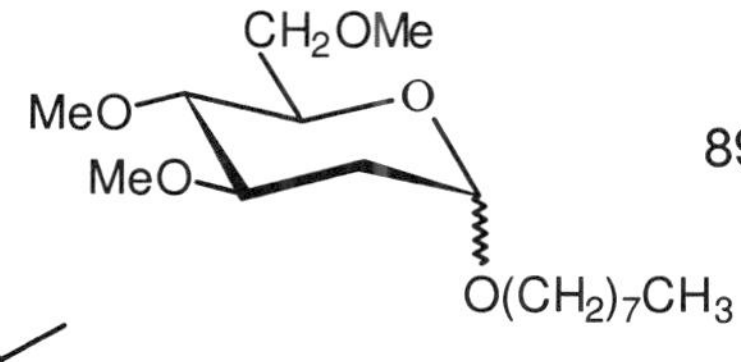	89	12
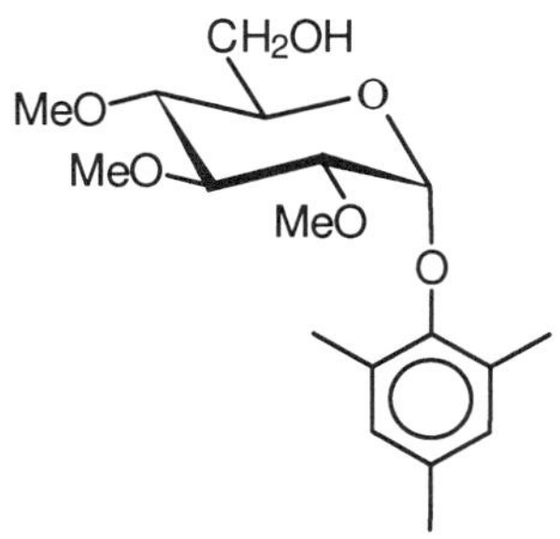	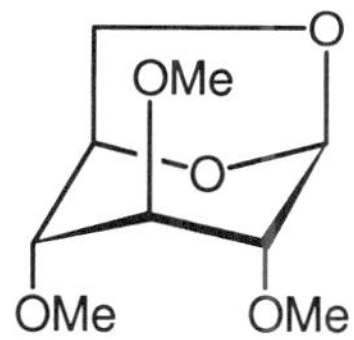	96	12

[a]All reactions were conducted in the presence of 1,4-dicyanonapthalene.

[b]Product yields depend upon the reaction conditions.

[c]In reference 11 levoglucosan also is a reported product. No yields are given.

[d]Although no cleavage products were isolated, starting material was not completely recovered.

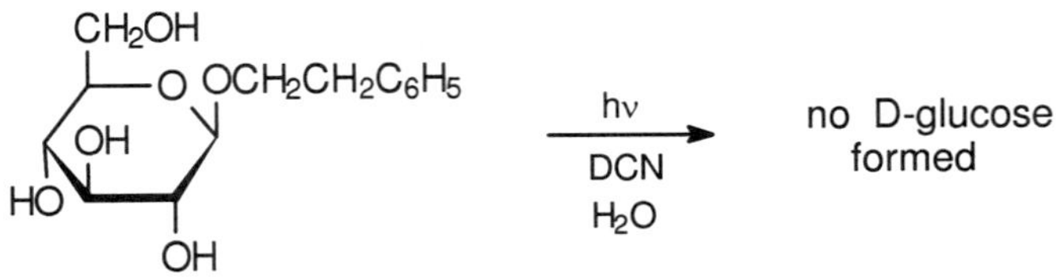

(2)

Equation 2

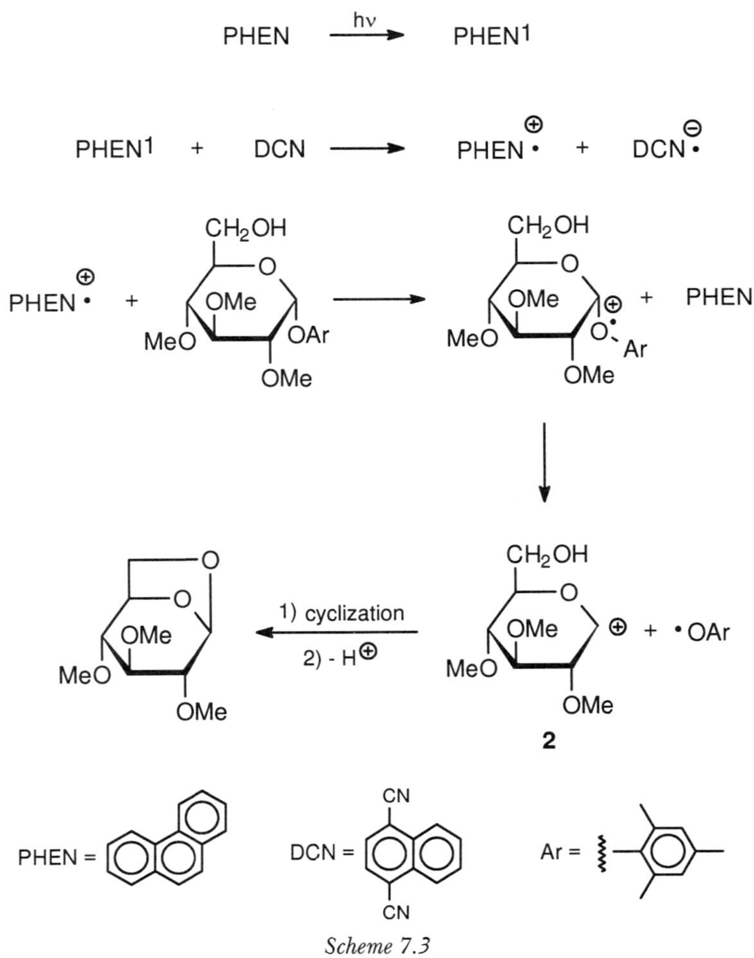

Scheme 7.3

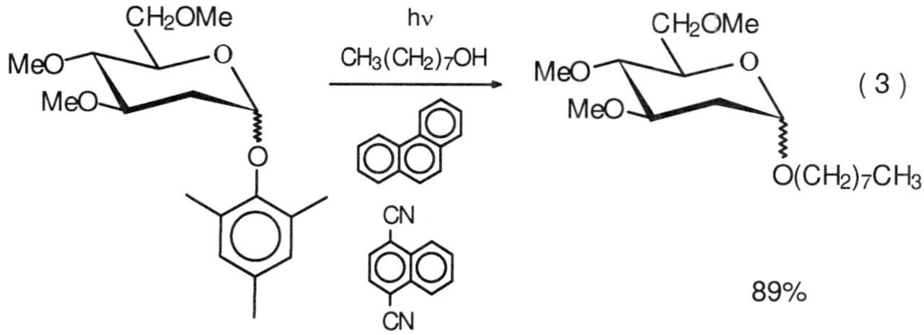

Equation 3

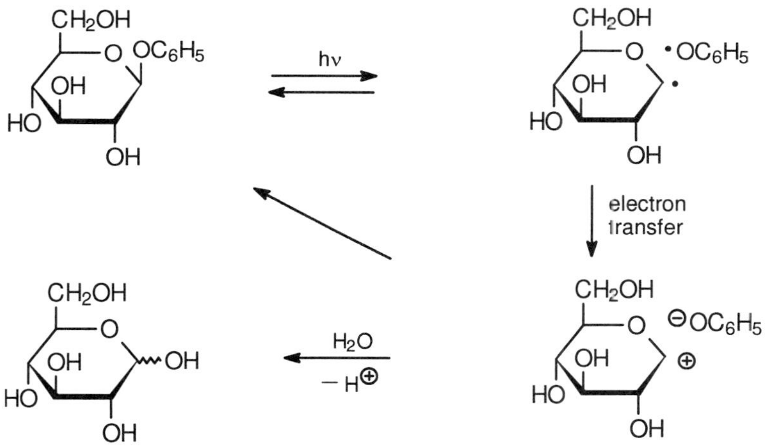

Scheme 7.4

between the 1-glycosyl and phenoxy radicals, the 1-glycosyl cation and phenoxide ion would be formed.[13] These ionic intermediates would produce the D-glucose and phenol observed upon photolysis in aqueous solution (Scheme 7.4). Since the opportunity for return to starting materials exists at both the radical pair and ion pair stages of the reaction, the observed, low quantum yield for this process is understandable.

Other Photochemical Reactions of Glycosides

Additional photochemical reactions that involve cleavage of glycosidic linkages are discussed in Chapters 5 and 16. Glycosidic cleavage arising from photolysis of uronic acids and esters is considered in Chapter 5. The photochemical reaction of carbohydrates protected as *o*-nitrobenzyl glycosides is discussed in Chapter 16.

Summary

The photochemistry of aryl glycosides has been studied under two types of conditions, direct irradiation and photochemical electron-transfer. In both instances the carbohydrate photoproducts appear to be formed from reaction of 1-glycosyl cations with nucleophiles such as water or an alcohol.

References

1. Phillips, G. O. *Advan. Carbohydr. Chem.* **1963**, *18*, 9.
2. McKelvey, R. D. *Carbohydr. Res.* **1975**, *42*, 187.
3. Tanret, G. *C. R. Acad. Sci.* **1935**, *201*, 1057.
4. Heidt, L. J. *J. Am. Chem. Soc.* **1939**, *61*, 2981.
5. Heidt, L. J. *J. Franklin Inst.* **1942**, 473.
6. Yamada, T.; Sawada, M.; Taki, M. *Arg. Biol. Chem.* **1975**, *39*, 909.
7. Filby, W. G.; Phillips, G. O.; Webber, M. G. *Carbohydr. Res.* **1976**, *51*, 269.
8. Kawaoka, K.; Khan, A. U.; Kerns, D. R. *J. Chem. Phys.* **1967**, *46*, 1842.
9. Timpa, J. D.; Legendre, M. G.; Griffin, G. W.; Das, P. K. *Carbohydr. Res.* **1983**, *117*, 69.
10. Timpa, J. D.; Griffin, G. W. *Carbohydr. Res.* **1984**, *131*, 185.
11. Davis, H. F.; Das, P. K.; Griffin, G. W.; Timpa, J. D. *J. Org. Chem.* **1983**, *48*, 5256.
12. Hashimoto, S.; Kurimoto, I.; Fujii, Y.; Noyori, R. *J. Am. Chem. Soc.* **1985**, *107*, 1427.
13. Kudryashov, L. I.; Nedoborova, L. I.; Chizhov, O. S.; Kochetkov, N. K. *Izv. Nauk SSSR, Ser. Khim.* **1975**, 722. (Chem. Abstr. **1975**, *83*, 43626.)

8

Alkenes

[2 + 2] Photocycloaddition Reactions

Introduction

Photochemical [2 + 2] cycloaddition reactions provide a simple pathway for forming four-membered rings between two unsaturated compounds (eq 1).[1] Because a cyclobutane ring readily can undergo further, nonphotochemical transformations, [2 + 2] addition has become an important step in the synthesis of a variety of natural products.[2] Cycloaddition reaction does not occur equally well between any two unsaturated compounds; in fact, as Gilbert and Baggott[3] have pointed out, "The vast majority of reported ($2\pi + 2\pi$) cycloadditions and certainly those that have been found to be useful in synthetic organic chemistry involve the addition of photoexcited α,β-unsaturated carbonyl compounds (enones) to ground-state alkenes or other nonabsorbing ethenes." This observation holds true in carbohydrate chemistry also where for all the reported [2 + 2] cycloaddition reactions involving carbon–carbon double or triple bonds, an α,β-unsaturated carbonyl compound is always one of the reactants (Table 8.1).

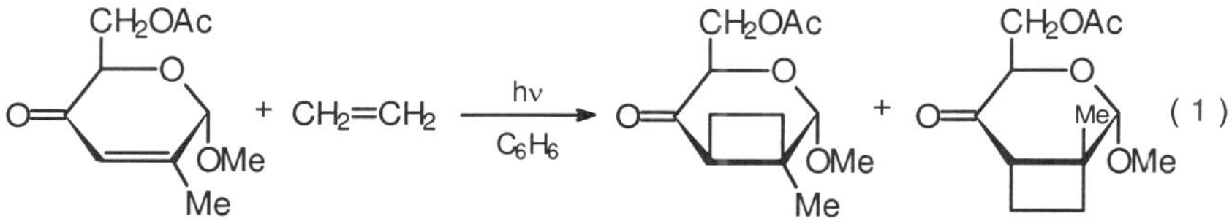

Equation 1

Table 8.1. Photocycloaddition Reactions.

Reactant	*Product*	*Percent Yield*	*Ref.*
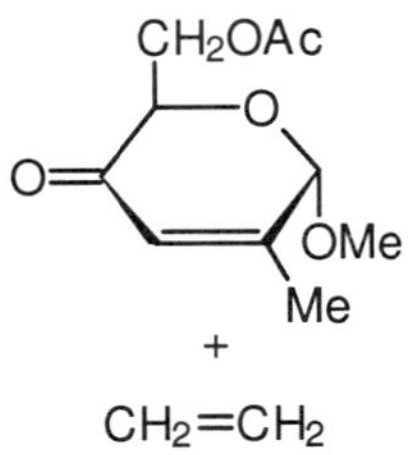	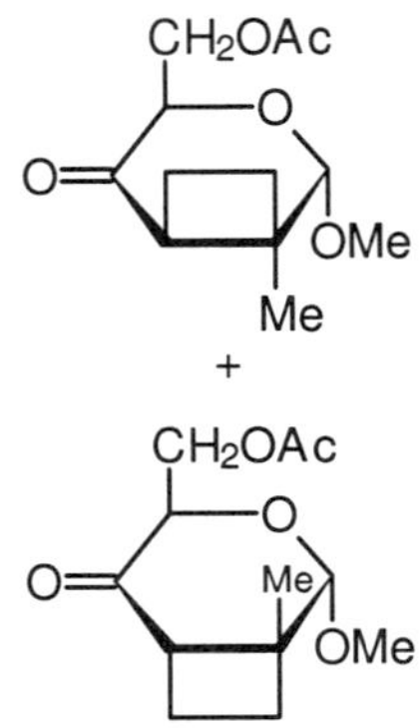	19[a]	1
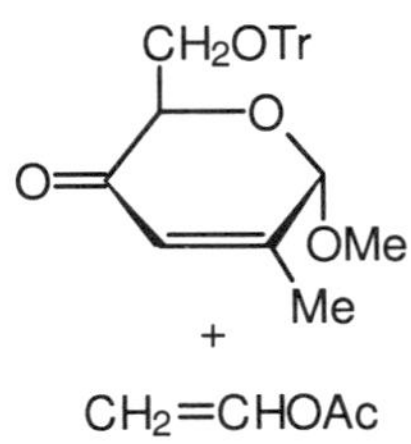	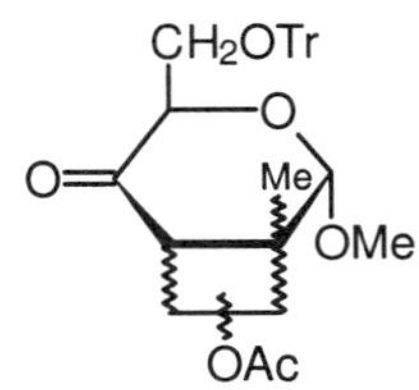	75	1
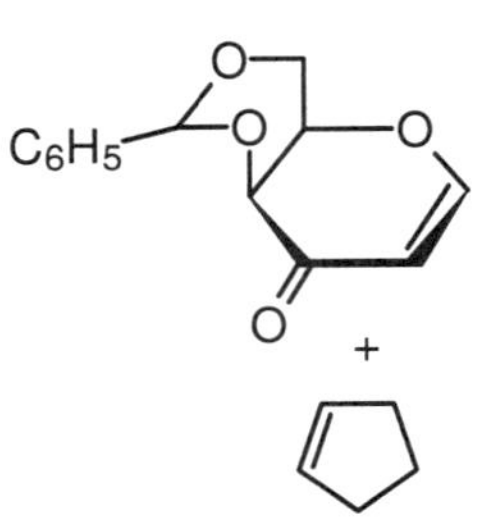	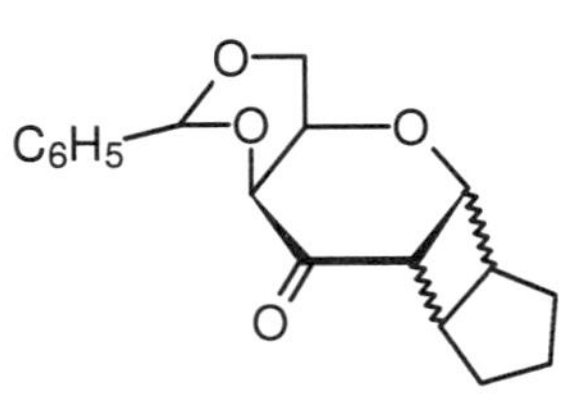	b	5
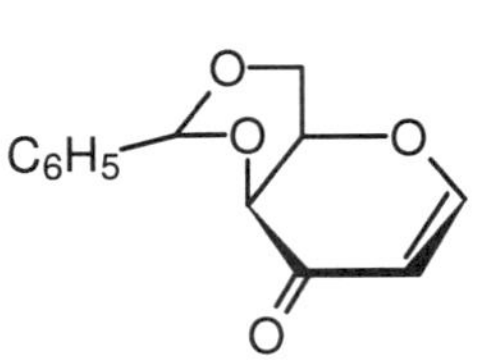	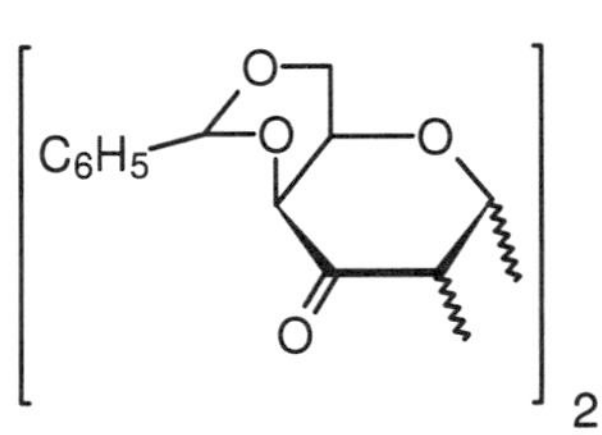	b	5

Continued on next page

Table 8.1. Photocycloaddition Reactions—*Continued*

Reactant	*Product*	*Percent Yield*	*Ref.*
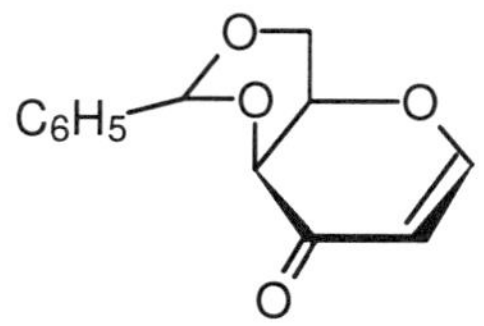 + 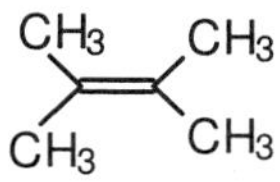	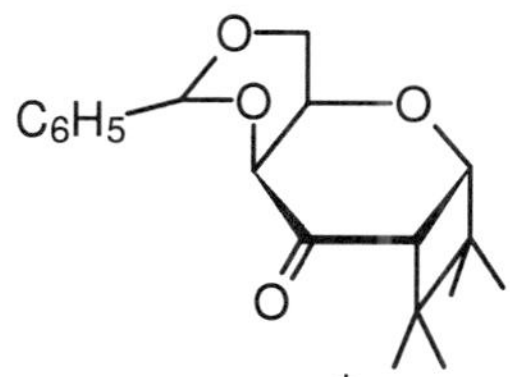 +	11	4
	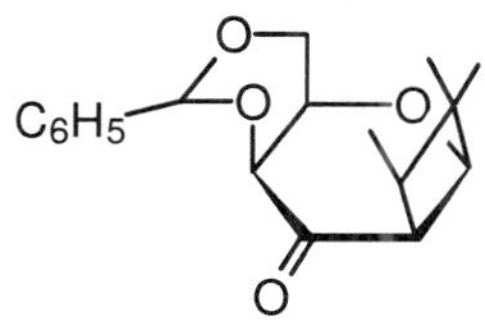 +	23	
	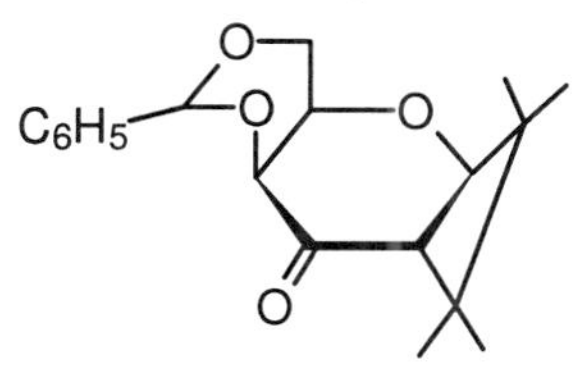 +	12	
	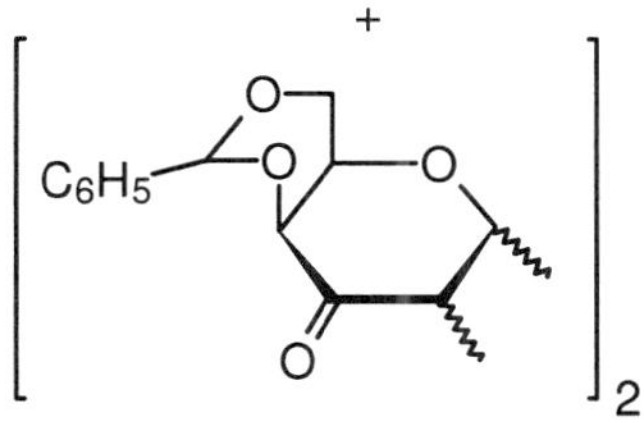	43	
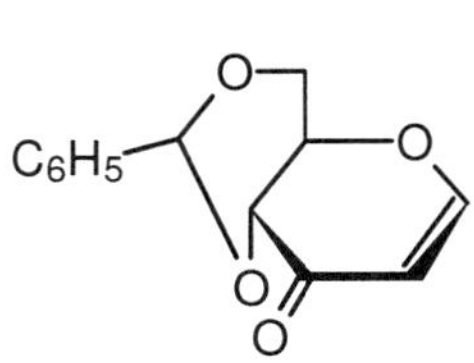 + 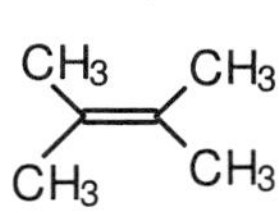	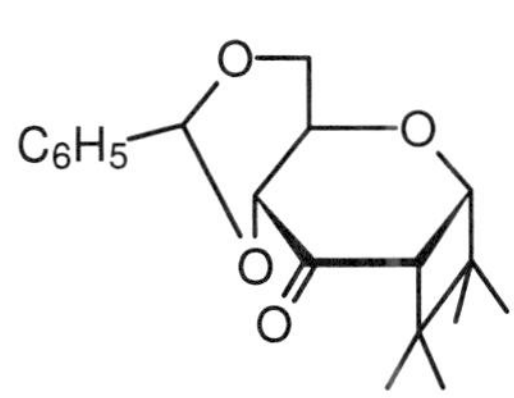 +	b	5
	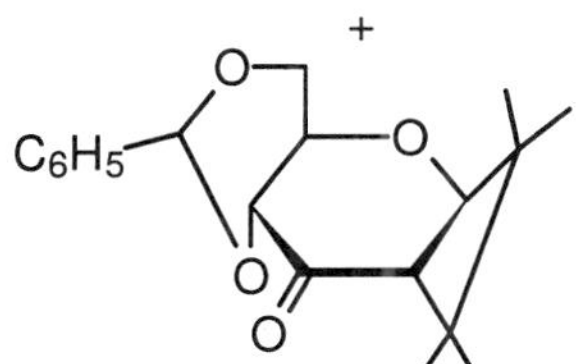	b	

Continued on next page

Table 8.1. Photocycloaddition Reactions—*Continued*

Reactant	*Product*	*Percent Yield*	*Ref.*
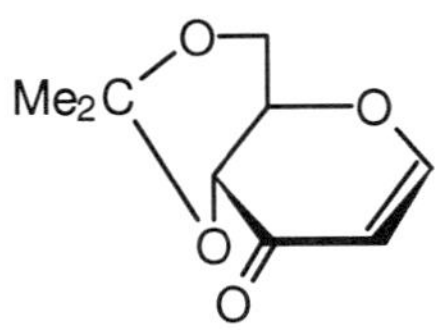 + HC≡CH	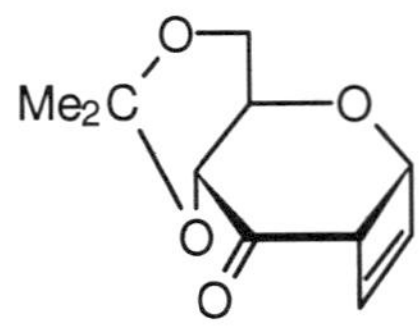	81	6
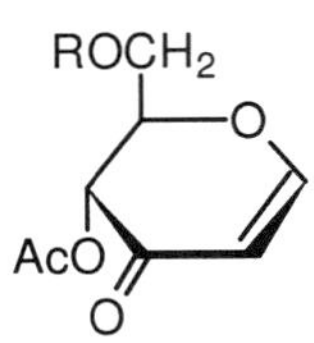 + HC≡CH	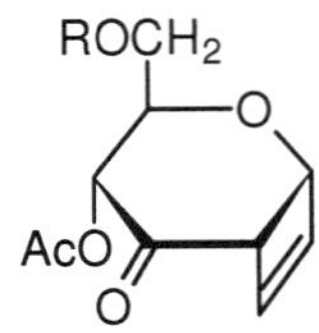		
R = Ac		82	7
R = $(C_6H_5)_3C$		75	6
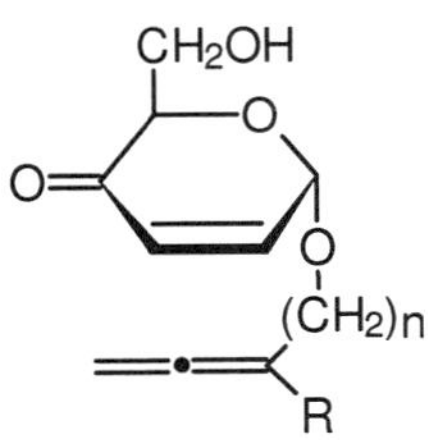	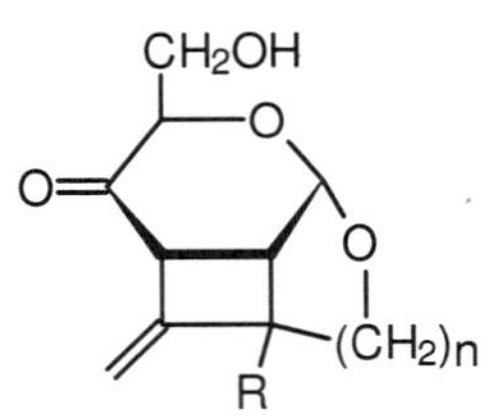		8
	n = 1, R = H	99	
	n = 1, R = Me	81	
	n = 2, R = H	46	
	n = 2, R = Me	60	
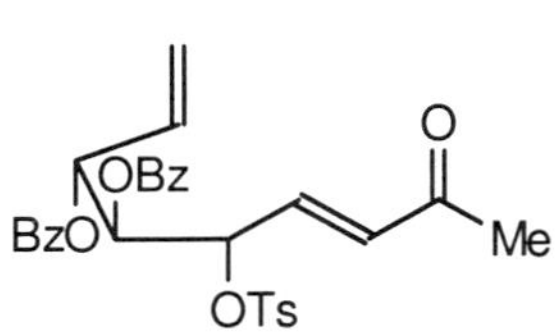	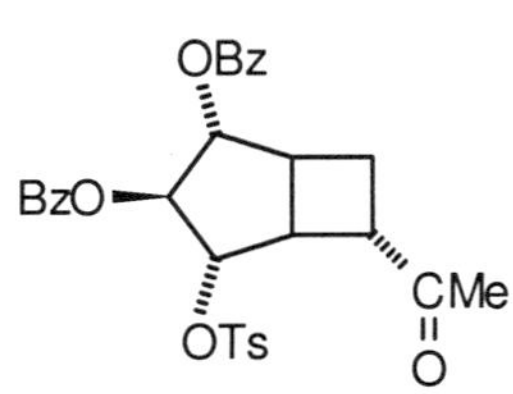	86	9

Continued on next page

Table 8.1. Photocycloaddition Reactions—*Continued*

Reactant	*Product*	*Percent Yield*	*Ref.*
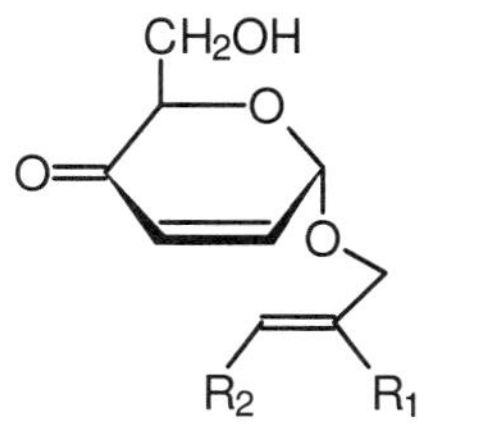	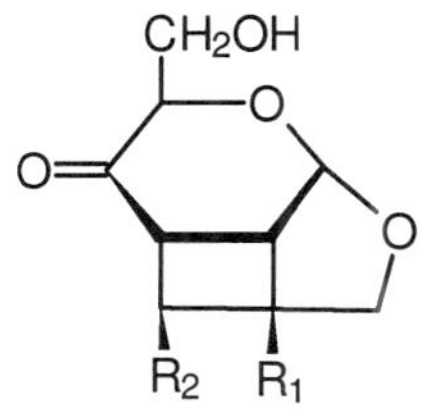		10[c]
$R_1 = R_2 = H$		95	
$R_1 = Me, R_2 = H$		93	
$R_1 = R_2 = (CH_2)_4$		62	
$R_1 = R_2 = (CH_2)_3$		75	
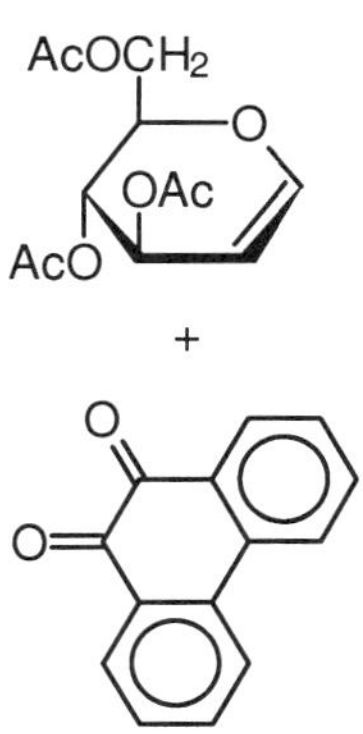	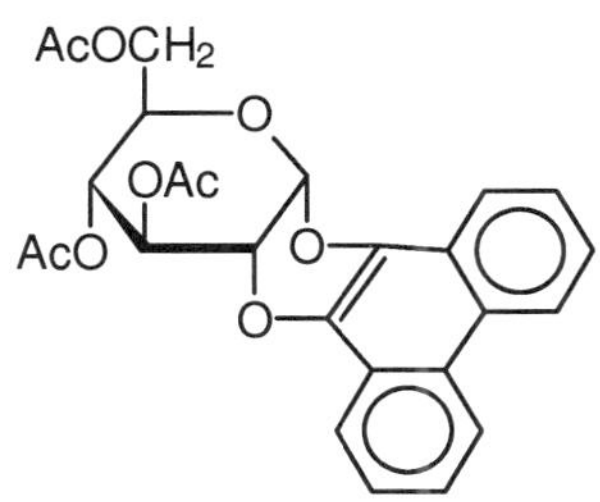		8

[a]This is a combined yield for both products.
[b]No yield was reported.
[c]The photoproducts experienced epimerization at C-5.

Reaction Mechanism

Direct Excitation

The mechanism for reaction between α,β-unsaturated carbonyl compounds and alkenes (or alkynes) has been extensively investigated and comprehensively reviewed on several occasions.[2,11–18] The widely accepted reaction mechanism is based upon a proposal published in 1964 by Corey et al.[19] and subsequently expanded upon by Loutfy and de Mayo.[20] These investigators suggested that the initially formed (n,π*) singlet excited state of an α,β-unsaturated carbonyl compound is transformed into the lowest triplet state by intersystem crossing (Scheme 8.1). (This triplet state also can be produced by sensitization.) In its lowest triplet state the enone forms an exciplex with a ground state alkene. The exciplex then reacts to give a 1,4-diradical that cyclizes to generate a four-membered ring (Scheme 8.2). Both the exciplex and the 1,4-diradical are assumed to be able to revert to the starting materials.

Although this mechanism is widely used, its acceptance is not universal. The primary point of contention is with the existence of the exciplex. The rationale for proposing the intermediacy of an exciplex was to explain the regioselectivity of the addition process. The thought was that in the exciplex the polarity of the excited enone [assumed to be in an (n,π*) state] and the polarity of the alkene would dictate the manner in which these reactants would be aligned with respect to each other (Figure 8.1). This alignment then would determine the regiochemistry of the products.

Schuster et al. have challenged this rationale by analyzing the available information and concluding that exciplexes are unnecessary for explaining the regioselectivity of product formation.[17] An important part of the information they considered came from the work of Hastings and Weedon,[21] who showed that, at least in some instances, regioselectivity in [2 + 2] photocyclo-

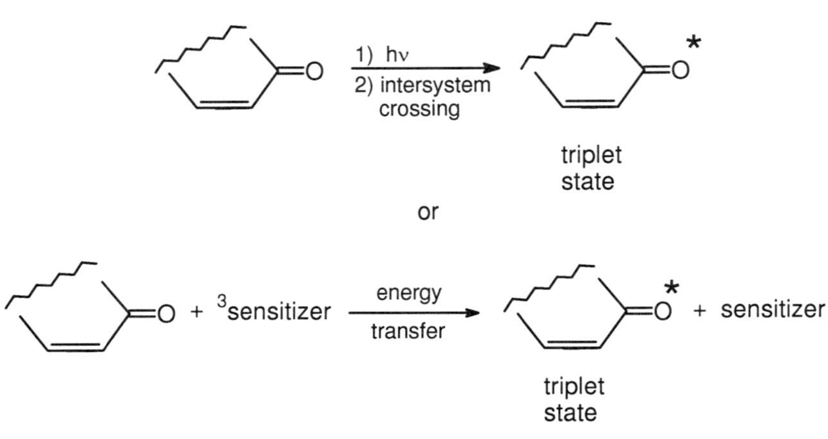

Scheme 8.1

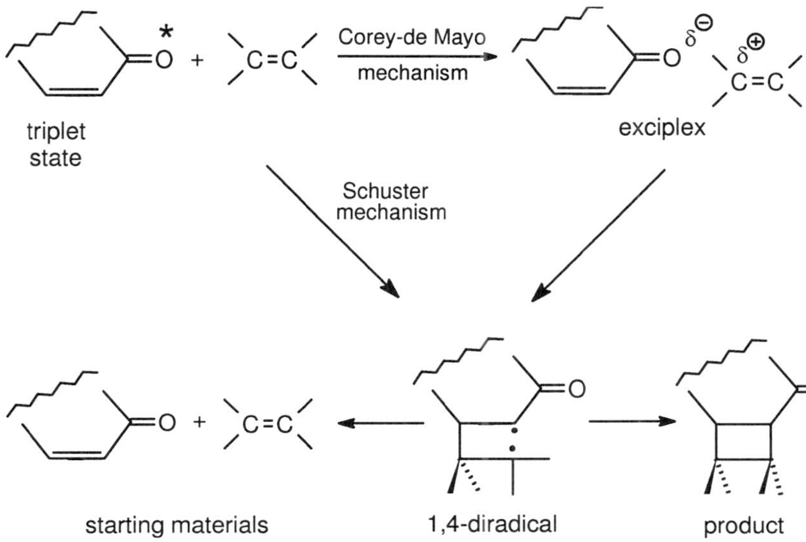

Scheme 8.2

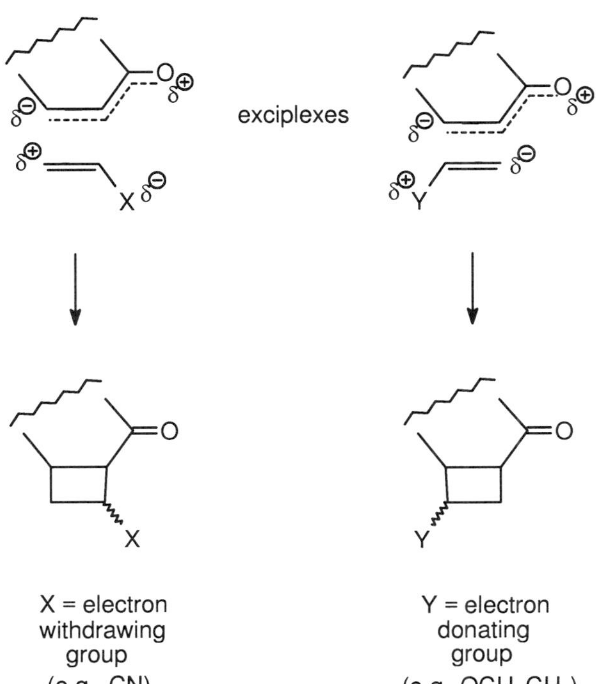

Figure 8.1. Alignment of reactants in proposed exciplexes

addition reaction was determined at the 1,4-diradical stage of reaction. The relative amounts of cyclobutanes formed depended upon which diradical (**A** or **B**, Scheme 8.3) was more likely to form a cyclobutane rather than cleave to regenerate the starting materials. Schuster et al. concluded from their analysis that "since no evidence *requiring* exciplex precursors to biradicals has been reported for these photocycloadditions, and since exciplexes are not necessary to explain the course of these reactions, we see no reason why exciplexes should continue to be invoked in mechanistic discussions of enone [2 + 2] photocycloaddition reactions."[17] Based upon the Schuster analysis, a reasonable assessment of the situation is that exciplexes should not be included as a part of the reaction mechanism for [2 + 2] photocycloaddition unless direct evidence is obtained for their existence.

Houk and co-workers have provided a different type of insight into the question of regioselectivity in cycloaddition reactions.[22] Their contribution is based on an interpretation of molecular orbital calculations related to the cycloaddition process. These investigators found that a twisted, (π,π*) enone triplet (Figure 8.2) was responsible for initiating the addition process. Furthermore, this triplet behaves as if it consists of an alkyl radical on the β-carbon

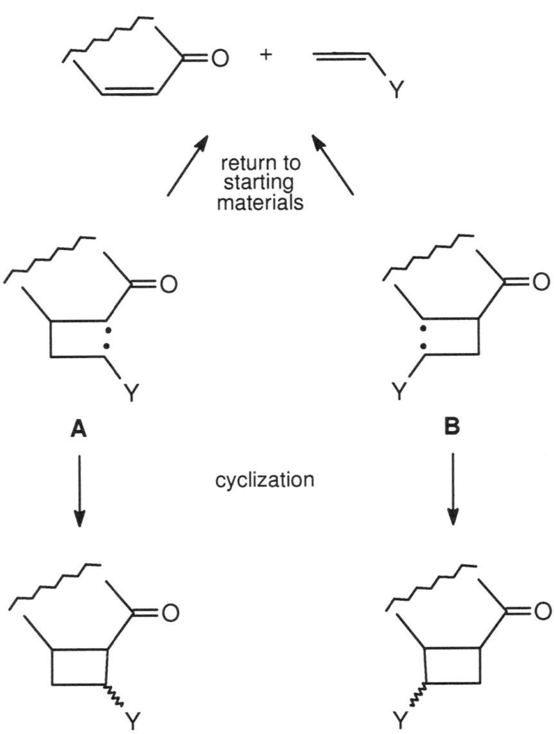

Scheme 8.3

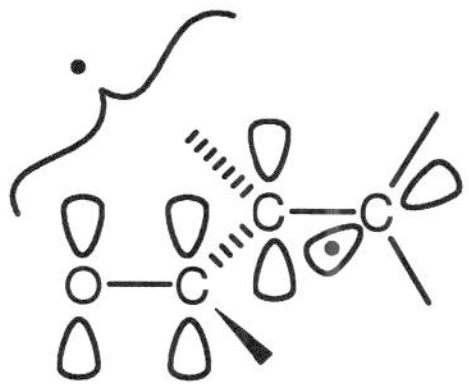

Figure 8.2. A twisted enone triplet

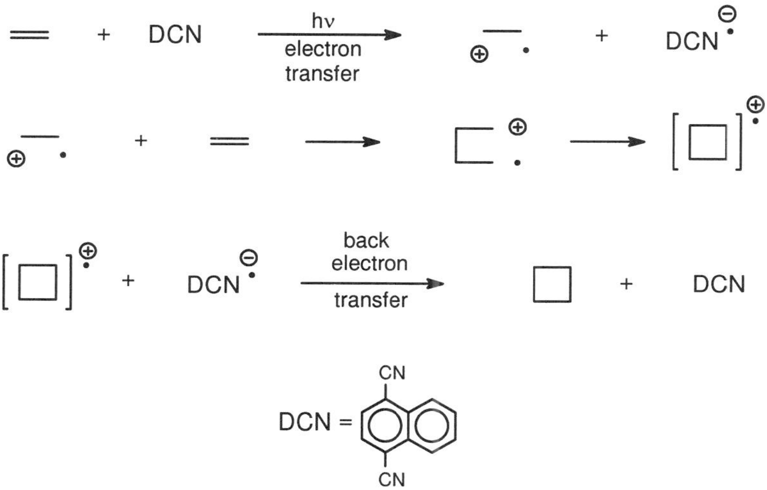

Scheme 8.4

atom linked to an α-acyl radical by a single bond. Houk points out that "alkyl radicals are nucleophilic and add rapidly to the less substituted terminus of electron-deficient alkenes. The α-acyl radicals are more electrophilic than alkyl radicals and add rapidly to the less substituted terminus of electron-rich alkenes... The polarity of the alkene directs the attack of the triplet and thus determines the regioselectivity of the reaction."[22] Chapter 9 contains further information about nucleophilic alkyl radicals and electrophilic alkyl radicals. See the section on "Effect of β-Substituents on Regioselectivity."

Electron-Transfer Reaction

Müller and Mattay have pointed out that [2 + 2] cycloaddition reactions also can occur by electron-transfer reaction.[14] The mechanism they proposed for this process, using 1,4-dicyanonaphthalene as an electron acceptor, is outlined in Scheme 8.4. Although there are no reported examples of [2 + 2] cycloadditon reactions of carbohydrates proceeding by photochemical electron transfer (PET), this type of reaction for carbohydrates is worthy of exploration since cycloaddition reactions occurring by PET can give different products from those formed by direct irradiation (Scheme 8.5).[16,23]

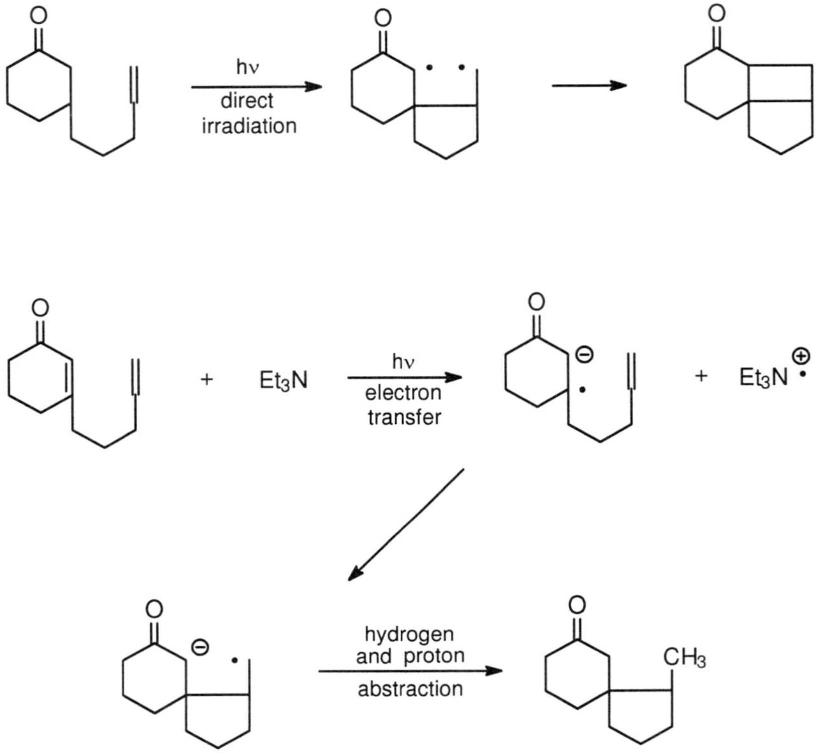

Scheme 8.5

Regioselectivity

Intermolecular Reaction

When an unsymmetrically substituted alkene adds to an α,β-unsaturated carbonyl compound, two regioisomers (head-to-head and head-to-tail) are possible (Scheme 8.6). Predicting the major product is not an easy task, however, since the identity of this product, according to the Schuster analysis, is determined by the manner in which the various 1,4-diradicals are partitioned between cyclization and return to starting material (Scheme 8.3). This partitioning is influenced by the structures of the reactants as well as other factors, including the solvent, temperature, reactant concentration, and presence of micelles.[15]

One generalization about regioselectivity that emerges from study of [2 + 2] photoaddition reactions is that electron-rich alkenes preferentially form head-to-tail isomers while electron-deficient ones often form head-to-head adducts (Scheme 8.6).[11,14,15] Another general feature arising from this infor-

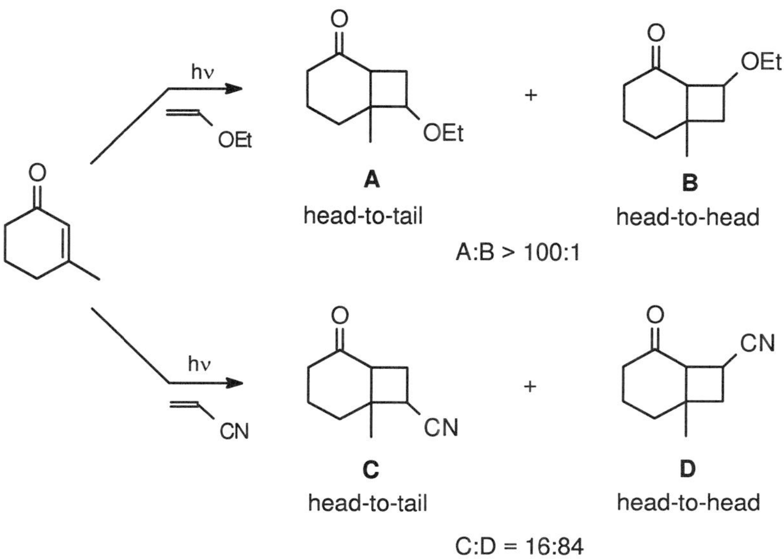

Scheme 8.6

mation is that even though one isomer may dominate, these reactions usually form mixtures of regioisomers. When attention is focused exclusively on carbohydrates, this second generalization is reinforced because the only reported intermolecular photocycloaddition of an unsymmetrical alkene to a carbohydrate produces an inseparable mixture of isomers (eq 2).[1] Otherwise, the question of regioselectivity in these reactions is not pertinent at present because all of the other reported reactions of alkenes adding to carbohydrates involve symmetrical alkenes (Table 8.1).

Intramolecular Reaction

In commenting on intermolecular photocycloaddition reactions, Oppolzer observed that the main impediment to the synthetic use of cycloaddition reactions is their generally poor regiochemical control.[2] He further noted that the regioselectivity of photochemical cycloaddition can be decidedly improved upon if the interacting π-systems are part of the same molecule. This fact is illustrated in the reaction shown in eq 3.[8] In addition to improved regioselectivity, another characteristic of intramolecular reaction is that whenever there is a choice between initial formation of a five-membered ring or a ring of a different size, the five-membered ring is preferred. This generalization, which was recognized simultaneously by two separate research groups, is known as "the rule of five".[24, 25] Maradyn and Weedon have trapped the diradicals arising from intramolecular cycloaddition and found that the diradicals predicted

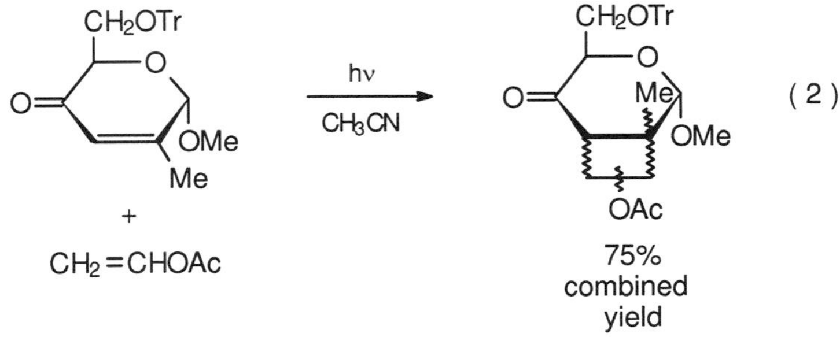

Equation 2

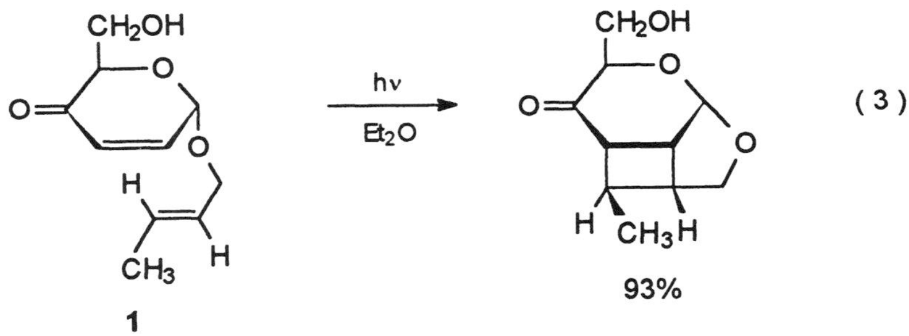

Equation 3

by the rule of five are formed faster than any others.[26] The reaction shown in Scheme 8.7 provides an example in which a carbohydrate cyclizes obeying the rule of five. While the examples of intramolecular photocycloaddition reactions involving carbohydrates are limited (Table 8.1), those that exist obey this rule and experience excellent regiochemical control.

Stereoselectivity

Photocycloaddition reactions present the possibility for generating a large number of stereoisomers. There is no simple means at present for predicting the stereoselectivity of this reaction; in fact, based upon analysis of a large number of reactions, Weedon has compiled a list of eight general statements, some of which contain multiple possibilities, designed to explain the type and number of stereoisomers formed from photocycloaddition reactions of enones to alkenes or alkynes.[14] Although the stereochemistry of the products from reactions of carbohydrates presents the same variety of possibilities, far fewer of these reactions have been investigated. With those that have been studied (Table 8.1), however, the many possibilities (e.g., cis and trans ring fusions, preferential addition to one face of a molecule) are apparent.

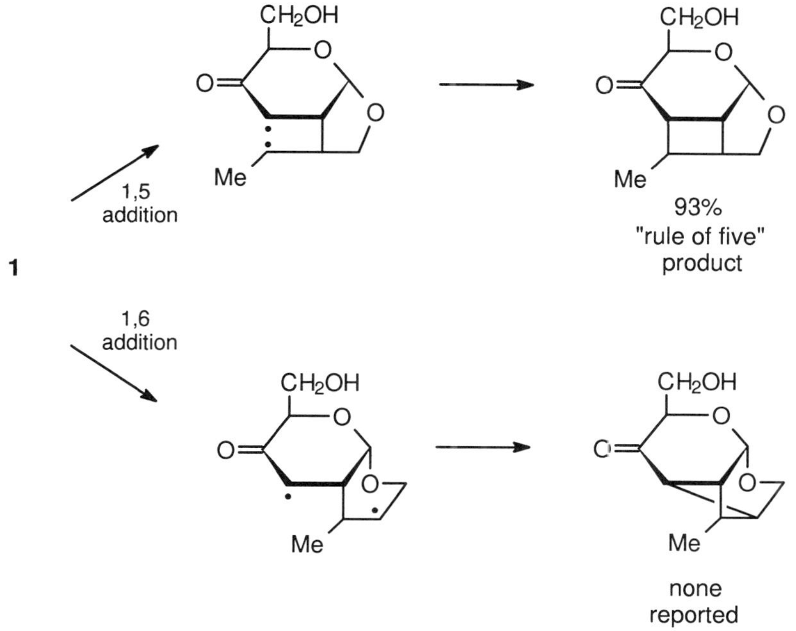

Scheme 8.7

Reaction with Singlet Oxygen

Ground and Excited States of Molecular Oxygen

The electronic configuration of molecular oxygen (dioxygen) in its ground state has the two highest energy electrons in two degenerate orbitals. In order to satisfy Hund's rule, the spins of these two electrons are unpaired; therefore, the ground state of molecular oxygen is a triplet (Figure 8.3). (The spectroscopic notation used for this state is $^3\Sigma$.) The first two excited states have paired spins and thus are both singlets ($^1\Delta$ and $^1\Sigma$) (Figure 8.3).[26,27] Between these two the $^1\Delta$ state is lower in energy and is the "singlet oxygen" ($^1O_2$) of importance in reactions of carbohydrates and other organic compounds. Singlet oxygen typically is generated by energy transfer from an excited sensitizer (eqs 4 and 5), a process that is thought to occur by electron exchange (also known as triplet–triplet annihilation) (Figure 8.4, eq 5).[27] Common sensitizers for this process are tetraphenylporphyrin, methylene blue, and Rose Bengal. Although singlet oxygen also can be generated in other ways (e.g., photochemical electron-transfer reaction[28] and thermal decomposition of endoperoxides[29]) for reactions involving carbohydrates the only photochemical technique that has been reported is the sensitization process.

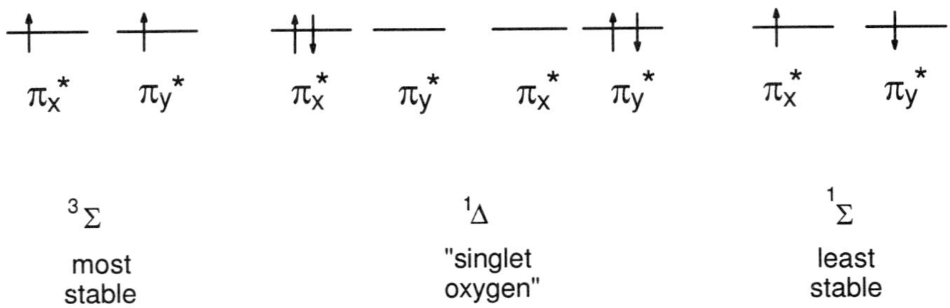

Figure 8.3. Electronic configuration of the highest populated energy levels for molecular oxygen.

$$\text{sensitizer} \xrightarrow[\text{2) intersystem crossing}]{\text{1) } h\nu} {}^{3}\text{sensitizer} \qquad (4)$$

Equation 4

$$^{3}\text{sensitizer} + {}^{3}O_2 \longrightarrow \text{sensitizer} + \underset{\text{singlet oxygen}}{{}^{1}O_2} \qquad (5)$$

Equation 5

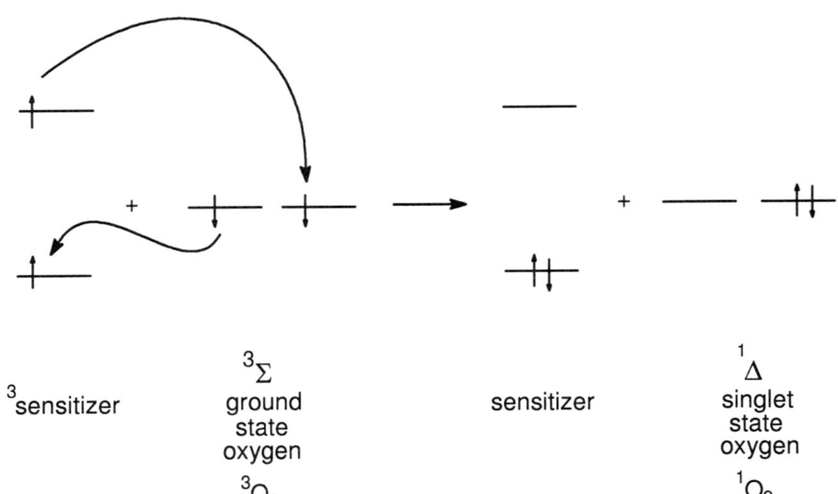

Figure 8.4. Formation of singlet oxygen by electron exchange

Basic Reaction Types

The three basic types of reaction that take place between singlet oxygen and unsaturated compounds are [2 + 2] addition, [4 + 2] addition, and the "ene" reaction. Only two of the three, the ene reaction (eq 6) and [4 + 2] addition (eq 7), are known when a carbohydrate is one of the reactants. Although the current discussion is about singlet oxygen, it is worth remembering that triplet (ground state) oxygen is an excellent radical scavenger and excited state quencher. It too is an important factor in some photochemical reactions of carbohydrates (see Chapter 5).

The Ene Reaction

A number of different mechanisms have been considered for the ene reaction,[30] which is sometimes referred to as the Schenck reaction.[31] It initially was thought to be a concerted process,[27] but the work of Hurst et al. established that reaction occurs in several steps.[28] These investigators found that the process begins with the reversible formation of an exciplex between the unsaturated substrate and singlet oxygen. This exciplex then irreversibly produces a perperoxide that decomposes to an intermediate whose structure depends upon the reaction conditions; for example, in polar solvents the intermediate is zwitterionic. This zwitterion then rearranges to the hydroperoxide photoproduct (Scheme 8.8).[32] Although the possibility of a concerted process has not been eliminated, further research has continued to support the perperox-

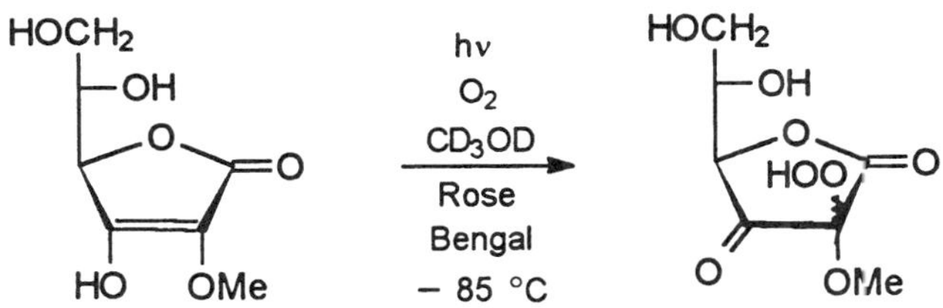

(6)

Equation 6

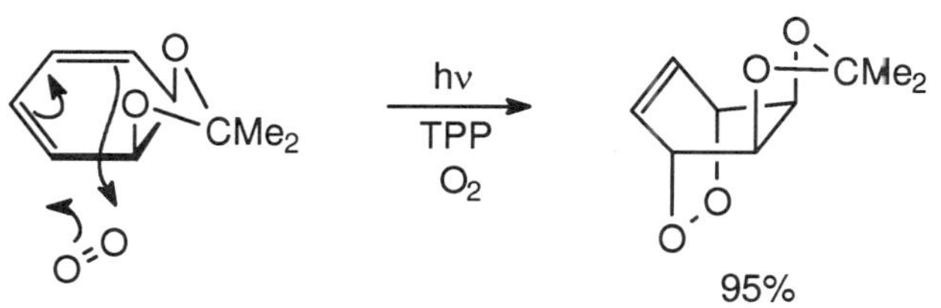

(7)

TTP = tetraphenylporphyrin

Equation 7

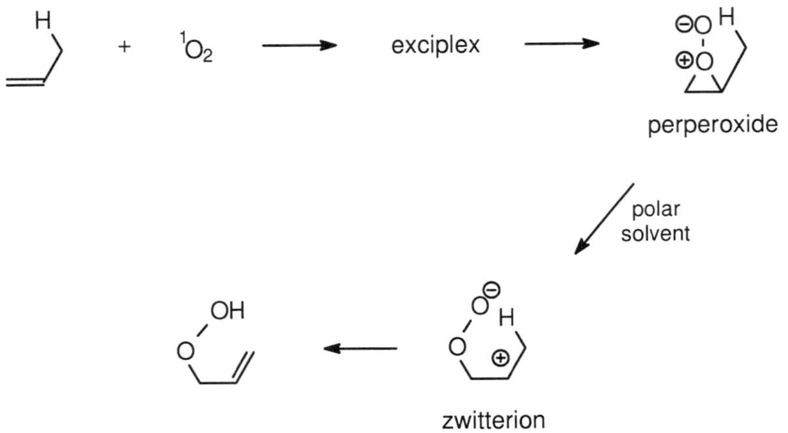

Scheme 8.8

ide mechanism.[33,34] The reported ene reactions involving carbohydrates are all of ascorbic acid and its derivatives and are given in Table 8.2.

[4 + 2] Cycloaddition

The addition of singlet oxygen to a diene, a process that has the characteristics of a Diels–Alder reaction, produces an endoperoxide (eq 7).[37–39] (Endoperoxides are often sufficiently unstable that they are difficult to isolate.) This [4 + 2] addition is generally considered to be a two-step process in which an exciplex initially forms between singlet oxygen and the diene prior to the cycloaddition reaction between these reactants. The stereoselectivity of the addition is sensitive to the substituents already present in the diene.[40] In eq 7, for example, singlet oxygen approaches the diene from the side of the molecule opposite the bulky isopropylidene group.[37–39] If the diene contains an allylic hydrogen atom, the possibility exists for both [4 + 2] addition and ene reaction to occur.[40]

In addition to the work of Kwon and Foote listed in Table 8.2, reactions of singlet oxygen that are related to carbohydrate chemistry have been reported by several research groups. Carless et al. demonstrated that photooxidation with singlet oxygen, which simultaneously creates two, carbon–oxygen bonds upon reaction with a diene, can be a valuable step in the synthesis of inositols (Scheme 8.9).[41] Carless and co-workers[41–45] and Shing and Tam[46] have applied this type of reaction with success in functionalizing a number of dienes. Diaz et al. have studied the addition of singlet oxygen to a variety of furans that contain carbohydrate substituents.[47–53] Photoproducts from these reactions are endoperoxides (Scheme 8.10) that generally are stable only at low temperature and rearrange to ring-opened structures at room temperature.[54] In reactions with furans the carbohydrate portion of the molecule does not appear to be involved in the photochemical process.

Table 8.2. Reactions of Unsaturated Carbohydrates with Singlet Oxygen.

Reactant	*Product*	*Percent Yield*	*Ref.*
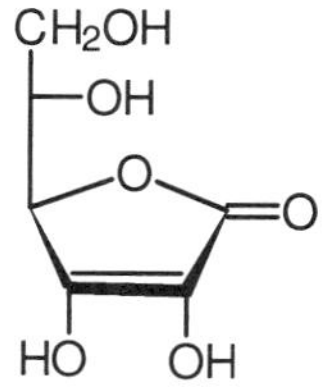	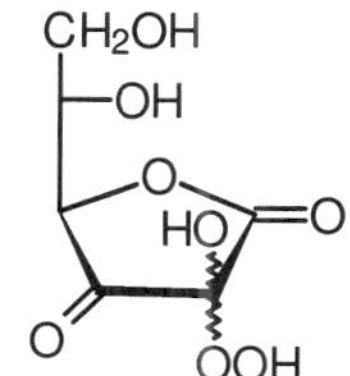 +	65[a,b]	35
	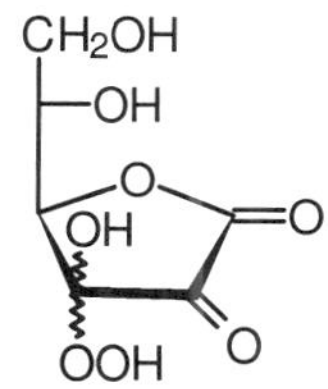	35[a,b]	
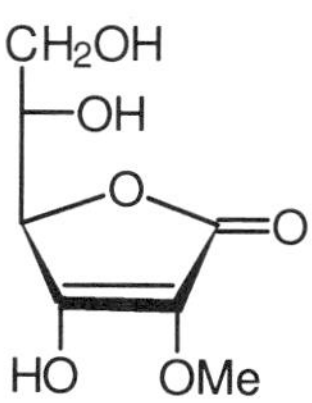	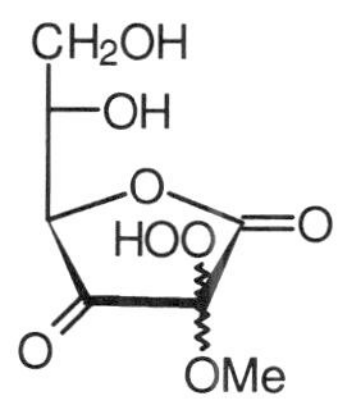	a,b,c	35
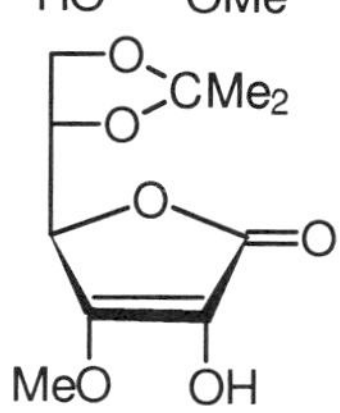	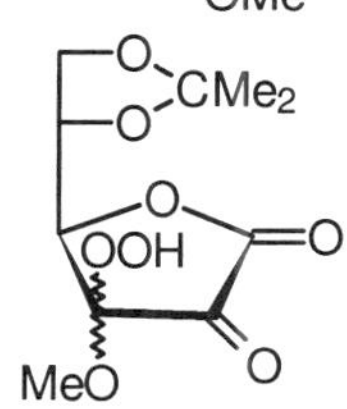	80[a,b]	36
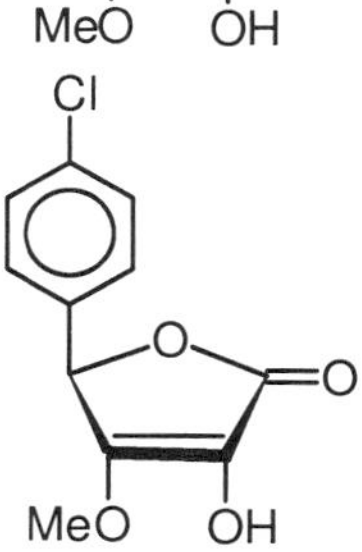	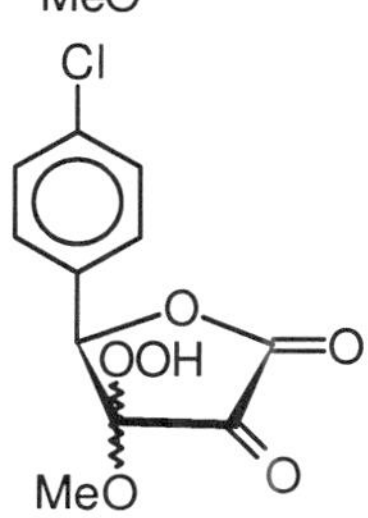	80[a,b]	36

[a]Reaction was conducted at -85 °C.
[b]Rose Bengal was used as a sensitizer.
[c]The yield was not reported.

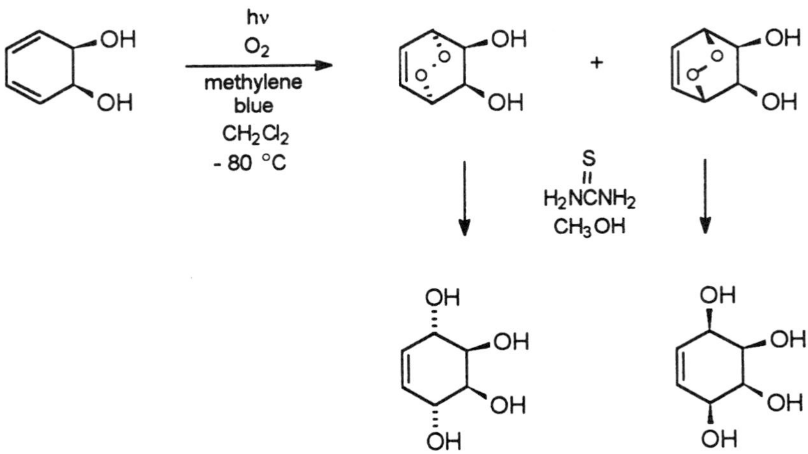

Scheme 8.9

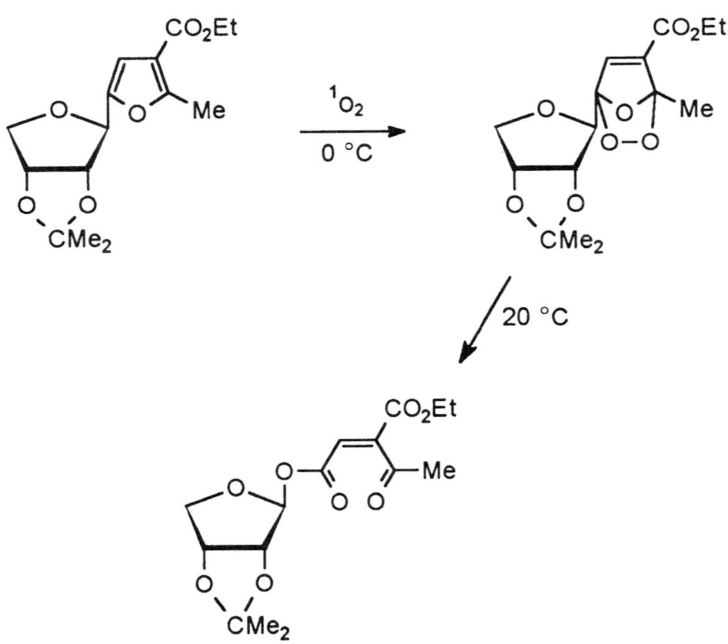

Scheme 8.10

Geometrical Isomerization

Acyclic Alkenes

Geometrical isomerization of a carbon–carbon double bond is one of the most general and well-documented reactions in organic photochemistry. Probably more is known about the mechanistic details of this type of reaction than any other photochemical process. (Gilbert and Baggot provide an excellent summary of the mechanistic understanding of this reaction.[55]) This reaction is less important and much less extensively studied in carbohydrate chemistry because acyclic compounds and medium to large ring systems—structural types where geometrical isomerization is readily achieved—are relatively rare in carbohydrates. Several photochemical, geometrical isomerization reactions of acyclic carbohydrates have been studied (Table 8.3).

Cyclic Alkenes

Photochemical, geometrical isomerization also can occur in cyclic systems but only if the ring is large enough to accommodate a *trans* double bond. Such isomerization is not possible for rings with fewer than six atoms. For rings with eight or more atoms (C, N, O) both *cis* and *trans* isomers are sufficiently stable that they can be isolated.[60] When a ring contains six or seven atoms, only the *cis* isomer can be isolated; however, extremely reactive intermediates, believed to be *trans* isomers, are formed during irradiation.[61–64] These intermediates react rapidly with alcohols at room temperature to form products consistent with ionic addition. An example of such a reaction is shown in eq 8.[65] Scheme 8.11 contains a proposed mechanism for this reaction; however, since no mechanistic studies have been conducted on this process, the mechanism is a tentative one.

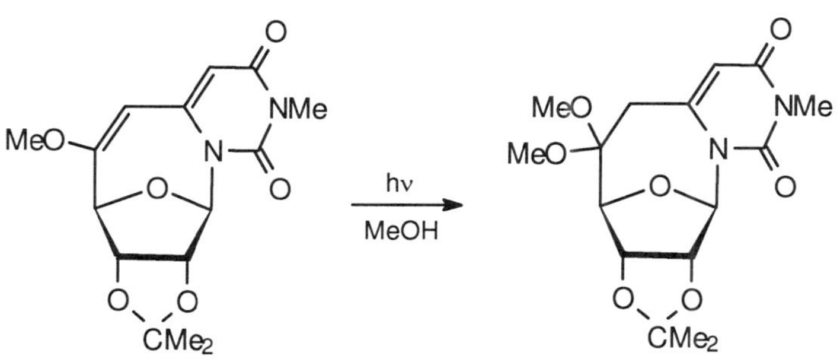

(8)

Equation 8

Table 8.3. Photochemical Geometrical Isomerization

Reactant	*Product*	*Percent Yield*	*Ref.*
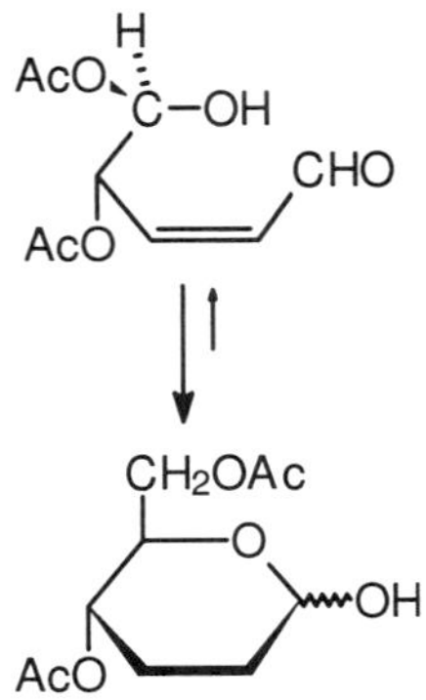	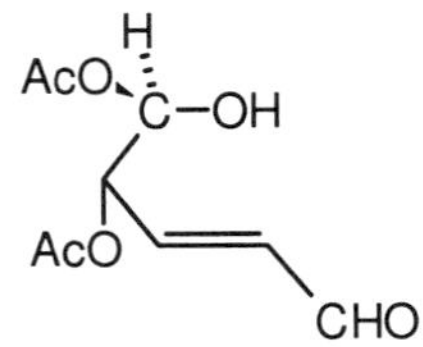	a	56, 57
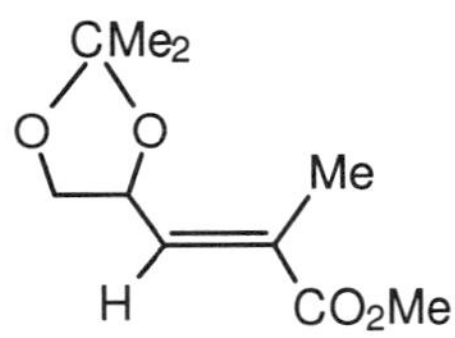	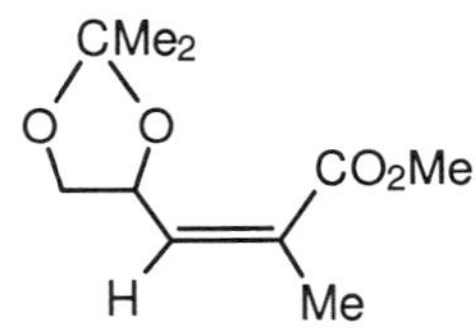	72[b]	58
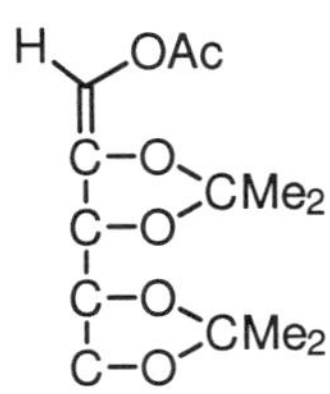	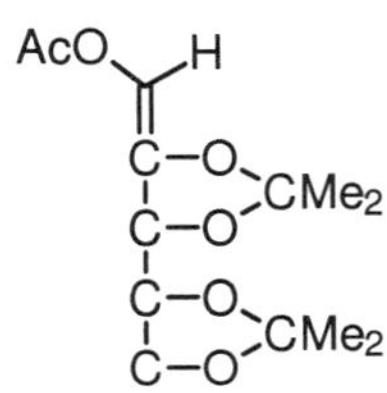	a	59
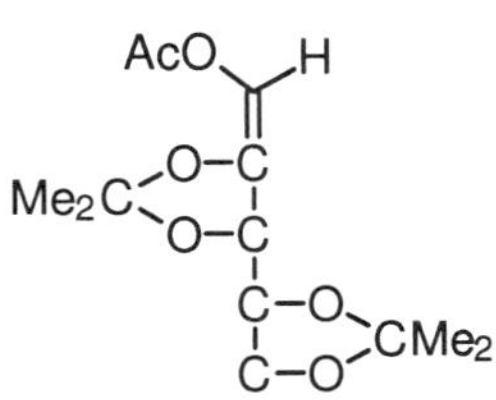	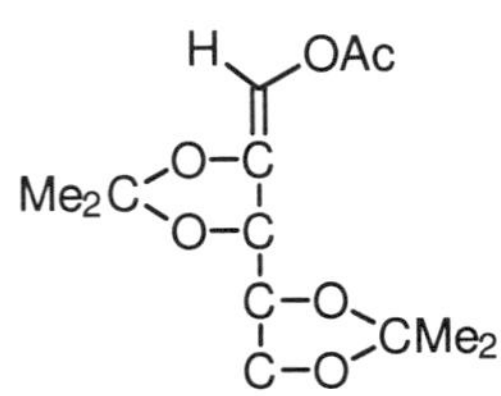	a	59

[a]The yield was not reported.

[b]The product shown was not isolated. It experienced a nonphotochemical cyclization after isomerization.

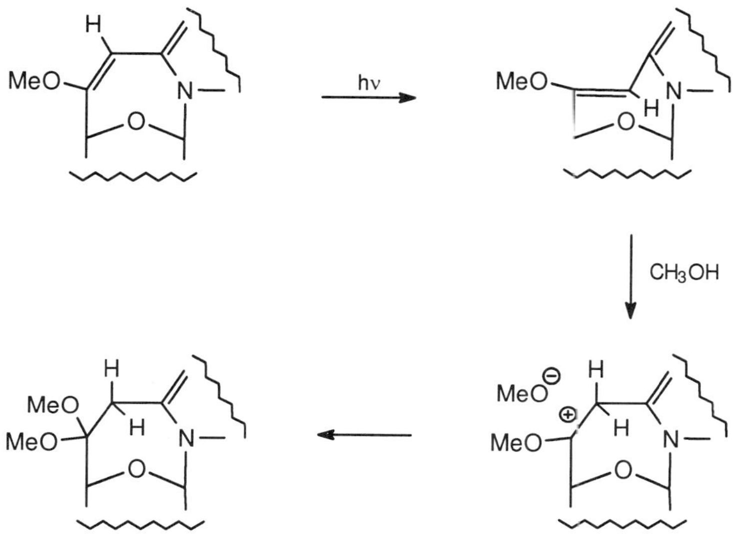

Scheme 8.11

Azodicarboxylates: [4 + 2] Cycloaddition

Leblanc and co-workers have studied [4 + 2] photocycloaddition reactions of furanoid and pyranoid glycals with azodicarboxylates.[66–70] These authors were interested in this reaction due to its potential in synthesizing new 2-amino-2-deoxy sugars. Mechanistic study has shown that the role of light in this reaction is to generate the *cis* isomer of the azo compound.[71] The *cis* isomer is highly reactive and adds to the glycal thermally at room temperature (Scheme 8.12).

Nitrene Cycloaddition Reactions

Photochemical reaction of an azide causes loss of nitrogen and formation of a nitrene that then can add to a carbon–carbon double bond.[72] When the double bond is part of a glycal, the azridine formed is unstable and fragments to give ring-opened compounds.[73] Kozlowska-Gramsz and Descotes have used this reaction for synthesis of aminodeoxy sugars (Scheme 8.13).[73,74]

Summary

α,β-Unsaturated carbonyl compounds undergo [2 + 2] photocycloaddition with alkenes to form cyclobutane derivatives (Table 8.1). Typically, many iso-

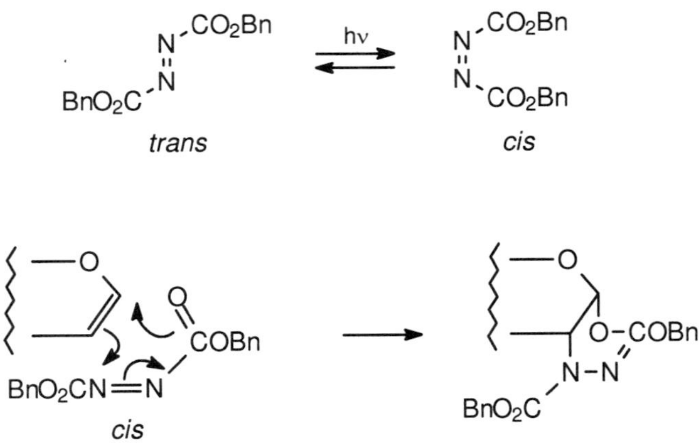

Scheme 8.12

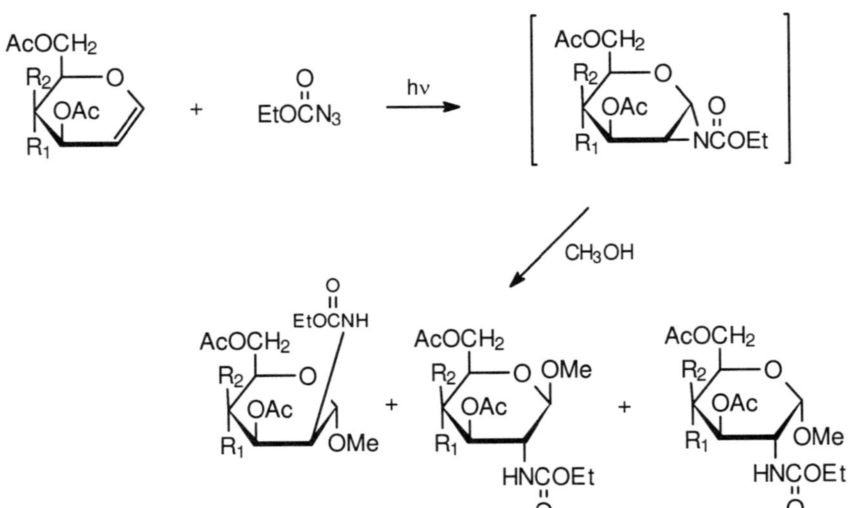

R_1 = H, R_2 = OAc	5%	26%	11%
R_1 = OAc, R_2 = H	4%	4%	34%

Scheme 8.13

mers are possible from such reactions, and it is generally difficult to predict what mixture of regio- and stereoisomers will be formed. Singlet oxygen reacts with unsaturated carbohydrates by an ene reaction to produce hydroperoxides (Table 8.2). Geometrical isomerization occurs in carbohydrates but it is not as common a process as with other organic compounds (Table 8.3).

References

1. Hicks, D. R.; Primeau, J. L.; Fraser-Reid, B. *Carbohydr. Res.* **1982**, *108*, 41.
2. Oppolzer, W. *Acc. Chem. Res.* **1982**, *15*, 135.
3. Gilbert, A.; Baggott, J. *Essentials of Molecular Photochemistry*; Blackwell Scientific: London, UK, 1991; p 271.
4. Collins, P. M.; Whitton, B. *Carbohydr. Res.* **1972**, *21*, 487.
5. Collins, P. M.; Whitton, B. R. *J. Chem. Soc., Perkin Trans. 1* **1973**, 1470.
6. Matsui, T.; Morooka, T.; Nakayama, M. *Bull. Chem. Soc. Jpn.* **1987**, *60*, 417.
7. Matsui, T.; Kawano, Y.; Nakayama, M. *Chem. Express* **1990**, *5*, 697.
8. Tenaglia, A.; Barillé, D. *Synlett* **1995**, 776.
9. Ferrier, R. J.; Prasit, P.; Gainsford, G. J.; Le Page, Y. *J. Chem. Soc., Perkin Trans. 1* **1983**, 1635.
10. Helferich, B.; von Gross, E. *Chem. Ber.* **1952**, *85*, 531.
11. Becker, D.; Haddad, N. *Org. Photochem.* **1989**, *10*, 1.
12. Hoffmann, N.; Buschmann, H.; Raabe, G.; Scharf, H.-D. *Tetrahedron* **1994**, *50*, 11167.
13. Baldwin, S. W. *Org. Photochem.* **1981**, *15*, 123.
14. Weedon, A. C. In *Synthetic Organic Photochemistry;* Horspool, W. M., Ed.; Plenum: New York, 1984; p 61.
15. Demuth, M.; Mikhail, G. *Synthesis* **1989**, 145.
16. Müller, F.; Mattay, J. *Chem. Rev.* **1993**, *93*, 99.
17. Schuster, D. I.; Lem, G.; Kaprinidis, N. A. *Chem. Rev.* **1993**, *93*, 3.
18. Schuster, D. I. In *The Chemistry of the Enones;* Patai, S.; Rappoport, Z.; Eds.; Wiley & Sons: Chichester, UK, 1989; p 623.
19. Corey, E. J.; Bass, J. D.; Le Mathieu, R.; Mitra, R. B. *J. Am. Chem. Soc.* **1964**, *86*, 5570.
20. Loutfy, R. O.; de Mayo, P. *J. Am. Chem. Soc.* **1977**, *99*, 3559.
21. Hastings, D. J.; Weedon, A. C. *J. Am. Chem. Soc.* **1991**, *113*, 8525.
22. Broeker, J. L.; Eksterowicz, J. E.; Belk, A. J.; Houk, K. N. *J. Am. Chem. Soc.* **1995**, *117*, 1847.
23. Bischof, E. W.; Mattay, J. *Tetrahedron Lett.* **1990**, *31*, 7137.
24. Srinivasan, R.; Carlough, K. H. *J. Am. Chem. Soc.* **1967**, *89*, 4932.
25. Liu, R. S. H.; Hammond, G. S. *J. Am. Chem. Soc.* **1967**, *89*, 4936.
26. Maradyn, D. J.; Weedon, A. C. *J. Am. Chem. Soc.* **1995**, *117*, 5359.
27. Turro, N. J. *Modern Molecular Photochemistry*; Benjamin/Cummings: Menlo Park, CA, 1978; p 583.
28. Mattay, J. *Synthesis* **1989**, 233.
29. DiMascio, P.; Sies, H. *J. Am. Chem. Soc.* **1989**, *111*, 2909.
30. Frimer, A. A. *Chem. Rev.* **1979**, *79*, 359.
31. Schenck, G. O.; Eggert, H.; Denk, W. *Ann. Chem.* **1965**, 177.
32. Hurst, J. R.; Wilson, S. L.; Schuster, G. B. *Tetrahedron* **1985**, *41*, 2191.
33. Orfanopoulos, M.; Smonou, I.; Foote, C. S. *J. Am. Chem. Soc.* **1990**, *112*, 3607.
34. Poon, T. H. W.; Pringle, K.; Foote, C. S. *J. Am. Chem. Soc.* **1995**, *117*, 7611.
35. Kwon, B.-M.; Foote, C. S. *J. Am. Chem. Soc.* **1988**, *110*, 6582.
36. Kwon, B.-M.; Foote, C. S.; Khan, S. I. *J. Am. Chem. Soc.* **1989**, *111*, 1854.
37. Sütbeyaz, Y.; Seçen, H.; Balci, M. *J. Chem. Soc., Chem. Commun.* **1988**, 1330.

38. Seçen, H.; Sütbeyaz, Y.; Balci, M. *Tetrahedron Lett.* **1990**, *31*, 1323.
39. Seçen, H.; Güttekin, S.; Sütbeyaz, Y.; Balci, M. *Synthetic Commun.* **1994**, *24*, 2103.
40. Clennan, E. I. *Tetrahedron* **1991**, *47*, 1343.
41. Carless, H. A. J.; Billinge, J. R.; Oak., O. Z. *Tetrahedron Lett.* **1989**, *30*, 3113.
42. Carless, H. A. J.; Oak, O. Z. *Tetrahedron Lett.* **1989**, *30*, 1719.
43. Carless, H. A. J.; Malik, S. S. *Tetrahedron Asymmetry* **1992**, *3*, 1135.
44. Carless, H. A. J.; Busia, K. *Tetrahedron Lett.* **1990**, *31*, 1617.
45. Carless, H. A. J.; Busia, K. *Carbohydr. Res.* **1992**, *234*, 207.
46. Shing T. K. M.; Tam, E. K. W. *Tetrahedron Asymmetry* **1996**, *7*, 353.
47. Aparicio, F. J. L.; Diaz, R. R.; Garcia, J. I.; Calvo-Flores, F. G. *Carbohydr. Res.* **1986**, *148*, 235.
48. Aparicio, F. J. L.; Diaz, R. R.; Lopez-Espinosa, M. T. P.; Rojas, A. M. P. *An. Quim., Ser. C* **1986**, *82*, 179.
49. Diaz, R. R.; Calvo-Flores, F. G.; Guardia, L. A.; Aparicio, F. J. L. *Carbohydr. Res.* **1989**, *191*, 209.
50. Aparicio, F. J. L.; Lopez-Espinosa, M. T. P.; Diaz, R. R. *An. Quim., Ser. C* **1984**, *80*, 215.
51. Aparicio, F. J. L.; Lopez-Espinosa, M. T. P.; Diaz, R. R. *An. Quim., Ser. C* **1984**, *80*, 156.
52. Aparicio, F. J. L.; Sastre, J. A. L.; Garica, J. I.; Diaz, R. R. *Carbohydr. Res.* **1984**, *132*, 19.
53. Aparicio, F. J. L.; Lopez-Espinosa, M. P.; Diaz, R. R. *Carbohydr. Res.* **1984**, *132*, 233.
54. K. Gollnick and A. Griesbeck, *Tetrahedron* **41**, 2057 (1985).
55. Gilbert, A.; Baggott, J. *Essentials of Molecular Photochemistry*; Blackwell Scientific: London, UK, 1991; p 230.
56. Fraser-Reid, B.; Radatus, B. *J. Am. Chem. Soc.* **1970**, *92*, 5288.
57. Card, P. J. *Carbohydr. Res.* **1982**, *104*, 338.
58. López-Herrera, F. J.; Sarabia-García, F.; M. S. Pino-González, M. S.; García-Aranda, J. F. *J. Carbohydr. Chem.* **1994**, *13*, 767.
59. Ducruix, A.; Pascard-Billy, C.; Eitelman, S. J.; Horton, D. *J. Org. Chem.* **1976**, *41*, 2652.
60. J. S. Swenton, *J. Org. Chem.* **34**, 3217 (1969).
61. P. J. Kropp, *Org. Photochem.* **1979**, *4*, 1.
62. Verbeek, J.; van Lenthe, J. H.; Timmermans, P. J. J. A.; Mackor, A.; and Budzelaar, P. H. M. *J. Org. Chem.* **1987**, *52*, 2955.
63. Inoue, Y.; Takamuku, S.; Sakurai, H. *J. Chem. Soc., Perkin Trans. 2* **1977**, 1635.
64. Inoue, Y.; Ueoka, T.; Kuroda, T.; Hakushi, T. *J. Chem. Soc., Perkin Trans. 2* **1983**, 983.
65. Otter, B. A.; Falco, E. A. *Tetrahedron Lett.* **1978**, 4383.
66. Leblanc, Y.; Labelle, M. In *Cycloaddition Reactions in Carbohydrate Chemistry;* Giuliano, R. M., Ed.; American Chemical Society: Washington, DC, 1992; p 81.
67. Fitzsimmons, B. J.; Leblanc, Y.; Rokach, J. *J. Am. Chem. Soc.* **1987**, *109*, 285.
68. Leblanc, Y.; Fitzsimmons, B. J.; Springer, J. P.; Rokach, J. *J. Am. Chem. Soc.* **1989**, *111*, 2995.
69. Leblanc, Y.; Fitzsimmons, B. J. *Tetrahedron Lett.* **1989**, *30*, 2889.
70. Grondin, R.; Leblanc, Y.; Hoogsteen, K. *Tetrahedron Lett.* **1991**, *32*, 5021.
71. Koerner von Gustorf, E.; White, D. V.; Kim., B.; Hess, D.; Leitich, J. *J. Org. Chem.* **1970**, *35*, 1155.
72. Dürr, H.; Kober, H. *Topics Curr. Chem.* **1976**, *66*, 89.
73. Kozlowska-Gramsz, E.; Descotes, G. *Tetrahedron Lett.* **1981**, *22*, 563.
74. Kozlowska-Gramsz, E.; Descotes, G. *Can. J. Chem.* **1982**, *60*, 558.

9

Addition of Radicals to Carbon–Carbon Double Bonds

Introduction

Photochemically promoted addition of organic radicals to unsaturated carbohydrates is an attractive and often-used process for the synthesis of extended- and branched-chain sugars. Two examples of situations in which this type of reaction has proven useful are the synthesis of pillarose, the sugar component of the antitumor agent pillaromycin,[1, 2] and the synthesis of dihydroshowdomycin, the hydrogenation product of the nucleoside antibiotic showdomycin.[3] Photochemically promoted addition reactions also have become part of new routes to cyclitol derivatives[4] and to annulated pyranosides.[5] Since a radical adding to an unsaturated carbohydrate may contain a nitrogen, sulfur, phosphorous, or halogen atom, photoaddition can be a convenient method for introducing any of these heteroatoms into a carbohydrate structure.

Photochemical reaction actually is only one of several methods for initiating the addition of a radical to an unsaturated carbohydrate. In a photochemically promoted reaction an excited ketone, typically acetone or benzophenone, abstracts a hydrogen atom from a compound present in the reaction mixture (often the solvent). This abstraction generates a radical that then adds to the unsaturated carbohydrate. Examples of addition reactions involving excited acetone and benzophenone are shown in eqs 1[6] and 2,[1, 2] respectively.

A problem arises at this point with respect to the requirements for a photochemical reaction of a carbohydrate. If the photochemical component of a reaction under consideration consists of excited acetone or benzophenone abstracting a hydrogen atom from an organic compound not a carbohydrate, should such a reaction be considered to be a photochemical reaction of a carbohydrate? The answer to the question probably is no; the reaction is better described as a radical addition that happens to be initiated photochemically.

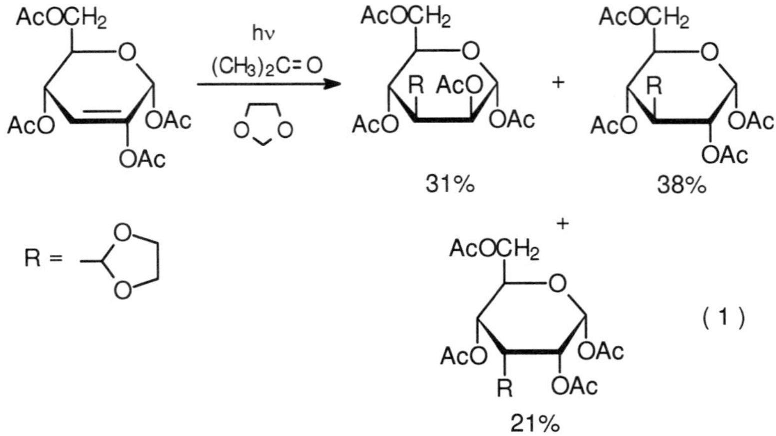

(1)

Equation 1

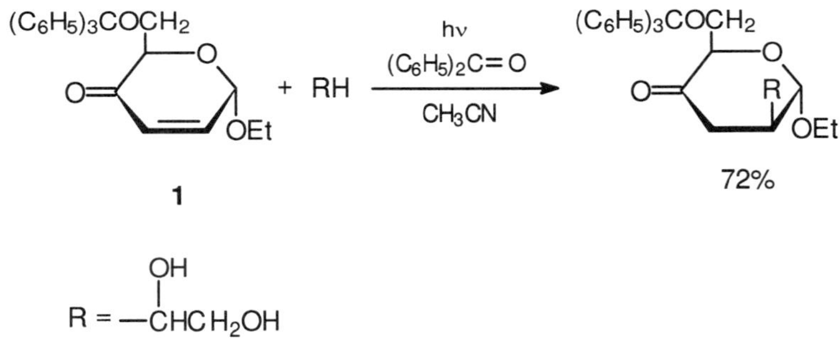

(2)

Equation 2

However, completely "passing over" such reactions does not seem advisable, since they represent some of the more important reactions in carbohydrate chemistry that involve photolysis. We have decided to adopt an intermediate position with respect to treatment of these reactions. Here, and in later chapters where this problem arises, photochemically initiated, radical addition reactions will be discussed as completely as other photochemical reactions, but tables of these reactions will not be included.

Reaction Mechanism

The mechanism for photochemically promoted radical addition to unsaturated carbohydrates has been studied by several research groups.[6–13] The most thorough of these investigations was conducted by Benko et al., who examined the addition of the hydroxymethyl radical to four α,β-unsaturated ketones (**1**–

4, Chart 9.1).[13] These investigations led to a number of mechanistically significant findings about photochemically promoted, radical-addition reactions. For example, the quantum yields for reaction of these compounds ranged from 0.14 to 0.18; consequently, there was inefficiency in radical formation or addition or both. Furthermore, the proposed intermediacy of the radical **6** (Scheme 9.1) in the reaction of **2** was supported by detection of **6** by ESR spectroscopy of the photolysis mixture. Since hydrogen-abstracting agents other than excited benzophenone (e.g., the *t*-butoxy radical) caused the same addition reaction to occur, the role of the excited ketone as a radical-generating agent became firmly established. Based upon these and other observations, including the regio- and stereoselectivity of the addition process, Benko et al. proposed the mechanism shown in Scheme 9.1.[13] These investigators also considered the alternate possibility that excited benzophenone was transferring energy to **2** and that the observed reaction actually was taking place from excited **2** (Scheme 9.2), but they rejected this alternative because the energy-transfer process would be an endothermic one and inefficient when compared to hydrogen abstraction.

Another argument against the energy-transfer option (Scheme 9.2) is that direct irradiation of **2** in methanol (no benzophenone present) produces none of the photoaddition product **5**.[13] Although this observation renders the energy-transfer option unlikely, it does not completely eliminate it. Since direct excitation produces a singlet excited state and energy transfer (from benzophenone) generates a triplet state, the possibility exists that these excited states have different reactivities; that is, the triplet state reacts to give **5** but the singlet state does not. Although such a situation would explain the observed results, it is not a likely scenario, since the lowest singlet excited state of **2** should undergo rapid intersystem crossing to the lowest triplet state.

Regioselectivity

Definition of α- and β-Substituents

Photochemically promoted radical addition to carbon–carbon double bonds is often a quite regioselective process. Giese has analyzed the addition of carbon

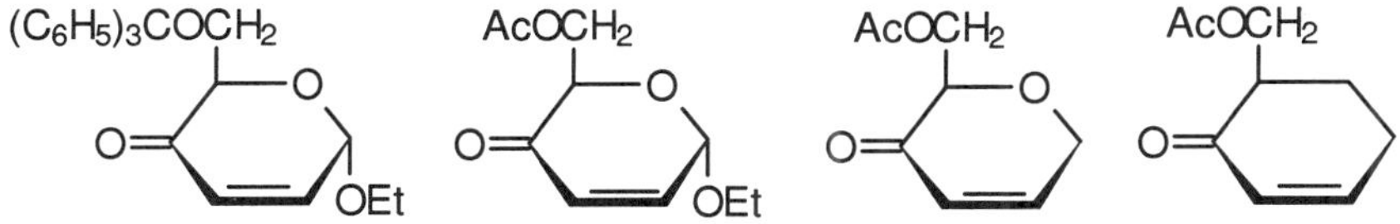

1 2 3 4

Chart 9.1. α,β-Unsaturated carbonyl compounds

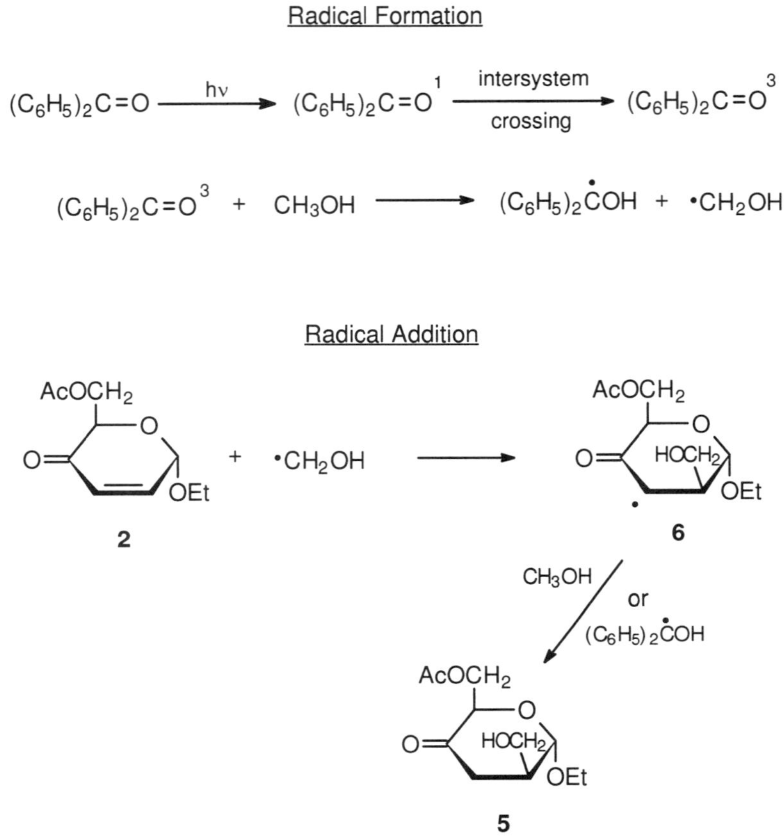

Scheme 9.1

radicals to alkenes and formulated several generalizations concerning how substituents effect reactivity.[14] In making these generalizations he has used the definitions shown in eq 3 for α and β substituents; that is, the substituent on the alkene carbon atom that becomes bonded to the radical center is the α-substituent. The β-substituent is attached to the carbon atom that becomes a radical center as a result of reaction.

Effect of β-Substituents on Regioselectivity

Drawing on work of other investigators and his own research, Giese concluded that β-substituents influence radical addition only by polar effects.[15–17] In explaining this observation he noted that alkyl radical addition to a double bond usually is quite exothermic because a π-bond is replaced by a σ-bond (i.e., a weaker bond by a stronger bond). According to Hammond's postu-

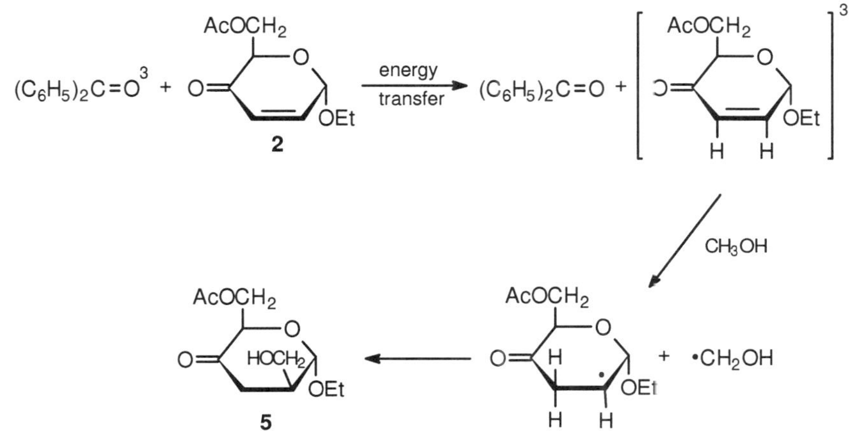

Scheme 9.2

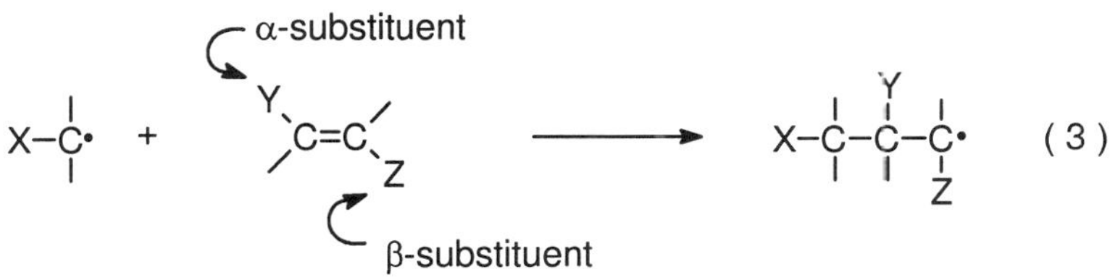

Equation 3

late,[18] the transition state for such a reaction should occur early in the reaction when bond making and bond breaking are just beginning. Under these conditions a β-substituent is far enough removed from the bonding atoms that its steric effects are inconsequential. Giese proposed that since β-substituents exert only a polar influence on free-radical addition reactions, their effect could be analyzed in terms of frontier orbital theory.

In discussing application of frontier orbital theory to the addition of radicals to alkenes, Giese summarized the salient points by recalling that "to a first approximation, this [frontier orbital] theory states that the energy differences between the highest occupied molecular orbitals (HOMO's) and the lowest unoccupied molecular orbitals (LUMO's) of the reacting species are decisive in determining rate variations.[19–21] The smaller the energy difference between these frontier orbitals, the larger is the stabilizing effect when the reactants approach one another. The frontier orbital of a free radical is the SOMO (singly occupied molecular orbital). Thus, the interaction between the SOMO of the free radical and the LUMO and HOMO of the alkene should play a decisive role in the interpretation and prediction of polar effects."[14]

Fleming[21] has described the basis for understanding the frontier orbital interactions mentioned by Giese[14] by showing that the SOMO of a radical interacts with both the HOMO and the LUMO of an alkene reactant (Figures 9.1 and 9.2). The SOMO–LUMO interaction clearly leads to stabilization because the orbital containing the single electron is lowered in energy as a result of interaction (Figure 9.1). SOMO–HOMO interaction also results in stabilization because there are two electrons in the lower energy orbital and only one in the higher energy orbital (Figure 9.2). Since the interaction between two orbitals increases as the orbitals become closer in energy, a radical with a high energy SOMO will react faster with a molecule having a low energy LUMO. (Compare Figures 9.1 and 9.3.) Likewise, a radical with a low energy SOMO will react more rapidly with a molecule having a high energy HOMO. (Compare Figures 9.2 and 9.4.)

Both Fleming[21] and Giese[14] have pointed out that introduction of an electron-donating group (e.g., an alkyl, alkoxy, or amino group) into an alkyl radical increases the SOMO energy. [This increase in SOMO energy can be attributed to the interaction between an orbital on carbon containing a single electron and a neighboring, nonbonding orbital on a heteroatom (Figure 9.5). The orbital interaction causes the radical to become more stable even though the SOMO increases in energy.] One consequence of this increase is that, as mentioned in the previous paragraph, SOMO–LUMO interaction becomes more important. If an electron-withdrawing substituent then is introduced into the alkene, the SOMO–LUMO energy difference becomes still smaller (an electron-withdrawing substituent lowers the LUMO energy[22]) and, consequently, the stabilizing SOMO–LUMO interaction becomes larger (Figure

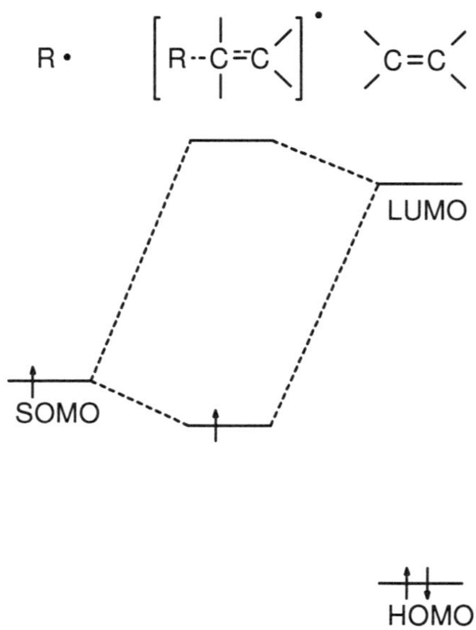

Figure 9.1. SOMO–LUMO interaction in radical addition to an alkene

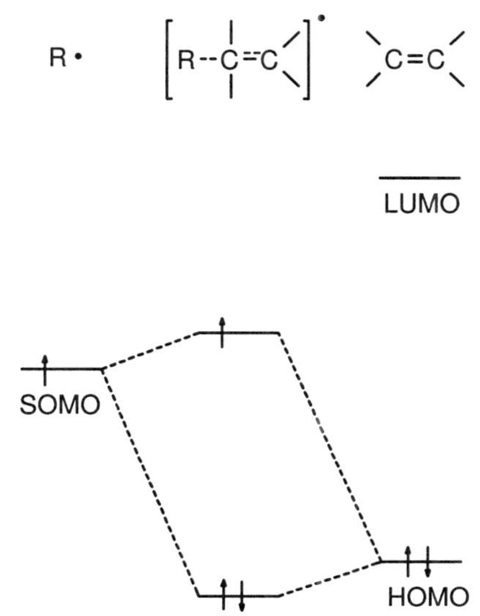

Figure 9.2. SOMO–HOMO interaction in radical addition to an alkene

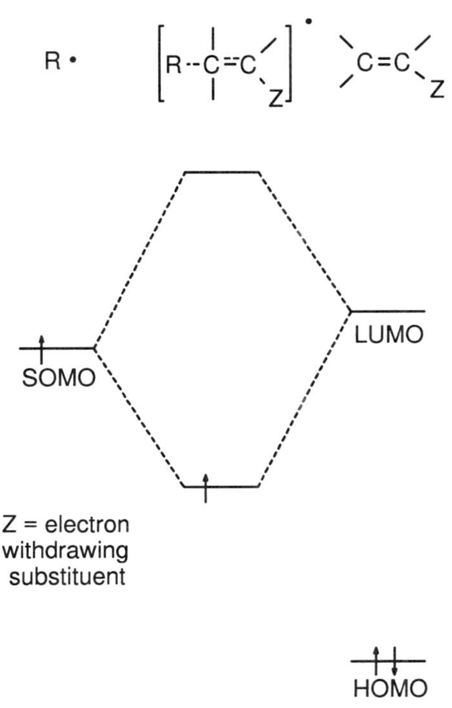

Figure 9.3. Interaction between SOMO and LUMO orbitals of similar energy

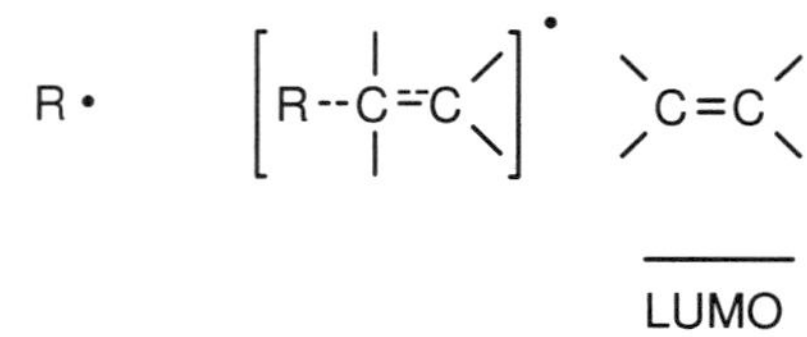

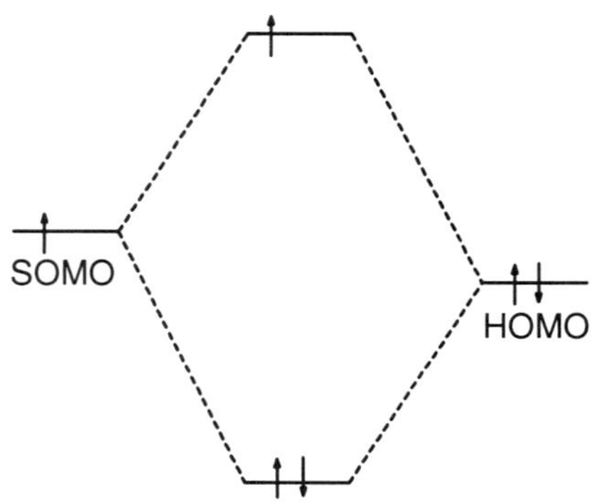

Figure 9.4. Interaction between SOMO and HOMO orbitals of similar energy

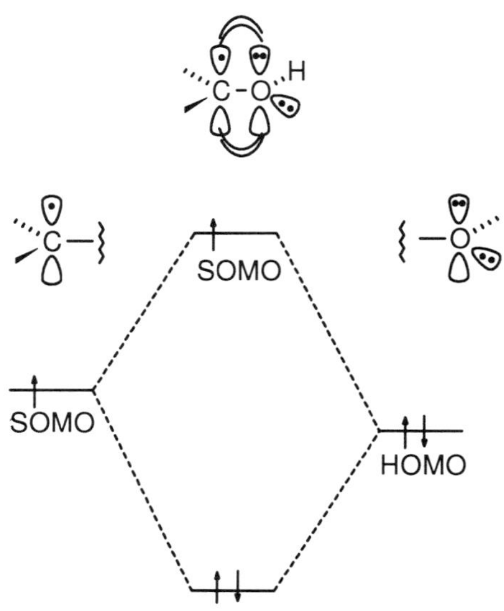

Figure 9.5. Effect on the SOMO of interaction with an oxygen atom

9.6). This stabilizing interaction increases the rate of reaction. Since reaction between the radical and the alkene is more rapid when the alkene is electron-deficient, the radical is described as being nucleophilic.

Factors that dramatically stabilize a radical can lower its SOMO energy to the point that the most important interaction is with the HOMO of the alkene (Figure 9.4). This situation is typical of nitrogen-, oxygen- and sulfur-centered radicals.[23] In these cases electron-withdrawing substituents on the double bond serve to increase the difference in energy between the frontier orbitals which, in turn, decreases the stabilizing interaction (Figure 9.7). Since electron-withdrawing substituents on the alkene reduce the rate of reaction with highly stabilized radicals, the radicals themselves are considered to be electrophilic.

Effect of α-Substituents on Regioselectivity

α-Substituents exert both polar and steric effects on radical-addition reactions. Steric effects are usually the more important of the two in determining regioselectivity when a substituent is directly attached to a reacting atom. The

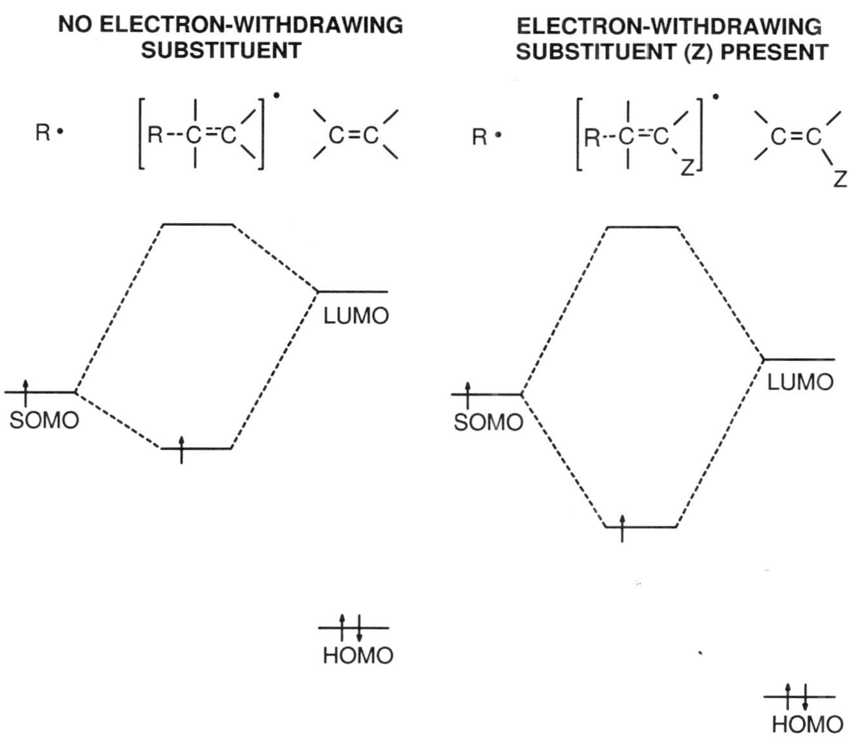

Figure 9.6. The stabilizing effect of an electron-withdrawing substituent on a SOMO energy level

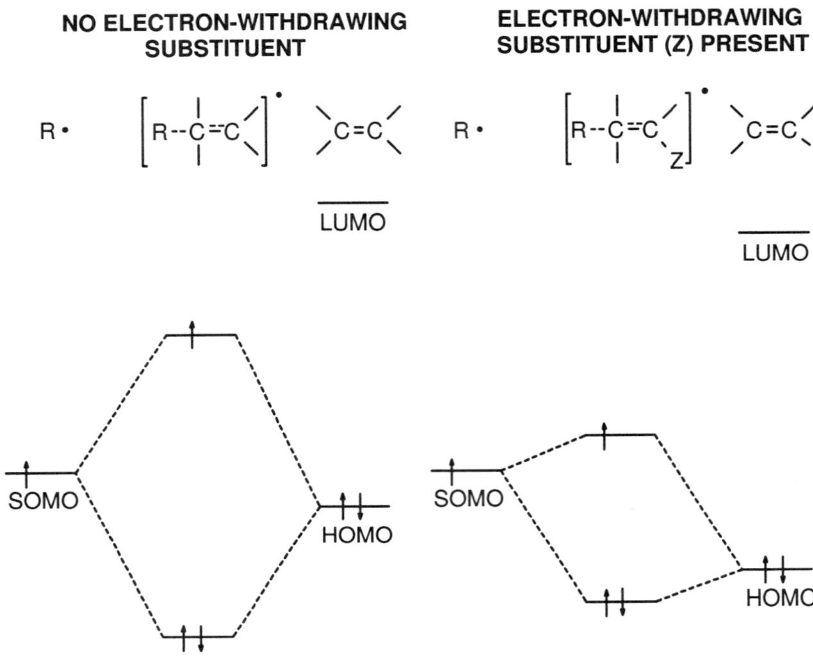

Figure 9.7. Reduction in stabilization caused by the introduction of an electron-withdrawing group

decisive influence of steric effects is overcome when the substituent is a carbonyl or nitro group. In these instances polar effects become dominant (see the section "Compounds with a Carbonyl or Nitro Group in Conjugation with a Double Bond").

Regioselective Addition to Unsaturated Carbohydrates

Compounds in Which a Double Bond Contains a Methylene Group

When a double bond in an unsaturated carbohydrate contains a methylene group, steric effects are the most important factors in determining regioselectivity. In these situations addition of a radical to the methylene group occurs in a highly regioselective fashion. Examples of this type of reaction involving a monosubstituted and a disubstituted double bond are shown in eqs 4[24] and 5,[8] respectively. These are two of many instances of photochemically initiated, radical addition to carbohydrates with methylene groups.[25–43]

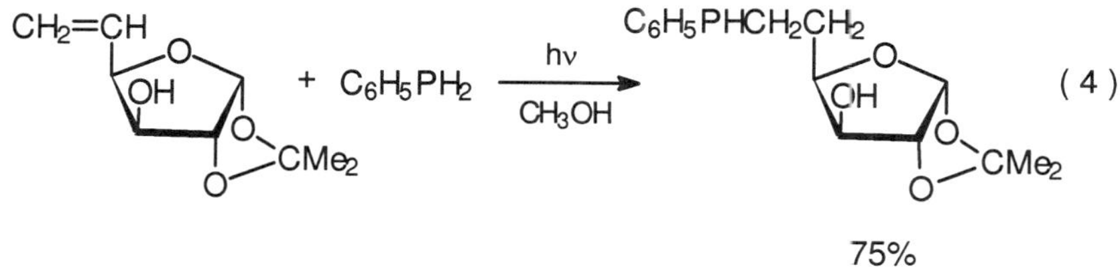

Equation 4

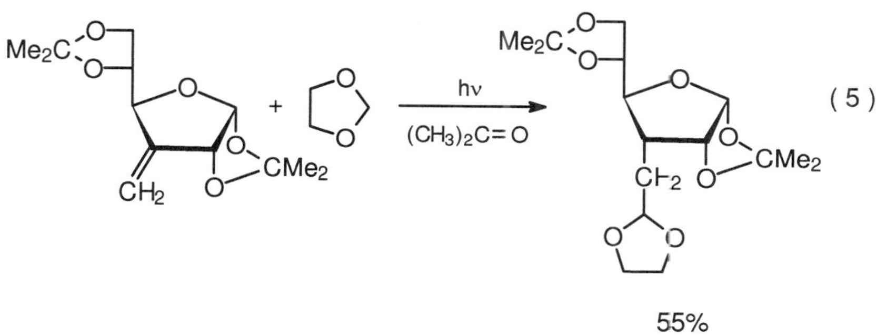

Equation 5

Compounds in Which a Heteroatom Is in Conjugation with a Double Bond

When a double bond has a substituent on each carbon atom and when one of these substituents contains a heteroatom (almost always an oxygen atom) in conjugation with the π-system, a carbon radical adds regioselectively to the carbon atom of the double bond with the heteroatom-containing substituent. The presence of the heteroatom polarizes the double bond in such a manner (Figure 9.8) that a nucleophilic carbon radical adds regioselectively to the electron-deficient carbon atom (eq 6).[12]

If the radical center in the approaching radical is located on a nitrogen, oxygen, sulfur, phosphorous, or halogen atom, the radical is generally electrophilic and radical addition occurs at the more electron-rich carbon atom of the double bond (Figure 9.8). This type of regioselectivity is illustrated by the reaction shown in eq 7.[27] The predicted pattern of selectivity, based upon electrophilicity and nucleophilicity of approaching radicals, is followed consistently by the reactions of unsaturated carbohydrates.

Compounds with a Carbonyl or Nitro Group in Conjugation with a Double Bond

When a carbon–carbon double bond is in conjugation with a carbonyl group, addition of a carbon radical occurs exclusively at the β position.[1,2,5,9,13,43–57]

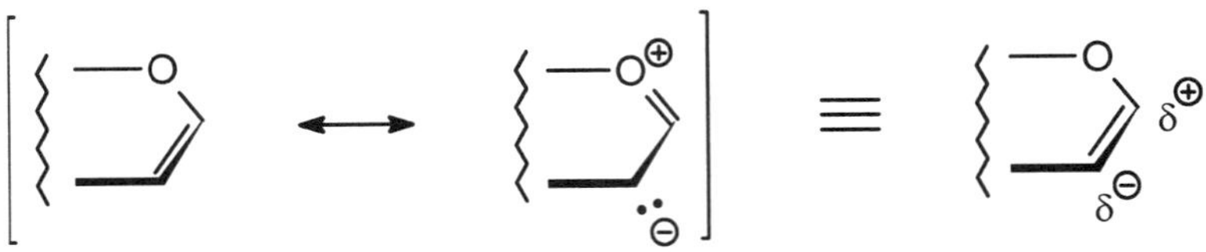

Figure 9.8. Polarization of a double bond in conjugation with an oxygen atom

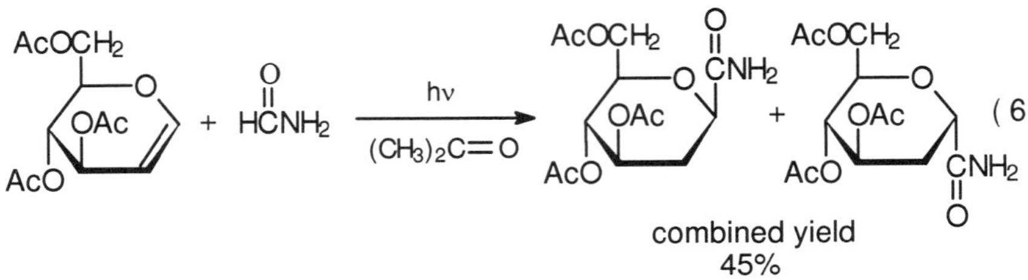

(6)

Equation 6

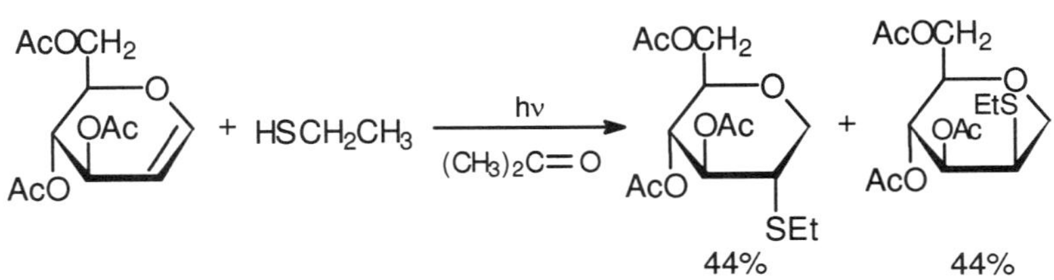

(7)

Equation 7

In reactions of this type polar effects are more important than steric effects. The dominance of polar effects is illustrated by the reaction shown in eq 8, where the addition of the hydroxymethyl radical is to the more electron-deficient carbon atom of the double bond (Figure 9.9), even though this atom is more sterically hindered.[47] This mode of addition is the result of the nucleophilic carbon radical adding in a "Michael-like" fashion to an α,β-unsaturated carbonyl compound (eq 2). A similar type of highly regioselective addition occurs when the double bond is in conjugation with a nitro group.[58–60]

Compounds with Trisubstituted Double Bonds

Since one of the sp^2-hybridized carbon atoms in a trisubstituted double bond is inherently less sterically hindered than the other, the direction of radical addition is determined primarily by steric effects and occurs at the less substituted alkene carbon atom (eq 9),[35] unless steric effects are overcome by a powerful polar substituent (eq 8).

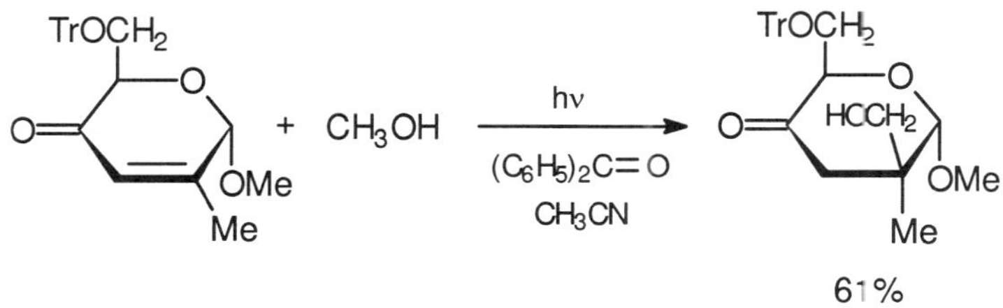

(8)

Equation 8

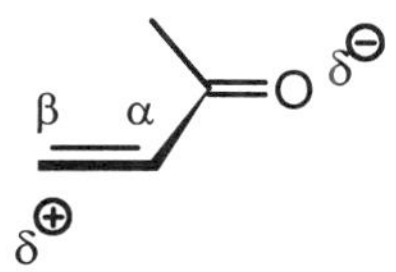

Figure 9.9. Polarization of an α,β-unsaturated carbonyl compound.

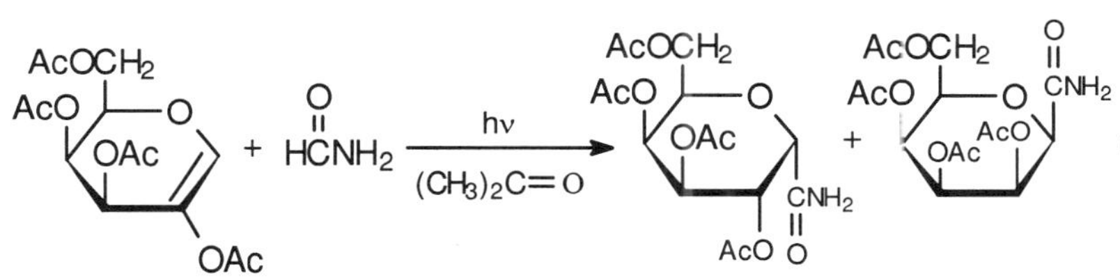

(9)

Equation 9

Compounds with Tetrasubstituted Double Bonds

The only reported photochemically promoted addition of a carbon radical to a tetrasubstituted double bond is the one shown in eq 10.[39] With four different substituents, each with different steric requirements, attached to the π system, it is not surprising that regioselective addition takes place. It also is not surprising that for such a complicated substitution pattern the regioselectivity of reaction is not amenable to a simple rationalization.

Stereoselectivity

Since radical addition to a double bond converts two sp^2-hybridized carbon atoms into ones that are sp^3-hybridized, new stereochemistry is established at each of these atoms (Scheme 9.3). The stereochemistry at the carbon atom bonding with the approaching radical (the α-carbon atom) is determined primarily by steric effects of the substituents at the radical center and those attached to the α-carbon atom. The importance of steric effects in determining stereochemistry is illustrated by comparison of the reactivity of compounds **2** and **3** (eq 11).[13] The presence of the ethoxy group in **2** forces the incoming

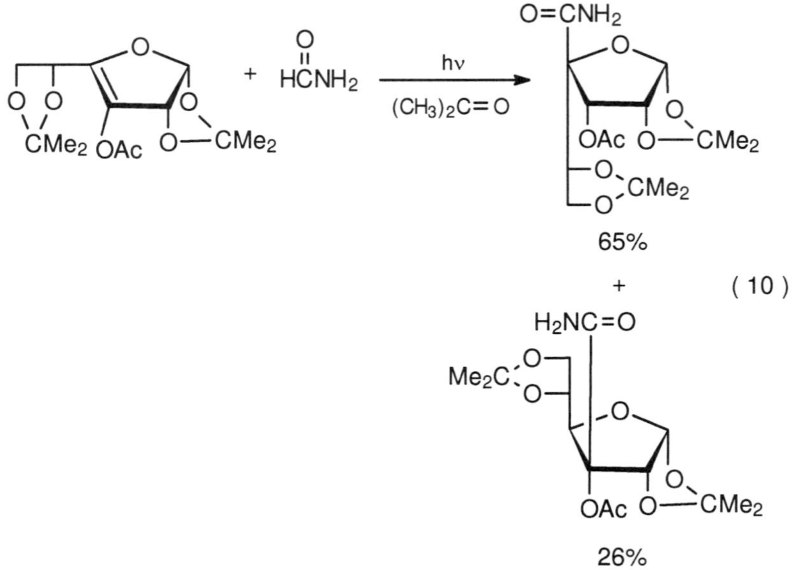

Equation 10

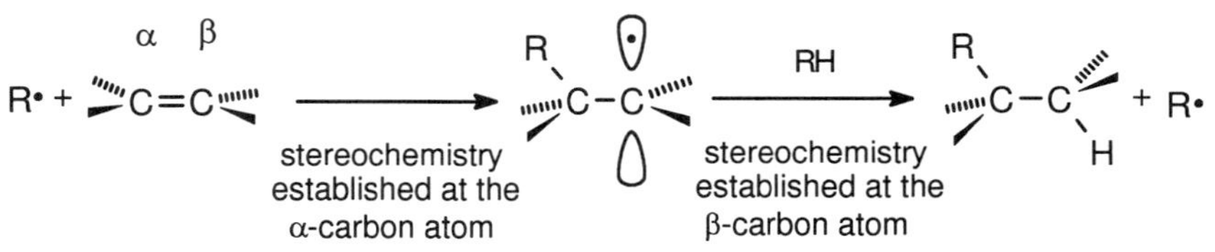

Scheme 9.3

hydroxymethyl radical to enter exclusively from the opposite side (β-face) of the molecule (eq 11, R = OEt). When the ethoxy substituent is absent (compound **3**), the stereoselectivity of this reaction is dramatically reduced, although a small amount of selectivity does remain (eq 11, R = H). The stereoselectivity still associated with this reaction (eq 11, R = H) is not attributable to the ring oxygen atom because replacement of the ring oxygen atom with a methylene group (eq 12) does not alter the relative amounts of stereoisomers formed.[13]

After initial reaction at the α-carbon atom, completion of the addition process requires transfer of an atom or group to the β-carbon atom (Scheme 9.3). The transfer is nearly always a hydrogen-atom transfer. The stereoselectivity of this phase of the addition process is determined by the β-substituents and characteristics of the reagent making the transfer to the β-carbon atom. Steric effects are dominant. Situations such as the formation of the less stable C-3 epimer **8** from photolysis of **7** are common because the radical **9** is approached by the hydrogen atom donor (dioxolane) from the sterically less hindered β-face of the radical (**9**) (Scheme 9.4).[8]

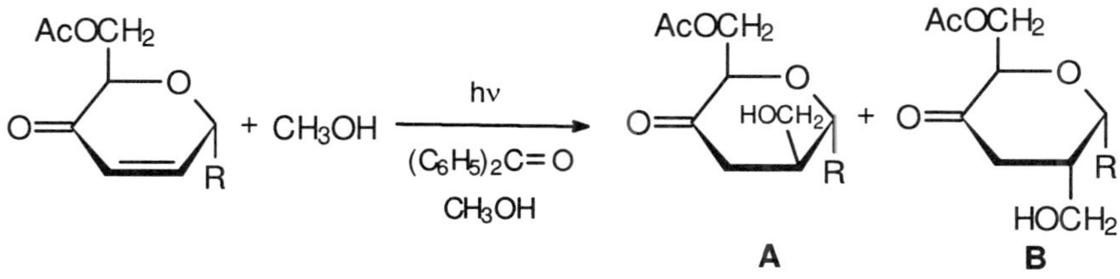

2 R = OEt A/B > 100 (11)

3 R = H A/B = 1.8

Equation 11

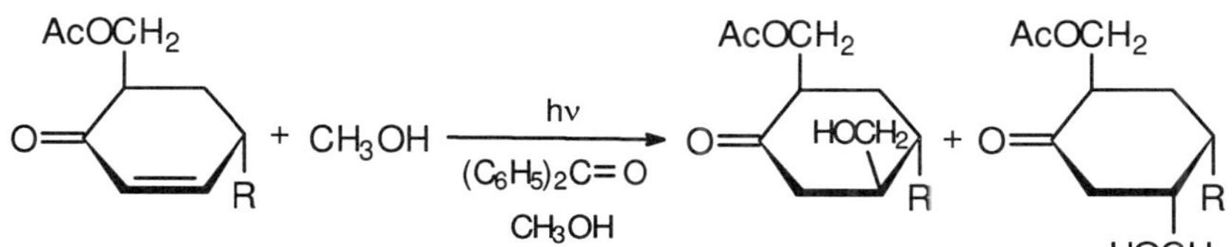

6 R = H

C C/D = 1.8 **D**

(12)

Equation12

Excited Ketones as Radical-Generating Agents

Since most reactions involving addition of a carbon radical to an unsaturated carbohydrate are initiated either by excited acetone or benzophenone, it is worthwhile to consider advantages, disadvantages, and problems that can arise from photolyses of these compounds. Benzophenone has the more intense and longer wavelength absorption [acetone, λ_{max} 265 (ϵ 11, isooctane); benzophenone, λ_{max} 350 (ϵ 100, hexane)];[61] consequently, when benzophenone is the light-absorbing agent, other compounds present in the reaction mixture are less likely to absorb the incident radiation. Absorption of light by a compound other than the hydrogen-abstracting agent can complicate the photolysis mixture by causing reactions that do not involve radical addition. A related advantage in the use of benzophenone is that it absorbs light at sufficiently long wavelengths that reaction takes place even when a Pyrex* filter is inserted between the light source and the reaction mixture. A Pyrex filter is valuable because it absorbs high energy radiation and thus insures that only less energetic light reaches the reaction mixture. Side reactions that are favored by absorption of high energy, UV light are reduced in importance.

*A Pyrex filter transmits only light of wavelength greater than 280 nm.[62]

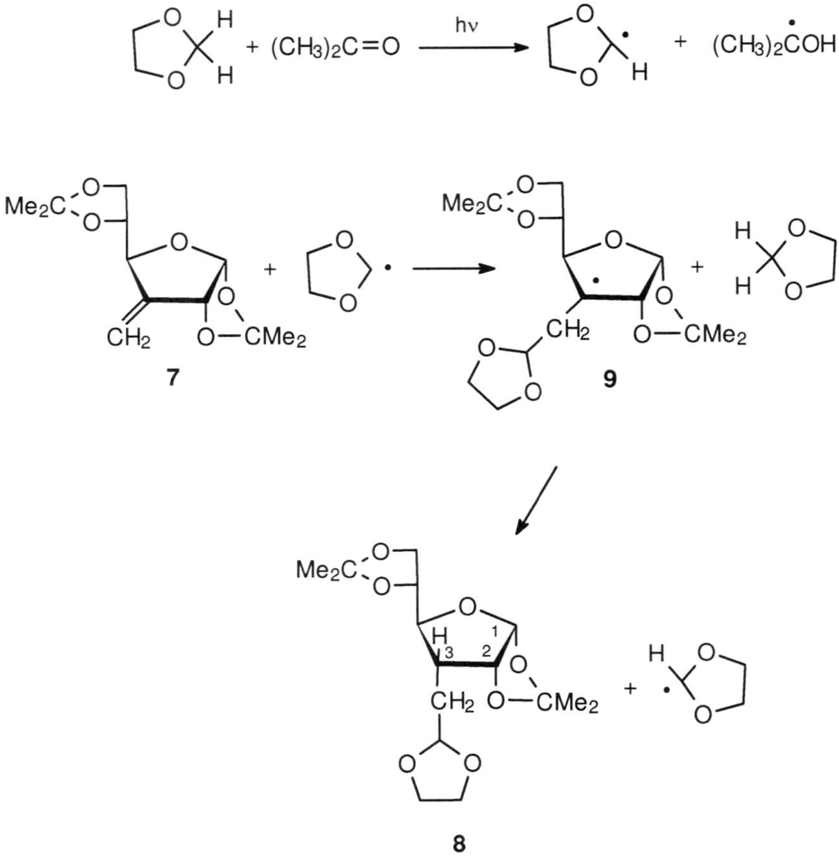

Scheme 9.4

Hydrogen abstraction by both acetone and benzophenone leads to formation of alcohols and pinacols (Scheme 9.5). Where acetone is concerned, these products are sufficiently volatile that they usually can be removed without resorting to chromatography. For benzophenone, however, the presence of the photoproducts benzhydrol and benzpinacol, along with unreacted benzophenone, presents a separation problem that complicates photoproduct purification.[63]

Participation of Organotin Compounds in Radical Addition Reactions

An alternative procedure for forming carbon radicals is to generate them from organotin radicals. Organotin radicals, which are produced by irradiation of organotin hydrides (and by other procedures), can initiate chain reactions such

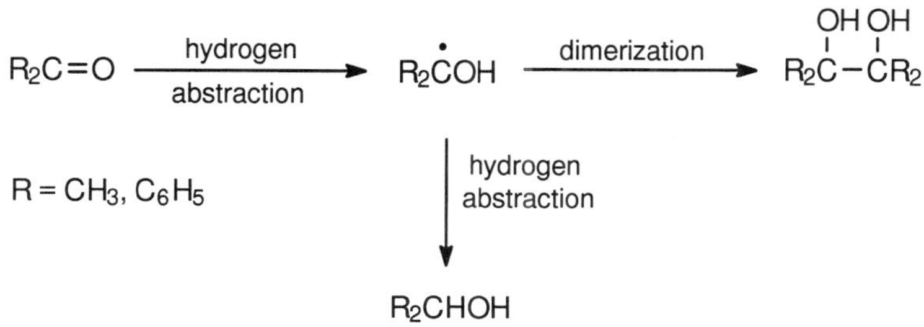

Scheme 9.5

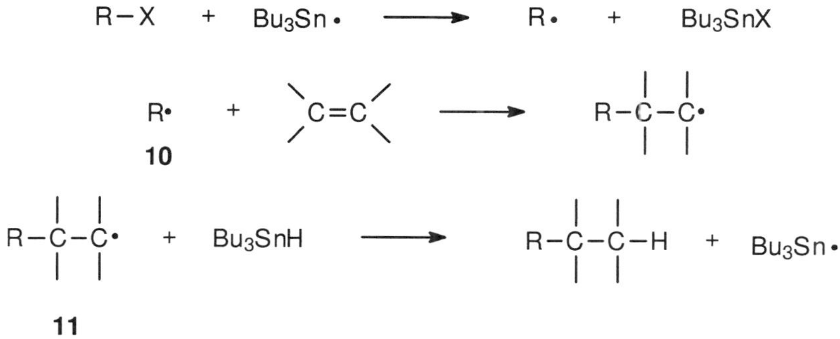

Scheme 9.6

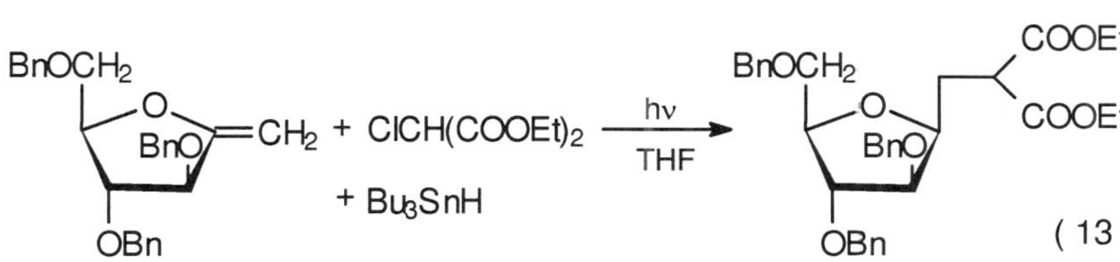

Equation 13

as that shown in Scheme 9.6 by abstracting a chlorine, bromine, or iodine atom from a halogenated organic compound to produce a carbon radical.[64] Carbon radicals formed in this fashion add to unsaturated carbohydrates (eq 13).[42]

Careful selection of the reactants and reaction conditions is necessary to insure that the radical **10** (Scheme 9.6) adds to the double bond in preference to forming RH by abstracting a hydrogen atom from a molecule of Bu_3SnH.[42] Although formation of RH is a competing process, for many radical–alkene combinations (e.g., a nucleophilic radical and an electron-deficient alkene) addition of **10** to the π-system competes effectively with hydrogen abstraction. If the reactants have been properly selected, the radical **11** (Scheme 9.6), produced by the addition process, is sufficiently different in reactivity from **10** (e.g., less nucleophilic and more sterically hindered) that hydrogen abstraction from Bu_3SnH is favored over addition to a second molecule of alkene.

Electron Transfer: An Alternative Pathway

Mann and Weymouth-Wilson reported a highly regio- and stereoselective, photochemical addition reaction to an α,β-unsaturated lactone that occurred in the absence of benzophenone and acetone (eq 14).[55] They proposed that this reaction was following an electron-transfer pathway and suggested the mechanism shown in Scheme 9.7.

Photochemically Promoted, Ionic Addition Reactions

Matsuura et al. discovered a group of photochemical addition reactions (e.g., eq 15) that appeared to result from formation of ionic intermediates,[65] but actually the products came from acid-catalyzed, addition reactions.[65–67] The

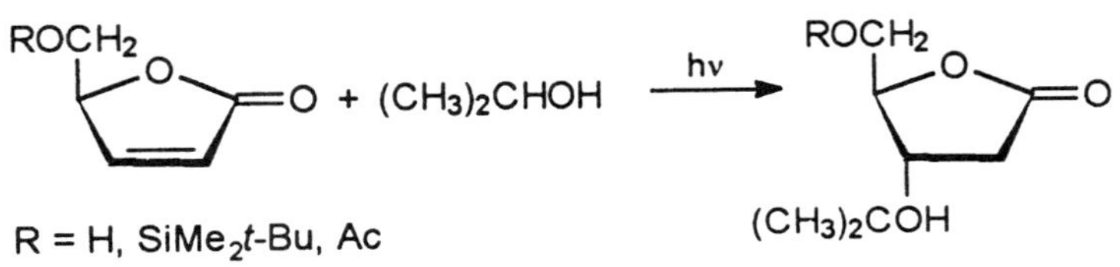

(14)

Equation 14

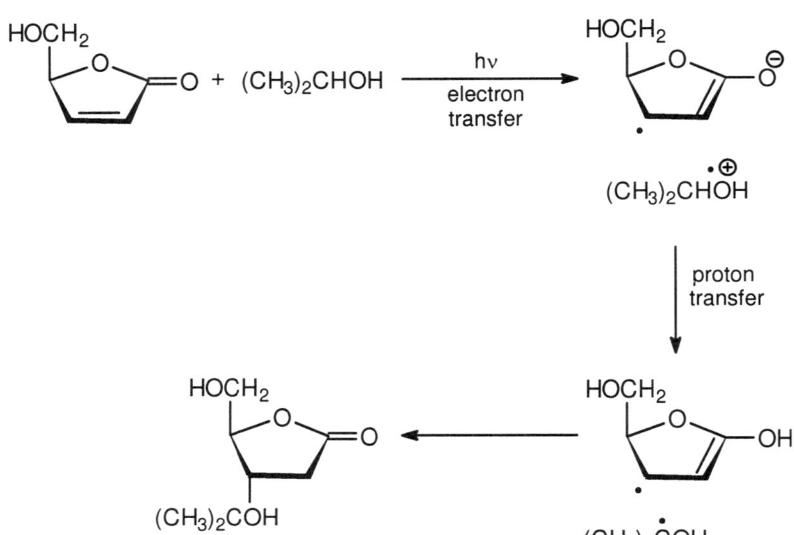

Scheme 9.7

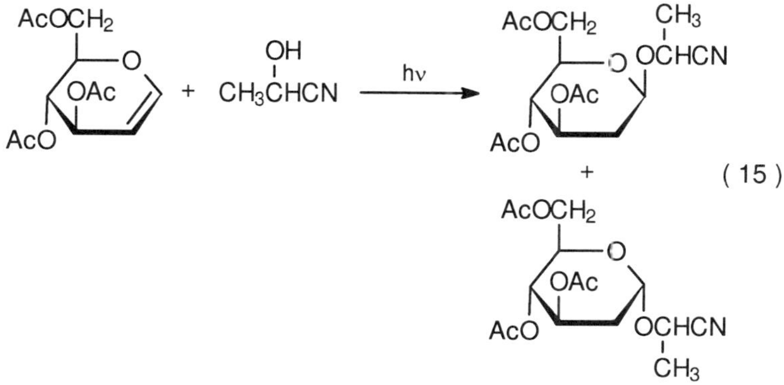

Equation 15

catalyst was not identified, but the authors demonstrated that it was an acidic material produced during photolysis.

Summary

Photochemically promoted addition of carbon radicals to unsaturated carbohydrates is a useful process for extending carbon atom chains and for synthesizing branched-chain sugars. When a radical contains a nitrogen, sulfur, or phosphorous atom, addition reaction provides a convenient method for introducing a heteroatom into a carbohydrate. The role of photochemistry in most of these reactions is to provide a means for generating carbon radicals. Radical formation typically is accomplished by hydrogen abstraction by excited acetone or benzophenone.

The regio- and stereoselectivity of addition reactions is influenced by steric and polar effects. The importance of steric effects in the regioselectivity of reaction is evident from the fact that a radical typically bonds to the less substituted carbon atom of a double bond. Polar effects determine the regioselectivity of reaction when a carbonyl or nitro group is in conjugation with the double bond or when the steric size of the groups attached to the double bond is similar.

References

1. Walker, D. L.; Fraser-Reid, B. *J. Am. Chem. Soc.* **1975**, *97*, 6251.
2. Fraser-Reid, B.; Walker, D. L. *Can. J. Chem.* **1980**, *58*, 2694.
3. Rosenthal, A.; Chow, J. *J. Carbohydr., Nucleosides, Nucleotides* **1980**, *7*, 77.
4. Othman, A. A.; Al-Masudi, N. A.; Al-Timari, U. S. *J. Antibiotics* **1978**, *31*, 1007.
5. Fitzsimmons, B. J.; Fraser-Reid, B. *J. Am. Chem. Soc.* **1979**, *101*, 6123.

6. Araki, Y.; Nishiyama, K.; Senna, K.; Matsuura, K.; Ishido,Y. *Carbohydr. Res.* **1978,** *64,* 119.
7. Rosenthal, A.; Ratcliffe, M. *Carbohydr. Res.* **1975,** *39,* 79.
8. Jewell, J. S.; Szarek, W. A. *Tetrahedron Lett.* **1969,** 43.
9. Udodong, U. E.; Fraser-Reid, B. *J. Org. Chem.* **1989,** *54,* 2103.
10. Rosenthal, A.; Shudo, K. *J. Org. Chem.* **1972,** *37,* 1608.
11. Rosenthal, A.; Ratcliffe, M. *Can. J. Chem.* **1976,** *54,* 91.
12. Rosenthal, A.; Zanlungo, A. *Can. J. Chem.* **1972,** *50,* 1192.
13. Benko, Z.; Fraser-Reid, B.; Mariano, P. S.; Beckwith, A. L. J. *J. Org. Chem.* **1988,** *53,* 2066.
14. Giese, B. *Angew. Chem. Int. Ed. Engl.* **1983,** *22,* 753.
15. Giese, B.; Meister, J. *Chem. Ber.* **1977,** *110,* 2558.
16. Giese, B.; Meister, J. *Angew. Chem. Int. Ed. Engl.* **1977,** *16,* 178.
17. Giese, B.; Mehl, W. *Tetrahedron Lett.* **1991,** *32,* 4275.
18. Hammond, G. S. *J. Am. Chem. Soc.* **1955,** *77,* 334.
19. Fukui, K. *Fortschr. Chem. Forsch.* **1970,** *15,* 1.
20. Fujimoto, H.; Yamabe, S.; Minato, T.; Fukui, K. *J. Am. Chem. Soc.* **1972,** *94,* 9205.
21. Fleming, I. *Frontier Orbitals and Organic Chemical Reactions;* John Wiley and Sons: New York, 1976; p 182.
22. Fleming, I. *Frontier Orbitals and Organic Chemical Reactions;* John Wiley and Sons: New York, 1976; p 120.
23. Mojelsky, T.; Chow, Y. L. *J. Am. Chem. Soc.* **1974,** *96,* 4549.
24. Whistler, R. L; Wang, C.-C.; Inokawa, S. *J. Org. Chem.* **1968,** *33,* 2495.
25. Horton, D.; Turner, W. N. *Chem. Ind.* **1964,** 76.
26. Horton, D.; Turner, W. N. *Carbohydr. Res.* **1966,** *1,* 444.
27. Matsuura, K.; Araki, Y.; Ishido, Y.; Satoh, S. *Chem. Lett.* **1972,** 849.
28. Araki, Y.; Matsuura, K.; Ishido, Y.; Kushida, K. *Chem. Lett.* **1973,** 383.
29. Matsuura, K.; Nishiyama, K.; Yamada, K.; Araki, Y.; Ishido, Y. *Bull. Chem. Soc. Jpn.* **1973,** *46,* 2538.
30. Matsuura, K.; Araki, Y.; Ishido, Y.; Mural, A.; Kushida, K. *Carbohydr. Res.* **1973,** *29,* 459.
31. Araki, Y.; Nishiyama, K.; Matsuura, K.; Ishido, Y. *Carbohydr. Res.* **1978,** *63,* 288.
32. Araki, Y.; Senna, K.; Matsuura, K.; Ishido, Y. *Carbohydr. Res.* **1978,** *60,* 389.
33. Matsuura, K.; Maeda, S.; Araki, Y.; Ishido, Y. *Tetrahedron Lett.* **1970,** 2869.
34. Araki, Y.; Senna, K.; Matsuura, K.; Ishido, Y. *Carbohydr. Res.* **1978,** *64,* 109.
35. Chmielewski, M.; BeMiller, J. N.; Cerretti, D. P. *J. Org. Chem.* **1981,** *46,* 3903.
36. Chmielewski, M.; BeMiller, J. N.; Cerretti, D. P. *Carbohydr. Res.* **1981,** *97,* C1.
37. Cerretti, P. *Carbohydr. Res.* **1981,** *94,* C10.
38. Kumamoto, K.; Yoshida, H.; Ogata, T.; Inokawa, S. *Bull. Chem. Soc. Jpn.* **1969,** *42,* 3245.
39. Rosenthal, A.; Ratcliffe, M. *Carbohydr. Res.* **1977,** *54,* 61.
40. Araki, Y.; Senna, K.; Matsuura, K.; Ishido, Y. *Carbohydr. Res.* **1978,** *65,* 159.
41. N. V. Bovin, S. É. Zurabyan, and A. Y. Khorlin, *Carbohydr. Res.,* **98,** 25 (1981).
42. Cipolla, L.; Liguori, L.; Nicotra, F.; Torri, G.; Vismara, E. *J. Chem. Soc., Chem. Commun.* **1996,** 1253.
43. Hicks, D. R.; Anderson, R. C.; Fraser-Reid, B. *Synthetic Commun.* **1976,** *6,* 417.
44. Walker, D. L.; Fraser-Reid, B.; Saunders, J. K. *J. Chem. Soc., Chem. Commun.* **1974,** 319.
45. Fraser-Reid, B.; Holder, N. L.; Yunker, M. B. *J. Chem. Soc., Chem. Commun.* **1972,** 1286.
46. Fraser-Reid, B.; Anderson, R. C.; Hicks, D. R.; Walker, D. L. *Can. J. Chem.* **1977,** *55,* 3986.
47. Fraser-Reid, B.; Holder, N. L.; Hicks, D. R.; Walker, D. L. *Can. J. Chem.* **1977,** *55,* 3978.
48. Udodong, U. E.; Fraser-Reid, B. *J. Org. Chem.* **1988,** *53,* 2132.
49. Fraser-Reid, B.; Hicks, D. R.; Walker, D. L.; Iley, D. E.; Yunker, M. B.; Tam, S. Y.-K.; Anderson, R. C. *Tetrahedron Lett.* **1975,** 297.
50. Buenger, G. S.; Marquez, V. E. *Tetrahedron Lett.* **1992,** *33,* 3707.
51. Wang, J.; Gossauer, A. *Helv. Chim. Acta* **1994,** *77,* 533.
52. Fraser-Reid, B. *Acc. Chem. Res.* **1975,** *8,* 192.
53. Parry, R. J.; Haridas, K.; De Jong, R.; Johnson, C. R. *Tetrahedron Lett.* **1990,** *31,* 7549.

54. Paulsen, H.; Koebernick, W. *Carbohydr. Res.* **1977**, *56,* 53.
55. Mann, J.; Weymouth-Wilson, A. *Synlett.* **1992**, 67.
56. Mann, J.; Weymouth-Wilson, A. C. *J. Chem. Soc., Perkin Trans. 1* **1994**, 3141.
57. Shim, S. C.; Shim, H. K.; Cho, T. H. *Chem. Lett.* **1986**, 19.
58. Sakakibara, T.; Nakagawa, T. *Carbohydr. Res.* **1987**, *163,* 239.
59. Sakakibara, T.; Takaide, A.; Seta, A. *Carbohydr. Res.* **1992**, *226,* 271.
60. Sakakibara, T.; Matuo, I.; Takaide, A.; Nakamura, T.; Seta, A. *Chem. Lett.* **1992**, 177.
61. Murov, S. L.; Carmichael, I.; Hug, G. L. *Handbook of Photochemistry,* 2nd ed.; Marcel Dekker: New York, 1993; pp 157, 169.
62. Murov, S. L.; Carmichael, I.; Hug, G. L. *Handbook of Photochemistry,* 2nd ed.; Marcel Dekker: New York, 1993; p 323).
63. Venkateswara, B.; Chan, R. J. B.; Moskowitz, N.; Fraser-Reid, B. *Bull. Soc. Chim. Fr.* **1993**, *130,* 428.
64. Neumann, W. P. *Synthesis* **1987**, 665.
65. Matsuura, K.; Araki, Y.; Ishido, Y.; Kainosho, M. *Chem. Lett.* **1972**, 853.
66. Matsuura, K.; Senna, K.; Araki, Y.; Ishido, Y. *Bull. Chem. Soc. Jpn.* **1974**, *47,* 1197.
67. Csaszar, J.; Bruckner, V. *Ann. Univ. Sci. Budapest* **1973,** 87 (1973). [*Chem. Abstr.* **1973**, *80,* 83437y]

10

Ethers

Introduction

Easily the most important photochemical reaction of carbohydrate-containing ethers is one that occurs when an *o*-nitrobenzyl group is removed from a carbohydrate by photolysis; in fact, irradiation of an *o*-nitrobenzyl ether is probably the most frequently conducted photochemical reaction in carbohydrate chemistry (eq 1). Although this deprotection process has advantages and disadvantages and needs to be discussed thoroughly, the discussion will not take place here but rather in Chapter 16 along with similar treatment of other reactions of nitro compounds.

Aside from photolysis of *o*-nitrobenzyl ethers, there are few photochemical reactions of carbohydrates in which an ether linkage is involved in the photochemical process. This is an understandable situation since ethers are ineffective at absorbing the UV light normally available for photochemical reaction ($\lambda > 200$ nm) unless an appropriate chromophore (e.g., an aromatic ring or a carbonyl group) is present in the molecule. For a typical aliphatic ether containing no additional light-absorbing groups, the only UV absorption at wavelengths longer than 200 nm arises from end absorption of a peak with a maximum in the 180–185 nm range.[1,2]

Benzyl Ethers

Since benzylation of a carbohydrate creates an ether with an effective, light-absorbing chromophore, photochemical reaction of such compounds becomes a distinct possibility. Angyal and Odier noted that when cyclitols protected as

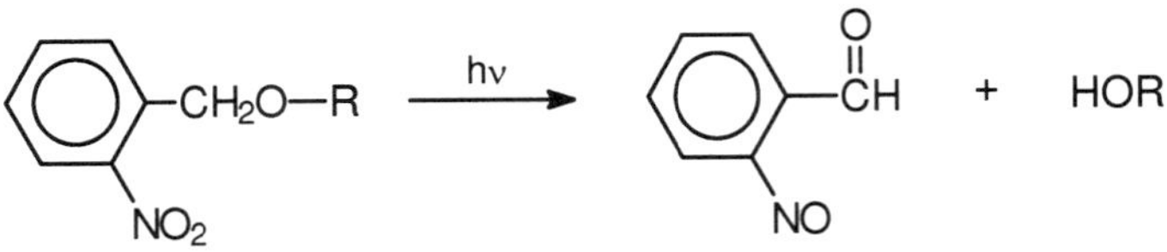

(1)

R = sugar moiety

Equation 1

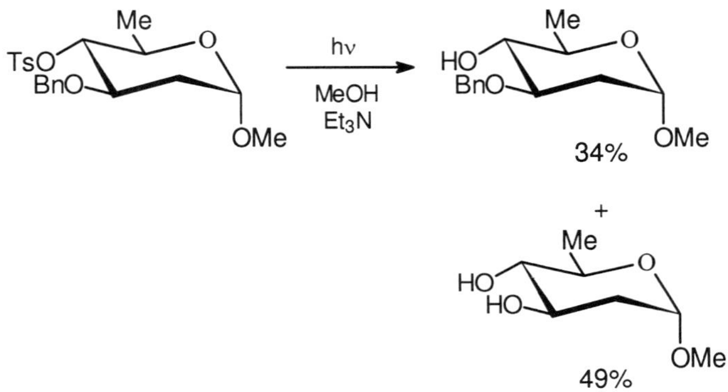

(2)

Equation 2

3,4-dimethoxybenzyl ethers were irradiated in aqueous hexamethylphosphoramide (HMPA), a distinct, but unspecified, photochemical reaction took place.[3] Binkley and Koholic found that irradiation of a benzyl-protected carbohydrate in a mixture of methanol and triethylamine caused the expected loss of tosyl protection and the unexpected removal of the benzyl group (eq 2).[4] The authors of both of these articles viewed benzyl ether photoreaction as an undesirable process that caused the unplanned loss of an otherwise quite stable protecting group.

Nishida et al. adopted a different view of benzyl ether photoreaction when they investigated the possibility that photolysis of benzyl ethers could be an alternative to traditional methods for removing *o*-benzyl protection.[5] These investigators found that photochemical deprotection occurred readily under electron-transfer conditions (eq 3) and proposed the mechanism shown in Scheme 10.1. In these reactions 9,10-dicyanoanthracene served as the electron acceptor, and magnesium perchlorate promoted reaction by reducing back electron transfer.[6,7] Liu and Binkley[8] also studied the photolysis of benzyl ethers in methanol in the presence of iodide ion, a different set of electron-transfer conditions (Scheme 10.2), and obtained results similar to those reported by Nishida et al. (Table 10.1).

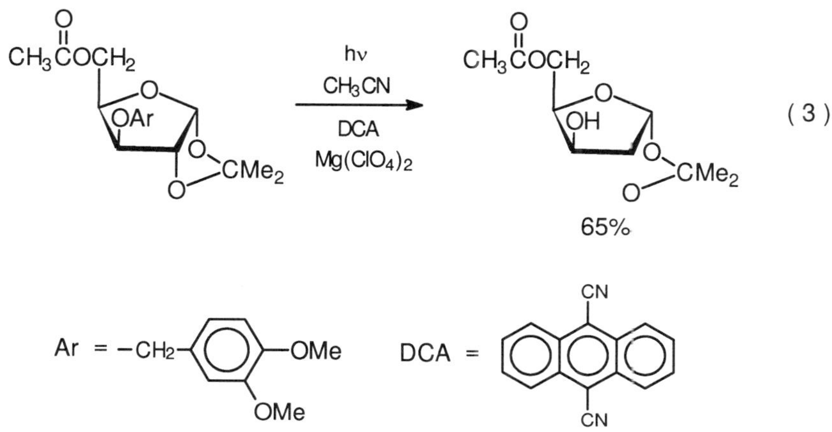

Equation 3

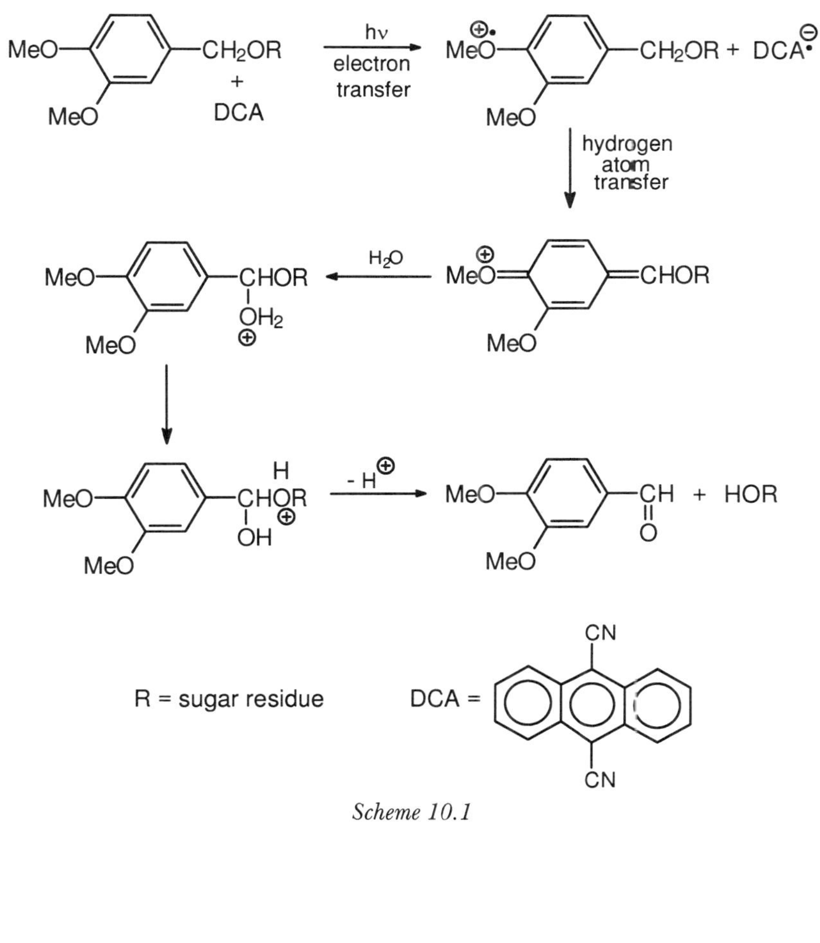

Scheme 10.1

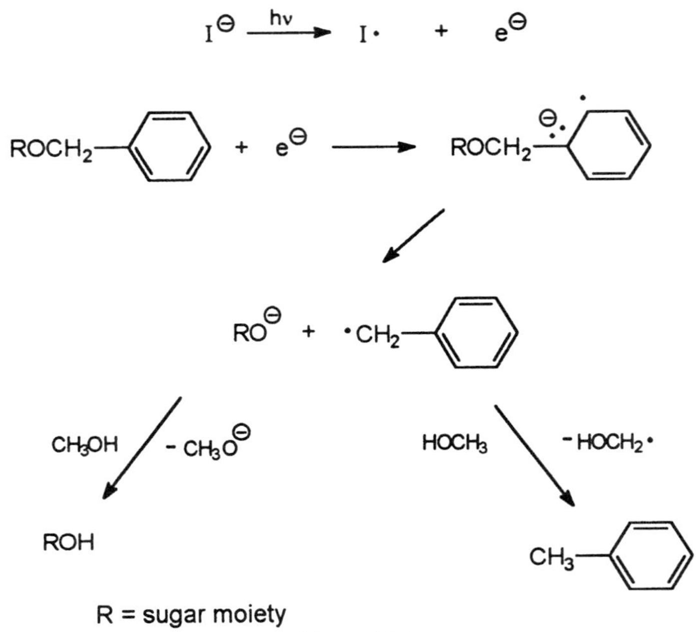

Scheme 10.2

Anhydro Compounds

If a light-absorbing group is in close proximity to a strained, cyclic ether, photochemical ring opening can take place. Such a reaction occurs when light is absorbed by the nucleoside **1** (eq 4).[9,10] Several other cleavage reactions of this type are found in Chapter 5, Table 5.3, where a light-absorbing keto group is adjacent to an oxirane ring.

Silyl Ethers

Pirrung and Lee have synthesized a type of silyl ether in which the silyl group can be removed by photolysis.[11] One of the ways that these researchers demonstrated the value of such a group was in the protection and photochemical deprotection of a thymidine derivative (eq 5).

Table 10.1. Photochemical Reaction of Benzyl Ethers.

Reactant	*Product*	*Percent Yield*	*Ref.*
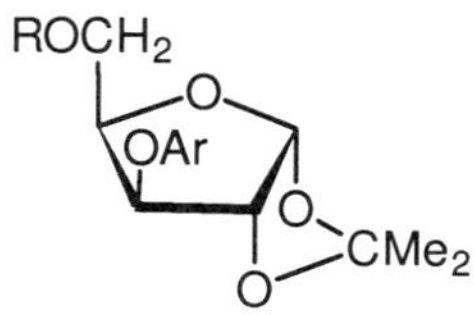	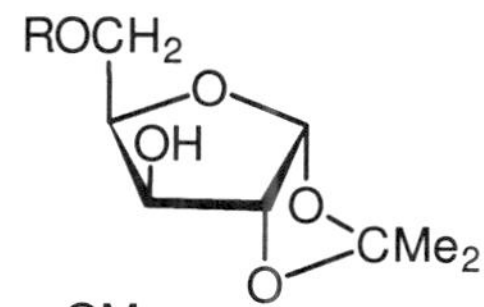		4[a]
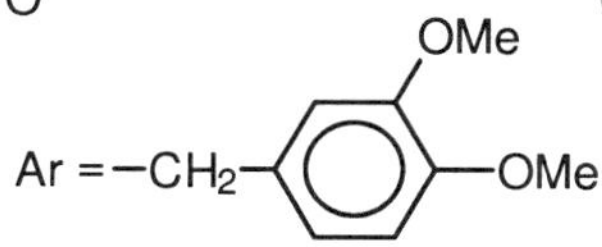			
R = *t*-BuMe$_2$Si		59	
R = CH$_3$CO		65	
R = C$_6$H$_5$CH$_2$		54	
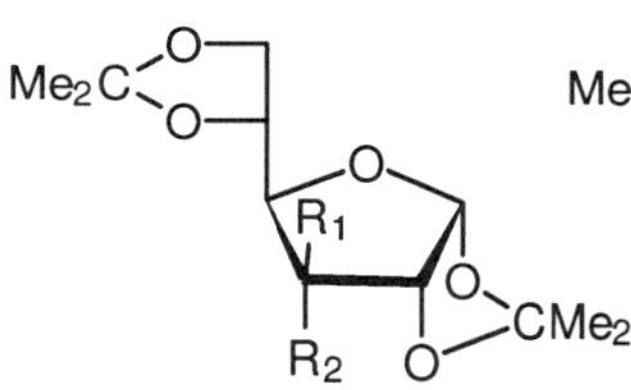	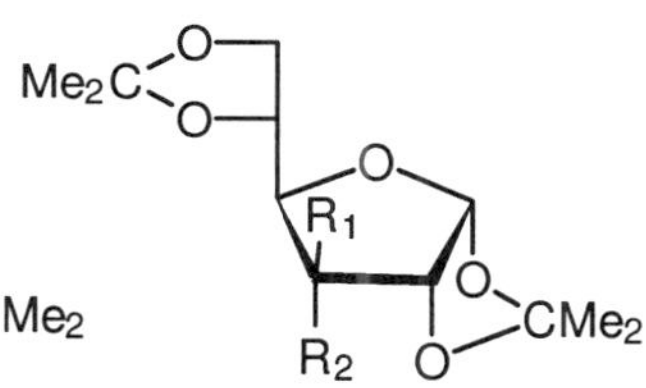		7
R$_1$ = OBn, R$_2$ = H	R$_1$ = OH, R$_2$ = H	85	
R$_1$ = H, R$_2$ = OBn	R$_1$ = H, R$_2$ = OH	65	
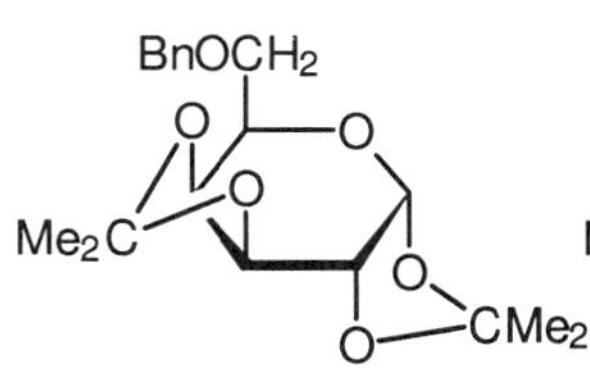	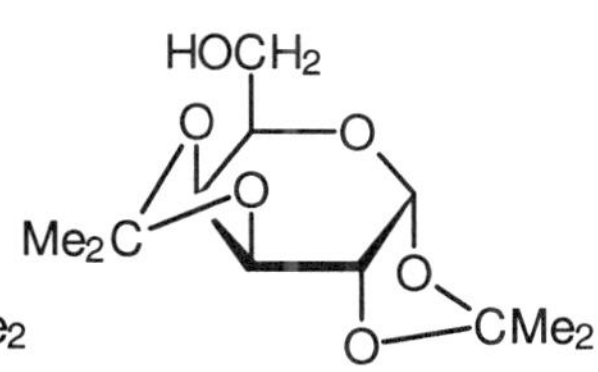	82	7
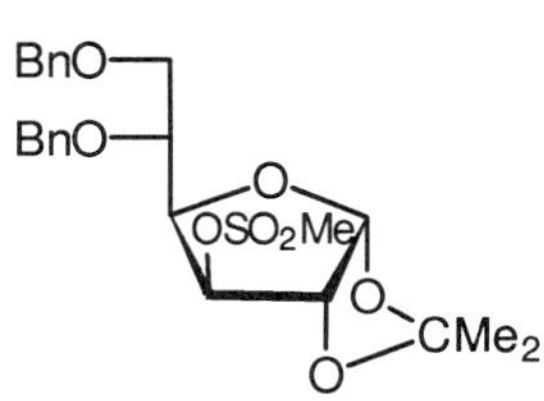	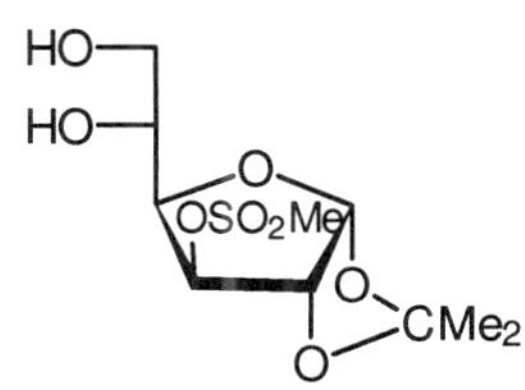	60	7

[a]Magnesium perchlorate and 9,10-dicyanoanthracene were added to the reaction mixture.

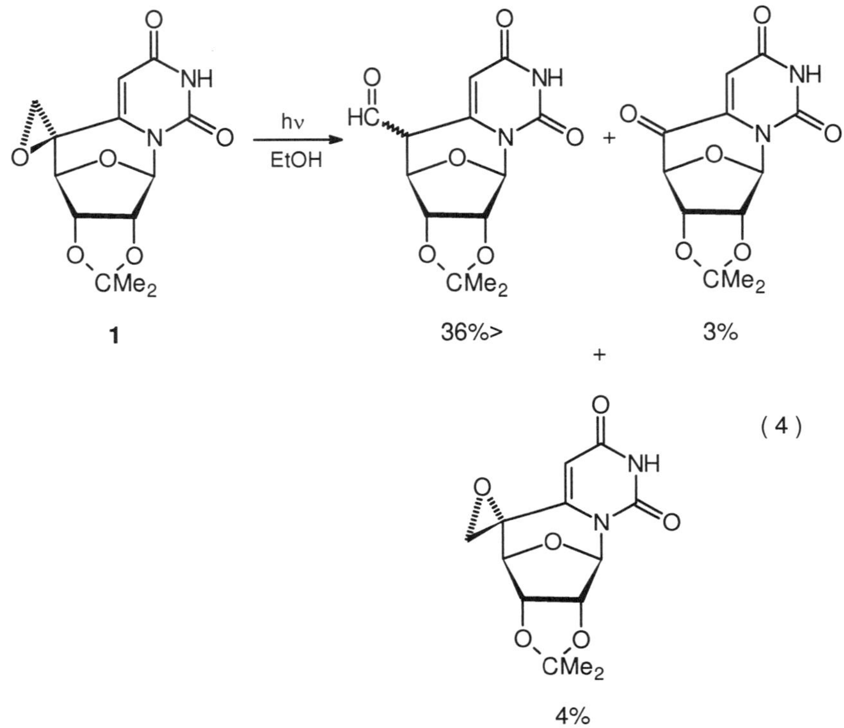

Equation 4

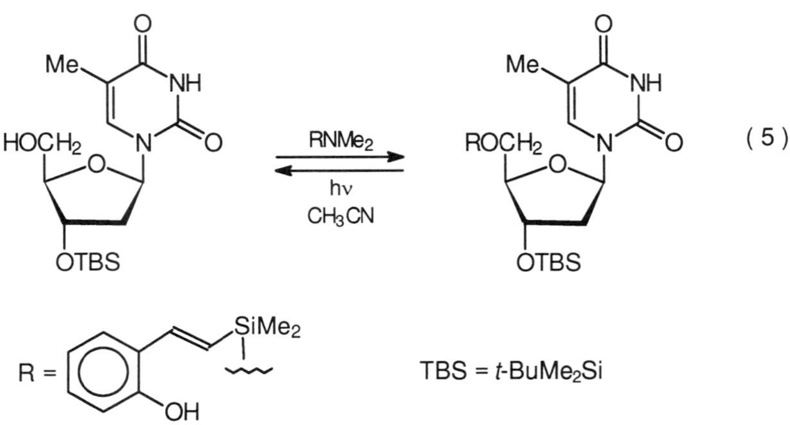

Equation 5

References

1. Gordon, A. J.; Ford, R. A. *The Chemists Companion;* John Wiley & Sons: New York, 1972; p 215.
2. Murov, S. L.; Carmichael, I.; Hug, G. L. *Handbook of Photochemistry;* 2nd ed., Marcel Dekker: New York, 1993; p 288.
3. Angyal. S. J.; Odier, L. *Carbohydr. Res.* **1982,** *101,* 209.
4. Binkley, R. W.; Koholic, D. J. *J. Org. Chem.* **1989,** *54,* 3577.
5. Nishida, A.; Oishi, S.; Yonemitsu, O. *Chem. Pharm. Bull.* **1989,** *37,* 2266.
6. Abe, M.; Oku, A. *J. Org. Chem.* **1995,** 60, 3065.
7. Mizuno, K.; Ichinose, N.; Otsuji, Y. *Chem. Lett.* **1985,** 455.
8. Liu, X.-G.; Binkley, R. W. *J. Carbohydr. Chem.* **1993,** *12,* 779.
9. Sasson, I. M.; Otter, B. A. *J. Org. Chem.* **1981,** *46,* 1114.
10. Sasson, I. M.; Otter, B. A. *J. Heterocyclic Chem.* **1987,** *24,* 1439.
11. Pirrung, M. C.; Lee, Y. R. *J. Org. Chem.* **1993,** *58,* 6961.

11

Unprotected Carbohydrates

Introduction

Early research on the photochemistry of carbohydrates was conducted primarily on compounds that contained no protecting groups. Study of such compounds was experimentally difficult because even in the early stages of reaction (20–30% conversion of the starting material) irradiation mixtures became quite complex due to the photochemical decomposition of the products initially formed. Without the assistance of modern chromatographic and instrumental techniques, the early researchers were forced to monitor changes in properties of reaction mixtures instead of actually identifying the photoproducts. Some of the properties monitored were formation of acid, release of gases (molecular hydrogen and carbon dioxide), and changes in optical rotation.[1,2]

Another difficulty that faced the early researchers was as fundamental as any problem in photochemistry can be; that is, there was uncertainty as to whether an unprotected carbohydrate contained any chromophore that could absorb UV light. Considerable controversy surrounded this topic prior to a set of definitive experiments by Phillips and Criddle that demonstrated that a definite, but quite weak, (n,σ*) absorption existed for unprotected carbohydrates.[3] These experiments are discussed in Chapter 1 and, in greater detail, in two review articles by Phillips.[1,2]

Photolysis of Unprotected Sugars

D-Glucitol

If it were necessary to designate a beginning point for modern carbohydrate photochemistry, it would be in 1963 when Phillips and co-workers published the first two papers[3,4] in a series[3–7] on photodegradation of carbohydrates. These workers began with a careful examination of the photochemistry of D-glucitol (D-sorbitol). They first established that D-glucitol absorbed UV radiation in the spectral region from 230 to 240 nm* and that this absorption was responsible for the observed photochemical reactions of this compound.[4] By using isotope dilution techniques, they positively identified the products from D-glucitol photolysis (eq 1) and showed that the complexity of the reaction was attributable to further photochemical decomposition of these products. Since some photoproducts absorbed light more efficiently than D-glucitol itself, secondary reaction became important after only a small amount of starting material had been consumed. Phillips proposed the stepwise degradation shown in Scheme 11.1 to account for the products formed and suggested that this type of process continued until all the carbon atoms in the starting material were converted into carbon dioxide.

It is worth recalling at this point that many unprotected carbohydrates in aqueous solution exist as equilibrium mixtures of cyclic and acyclic structures (see Scheme 1.3 in Chapter 1). This is true for each of the three degradation products shown in Scheme 11.1 and many of the other compounds discussed in this chapter. It would be cumbersome to show in each case all of these various forms in equations, schemes, and tables; therefore, only one of the possible forms (usually an acyclic one) typically will be used to represent a carbohydrate in this chapter.

Phillips did not propose a detailed mechanism for each of the steps shown in Scheme 11.1, but some of this detail is available from information on radical reactions of carbohydrates. Von Sonntag and co-workers have shown that radicals derived from sugars are efficiently scavenged by molecular oxygen and that the peroxy radicals thus produced readily eliminate •OOH.[8–10] This sequence of reactions provides an understandable pathway for formation of D-glucose from D-glucitol (Scheme 11.2) and D-gluconic acid from D-glucose (Scheme 11.3). Elimination of a peroxy radical from D-gluconic acid also is involved in its decarboxylation to give D-arabinose (Scheme 11.4), the final step in the reaction sequence shown in Scheme 11.1.

In the reactions pictured in Schemes 11.2–11.4 the radicals involved could arise either from photolysis or from hydrogen abstraction reactions. Once photolysis begins, radicals capable of hydrogen abstraction are present in solution. This fact follows from the demonstration by Phillips and Rickards that irradiation of D-glucose in the presence of oxygen produces hydrogen peroxide.[6] Photolysis of hydrogen peroxide is an excellent method for generating hydroxy radicals, intermediates that are effective at abstracting hydrogen atoms from sugars.[8–10]

*The molar absorption coefficient was determined to be 0.2 at 254 nm and rose to a value of 3.0 at 200 nm. The absorption was assigned to formation of an (n,σ *) excited state.[1-3]

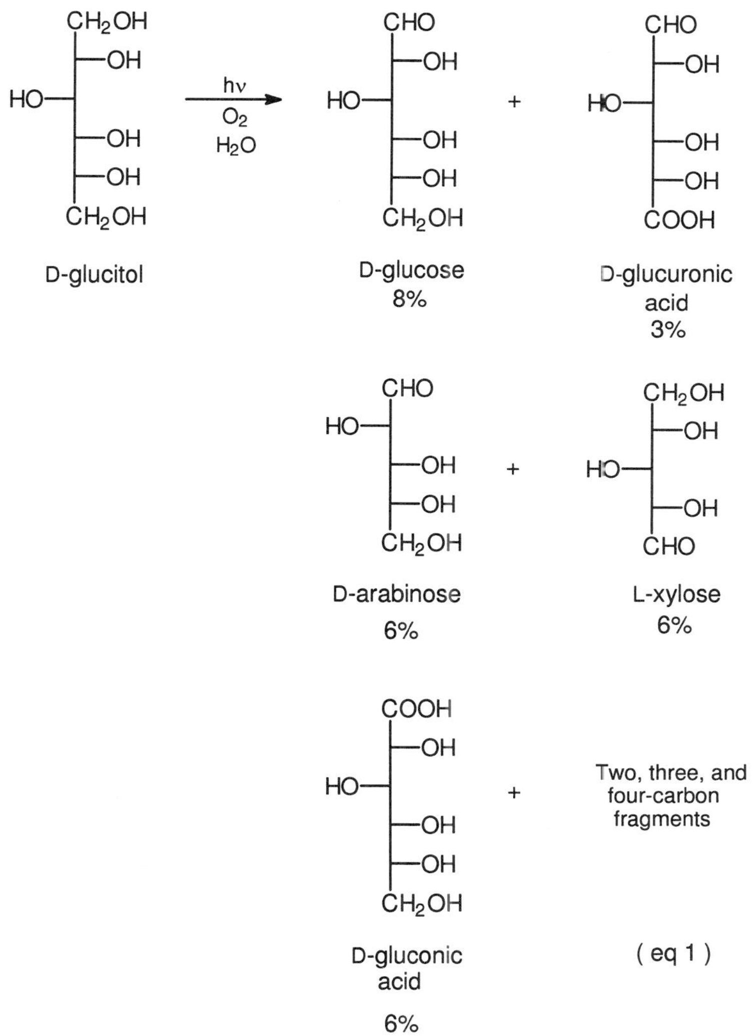

hν
O_2
H_2O
D-glucitol
D-glucose
8%
D-glucuronic acid
3%
D-arabinose
6%
L-xylose
6%
D-gluconic acid
6%
Two, three, and four-carbon fragments
(eq 1)

Equation 1

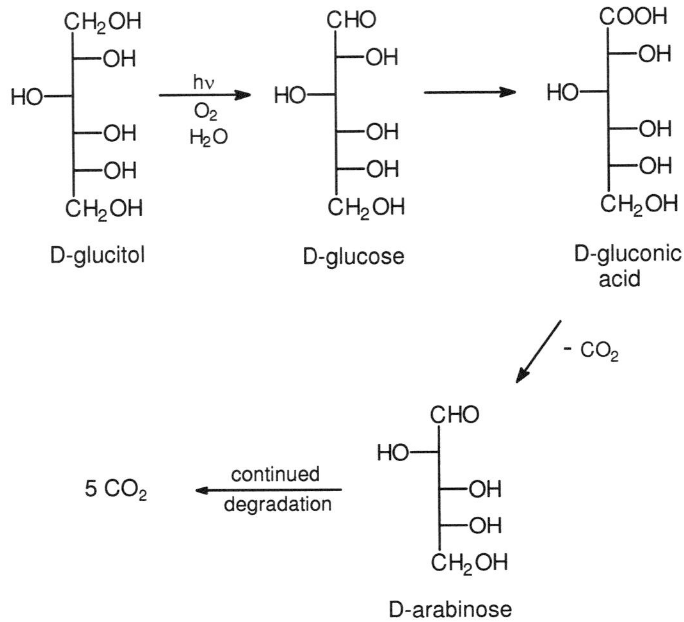

Scheme 11.1

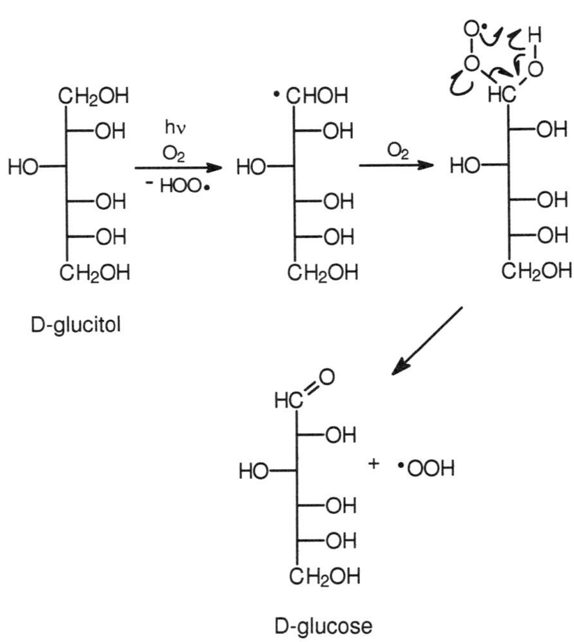

Scheme 11.2

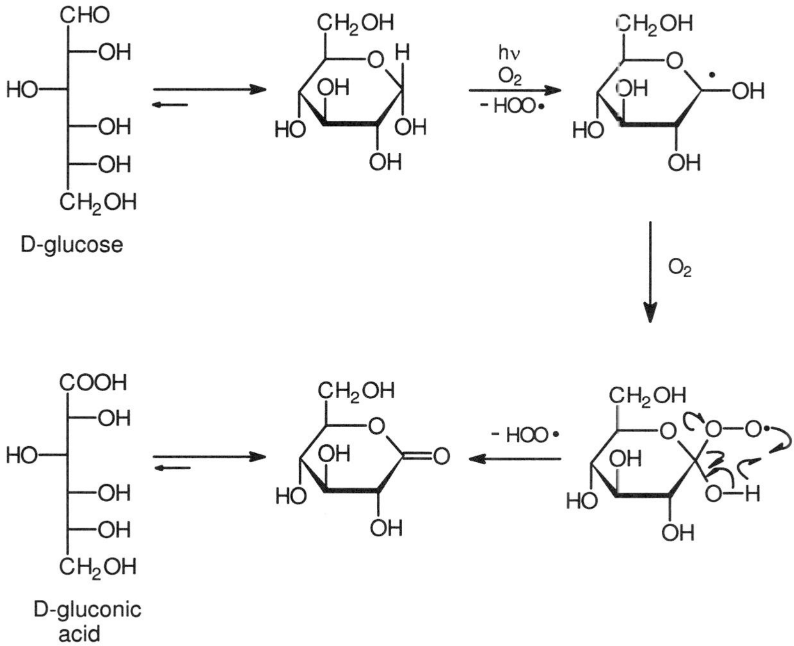

Scheme 11.3

Degradation reactions of the type shown in Schemes 11.2–11.4 also should initiate another degradation sequence beginning at the "other end" of the D-glucitol molecule. Phillips and Criddle proposed that such a sequence was responsible for the formation of L-xylose (eq 1).[3] (The intermediate hexose in this sequence is L-gulose, which was indicated by paper chromatography to be present in the reaction mixture but was not isolated.) Also, photochemical reaction could begin at one of the secondary carbon atoms and result in formation of ketoses and smaller chain fragments. Considering the number of possibilities, it is not surprising that the yield of any single product is quite low and that photolysis mixtures are exceedingly complex.

Phillips and co-workers also reported evidence for formation of an additional photoproduct from photolysis of D-glucitol.[3,4] This new compound was not isolated or identified but exhibited a characteristic UV absorption (λ_{max} 265 nm) in basic solution. This UV absorption had been observed earlier by other investigators, in particular Laurent,[11,12] who suggested several possible structures (including the correct one), even though the only information he had was the UV absorption of this material and the fact that the same absorption appeared from irradiation of a number of monosaccharides. The identity of this compound ultimately was established by Scherz to be malonaldehyde.[13] A possible mechanism for malonaldehyde formation is shown in Scheme 11.5.

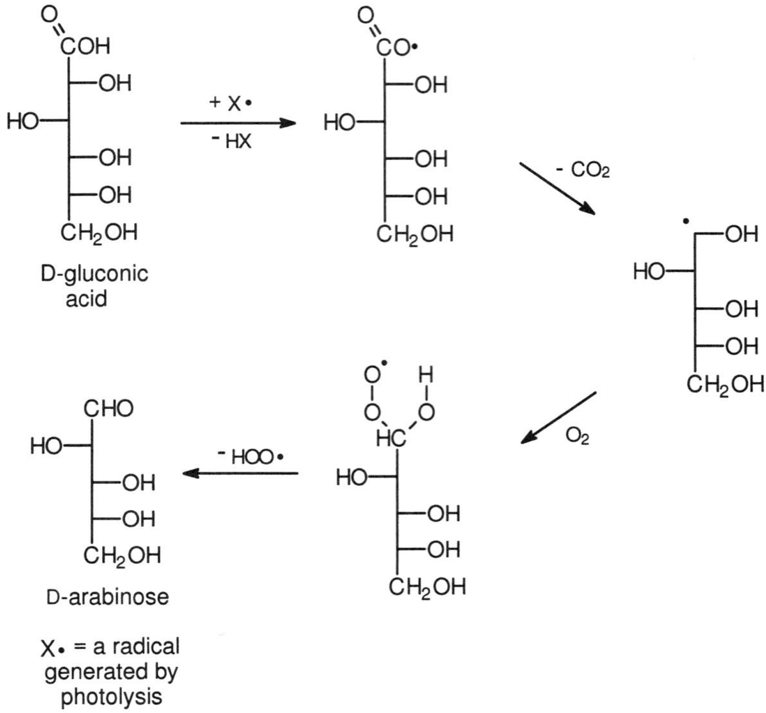

Scheme 11.4

D-Fructose

The equilibrium between cyclic and acyclic structures of monosaccharides such as D-glucose and D-fructose is known to favor the cyclic forms. (Equilibrium concentrations of carbonyl-containing forms of monosaccharides are discussed in Chapter 1.) If there is sufficient carbonyl-containing constituent present to absorb the light necessary for reaction, typical carbonyl photochemical reactions will take place. This condition is not met by D-glucose, which at equilibrium contains less than 0.024% of the *aldehydo* form[14] and does not have photochemical reactivity characteristic of carbonyl compounds. In contrast, even though the equilibrium concentration of the keto component in D-fructose is only 0.7–0.8%,[15,16] this concentration is sufficient for D-fructose to exhibit characteristic reactivity of a carbonyl compound; that is, it experiences Norrish Type I reaction upon photolysis and yields products arising from decarbonylation and subsequent radical reactions (eq 2).[15] (The mechanism for photoreaction of D-fructose is discussed, along with other compounds that undergo Norrish Type I reaction, in Chapter 2.)

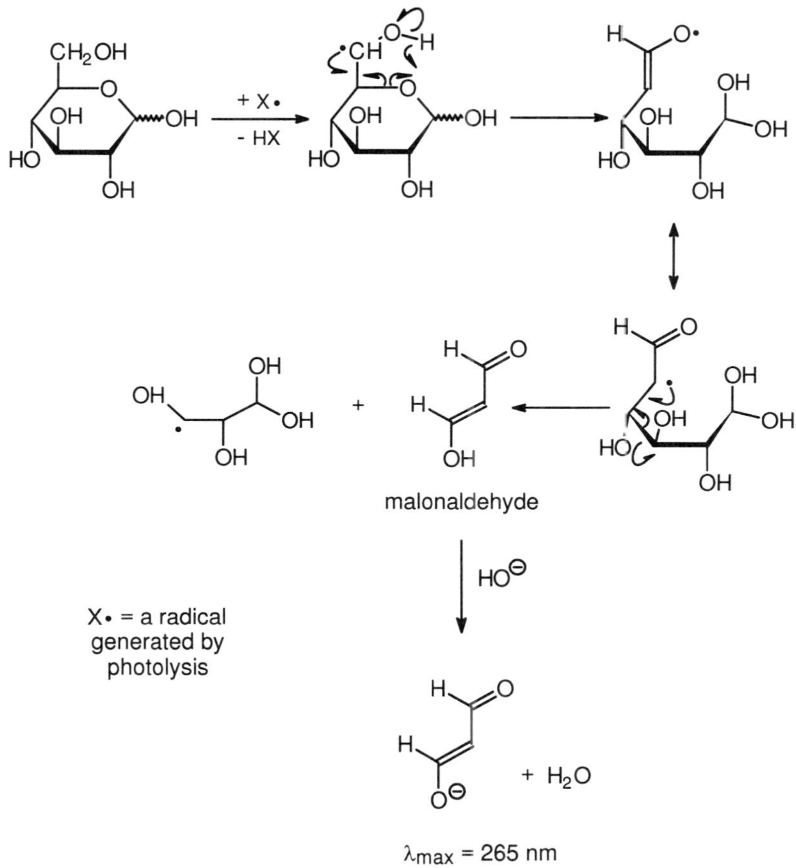

Scheme 11.5

D-Glucose

Phillips and Rickards conducted the same type of thorough investigation of D-glucose that had been done earlier for D-glucitol.[6] Their findings were consistent with those obtained from photolysis of D-glucitol; that is, the photolysis mixture became quite complex when conversion of starting material exceeded 10%, but at low conversions D-gluconic acid and D-arabinose accounted for nearly all product formation (eq 3).

Other Sugars

As mentioned in the introduction to this chapter, early photolyses of carbohydrates did not result in photoproduct identification. For this reason, these

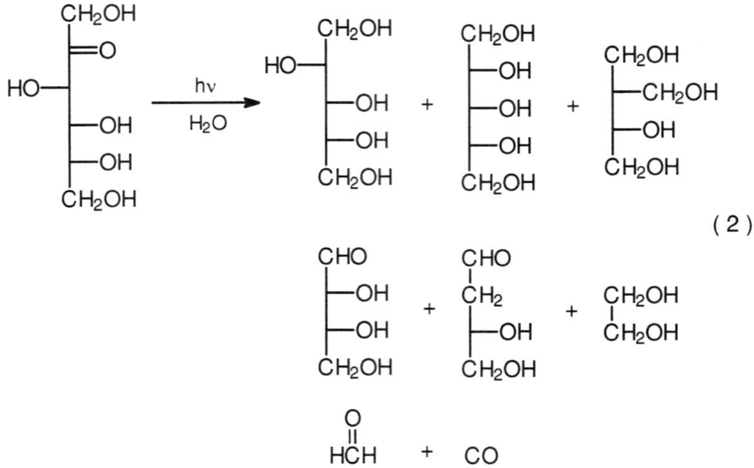

Equation 2

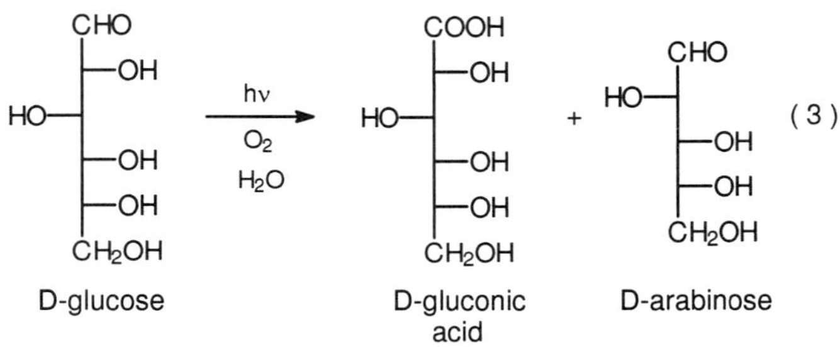

Equation 3

reactions are not discussed in this book. The interested reader is directed to references 1–3, where these investigations are described and references to the original articles are found. When consideration is restricted to compounds in which product identification took place, the several sugars, in addition to those already discussed, that need to be included are listed in Table 11.1. (The reactions of these compounds are analogous to those just described.) Some of these compounds also are mentioned in other places in this book. The photolysis of D-fructose is discussed in Chapter 2, and ascorbic acid is one of the compounds described in Chapter 8 as undergoing reaction with singlet oxygen.

Sugar Phosphates

Halmann and co-workers studied the photolysis of the disodium salts of α-D-glucose 6-phosphate and α-D-glucose 1-phosphate (Table 11.2) and found

Table 11.1. Photolysis of Unprotected Sugars and Related Compounds[a]

Reactant	*Product and Percent Yield*	*Ref.*
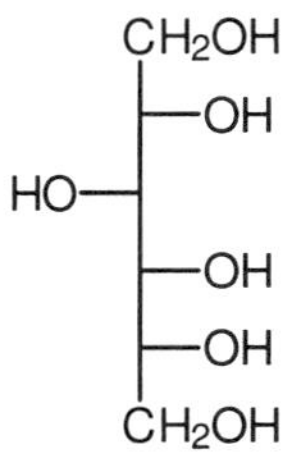	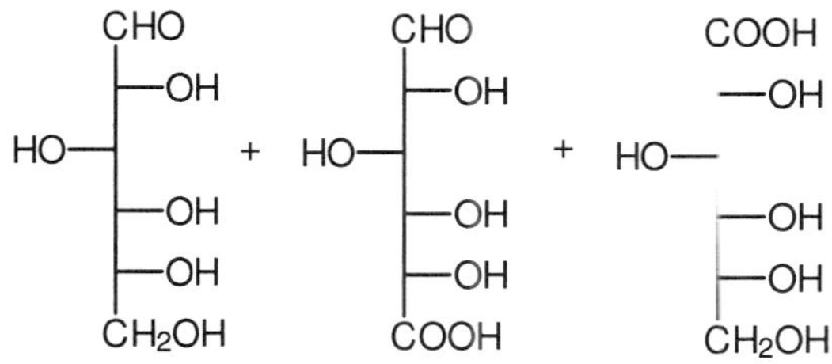 8% 3% 6% 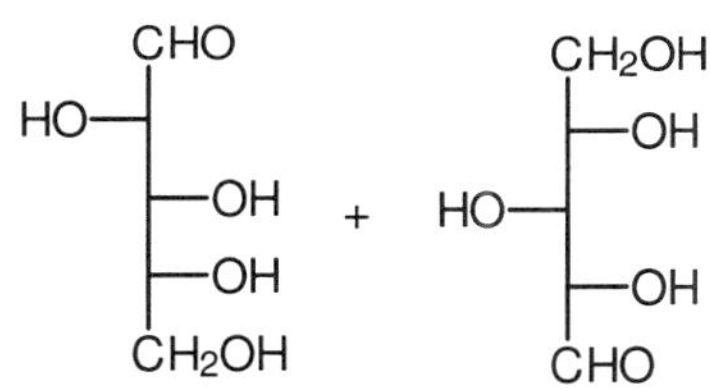 6% 6%	3, 4, 5[b,c]
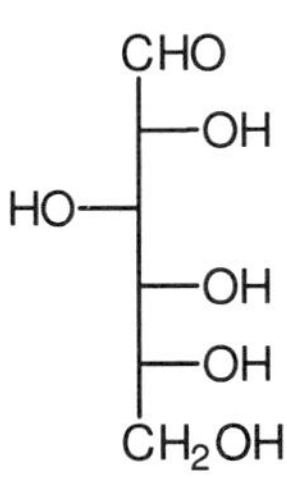	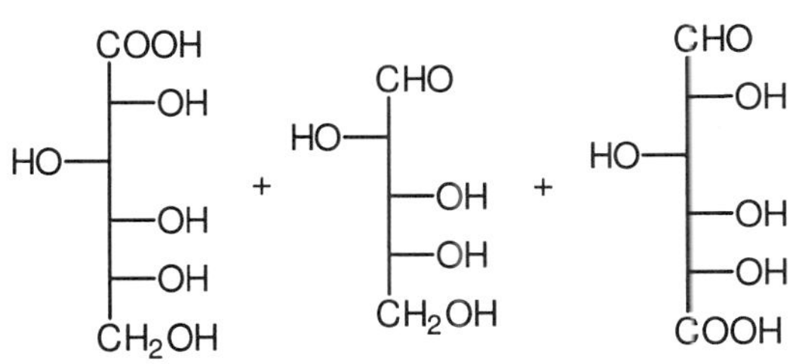 13% (0.5%) 7% (13%) 2% (0.6%) two-carbon compounds 16% (22%) + three-carbon compounds 1% (8%)	6, 7[d], 17[e]
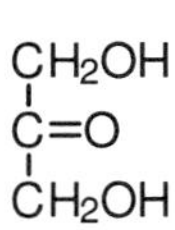	 80%	18

Continued on next page

Table 11.1. Photolysis of Unprotected Sugars and Related Compounds[a]—*Continued*

Reactant	*Product and Percent Yield*	*Ref.*
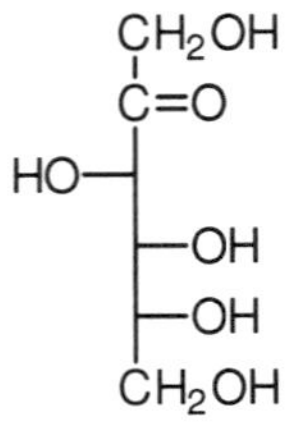	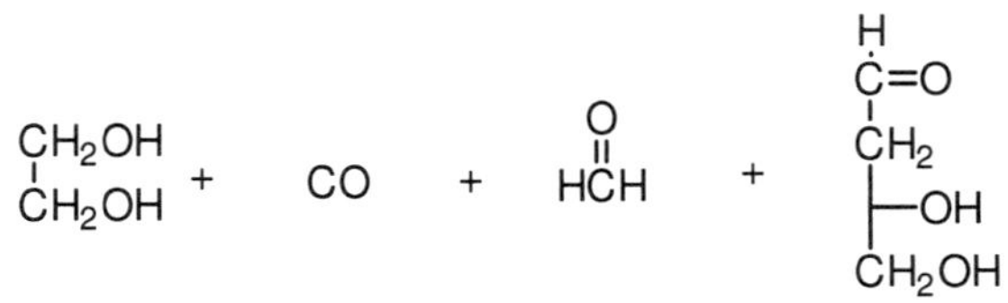 0.03 (0.00) 0.2 (0.02) 0.05 (0.06) 0.035 (0.00) 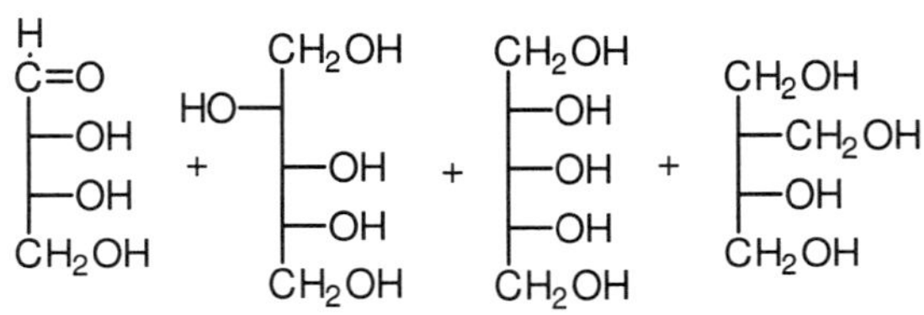 0.03 (0.08) 0.02 (0.00) 0.02 (0.00) 0.02 (0.00)	19[f]
$CHO-CH_2OH$	$CO + H_2C{=}O + CH_3OH + HC(=O)CH_2C(=O)H$	20[g]
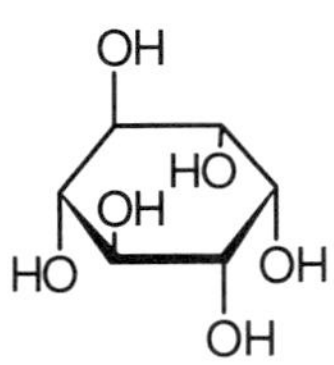	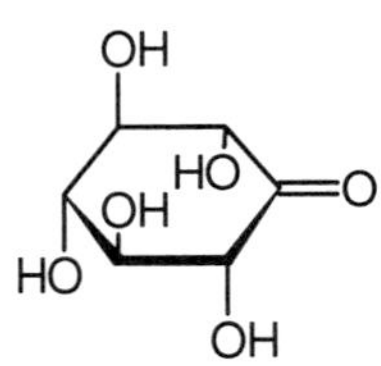	21
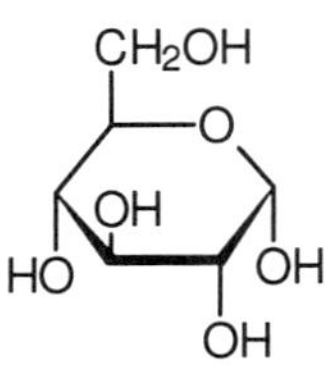	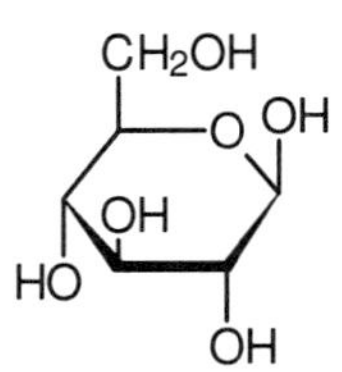	22, 23[g,h]

Continued on next page

Table 11.1. Photolysis of Unprotected Sugars and Related Compounds[a]—*Continued*

Reactant	*Product and Percent Yield*	*Ref.*
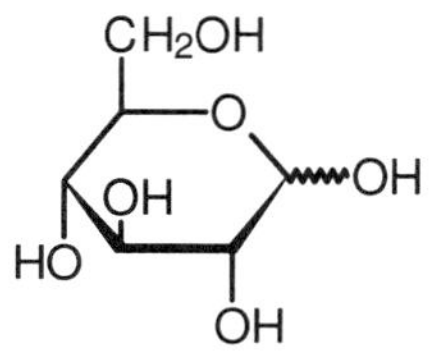 + $H_2N(CH_2)_4CH(NH_2)COOH$	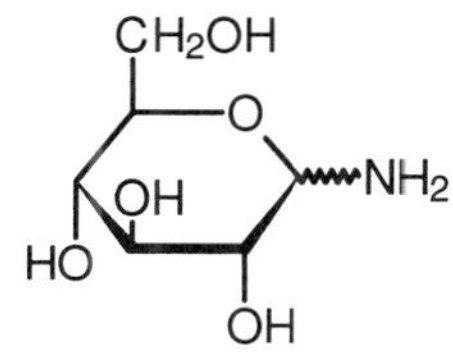 2.5% + 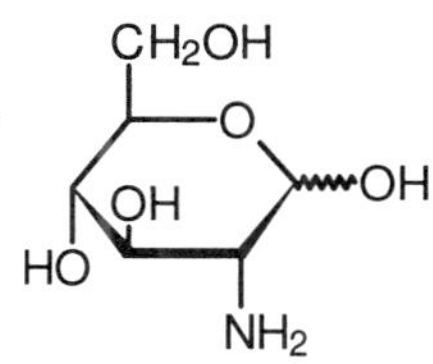 0.5%	249,i

[a]Reaction was conducted in the presence of oxygen, unless otherwise noted.

[b]In reference 5 reaction was achieved by sensitization with sodium anthraquinone-2-sulfonate and with sodium anthraquinone-2,6-disulfonate.

[c]Photosensitization was conducted on D-mannitol also.

[d]The numbers in parenthesis are from reference 7 and represent a reaction carried further toward completion.

[e]In this paper the focus was on conversion of D-glucose into D-gluconic acid.

[f]Structures for products with Φ>0.02 are included. Numbers without parenthesis are quantum yields determined in an inert atmosphere. Numbers in parenthesis are for quantum yields determined in the presence of oxygen.

[g]Photolysis was conducted in an inert atmosphere.

[h]The reaction solvent was methylsulfoxide and either benzophenone or acetophenone was present as a sensitizer.

[i]Reaction also was conducted in the presence of glycine.

Table 11.2. Photolysis of Sugar Phosphates[a]

Reactant	*Product*	*Ref.*
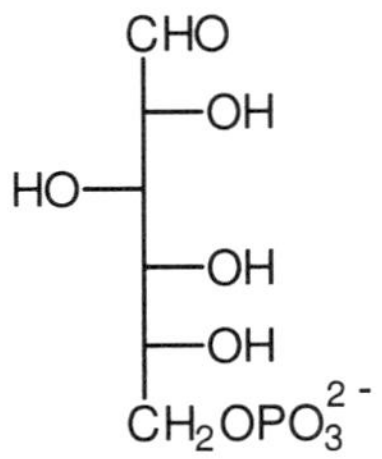	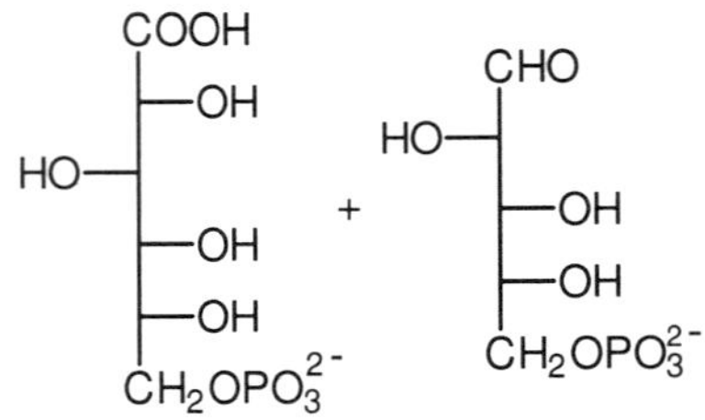	25
	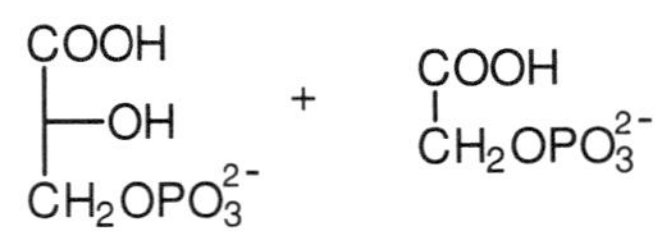	
	HPO_4^{2-} + CO_2	
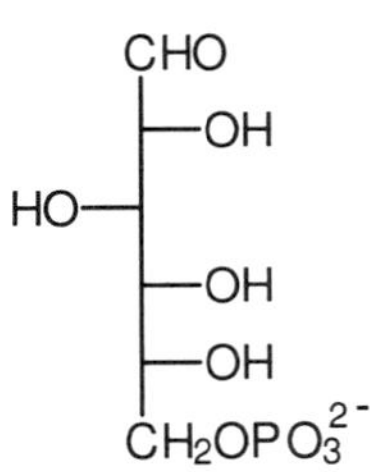	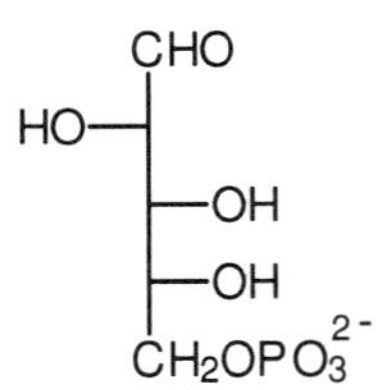 + HPO_4^{2-} + CO_2	25[b]
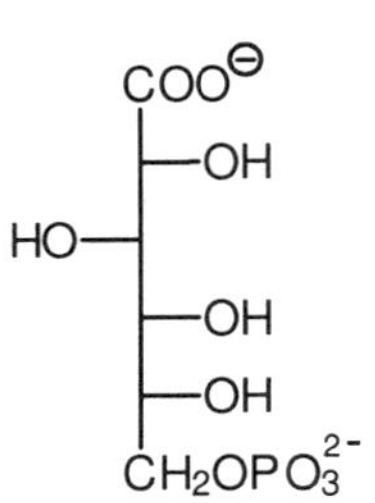	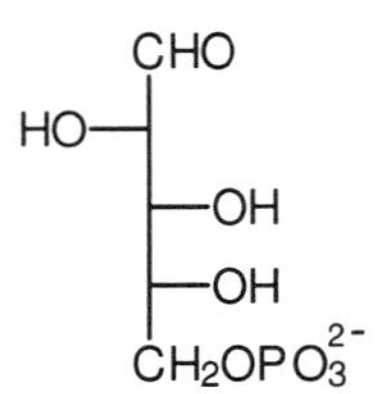	25

Continued on next page

Table 11.2. Photolysis of Sugar Phosphates[a]—*Continued*

Reactants	*Product and Percent Yield*	*Ref.*
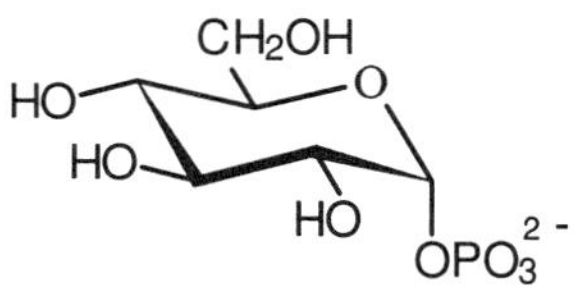 **2**	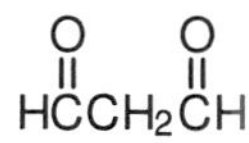 0.5% + HPO_4^{2-} 38% CO_2 2.8% + CO 2.2%	26
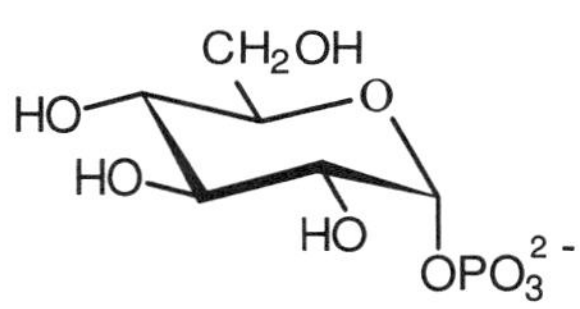 **2**	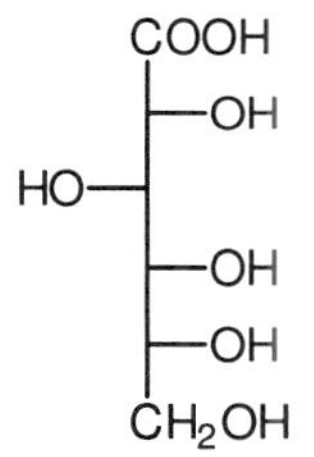 + 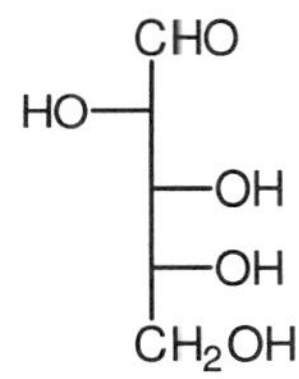 CO_2 + 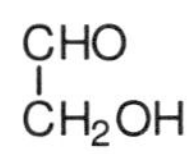	26[b]
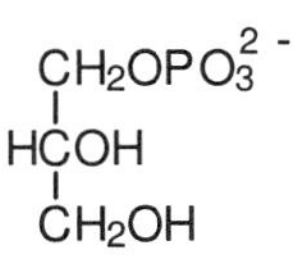	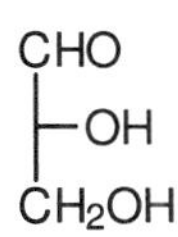 10% + 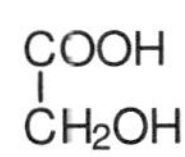 9% CO_2 + 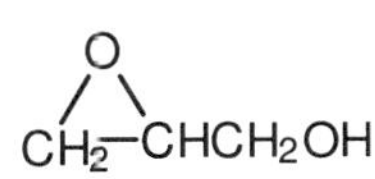 40%	27
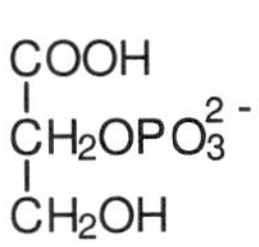	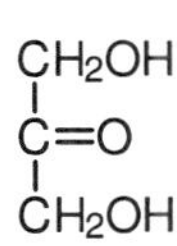 10% + 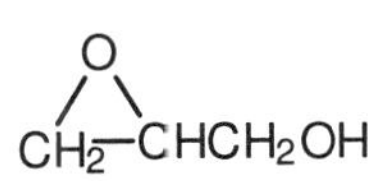 40% CO_2 +  9%	27

[a]Reaction was conducted under oxygen, unless otherwise noted.
[b]Reaction was conducted in an inert atmosphere.

that the most important reaction is the release of the orthophosphate anion (HPO_4^{2-}).[25,26] Aside from generating orthophosphate, however, photolyses of these compounds causes reactions analogous to those observed for D-glucose itself; for example, α-D-glucose 6-phosphate forms D-arabinose 5-phosphate (eq 4) by a stepwise degradation process similar to that shown in Scheme 11.1. Extended irradiation leads to extensive molecular fragmentation.

Metal Complexes of Unprotected Sugars

Complexation between an unprotected carbohydrate and a metal ion causes several important changes to occur in the photochemistry of the carbohydrate. The first notable change is that any uncertainty associated with absorption of light by an unprotected sugar[3–5] is no longer of concern because when complexed with a metal ion, a typical carbohydrate absorbs strongly in the UV region of the spectrum. For example, Araki and co-workers demonstrated that the reddish complex formed between D-glucose[28] or D-fructose[29] and iron(III) chloride begins absorbing light between 450 and 500 nm and has a maximum near 340 nm. Complexation changes the lowest energy excited singlet state from (n,σ*) in an uncomplexed carbohydrate[3, 4] to one involving internal electron transfer in a complexed compound.[29] This change in light absorption and excited state is responsible for a substantial reduction in complexity of the photochemical reaction mixture because secondary reactions arising from excitation of photoproducts are dramatically reduced. For example, irradiation of the complex formed between D-glucose and iron(III) chloride produces a 61% yield of 4-*O*-formyl-D-arabinose (eq 5).[30] This reaction stands in contrast to the variety of products formed in low yields from direct irradiation of D-glucose by itself (Table 11.1).

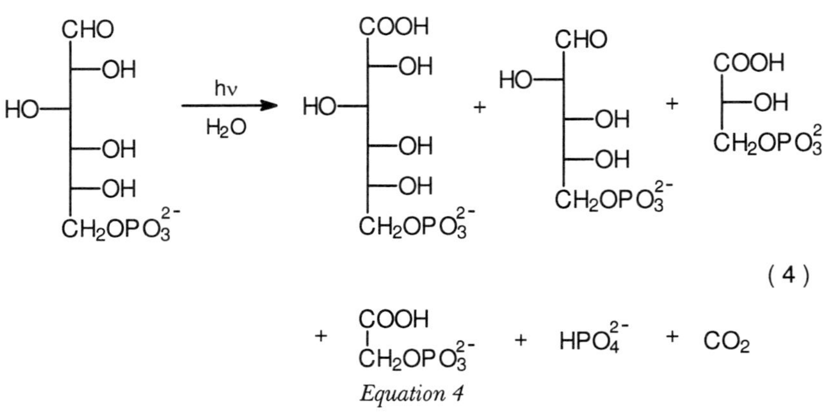

(4)

Equation 4

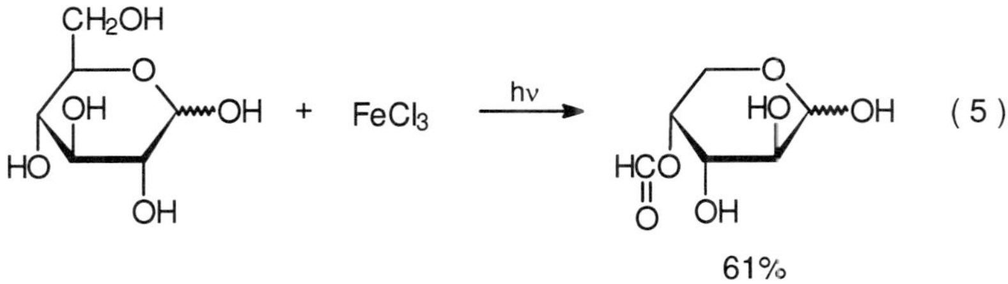

Equation 5

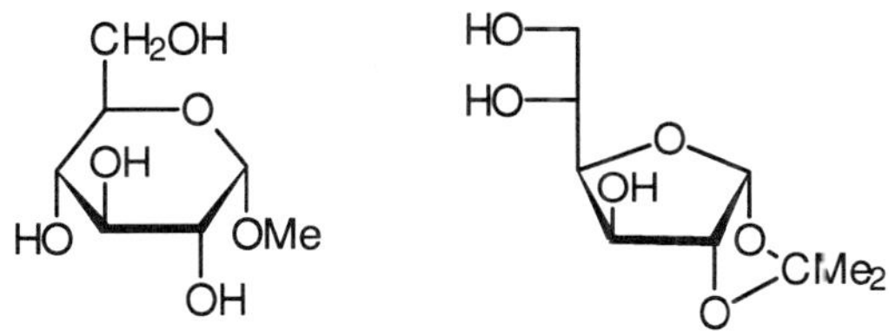

1 2

Chart 11.1. D-Glucose Derivatives

Ichikawa, Sato and co-workers have proposed that the formation of one major product from photolysis of D-glucose in the presence of iron(III) chloride occurs because the only complex that forms to a significant extent involves the unprotected, anomeric hydroxyl group.[30,31] In support of this proposal they found that methyl α-D-glucopyranoside (**1**) and 1,2-*O*-isopropylidene-α-D-glucofuranose (**2**) did not undergo reaction (Chart 11.1). They proposed that the photolysis of the complex between D-glucose and iron(III) chloride produces an open-chain intermediate (**3**) that then cyclizes to give 4-formyl-D-arabinose (Scheme 11.6). Although it has been suggested that photochemical electron transfer is taking place,[30] the details of this mechanism, particularly those relating to the photochemical changes that occur immediately after excitation of the complex, remain unclear and are in need of additional investigation.

The reported photolyses of complexes between iron(III), manganese(II), or titanium(IV) and a carbohydrate are listed in Table 11.3. In general, the reactions of iron(III) and manganese(II) complexes are quite similar in that fragmentation to a shorter chain sugar occurs as a result of cleavage of a bond involving the anomeric carbon atom. Titanium(IV) is different from the other two ions because it favors complexation of a primary hydroxyl group. Since the only carbohydrates complexed with titanium(IV) that have been investigated are aldohexoses, complexation necessarily involved the hydroxyl group at C-6. Photochemical cleavage of such complexes breaks the bond between C-5 and C-6. In order for complexation involving both O-5 and O-6 to occur, the monosaccharides must be in a hexofuranosyl form; consequently, photo-

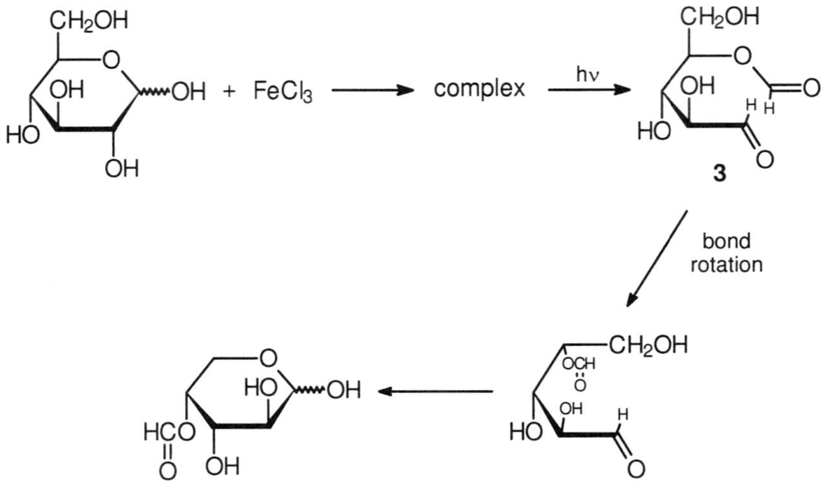

Scheme 11.6

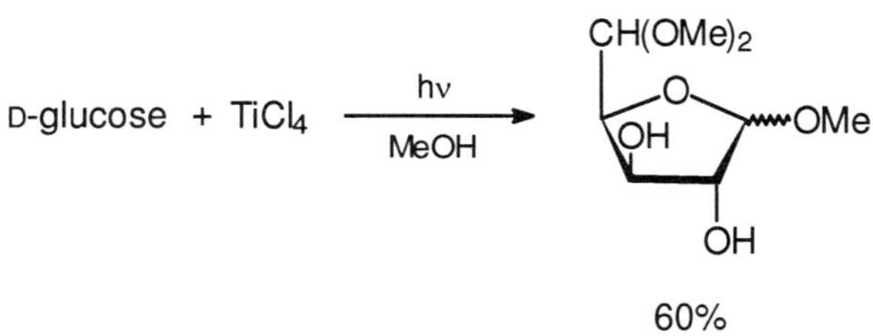

(6)

Equation 6

lysis produces five-membered ring compounds (eq 6).[36] When titanium(IV) chloride is used with methanol as the solvent, all carbonyl groups are converted into the corresponding methyl acetals.

Carbohydrate Polymers

Carbohydrate polymers have been the subject of a number of photochemical investigations. Most of this study has focused on changes in properties of the polymers being irradiated.[1,2] Little information exists about the specific molecular structure of the photoproducts and even less is known about reaction mechanisms. Since the data available about photolysis of carbohydrate polymers generally does not provide a basis for detailed mechanistic understanding of their photochemistry, the photochemical reactions of these polymers will not be discussed in this book. The reviews by Phillips[1,2] are excellent sources for existing information and for references to the primary literature.

Table 11.3. Reactions of Unprotected Sugars Complexed with Metal Ions[a]

Reactant	*Product*		*Percent Yield*	*Ref.*
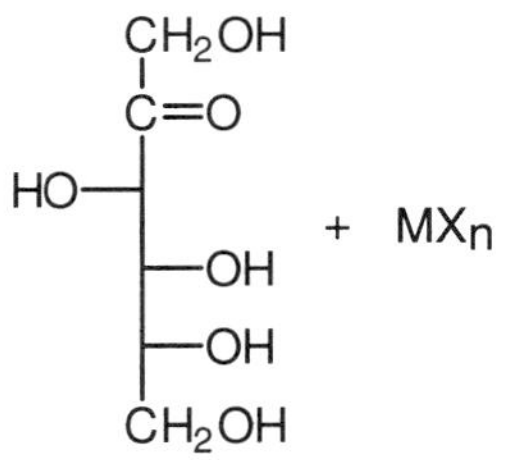	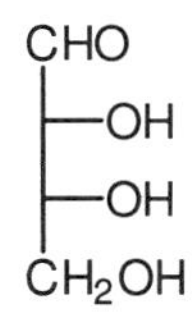			
$MX_n = FeCl_3$			82-88 60-74	28, 32 33
$MX_n = MnCl_2$			57 56	33 34
$MX_n = Fe(NO_3)_3$			85	28
$MX_n = FeCl_2$			73	34
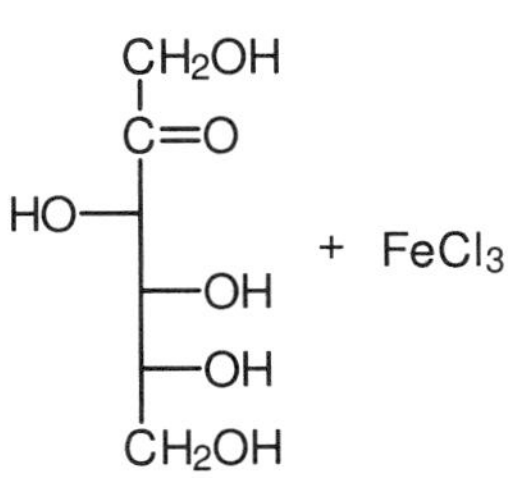	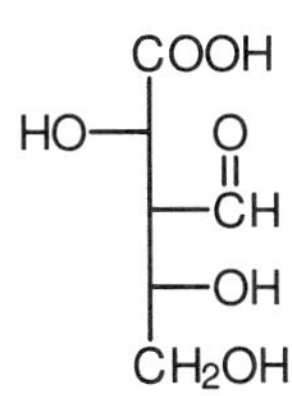		52[b]	30
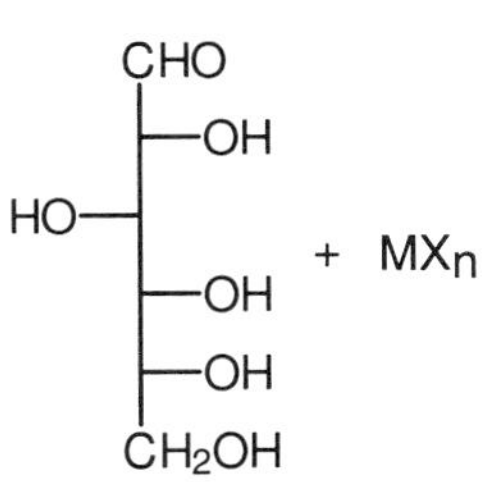	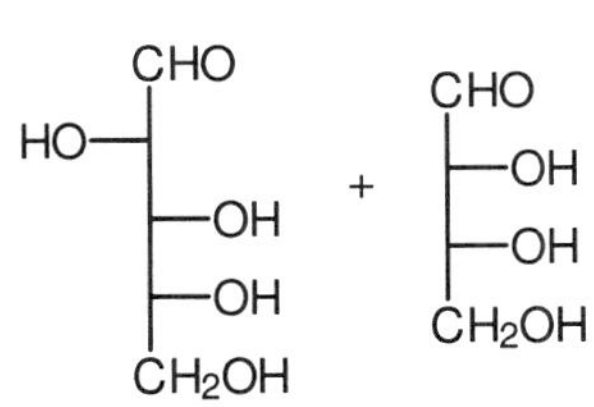			
$MX_n = MnCl_2$	20%			33
$MX_n = FeCl_3$	28-50%	50-68%		28

Continued on next page

Table 11.3. Reactions of Unprotected Sugars Complexed with Metal Ions[a]—*Continued*

Reactant	*Product*	*Percent Yield*	*Ref.*
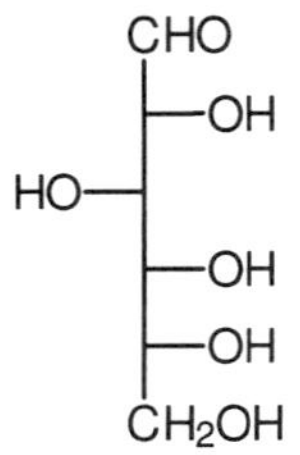 + MX_n	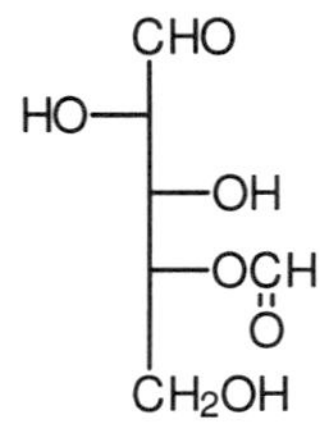		
MX_n = $FeCl_3$		38[b,c] 45[b,c]	30, 31 35
MX_n = $Fe(OTf)_3$		40[b,c,d] 44[b,c,d]	30 35
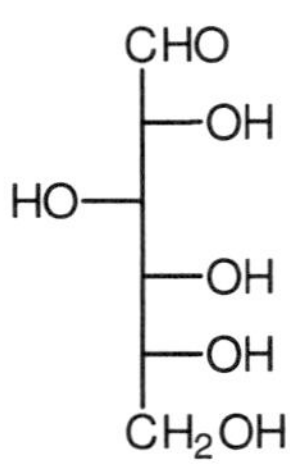 + $TiCl_4$ + CH_3OH	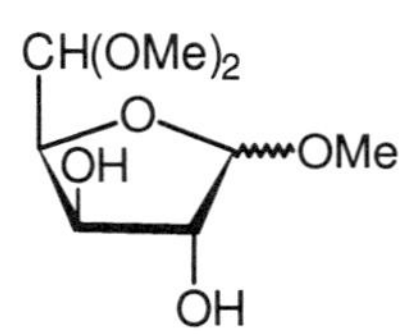	60	36
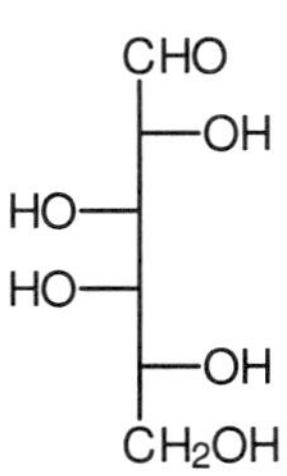 + $TiCl_4$ + CH_3OH	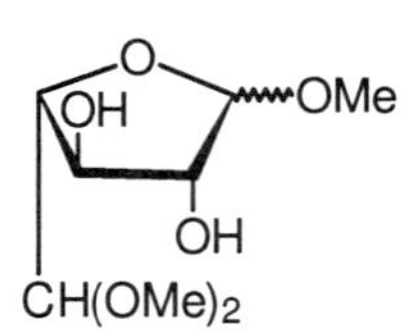	e	36
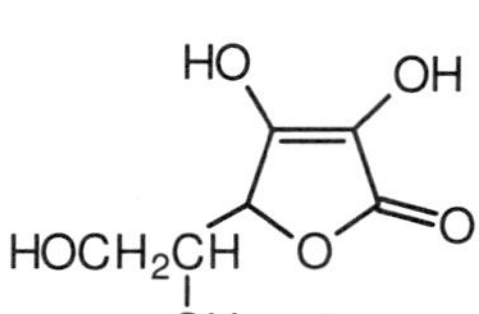 + O_2 + Ru^{2+}	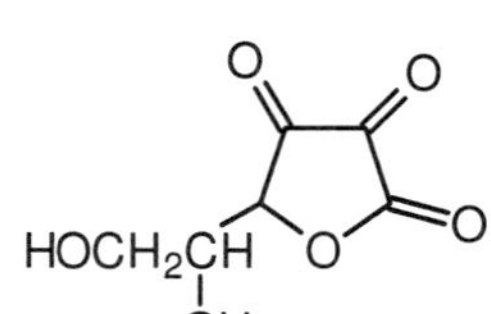 + H_2O_2	e	37

Continued on next page

Table 11.3. Reactions of Unprotected Sugars Complexed with Metal Ions[a]—*Continued*

Reactant	*Product*	*Percent Yield*	*Ref.*
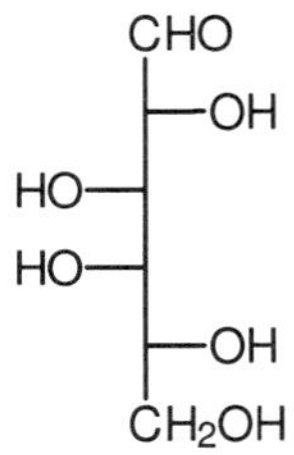 + MXn	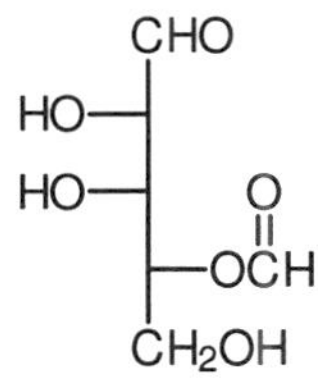		
MXn = $FeCl_3$		35[b,c] 40[b,c]	30, 31 35
MXn = $Fe(OTf)_3$		56[b,c,d]	35
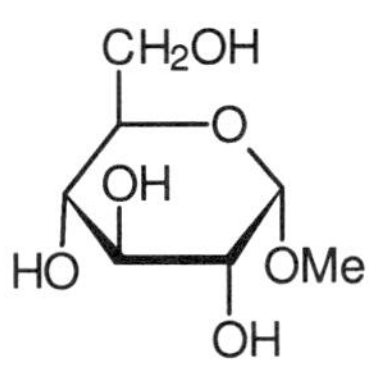 + MXn			
MXn = $TiCl_4$		e	38
MXn = $FeCl_3$		e	31
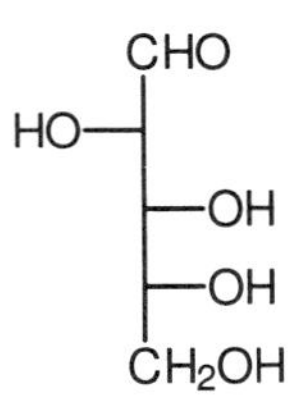 + $FeCl_3$	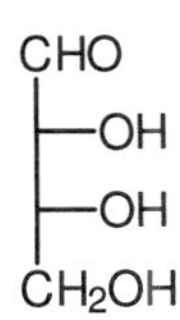	89	28
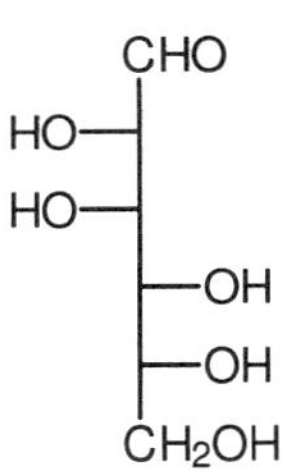 + $FeCl_3$	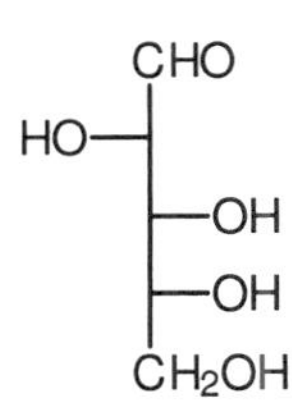 28%	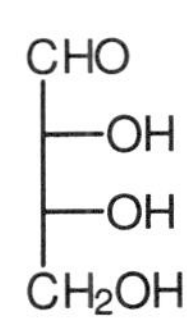 45%	28

Continued on next page

Table 11.3. Reactions of Unprotected Sugars Complexed with Metal Ions[a]—*Continued*

Reactant	*Product*	*Percent Yield*	*Ref.*
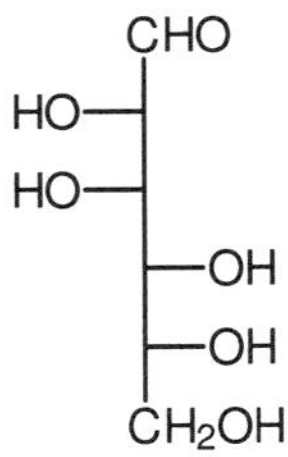 + MX_n	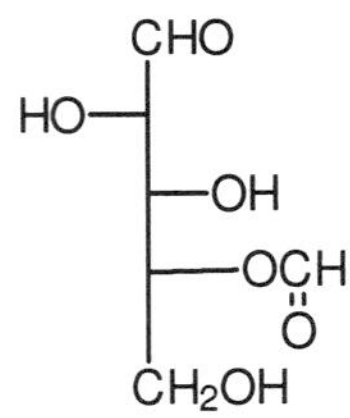		
MX_n = $FeCl_3$		17[b,c]	30, 31, 35
MX_n = $(OTf)_3$		57[b,c,d]	35
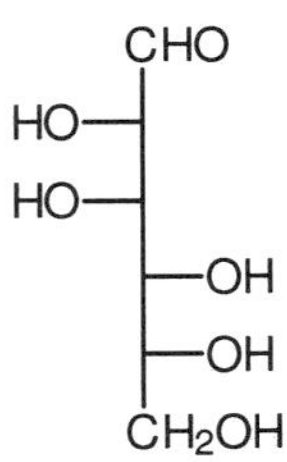 + $TiCl_4$ + CH_3OH	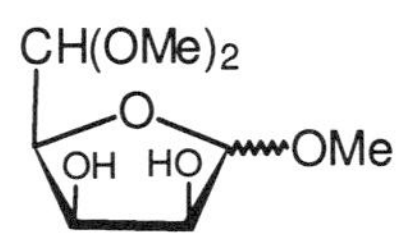	50[b,c]	38
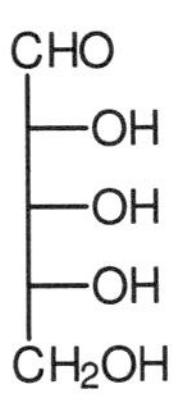 + $FeCl_3$	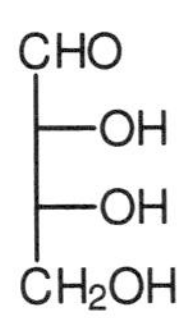 52% 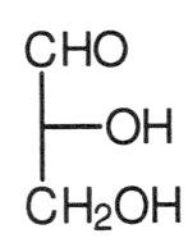 23%		28
 + OsO_4 + $Ba(ClO_3)_2$	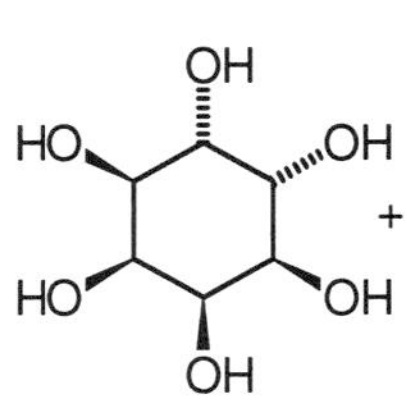 31% + 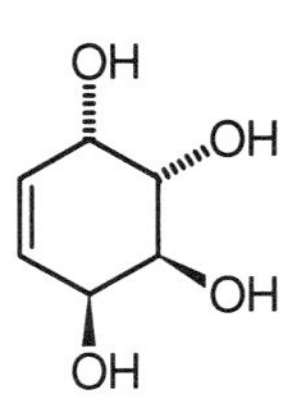 5%		39

Continued on next page

Table 11.3. Reactions of Unprotected Sugars Complexed with Metal Ions[a]—*Continued*

Reactant	*Product*	*Percent Yield*	*Ref.*
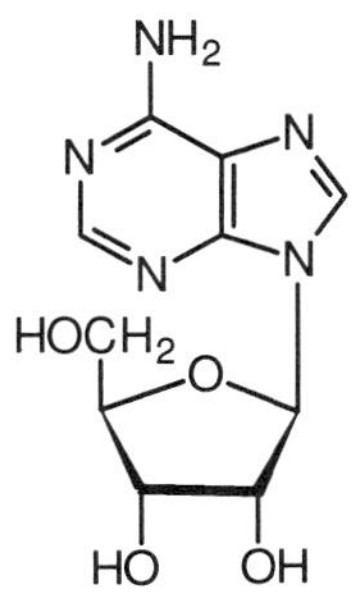	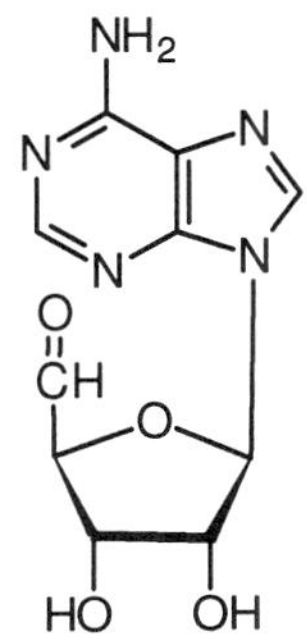	25 30[f]	40
	+ 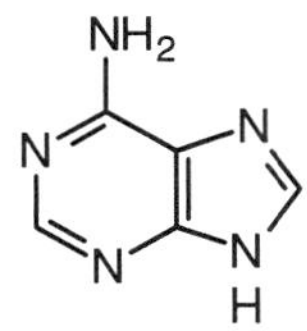	63 50[f]	

[a]Unless otherwise noted, reactions were conducted in the presence of oxygen.

[b]Pyridine was the reaction solvent.

[c]Considerable unreacted starting material was recovered.

[d]OTf = OSO_2CF_3.

[e]There was no reaction.

[f]Reaction was conducted in an inert atmosphere.

$$(C_6H_{12}O_6)_n + 6n\,H_2O \xrightarrow[\substack{RuO_2 \\ TiO_2 \\ Pt}]{\substack{h\nu \\ H_2O}} 6n\,CO_2 + 12n\,H_2 \quad (7)$$

Equation 7

Photochemical Conversion of Carbohydrates into Molecular Hydrogen and Carbon Dioxide

Kawai and Sakata discovered that photolysis of sucrose, starch, or cellulose in aqueous solution in the presence of a suspension of finely divided RuO_2, TiO_2 and Pt produces molecular hydrogen and carbon dioxide (eq 7).[41,42] When this reaction was carried to completion, no other products were detected. Also, reaction could be conducted in basic solution to release only hydrogen gas and form Na_2CO_3 and $NaHCO_3$. Kawai and Sakata proposed that the $RuO_2/TiO_2/$ Pt powder acts as an electrochemical microcell. They suggested that light of energy larger than the band gap of TiO_2 produces electron-hole pairs. The separated electrons and holes cause redox reactions of the carbohydrates at the surface of the particles.

Summary

Photolysis of unprotected carbohydrates in the presence of oxygen results in formation of complex reaction mixtures (Table 11.1) due to the rapid, sequential breakdown of the primary photoproducts (Scheme 11.2). Sugar phosphates react in a manner similar to that of unprotected sugars except that the major reaction is release of the orthophosphate anion (Table 11.2). Reaction of unprotected carbohydrates with iron(III), manganese(II), and titanium(IV) ions produces complexes that readily absorb UV light and easily undergo photochemically promoted, oxidation–reduction reactions that fragment the carbon skeleton (Table 11.3).

References

1. Phillips, G. O. In *The Carbohydrates*; Pigman, W.; Horton, D., Eds.; Academic Press: New York, 1980; Vol 1B, p 1217.
2. Phillips, G. O. *Adv. Carbohydr. Chem.* **1963**, *18*, 9.
3. Phillips, G. O.; Criddle, W. J. *J. Chem. Soc.* **1963**, 3984.
4. Phillips, G. O.; Barber, P. *J. Chem. Soc.* **1963**, 3990.
5. Phillips, G. O.; Barber, P.; Rickards, T. *J. Chem. Soc.* **1964**, 3443.

6. Phillips, G. O.; Rickards, T. *J. Chem. Soc. (B)* **1969**, 455.
7. Phillips, G. O.; Moody, G. J. *J. Chem. Soc.* **1960**, 3398.
8. Bothe, E.; Schulte-Frohlinde, D.; von Sonntag, C. *J. Chem. Soc., Perkin Trans. 2* **1978**, 416.
9. Bothe, E.; Schuchmann, M. N.; Schulte-Frohlinde, D.; von Sonntag, C. *Photochem. Photobiol.* **1978**, *28*, 639.
10. von Sonntag, C. *Adv. Carbohydr. Chem. Biochem.* **1980**, *37*, 7.
11. Laurent, T. C.; Wertheim, E. M. *Acta Chem. Scand.* **1952**, *6*, 678.
12. Laurent, T. C. *J. Am. Chem. Soc.* **1956**, *78*, 1875.
13. H. Scherz, *Carbohydr. Res.* **14**, 417 (1970).
14. Pigman, W.; Anet, E. F. L. J. In *The Carbohydrates*; Pigman, W.; Horton, D. Eds.; Academic Press: New York, 1972; Vol IA, p 165.
15. Funcke, W.; von Sonntag, C.; Triantaphylides, C. *Carbohydr. Res.* **1979**, *75*, 305.
16. Hayward, L. D.; Angyal, S. J. *Carbohydr. Res.* **1977**, *53*, 13.
17. Degelmann, H.-P.; Emig, G.; Hofmann, H. *Zuckerindustrie* **1982**, *107*, 117.
18. Steeken, S.; Jaenicke-Zauner, W.; Schulte-Frohlinde, D. *Photochem. Photobiol.* **1975**, *21*, 21.
19. Triantaphylides, C.; Schuchmann, H.-P.; von Sonntag, C. *Carbohydr. Res.* **1982**, *100*, 131.
20. Beeby, A.; Mohammed, D. B. H.; Sodeau, J. R. *J. Am. Chem. Soc.* **1987**, *109*, 857.
21. Criddle, W. J.; Jones, B.; Ward, E. *Chem. Ind.* **1967**, 1833.
22. Han, S. D.; Lee, T. Y.; Chae, W. K. *Bull. Korean Chem. Soc.* **1986**, *7*, 251.
23. Kang, B. I.; Chae, W. K. *Bull. Korean Chem. Soc.* **1990**, *11*, 183.
24. Donor, L. W.; Balicki, R.; Whistler, R. L. *Carbohydr. Res.* **1976**, *47*, 342.
25. Triantaphylides, C.; Halmann, M. *J. Chem. Soc., Perkin Trans. 2* **1975**, 34.
26. Trachtman, M.; Halmann, M. *J. Chem. Soc., Perkin Trans. 2* **1977**, 132.
27. Greenwald, J.; Halmann, M. *J. Chem. Soc., Perkin Trans. 2* **1972**, 1095.
28. Araki, K.; Sakuma, M.; Shiraishi, S. *Bull. Chem. Soc. Jpn.* **1984**, *57*, 997.
29. Araki, K.; Shiraishi, S. *Bull. Chem. Soc. Jpn.* **1986**, *59*, 3661.
30. Ichikawa, S.; Tomita, I.; Hosaka, A.; Sato, T. *Bull. Chem. Soc. Jpn.* **1988**, *61*, 513.
31. Ichikawa, S.; Sato, T. *Tetrahedron Lett.* **1986**, *27*, 45.
32. Araki, K.; Sakuma, M.; Shiraishi, S. *Chem. Lett.* **1983**, 665.
33. Araki, K.; Shiraishi, S. *Chem. Lett.* **1984**, 267.
34. Araki, K.; Shiraishi, S. *Bull. Chem. Soc. Jpn.* **1986**, *59*, 229.
35. Hosaka, A.; Ichikawa, S.; Shindo, H.; Sato, T. *Bull. Chem. Soc. Jpn.* **1989**, *62*, 797.
36. Sato, T.; Takahashi, K.; Ichikawa, S. *Chem. Lett.* **1983**, 1589.
37. Tsukahara, K.; Wada, Y.; Kimura, M. *Bull. Chem. Soc. Jpn.* **1991**, *64*, 908.
38. Ichikawa, S.; Takahashi, K.; Tanaka, M.; Sato, T. *Bull. Chem. Soc. Jpn.* **1988**, *61*, 505.
39. Motherwell, W. B.; Williams, A. S. *Angew. Chem. Int. Ed. Engl.* **1995**, *34*, 203.
40. Araki, K.; Shiraishi, S. *Photochem. Photobiol.* **1987**, *45*, 157.
41. Kawai, T.; Sakata, T. *Nature* **1980**, *286*, 474.
42. Sakata, T.; Kawai, T. *Nouv. J. Chim.* **1981**, *5*, 279.

12

Sulfonic Acid Derivatives

Introduction

Interest in the photochemistry of sulfonic acid esters dates back to the 1960's when Jeger, Schaffner, and co-workers reported that photolysis of α-sulfonyloxyketones and α,β-epoxyketones produced characteristic rearrangement and elimination reactions (eq 1).[1,2] Product yields were sufficiently good for Schaffner to suggest that photolyses of these compounds could become "potent synthetic tools of wide applicability".[3] This prediction was realized for α,β-epoxyketones as their photochemistry became an important area of research, but the proposed role for α-sulfonyloxyketones failed to materialize. Nevertheless, interest had been created in sulfonate photochemistry and was reinforced by the discovery of Stratenus and Havinga that aromatic tosylates experienced photo-Fries rearrangement (eq 2).[4] Although irradiation of a variety of sulfonic acid esters has occurred since these initial discoveries, nowhere has photochemical reaction of sulfonates become as important as in carbohydrate chemistry. When carbohydrate chemists began to investigate sulfonate photolyses, they found that these reactions met a significant, existing need.[5]

In 1968 Zen et al. reported that irradiation of carbohydrate *p*-toluenesulfonates caused replacement of the tosyl group with a hydrogen atom (eq 3).[6] This result provided the missing element needed for an improved protection–deprotection sequence for hydroxyl groups. Because it was well established that tosyl groups could be introduced easily into carbohydrates and that these tosyl-protected compounds were stable under a variety of reaction conditions, photolysis proved to be the mild deprotection process required to make tosyl group introduction and removal a synthetically useful, protection–deprotection sequence (Scheme 12.1). Subsequent research showed that the sodium methoxide used in this photochemical reaction could be replaced by a tertiary

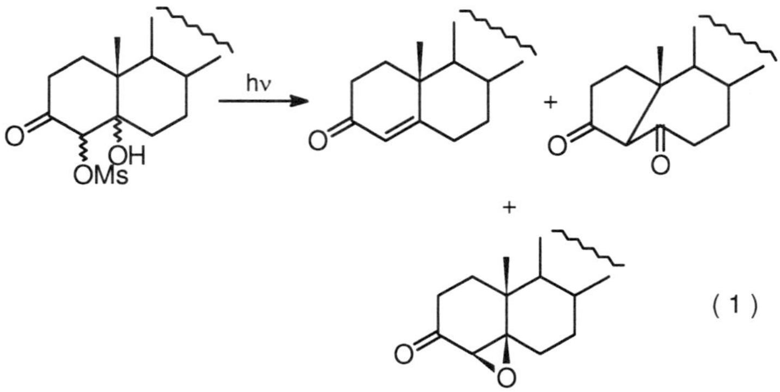

(1)

Equation 1

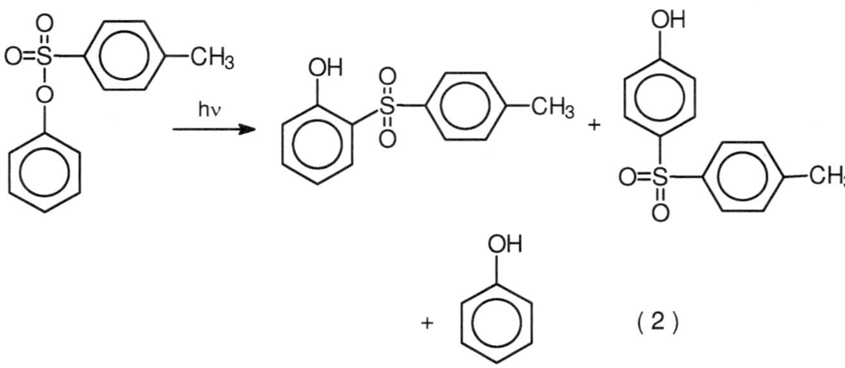

(2)

Equation 2

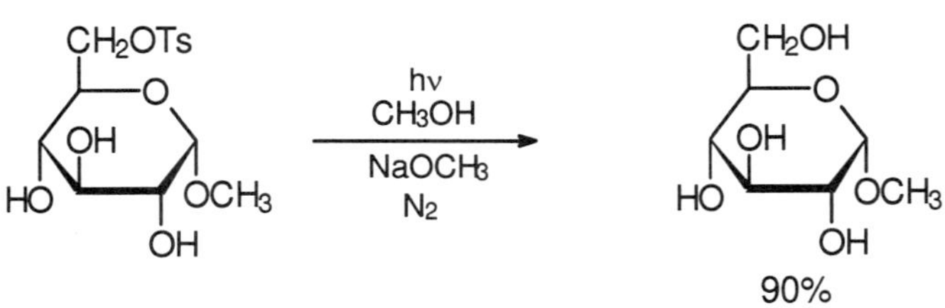

(3)

Equation 3

Scheme 12.1

amine. This finding "singled out" photolysis as a process superior to existing detosylation methods (e.g., reaction with lithium aluminum hydride,[7] Raney nickel,[8] sodium in liquid ammonia,[9] and the sodium naphthalene anion radical[10]), all of which required strongly basic or otherwise vigorous reaction conditions.

Once the usefulness of tosylate photochemical reaction became evident, investigation of the mechanism for this process began. Initial studies provided considerable insight and "paved the way" for later contributions that clarified the details of the reaction process. Later mechanistic work (see the section "Reaction Mechanisms") established reaction conditions that allowed broader application of this protection–deprotection sequence. Since the time of the initial investigation by Zen et al.,[6] a variety of sulfonic acid esters have been irradiated. The majority of the research conducted on these compounds has been on esters of *p*-toluenesulfonic acid.

p-*Toluenesulfonates*

Absorption of Ultraviolet Light by *p*-Toluenesulfonates

p-Toluenesulfonates have intense absorption in the UV with a maximum at 220 nm ($\in$ 10,000).[11] They also exhibit broad, much weaker absorption between 240 and 280 nm with a typical 0-0 absorption band at 272 nm;[12] thus, in order for the *p*-tolylsulfonyl group to become excited by direct irradiation, light of wavelength less than 300 nm is necessary.

Photochemical Deprotection of *p*-Toluenesulfonates

Effect of Structure on Reactivity

Table 12.1 contains a list of the carbohydrate tosylates that have been deprotected by photochemical reaction. Several observations about photochemical detosylation are possible from examining the entries in this table. First, photochemical reaction does not depend on the type of ring system (pyranoid or furanoid) that contains the tosyloxy group. Also, primary and secondary tosyloxy groups have similar reactivity. Compounds with two sulfonyloxy groups react in the same manner as those with only one. Finally, there are no reported examples of irradiation of tertiary or anomeric tosylates.

Effect of Added Reagents

Irradiation in the Presence of Sodium Methoxide. A significant characteristic of the tosyl protection–deprotection sequence is that it can be used in the presence of a variety of groups commonly found in carbohydrates. The originally reported reaction conditions,[6] which called for irradiation in a methanolic

Table 12.1. Photochemical Reactions of Carbohydrate Tosylates[a]

Reactant	*Product*	*Percent Yield*	*Ref.*
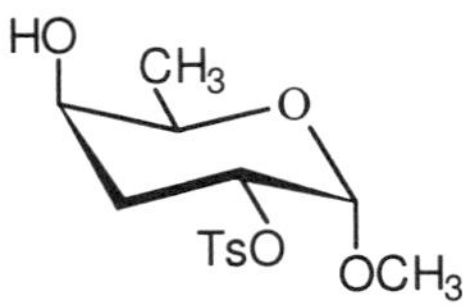	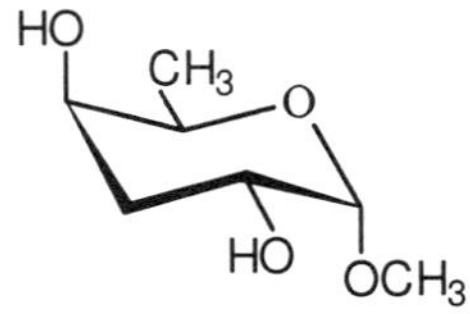	90	13
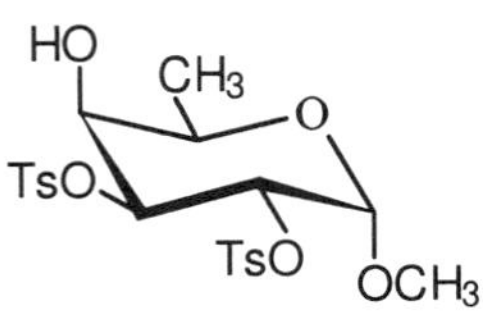	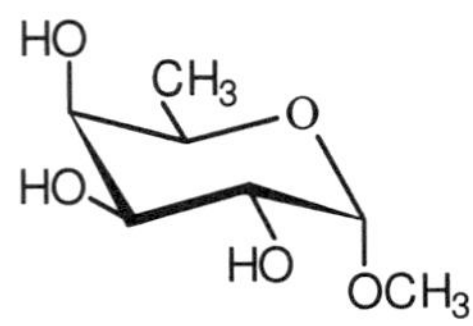	85	13
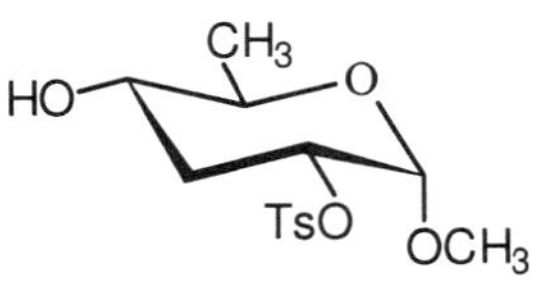	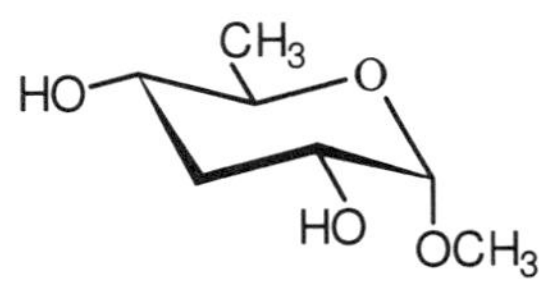	95	13
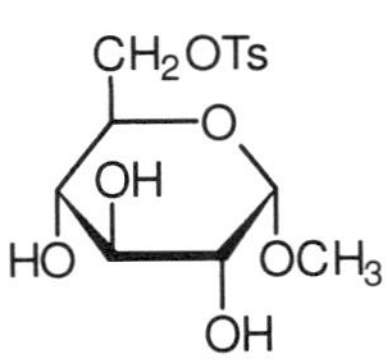	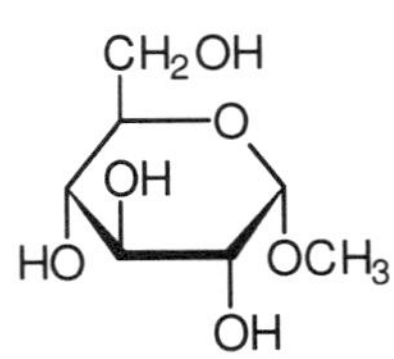	90	6
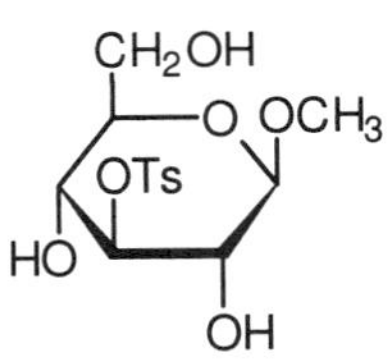	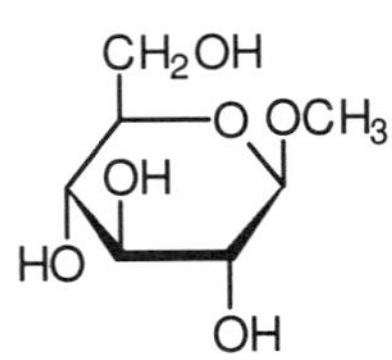	80	6
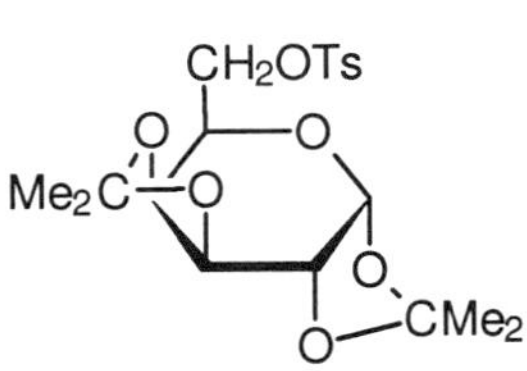	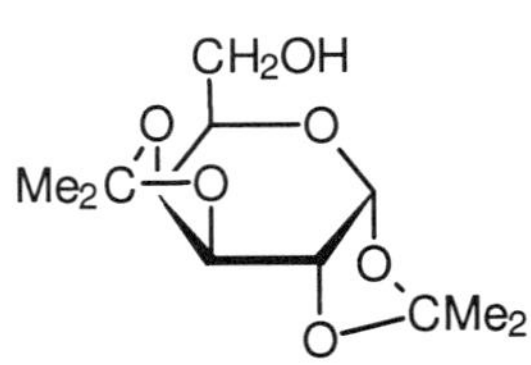	100 100	6 12,14

Continued on next page

Table 12.1. Photochemical Reactions of Carbohydrate Tosylates[a]—*Continued*

Reactant	*Product*	*Percent Yield*	*Ref.*
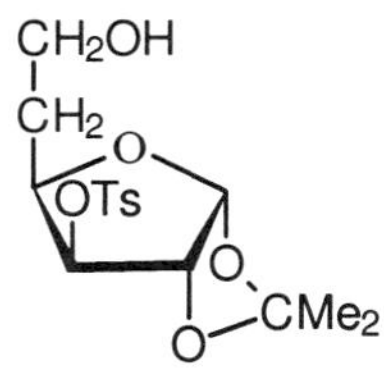	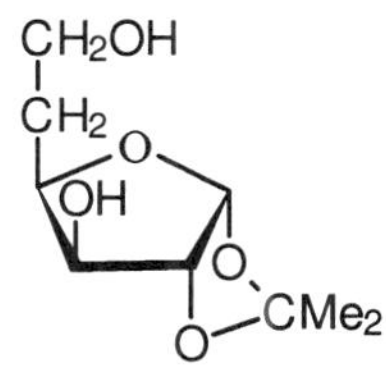	50	15
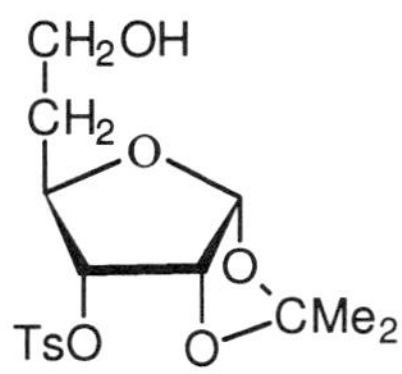	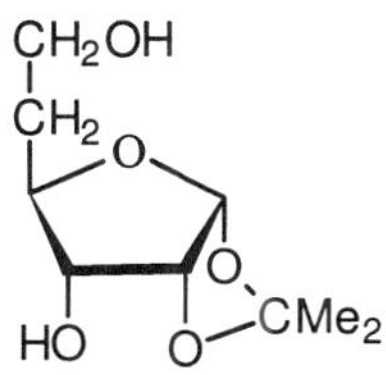	91	16
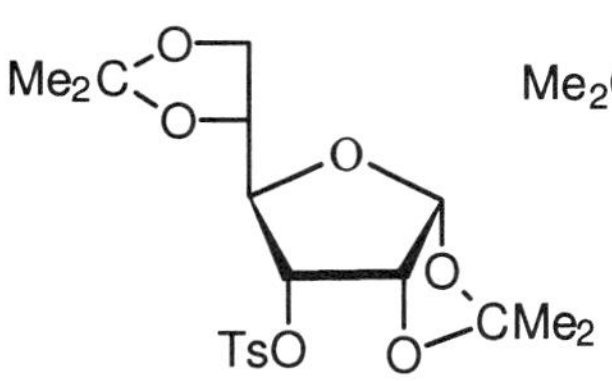	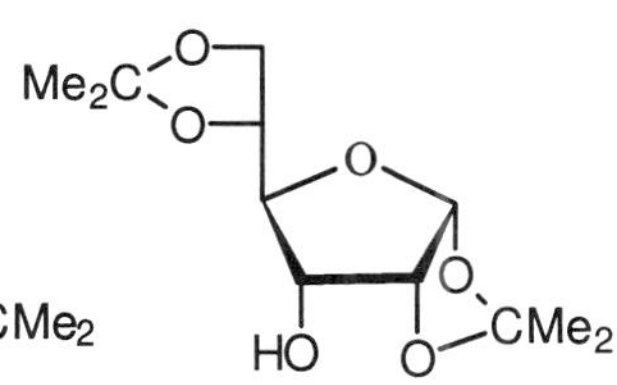	96 65[b]	12,14 17,18
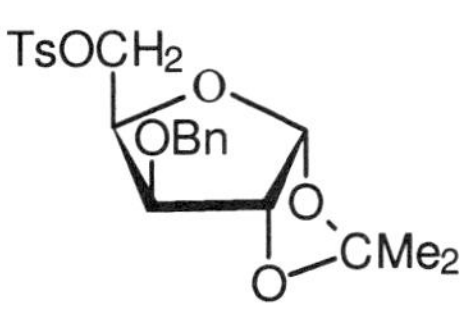	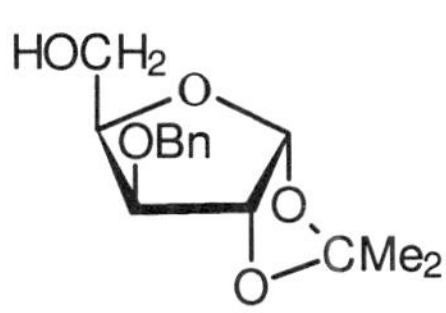	59[c] 68[d]	17,18 17,18
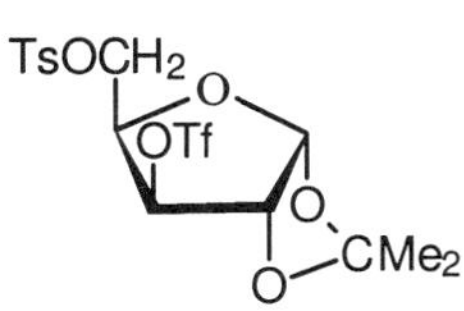	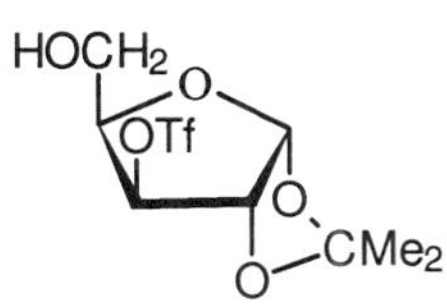	70	19
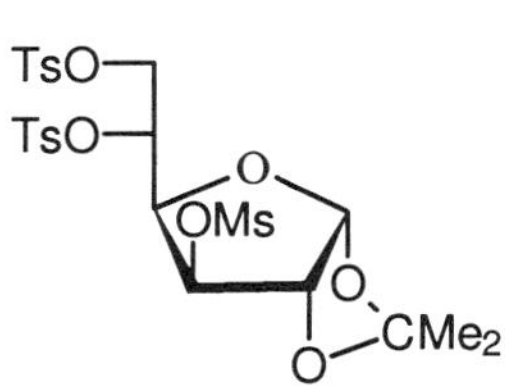	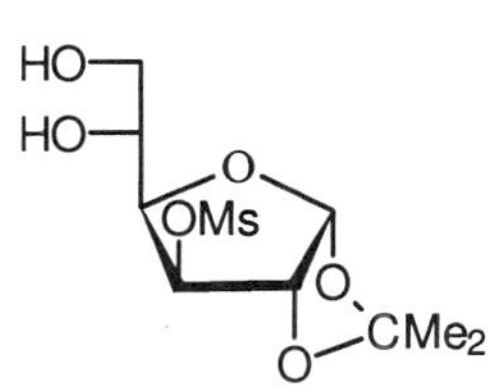	100	20, 21

Continued on next page

Table 12.1. Photochemical Reactions of Carbohydrate Tosylates[a]—*Continued*

Reactant	*Product*	*Percent Yield*	*Ref.*
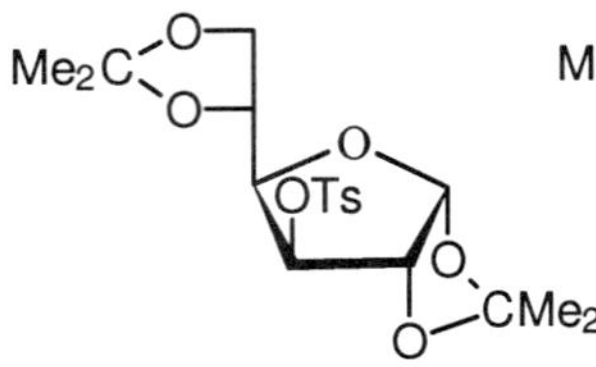	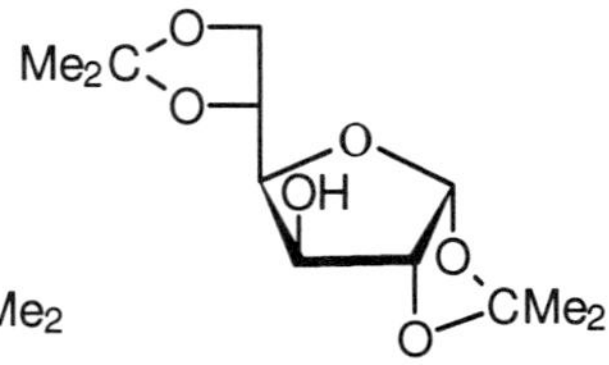	100	6
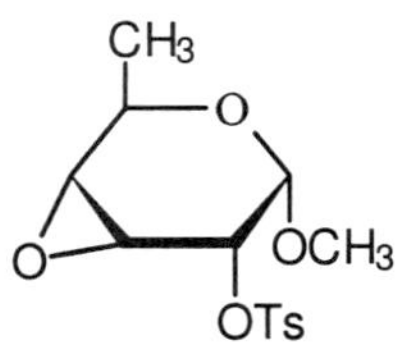	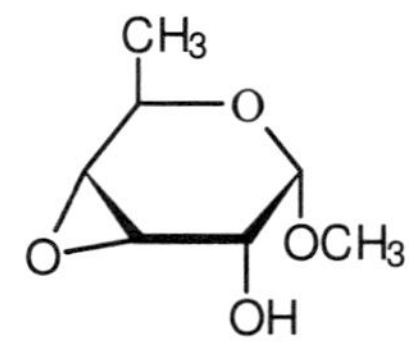	60	25
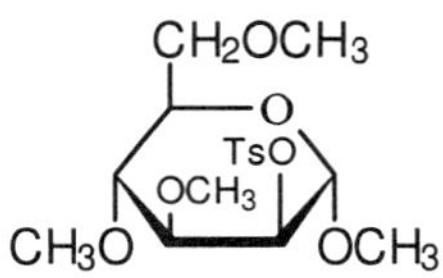	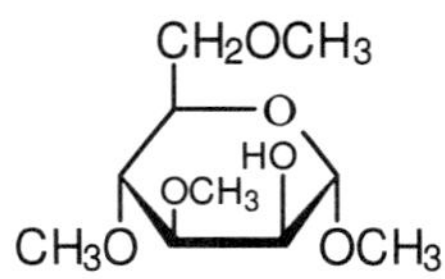	87	26
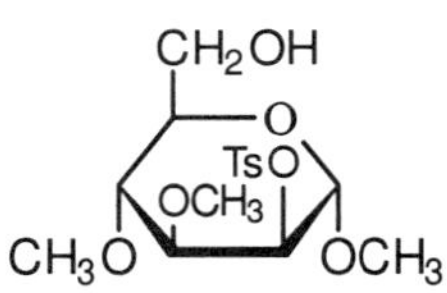	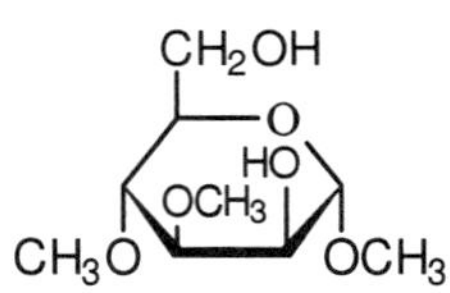	62	26
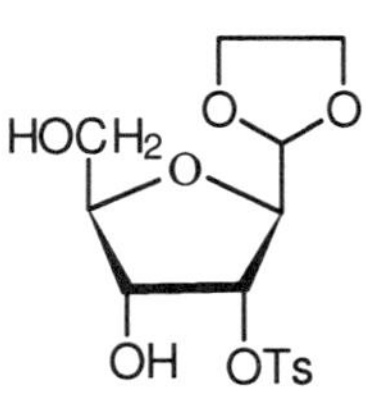	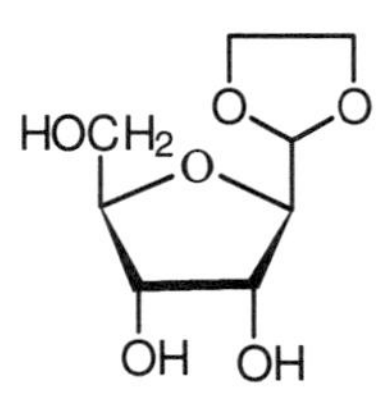	70	27
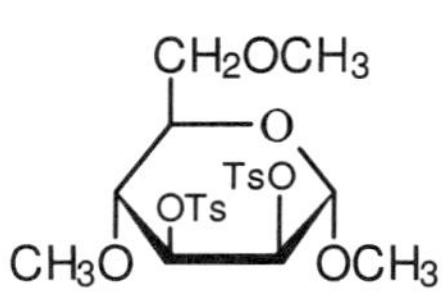	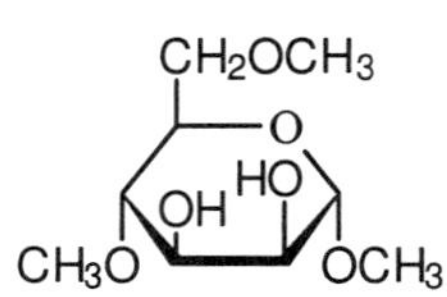	61	28

Continued on next page

Table 12.1. Photochemical Reactions of Carbohydrate Tosylates[a]—*Continued*

Reactant	*Product*	*Percent Yield*	*Ref.*
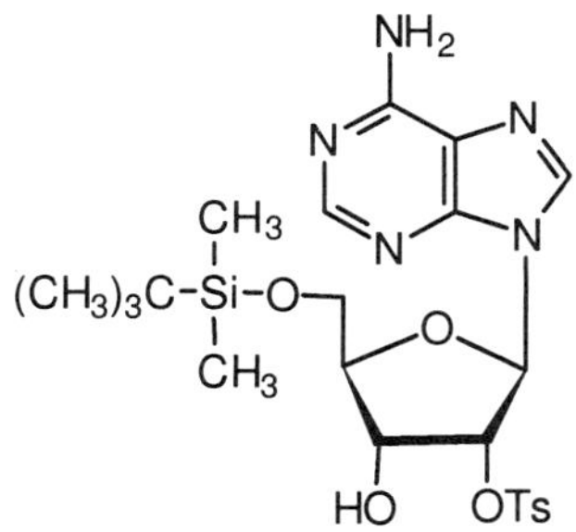	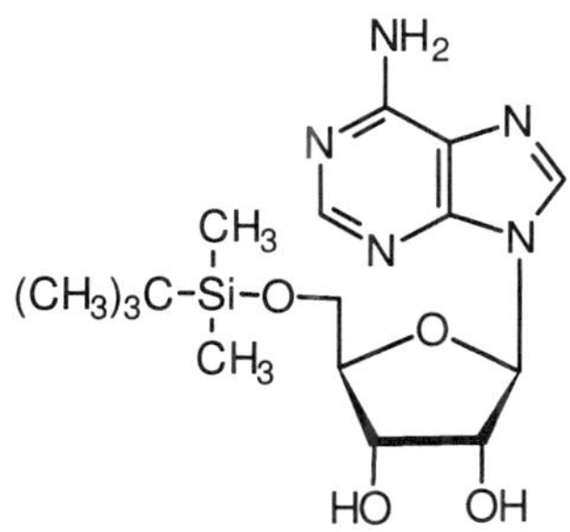	76[b] 0[c]	17,18 17,18
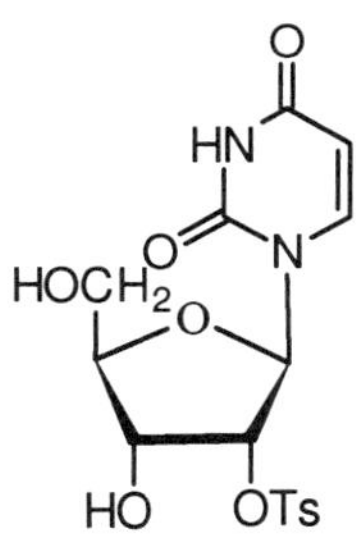	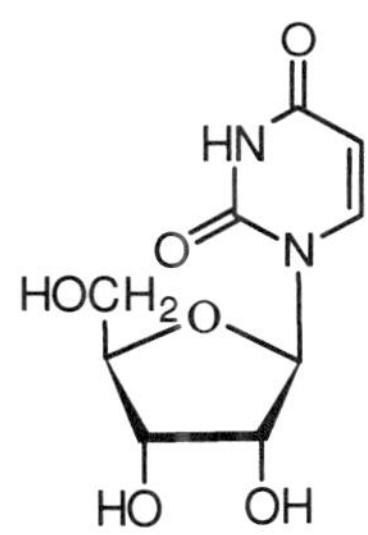	84[b] 0[c]	17,18 17,18
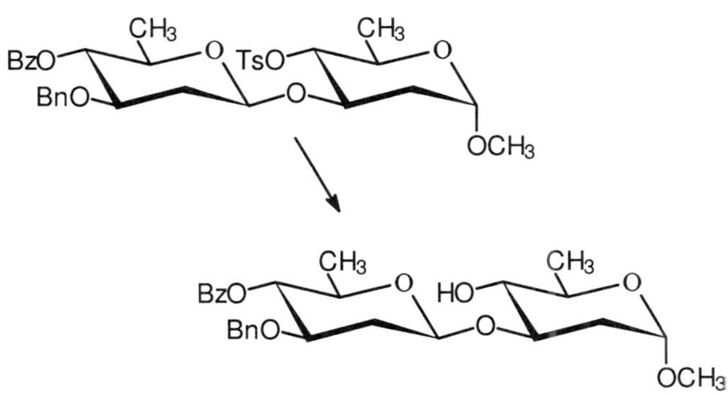		80	29
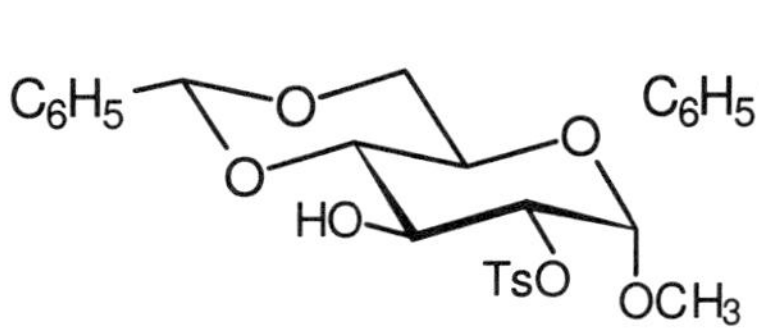	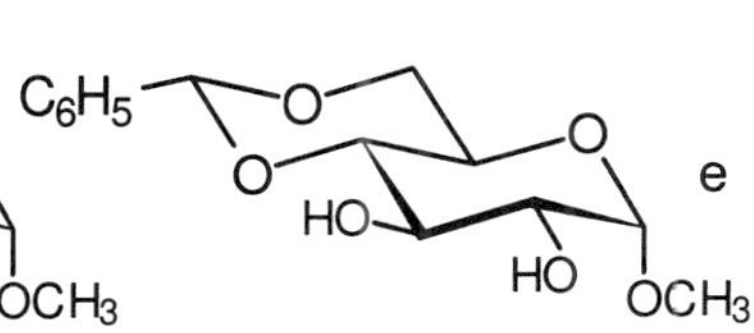	e	30

Continued on next page

Table 12.1. Photochemical Reactions of Carbohydrate Tosylates[a]—*Continued*

Reactant	*Product*	*Percent Yield*	*Ref.*
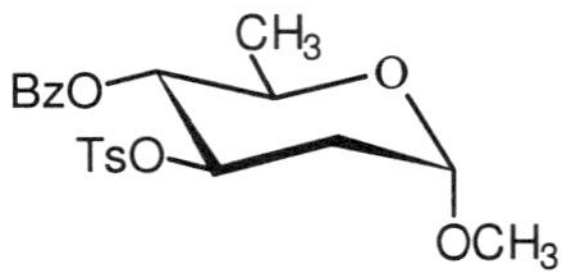	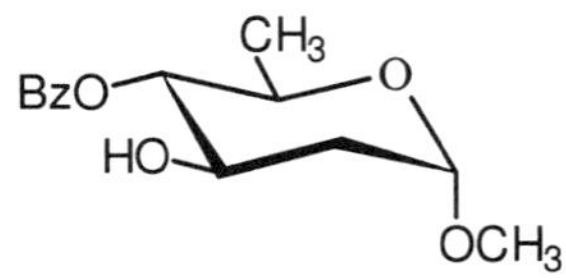	91	29
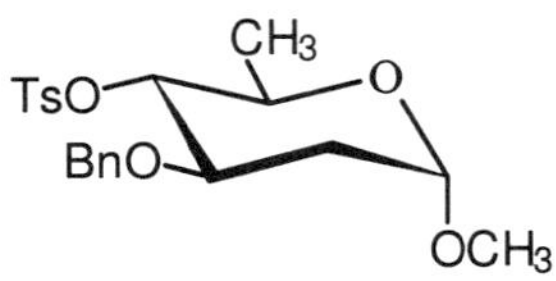	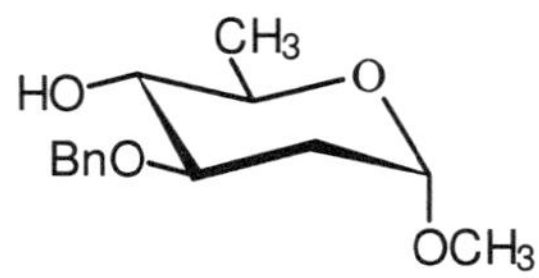	90	29
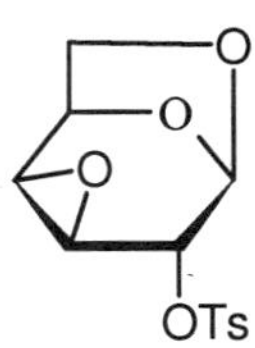	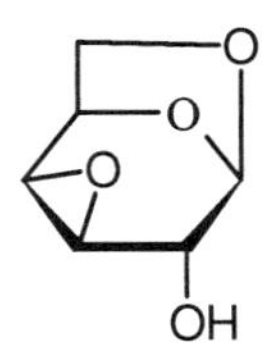	80	22, 23, 24
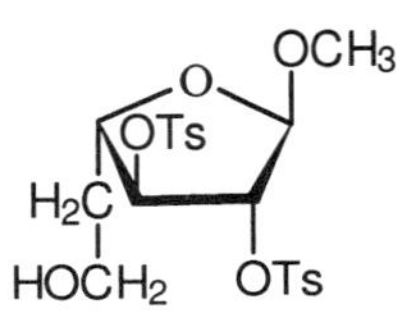	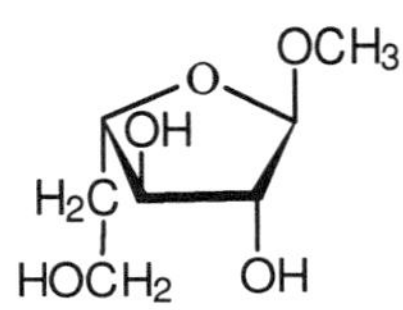	63	15
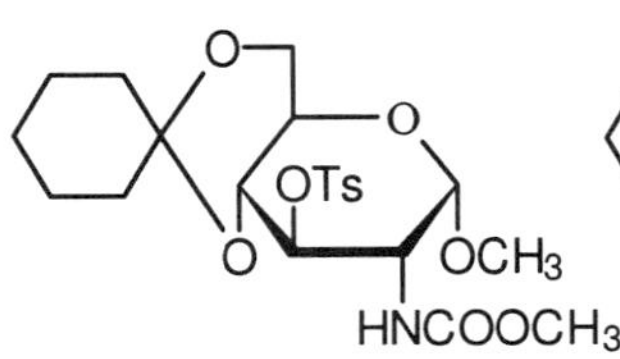	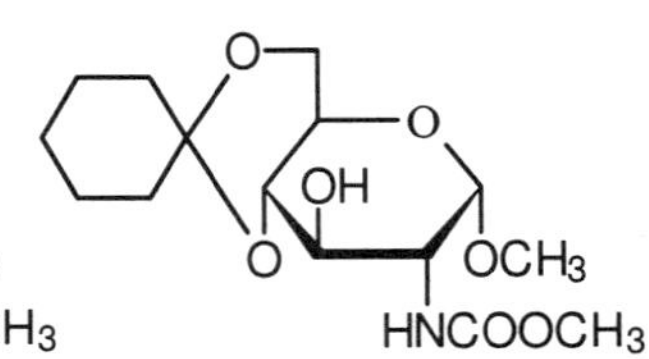	91	31

[a]Reactions were conducted under an atmosphere of either nitrogen or argon.

[b]3-(4,8-Dimethoxynapthyl)propanonic acid and hydrazine were added to the reaction mixture.

[c]1,4-Diazabicyclo[2.2.2]octane was added to the reaction mixture.

[d]1,5-Dimethoxynapthene and hydrazine were added to the reaction mixture.

[e]No yield was reported.

solution of sodium methoxide (eq 3), remained the standard conditions for many years. (Sodium methoxide was added because reactions in neutral or acidic solutions were slower and produced undesirable, foul-smelling products.) The problem with these reaction conditions was that they caused nonphotochemical transformations to occur in base-sensitive molecules. Reaction conditions, therefore, effectively limited the use of tosyl protecting groups to those compounds that were stable in the presence of a strong base. This restriction represented a severe limitation on the use of the protection–deprotection sequence; nevertheless, compounds that met the stability requirement included those containing either an isopropylidene, anhydro, methoxy, benzylidene, carbamoyl, cyclohexylidene, or mesyloxy group (Table 12.1).

Irradiation in the Presence of Tertiary Amines. A significant improvement in the conditions for photochemical deprotection was reported in 1987 by Masnovi et al.,[12] who determined that the strong base originally used during tosylate photolysis could be replaced by a tertiary amine. This alteration in the reaction conditions expanded the list of compounds that could be photochemically detosylated (without causing other changes in the molecules being irradiated) to include those with base-sensitive groups; for example, removal of the tosyl group from compound **1** did not cause loss of the benzoyl group (eq 4).[29]

Irradiation in the Presence of Electron-Transfer Sensitizers. Even though the replacement of sodium methoxide with a tertiary amine significantly increased the variety of compounds that could be deprotected photochemically without other reactions taking place, a notable group of tosylates remained for which photochemical deprotection was unsuccessful. These molecules contained chromophores capable of competing with the tosyloxy group for the incident light. The most important members of this class were tosyloxy-containing nucleosides. When these nucleosides were irradiated directly, no simple detosylation products were isolated; in contrast, deprotection did take place under the sensitized electron-transfer conditions developed by Nishida and co-workers (eq 5).[17,18] In sensitized electron transfer the incident light is not absorbed by the tosylate itself but by a sensitizer [in this case 3-(4,8-dimethoxynaphthyl)propanoic acid, DMNP]. The sensitizer transfers an electron directly to the tosyl group to form a radical anion that produces the deprotected nucle-

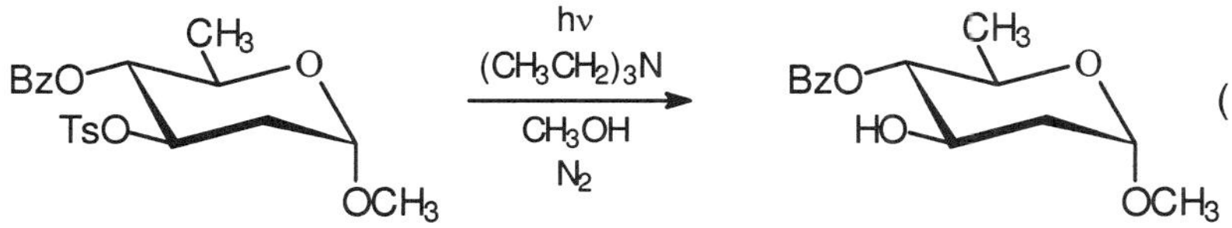

1 91% (4)

Equation 4

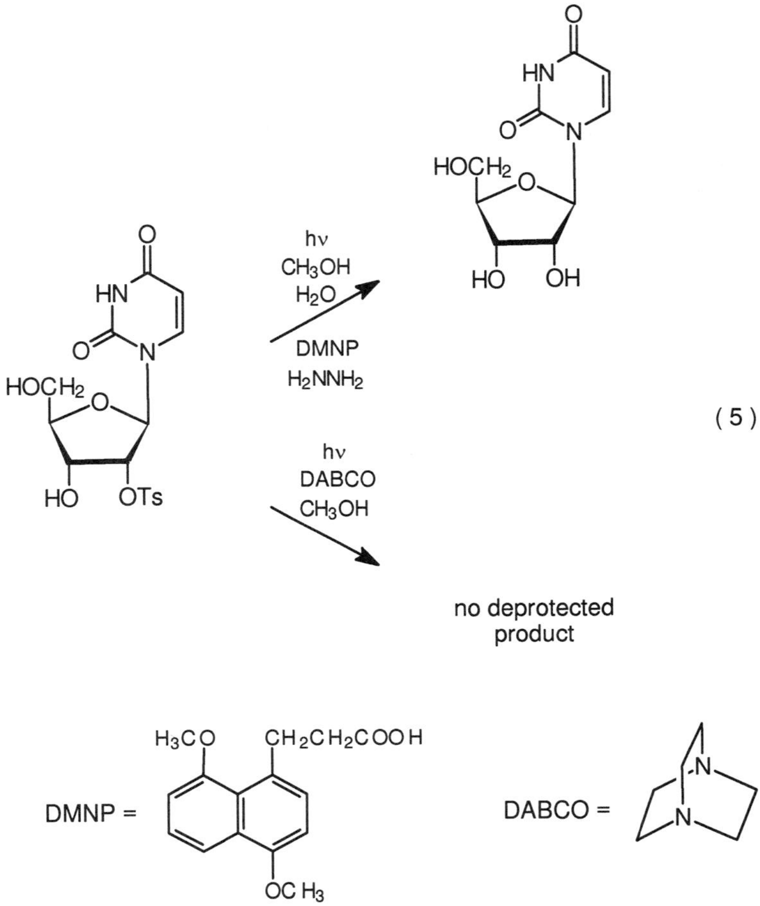

(5)

Equation 5

oside. (Sensitized electron transfer is discussed further in the section "Reactions in the Presence of Electron-Transfer Sensitizers.)

Reactive Deprotection of α-Tosyloxyketones

Although almost all carbohydrate tosylates undergo simple deprotection upon photolysis, α-tosyloxyketones experience a different reaction; that is, reductive deprotection takes place. When, for example, Szarek and Dmytraczenko irradiated the tosylate **2**, the deoxy sugar **3** was formed (eq 6).[30] These authors proposed that the photoreduction process was initiated by a hydrogen atom abstraction by the excited carbonyl group (Scheme 12.2). In support of this proposal, they found that photoreduction required the carbonyl group to be

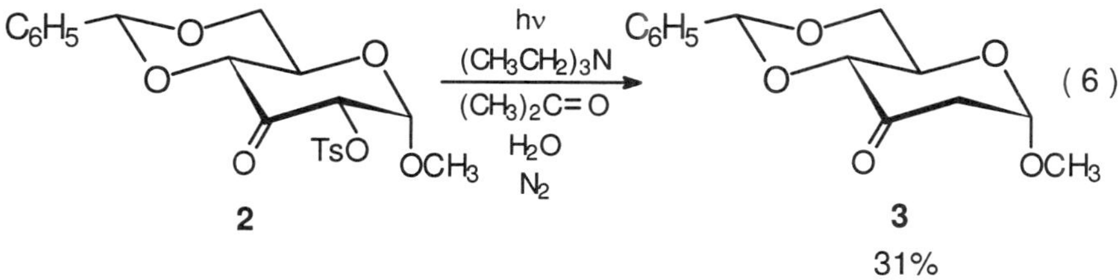

Equation 6

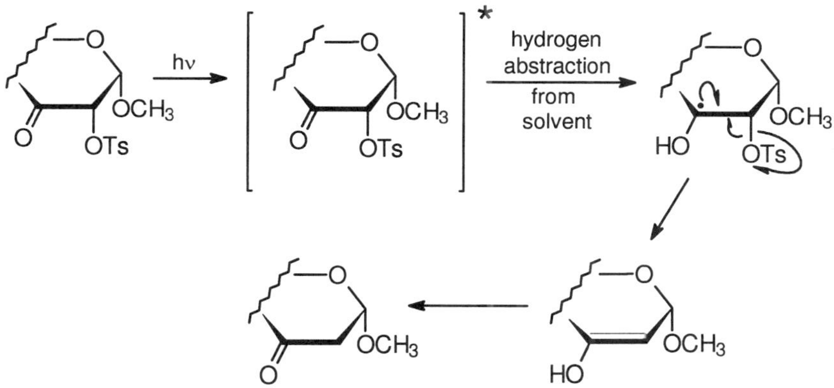

Scheme 12.2

present in the molecule because when it was absent, normal photodeprotection occurred (eq 7).[30] (Reductive photodeprotection of α-tosyloxyketones also may occur by an electron-transfer reaction similar to that discussed in Chapter 5 for α-acetoxyketones.)

Reaction Mechanisms

Reaction in the Absence of Electron-Transfer Agents

Mechanistic understanding of tosylate photolysis has developed in a slow but steady fashion during the past 25 years. Pète and co-workers initiated mechanistic investigations in 1971 when they published two communications that described photohydrolysis (eq 8) and photofragmentation (Scheme 12.3) as possible reaction pathways.[32,33] These researchers favored the fragmentation process primarily because toluene was produced during photolysis. The toluene, they reasoned, came from the *p*-tolylsulfonyl radical (Scheme 12.3). They also determined that photochemical reaction was relatively insensitive to the polarity of the solvent and argued that this finding was inconsistent with photohydrolysis, which they believed would involve ionic intermediates. These

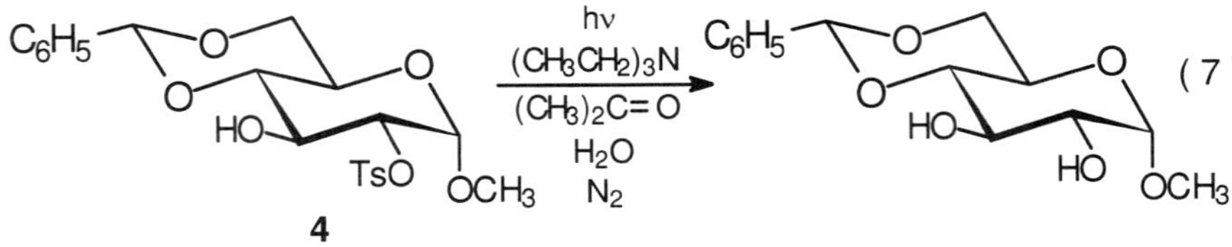

Equation 7

$$ArSO_2OR \xrightarrow[(CH_3CH_2)_2O,\ H_2O]{h\nu} ArSO_2OH + HOR \quad (8)$$

Equation 8

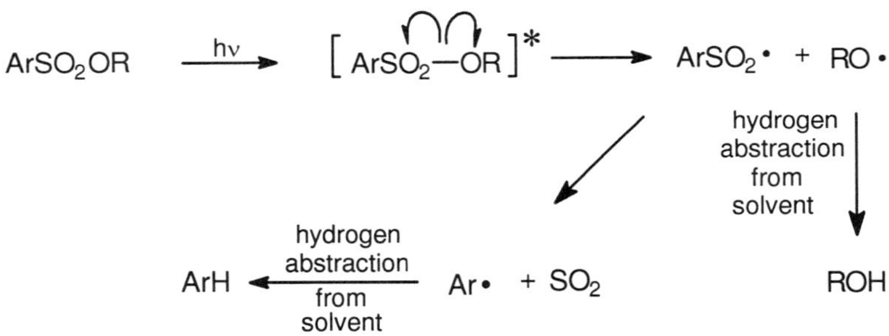

Scheme 12.3

workers also conducted qualitative experiments designed to identify the excited state responsible for reaction. They determined that the photochemical process was sensitized by benzene, but not by acetone, and quenched by piperylene. This information by itself, however, was deemed insufficient to make an excited state assignment.

Several years later Izawa and Kuromiya, who also were interested in the nature of the excited state responsible for sulfonate photoreaction, observed that 1,3-pentadiene quenched to the same extent both fluorescence (a singlet state phenomenon) and disappearance of starting material.[34] They concluded from this finding that the same excited state was responsible for both of these processes and, therefore, that reaction was occurring from an excited singlet state. These authors also considered several possible reaction mechanisms and argued that the one proposed by Pète and Portella (Scheme 12.3) was unlikely because the fragmentation of the arylsulfonyl radical into an aryl radical and sulfur dioxide should be decidedly endothermic. As an alternative they proposed a reaction mechanism in which an excited singlet state cleaved a carbon–sulfur bond to separate the tosyl group from the remainder of the molecule (Scheme 12.4).

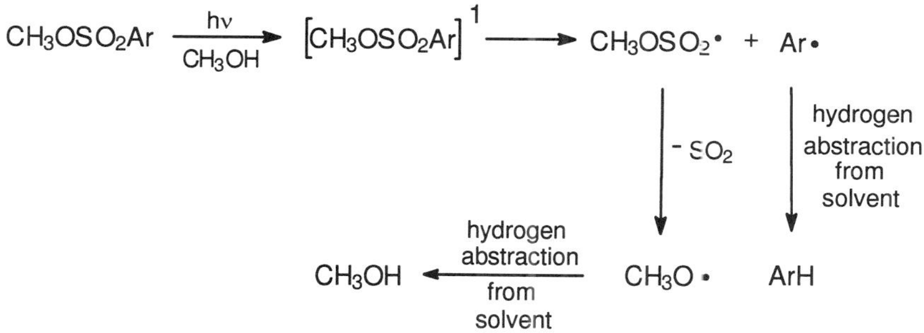

Scheme 12.4

Izawa and Kuromiya[34] also considered the possibility of electron transfer from an amine to an excited sulfonate ester, but they rejected such a mechanism because photolysis in the presence of triethylamine did not enhance the disappearance of starting material. As subsequent investigations revealed, their decision to look only at the disappearance of starting material and to analyze photolysis mixtures spectrophotometrically was an unfortunate choice. Irradiations involving amines can produce absorbing photoproducts;[16] thus, starting material could have been disappearing more efficiently in the presence of triethylamine than in its absence, but loss of starting material could have been masked by absorbing products derived from amine photolysis.

The mechanism for tosylate photoreaction was revisited by Pète and Portella in 1980,[11] nine years after their original publications on this topic.[32,33] In this more recent paper they abandon their earlier mechanistic proposal (Scheme 12.3) in favor of that suggested by Izawa and Kuromiya (Scheme 12.4).[34] Pète and Portella considered the mechanism shown in Scheme 12.4 to be the best choice for reaction taking place in neutral solution. By 1980, however, they had accumulated evidence favoring a different mechanism for reaction in the presence of base.

Reaction in the Presence of Electron Donors

In their 1980 publication on sulfonate ester photolysis Pète and Portella made several observations about reaction in the presence of hydroxide ion.[11] Perhaps the most interesting of these was that a base could quench tosylate fluorescence and simultaneously increase the quantum yield of the photochemical reaction. This finding was the first demonstration that hydroxide ion actually promoted reaction and that this ion was interacting directly with the excited singlet state of a tosylate. Pète and Portella, however, concurred with the earlier conclusion[34] that electron transfer was not taking place. Their basis for rejecting this possibility was that an added electron donor (triethylamine) had no effect on photolyses conducted in cyclohexane. It is not clear why they chose cyclohexane rather than methanol, the typical reaction solvent, but it is clear that photolysis in a solvent of low polarity can reduce or eliminate product formation in an electron-transfer reaction by favoring back electron transfer (Scheme 12.5).[35] Abandoning the possibility of an electron-transfer reaction caused these investigators to propose a complicated reaction mechanism to rationalize the fluorescence-quenching experiments; thus, although fluorescence quenching was difficult for Pète and Portella to explain, it became a useful observation later when others reconsidered the question of electron transfer.

In 1985 Binkley and Koholic conducted an experiment that provided a mechanistic tool for the study of tosylate photoreaction.[36] These investigators found that the D-*allo* radical **5**, produced by irradiation of a carbohydrate nitrate, experienced quantitative isomerization to its D-*gluco* epimer **6** (Scheme 12.6). This result was useful because the accepted mechanism for tosylate photoreaction at that time (Scheme 12.4) predicted the formation of the same radical (**5**) from photolysis of the tosylate **8**. Irradiation of **8**, therefore, should have given the deprotected, epimerized alcohol **9** (eq 9). When **8** was photolyzed, however, no epimerization took place (eq 10); thus, an alkoxy radical was not

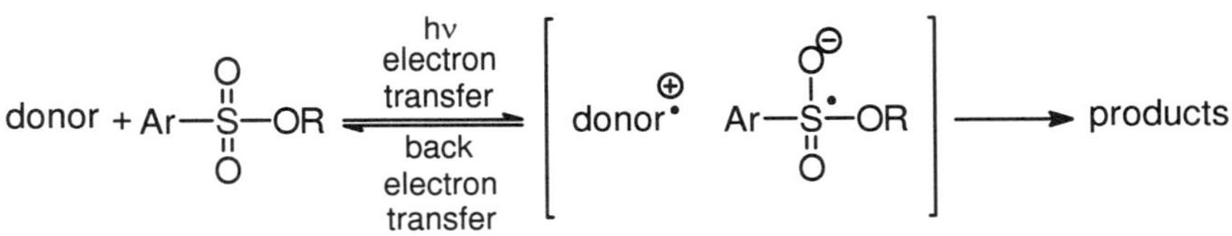

Scheme 12.5

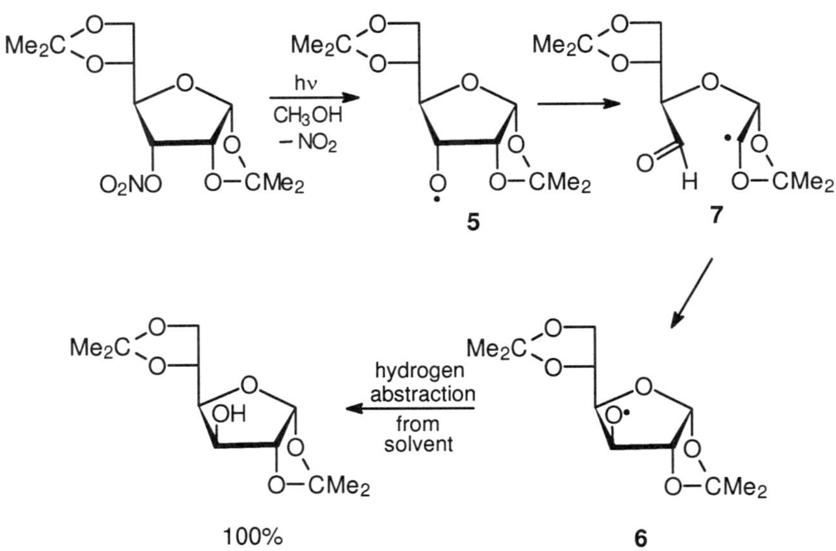

Scheme 12.6

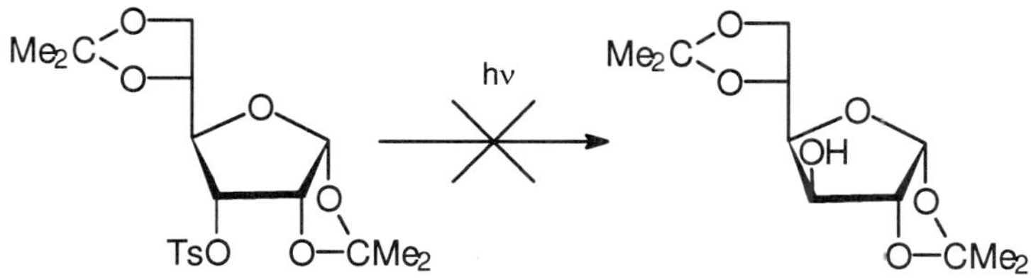

8 **9**

(9)

Equation 9

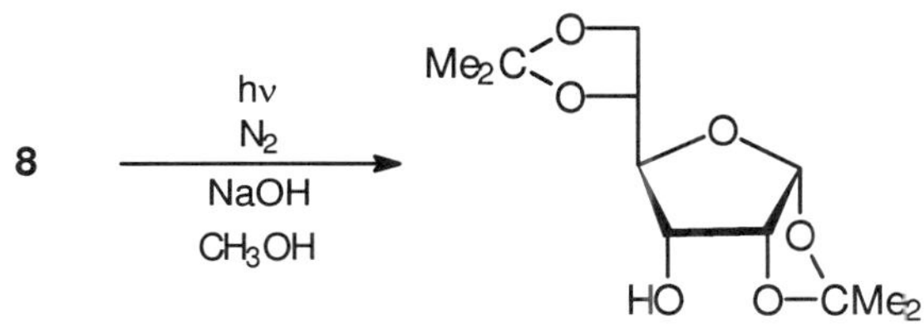

(10)

Equation 10

generated by photolysis. These workers explained this result by proposing a process in which an electron-transfer reaction produced a radical anion that fragmented to form an alkoxide ion and a sulfonyl radical (Scheme 12.7).[12,14]

By using flash photolysis techniques, Masnovi et al. showed that electron transfer was taking place between an excited tosylate and an electron donor.[12] They did this by observing directly the radical cation $D^{\bullet +}$ produced by transfer of an electron to an excited tosylate from the tertiary amine DABCO (1,4-diazabicyclo[2.2.2]octane) (Scheme 12.7). The *p*-tolylsulfonyl radical **10** was observed in the same experiment. Detection of these intermediates demonstrated that electron transfer was occurring under typical reaction conditions.

In addition to providing evidence for *p*-tolylsulfonyl radical (**10**) formation from tosylate photolysis, flash photolysis experiments also showed that **10** was rapidly and efficiently removed from solution by oxygen.[14] Preparative irradiations demonstrated that this deliberate addition of oxygen to the photolysis mixture reduced the amount of toluene produced and caused the *p*-toluenesulfonate anion to be formed (Scheme 12.8). This experiment showed that the *p*-tolylsulfonyl radical (**10**) was converted into toluene during reaction but left unanswered the question of how this conversion was taking place. The most direct path between **10** and toluene, loss of sulfur dioxide (Scheme 12.3), already had been determined to be problematic because such a transformation should be quite endothermic.[34] A link was established between the sulfonyl radical **10** and toluene when tosylate photochemical reaction was conducted to partial completion. Under these conditions the *p*-toluenesulfinate anion was present in the reaction mixture, although it was absent in reactions carried to completion. The explanation for this result was that the *p*-tolylsulfonyl radical

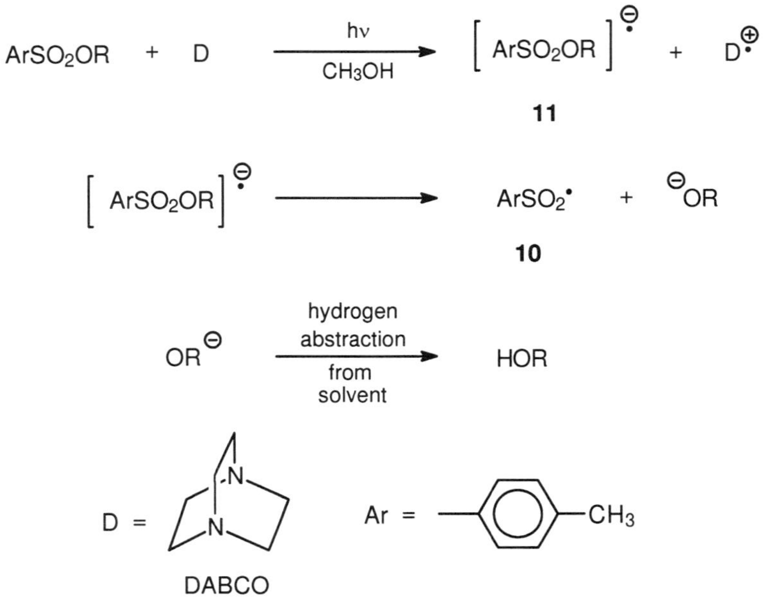

Scheme 12.7

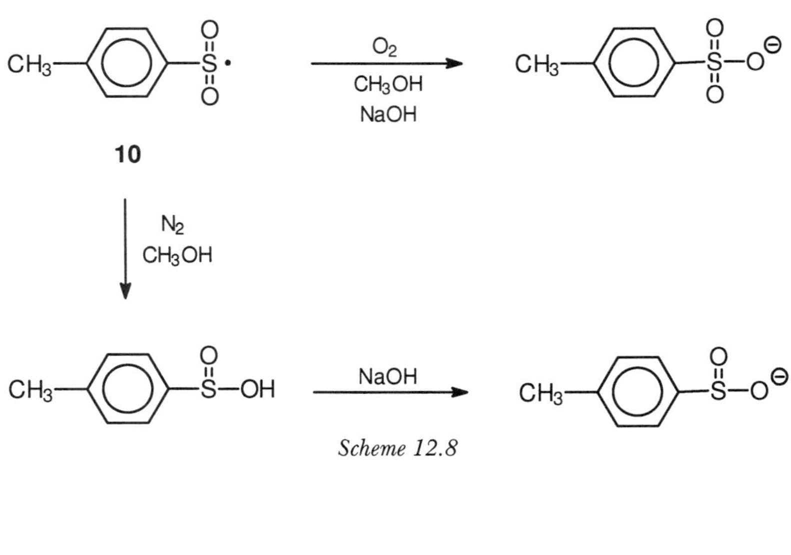

Scheme 12.8

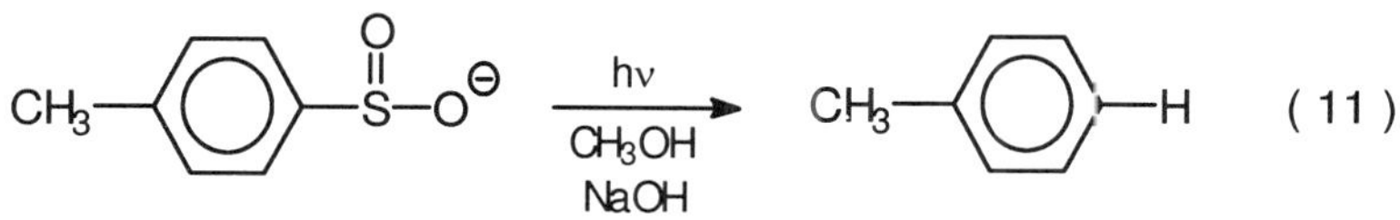

Equation 11

(in the absence of oxygen) was abstracting a hydrogen atom from methanol to give *p*-toluenesulfinic acid, which was neutralized immediately to form the *p*-toluenesulfinate anion (Scheme 12.8). Photolysis of the *p*-toluenesulfinate ion then produced toluene (eq 11).[14,37]

Reaction in the Presence of Electron-Transfer Sensitizers

One expected consequence of the electron-transfer mechanism shown in Scheme 12.7 is that a radical anion such as **11**, regardless of its method of formation, should fragment to give the *p*-tolylsulfonyl radical (**10**) and an alkoxide ion.* When **11** actually is generated in other ways (e.g., electrochemical reduction,[38–40] reaction with arene radical anions,[41–43] or alkali metal reductions[44]), the predicted behavior is observed; that is, an alkoxide ion is formed during detosylation. It is not difficult to imagine, therefore, that a photochemical process different from that outlined in Scheme 12.7, but still capable of generating **11**, could provide another means for photochemical deprotection. Nishida and co-workers have developed such an alternate procedure in which a tosylate is converted into its radical anion (**11**) by accepting an electron from an excited, electron-rich donor (Scheme 12.9).[17,18] Selection of 1,5-dimethoxynaphthalene (**12**) as the donor (and use of an appropriate light filter) makes it possible for the incident light to be absorbed entirely by the sensitizer; thus, detosylation takes place photochemically without ever forming an excited state of the tosylate.

Electron-transfer sensitization has both advantages and disadvantages when compared to direct excitation. Direct excitation can cause reactions other than the desired deprotection to occur, but these unwanted reactions sometimes can be avoided by electron-transfer sensitization. The advantages of sensitization are most apparent for compounds with chromophores that compete with the tosyl group for absorption of light. One instance in which this competition takes place is with tosyloxy-containing nucleosides, which can be deprotected by photolysis in the presence of electron-transfer sensitizers, but experience other reactions upon direct excitation (eq 5).[17,18] The primary disadvantage to this sensitization technique is that it requires the addition of a coreductant (e.g., hydrazine or sodium borohydride) to the photolysis mixture in order for efficient reaction to occur. Although a coreductant does not necessarily cause problems, coreductants generally are undesirable because their presence in the photolysis mixture creates an opportunity for reactions other than deprotection to take place.

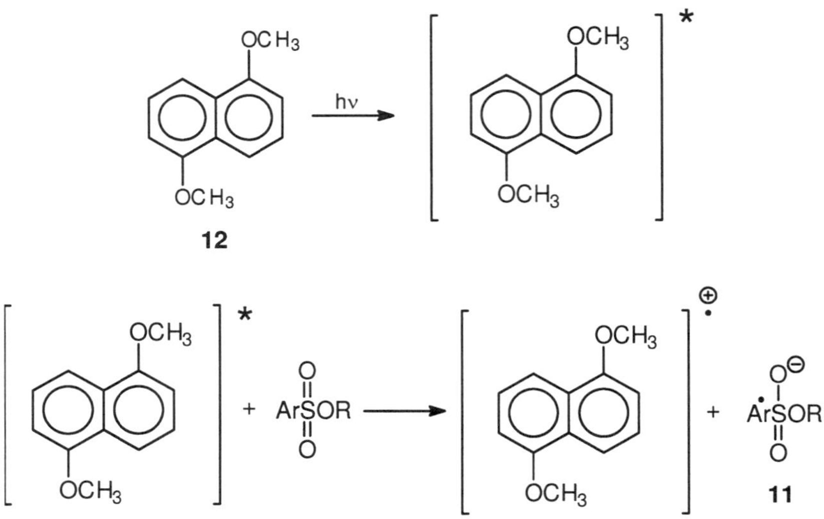

Scheme 12.9

Reaction in the Absence of Solvent

Negative-ion chemical-ionization mass spectrometry (NICI-MS) provides a means for creating the radical anion **11** in the absence of a solvent and observing its reaction under these conditions. This type of observation provides information about the importance of the solvent to the photochemical reaction. Examination of the NICI-MS spectra of tosylates shows that the presence of solvent is decisive in determining the reaction pathway to be followed; thus, when dissolved in methanol, **11** reacts by cleaving its oxygen–sulfur bond (Scheme 12.10, path A), but in the absence of a solvent a carbon–sulfur bond fragments (Scheme 12.10, path B).[45] Hehemann et al. suggested that the extremely high energy of the unsolvated alkoxide ion makes its formation difficult without solvent present. In solution, however, the alkoxide ion is stabilized sufficiently by solvent molecules to make reaction by path A (Scheme 12.10) possible.

Methanesulfonates

Methanesulfonates do not have significant UV absorption at wavelengths longer that 200 nm and, therefore, it is not surprising that direct photolyses of these compounds returns only unreacted starting material.[11,32] Even though direct irradiation is unproductive, photochemically promoted deprotection does occur in an indirect fashion. Irradiation of mesylates in the presence of hexamethylphosphoramide (HMPA), a compound that is capable of absorbing

Scheme 12.10

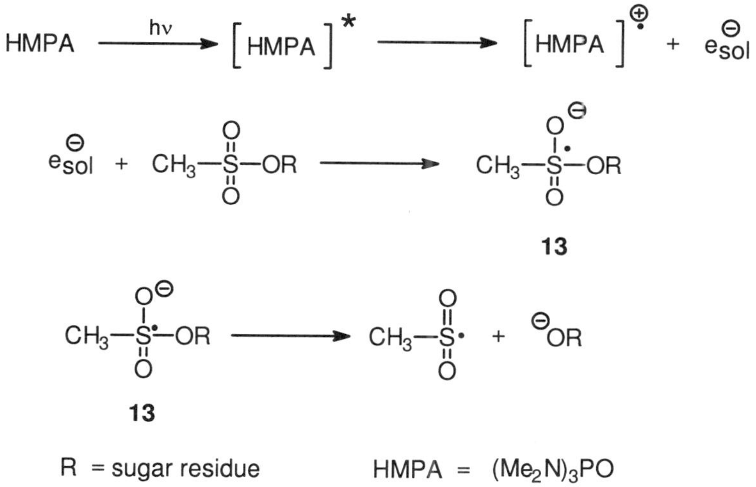

Scheme 12.11

UV light and then transferring an electron to a methanesulfonate (Scheme 12.11), creates the radical anion **13**.[46] [Scharf and co-workers[47] have shown that electron transfer from HMPA is a two-step process in which excited HMPA ionizes to produce a solvated electron that then is captured by an electron acceptor (in this case a mesylate).] The intermediate **13**, which is analogous to the radical anion **11** (Scheme 12.7), generated by *p*-toluenesulfonate photolysis, then fragments to give the anion of the deprotected sugar (Scheme 12.11).

Liu and Binkley reported an alternate set of reaction conditions for mesylate deprotection.[20,21] These conditions depend upon excited iodide ion as the electron source for forming the radical anion **13**. Upon irradiation, iodide ion ejects an electron into solution[48] where a methanesulfonate captures it to give **13**, the precursor of the deprotected sugar (Scheme 12.12).[20,21] Table 12.2 contains a list of the mesylates that have been investigated.

Compounds **14–17** (Chart 12.1) preferentially lose groups other than the mesyl group upon photolysis. It is not difficult to understand why this occurs. The aromatic chromophores in **14–17** absorb essentially all of the incident light, making it impossible for iodide ion to become excited and promote reaction. Direct excitation of compounds **14–17** leads to loss of the light-absorbing

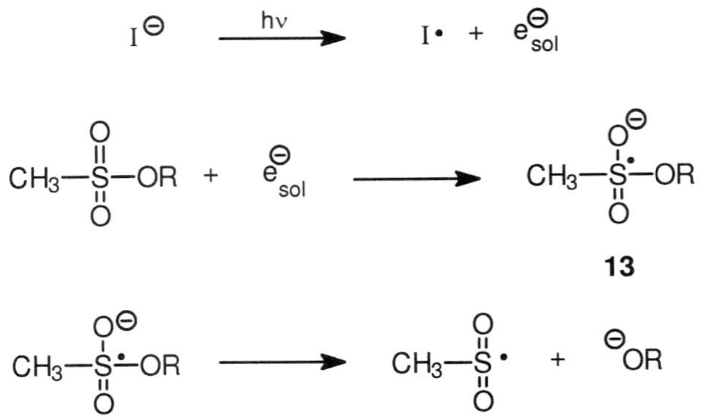

R = carbohydrate residue

Scheme 12.12

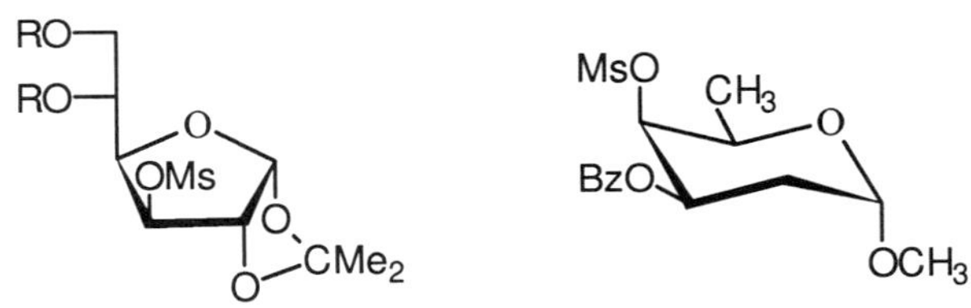

14 R = Bz

15 R = Bn

16 R = Ts

17

Chart 12.1

groups. Since the reactions of these compounds do not involve mesylate photolysis, they will not be discussed at this time, but they are considered at other places in this book. Benzyl-protected carbohydrates are discussed in Chapter 10 (and described, along with benzoyl-protected compounds, in reference 21). Photoremoval of a tosyl group was considered earlier in this chapter. Since undesired, photochemical loss of groups with aromatic chromophores does occur during mesylate irradiation, it is worth remembering that a tosyl group can be removed photochemically under conditions that permit benzyl and benzoyl groups to remain in place.

Trifluoromethanesulfonates

Photolysis of trifluoromethanesulfonates (triflates) takes a somewhat different course from that of other sulfonates. Although deprotection does occur under

Table 12.2. Photochemical Reactions of Carbohydrate Mesylates[a]

Reactant	*Product*	*Percent Yield*	*Ref.*
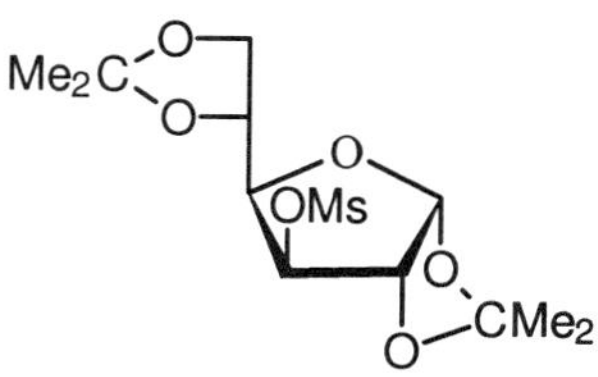	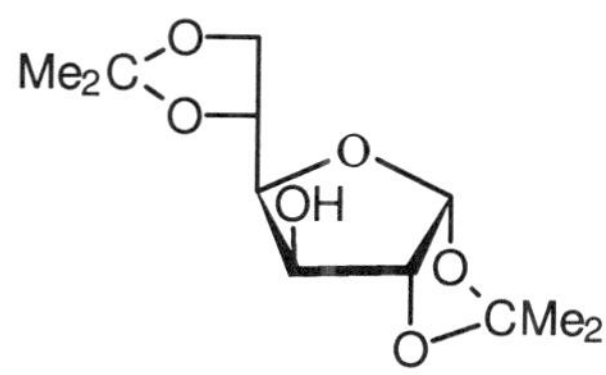	90	20, 21
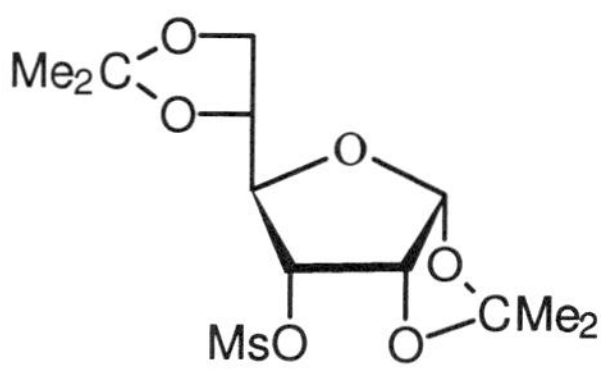	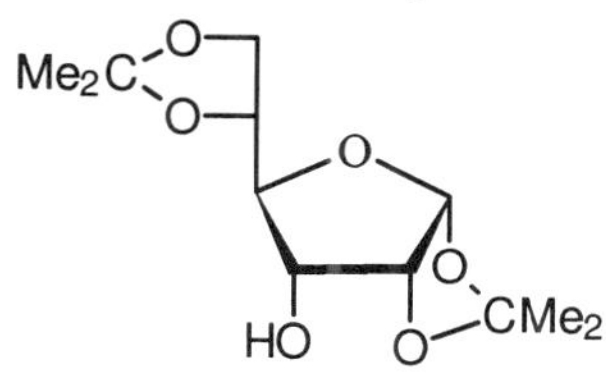	61	20, 21
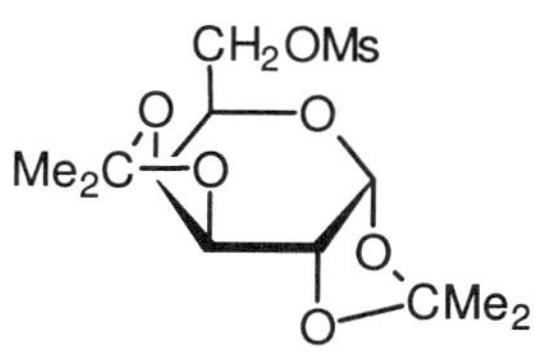	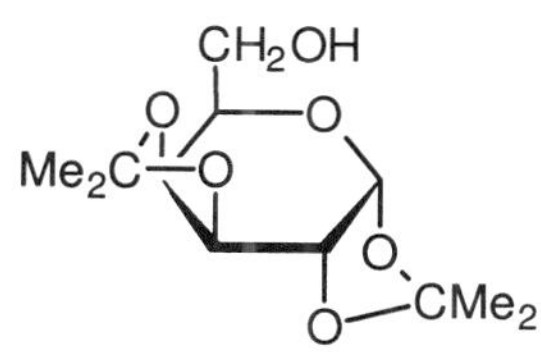	64	20, 21
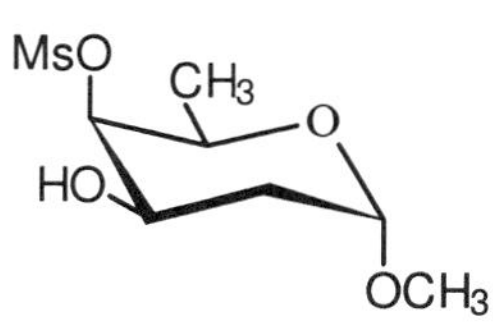	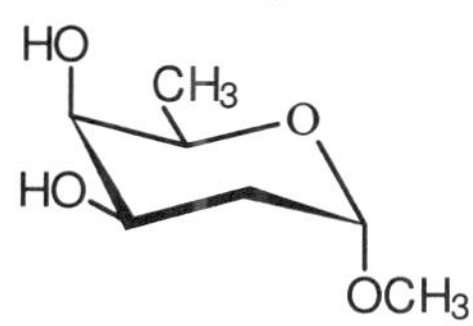	74	20, 21
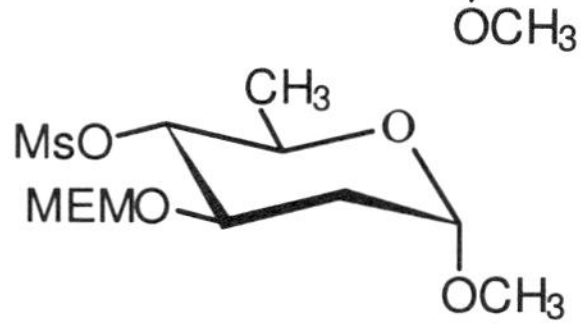	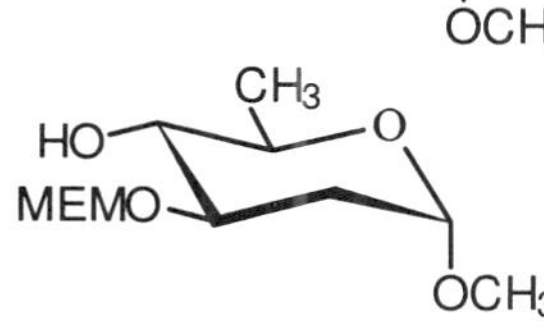	70[b]	20, 21
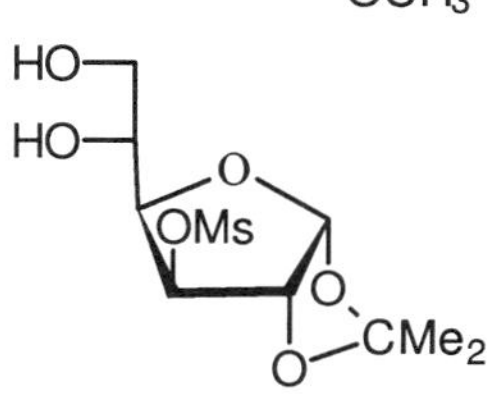	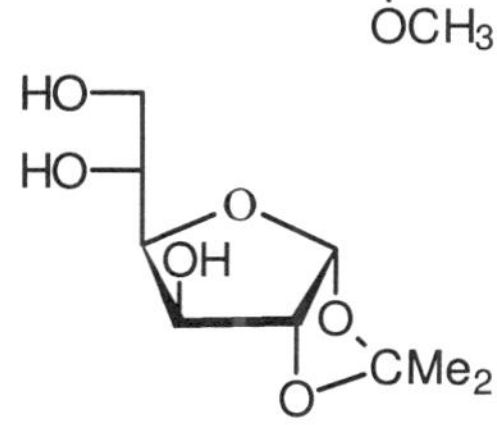	60	20, 21
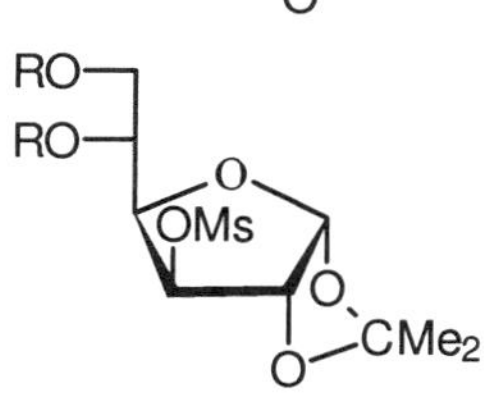	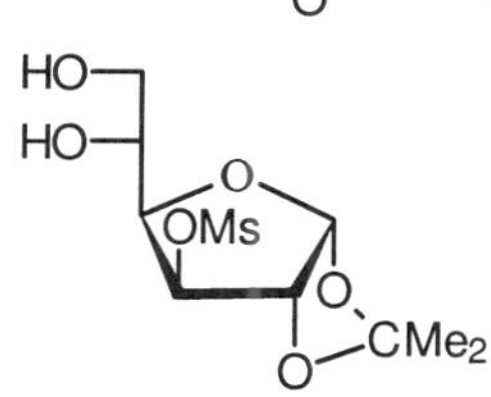	100(R=Bz) 60(R=Bn) 100(R=Ts)	21

Continued on next page

Table 12.2. Photochemical Reactions of Carbohydrate Mesylates[a]—*Continued*

Reactant	*Product*	*Percent Yield*	*Ref.*
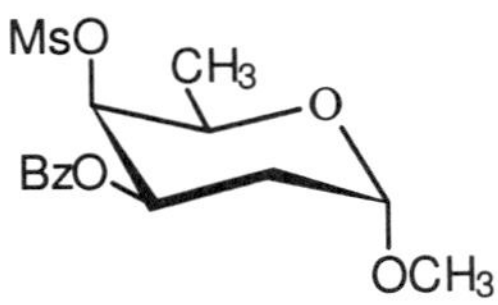	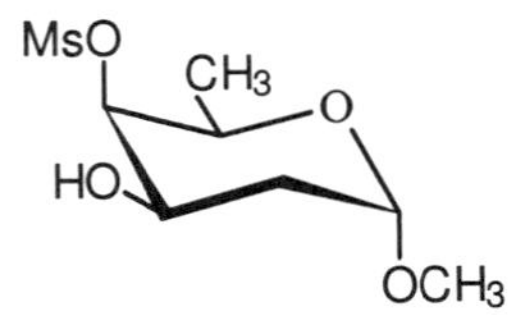	98	20
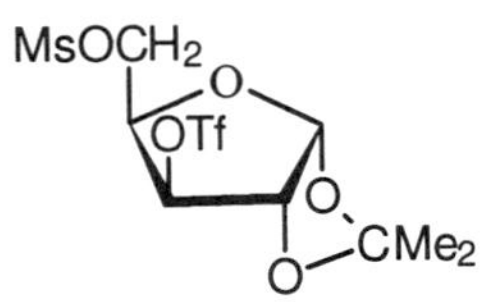	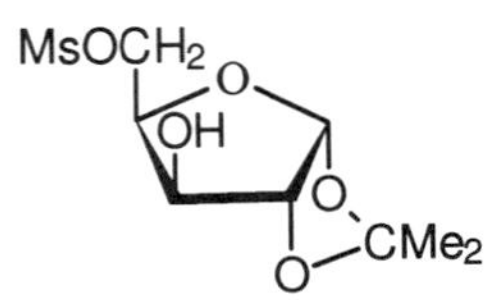	51	19
	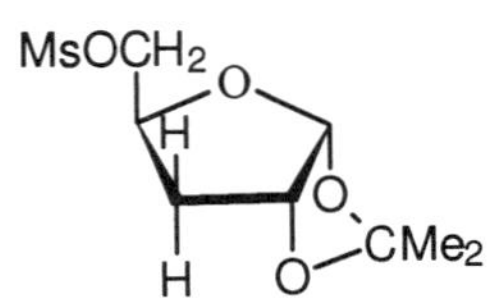	47	
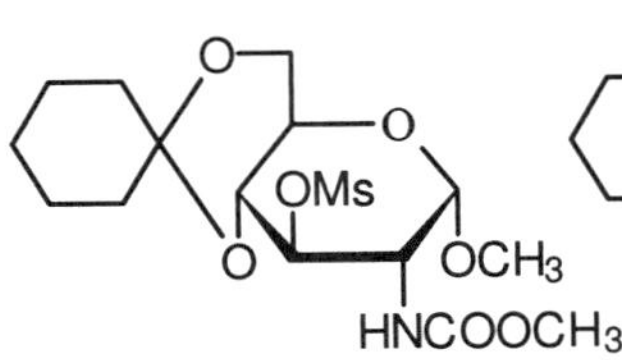	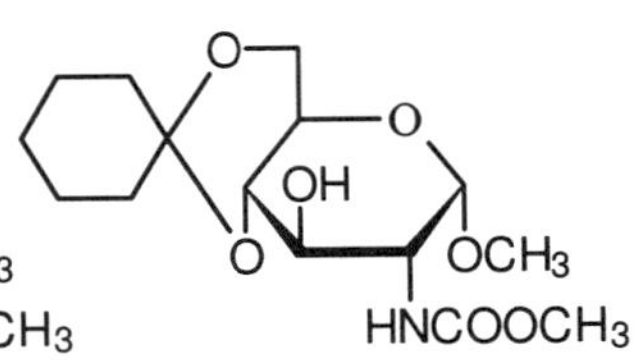	86	31[c]

[a]Photolyses were conducted under nitrogen in a methanolic solution of potassium iodide and sodium bicarbonate.

[b]MEM = 2-methoxyethoxymethyl.

[c]Reaction was conducted in HMPA.

some conditions, it is always accompanied by deoxy sugar formation, which sometimes is the only reaction observed (Table 12.3).[19,49] As with methanesulfonates, electron transfer to a triflate does not take place by direct excitation; rather, photolysis in the presence of HMPA[31] or iodide ion[19] produces the radical anion **18** indirectly (Scheme 12.13). Since **18** forms under the same conditions as its methylsulfonyl counterpart **13**, the mechanisms for formation of **18** presumably are the same as those leading to **13** (Schemes 12.11 and 12.12). Unlike **13**, however, the radical anion **18** has two reaction pathways open to it. One of these, the normal deprotection process, involves oxygen–sulfur bond cleavage (Scheme 12.13, path a), while the other requires cleavage of a carbon–oxygen bond and produces a deoxy sugar (Scheme 12.13, path b); thus, the highly electron-withdrawing trifluoromethyl group opens a new pathway (path b) for sulfonate photoreaction.[19] This pathway probably exists only because it leads to the extremely stable triflate anion (Scheme 12.13).

Pentafluorobenzenesulfonates

Duan et al. found that photolysis of pentafluorobenzenesulfonates (pentaflates) results not only in typical desulfonylation reaction but also in a photochemical rearrangement (Scheme 12.14, Table 12.4).[50] The first step in both of these processes is an electron transfer between triethylamine and the excited pentaflate. The deprotection process takes place in a manner similar to that described for tosylates (Scheme 12.7). The rearrangement occurs by attack of one of the oxygen atoms of the radical anion on the highly electron-deficient aromatic ring to form a sulfurous acid ester (Scheme 12.14).

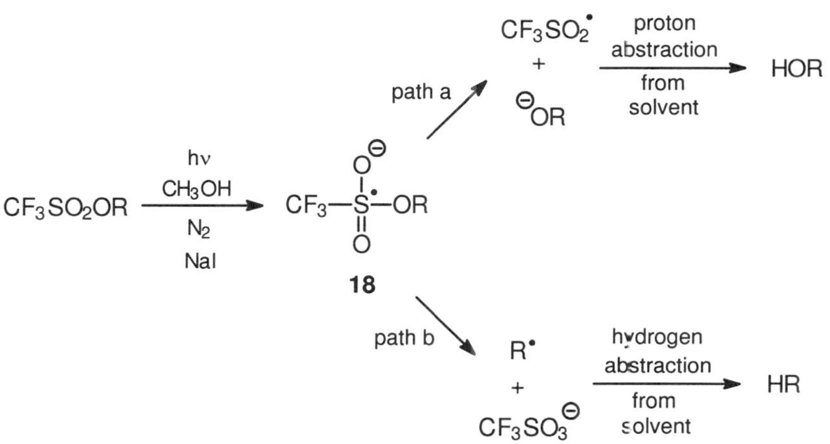

Scheme 12.13

Table 12.3. Photochemical Reactions of Carbohydrate Trifluoromethanesulfonates (Triflates)

Reactant	*Product and Percent Yield*	*Ref.*
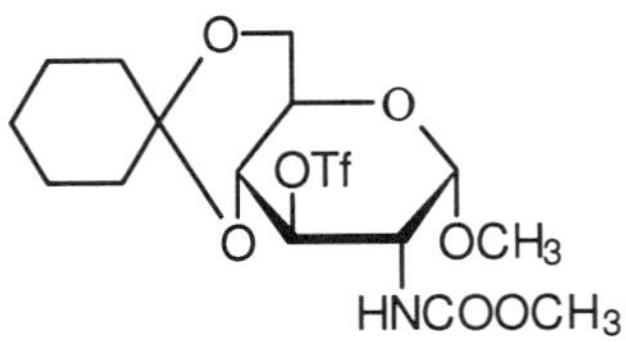	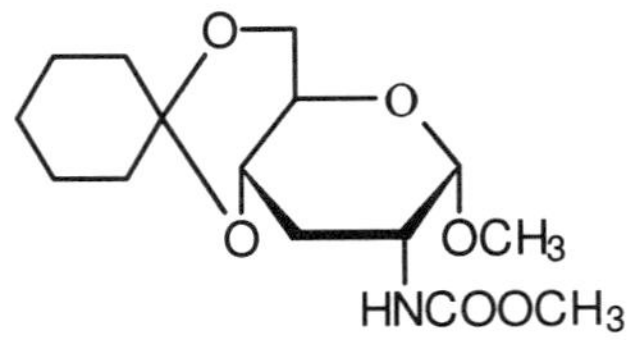 78%	49a
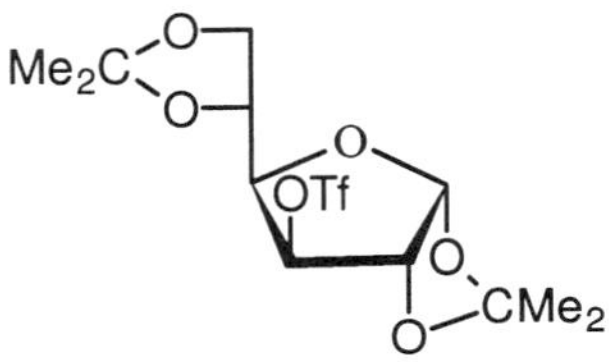	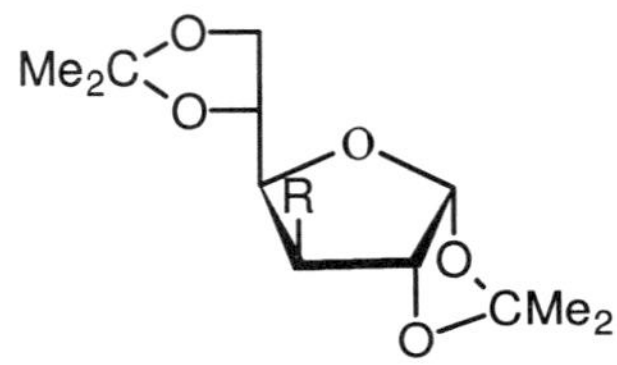 R = H 64% R = OH 0%	49a
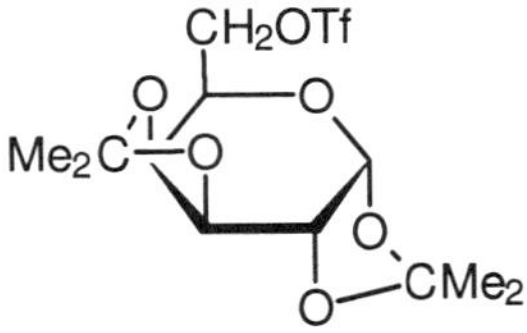	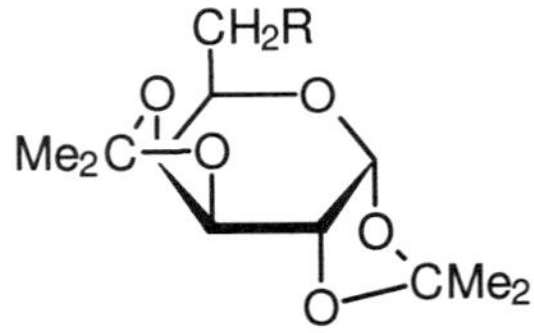 R = H 21% R = OH 31%	19b
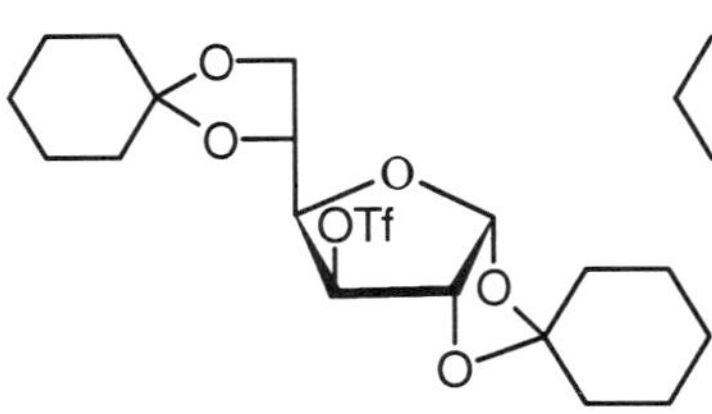	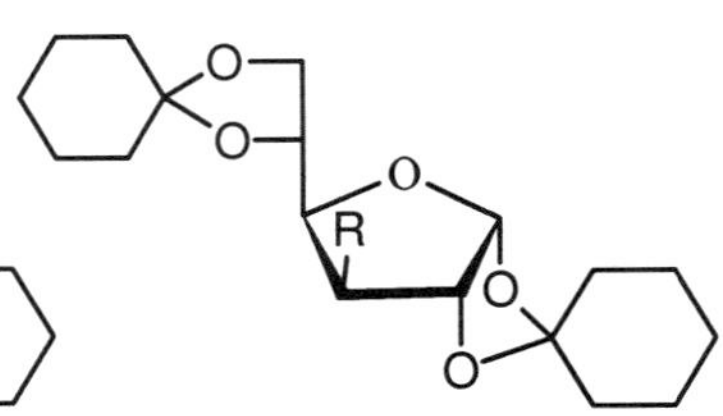 R = H 48% R = OH 38%	19b

Continued on next page

Table 12.3. Photochemical Reactions of Carbohydrate Trifluoromethanesulfonates (Triflates)—*Continued*

Reactant	*Product and Percent Yield*		*Ref.*
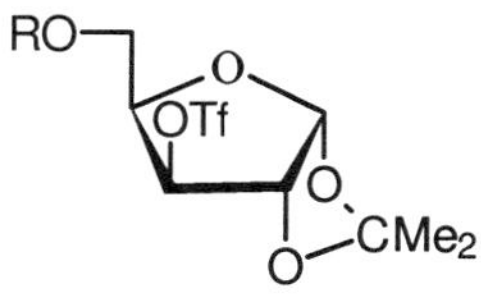	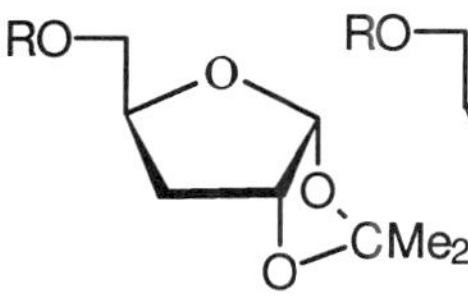	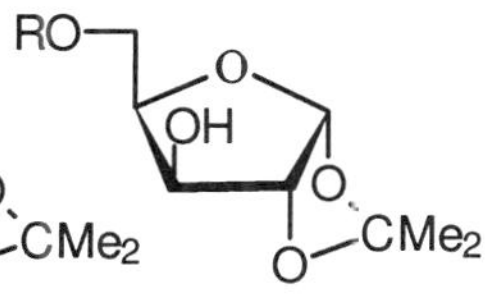	19[b,c]
R = MEM	31%	57%	
R = Ms	47%	51%	
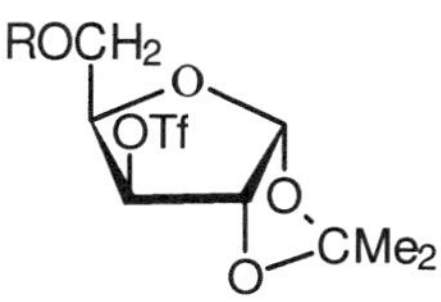	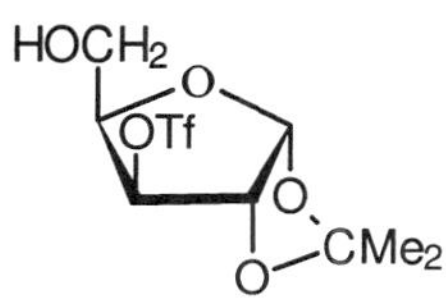		19[b]
R = Ts	70%		
R = Bz	83%		
R = Bn	77%		

[a]Photolyses were conducted in HMPA/H_2O.
[b]Photolyses were conducted in a methanolic solution of KI and $NaHCO_3$.
[c]MEM = 2-methoxyethoxymethyl.

Table 12.4. Photolysis of Pentafluorobenzenesulfonates (Pentaflates)[a]

Reactant	*Product and Percent Yield*	*Ref.*
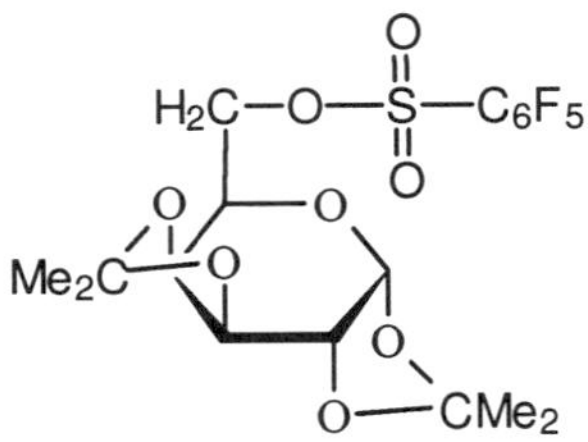	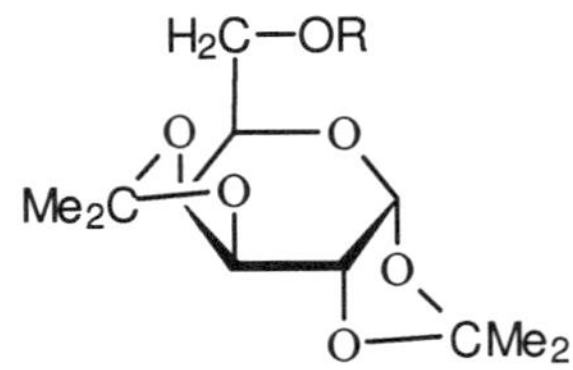 R = H 35% R = $C_6F_5OS{=}O$ 47%	50
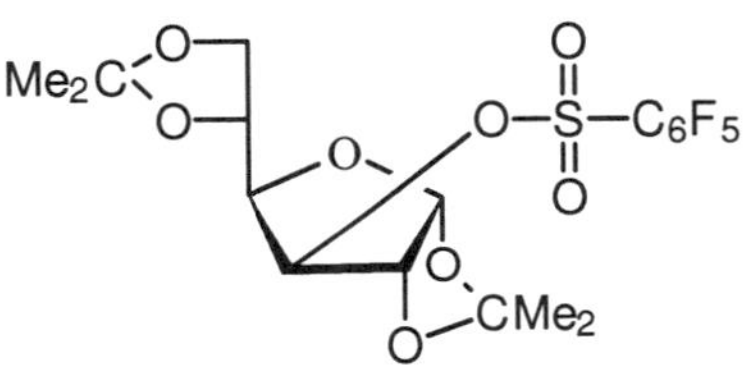	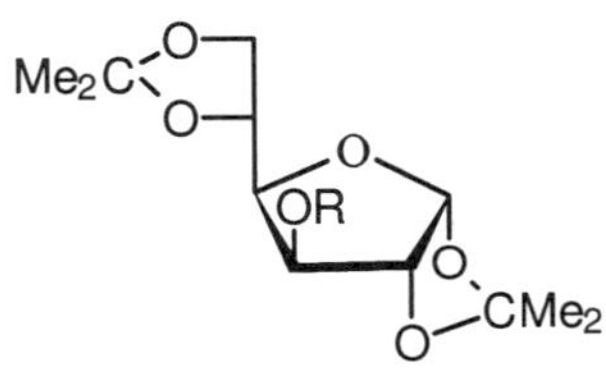 R = H 19% R = $C_6F_5OS{=}O$ 77%	50
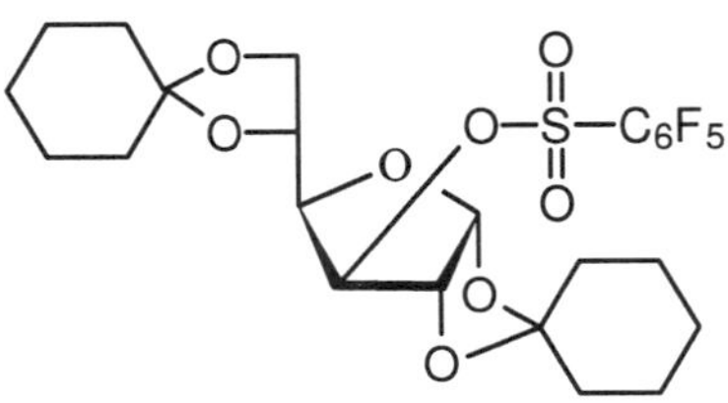	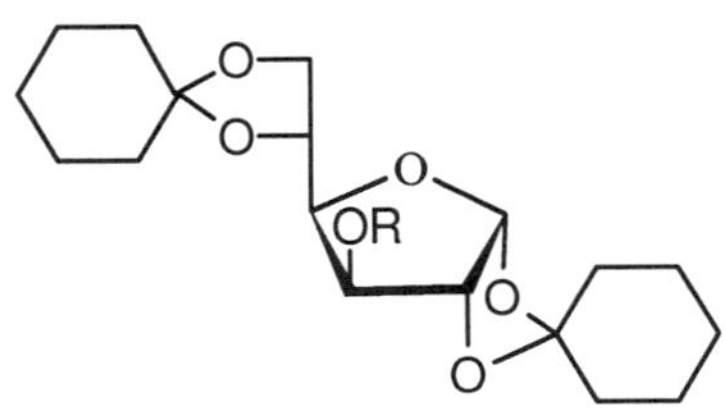 R = H 18% R = $C_6F_5OS{=}O$ 74%	50
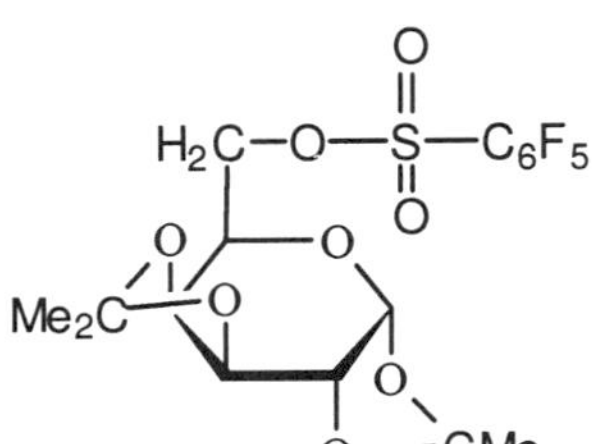	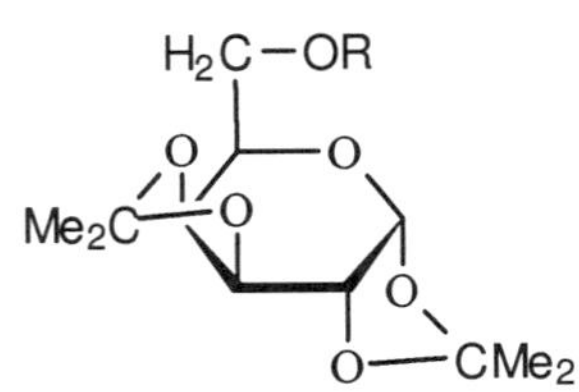 R = H 35% R = $C_6F_5OS{=}O$ 47%	50

[a]Photolyses were conducted in methanol containing triethylamine.

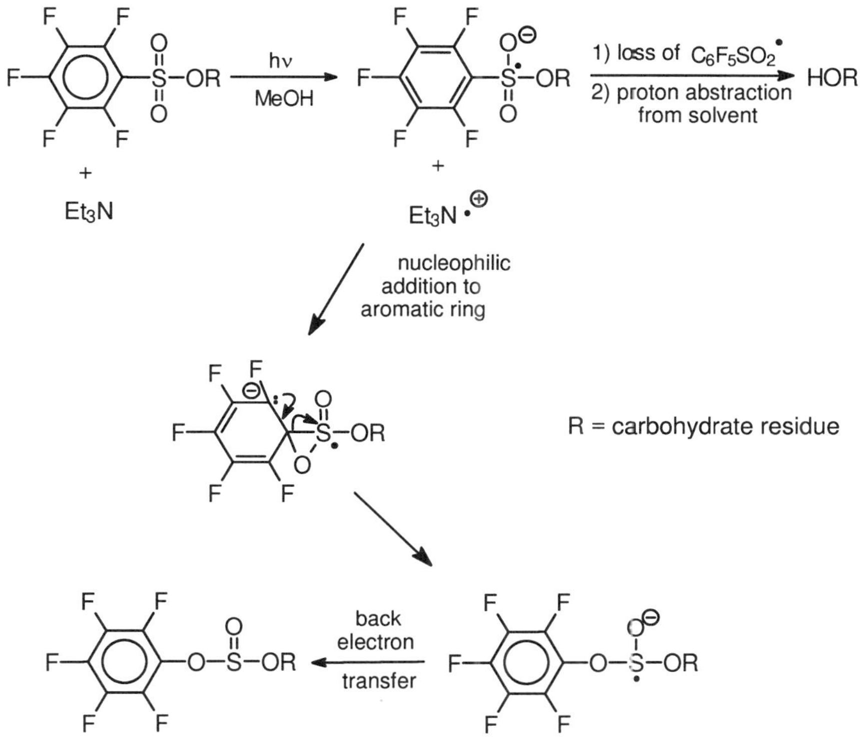

Scheme 12.14

Sulfonamides

In a reaction that parallels the photodeprotection of sulfonic acid esters, photolysis of sulfonamides produces the corresponding deprotected amines. Although this reaction is recognized as an effective process for removal of a tosyl group from an amine or amino acid,[5] its use in carbohydrate chemistry is rare. There are only two instances in which photolysis of a sulfonamide has been employed to remove a tosyloxy group to form an aminodeoxy sugar. Matsumoto et al. found that irradiation of compounds **19** and **20** in the presence of 1,5-dimethoxynaphthalene (**12**) gave the corresponding aminodeoxy sugars in high yield (eq 12 and 13).[51] Yonemitsu and co-workers previously had established in noncarbohydrate systems that irradiation of tosylamides with 1,5-dimethoxynaphthalene produced amines.[52,53] These authors proposed an electron-transfer mechanism for this process that is analogous to the one shown in Scheme 12.9 for tosylate photoreaction.

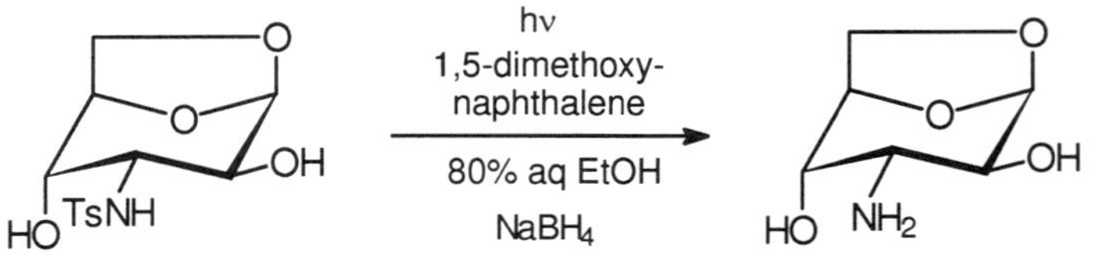

(12)

19 87%

Equation 12

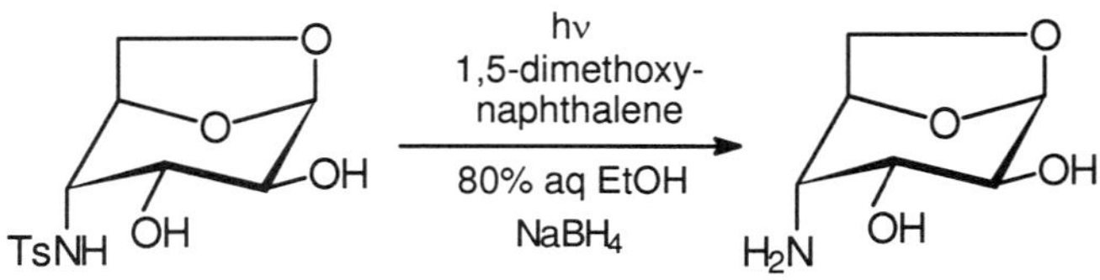

(13)

20 83%

Equation 13

Summary

The *p*-tolylsulfonyl group is an excellent protecting group. It is introduced and photochemically removed from carbohydrates under mild conditions in the presence of a wide variety of other protecting groups (Table 12.1). Upon direct irradiation in the absence of electron-transfer agents, photofragmentation occurs as shown in Scheme 12.4, but in the presence of electron donors, a radical anion is formed that reacts as shown in Scheme 12.7.

Other sulfonates also experience photochemical reaction. Mesylates react upon irradiation in the presence of either iodide ion (Scheme 12.12) or HMPA (Scheme 12.11) in a process thought to be mechanistically analogous to the reaction of tosylates. Triflates react under the same conditions but along two pathways; one leads to sulfonyl group loss (the typical sulfonate photoreaction) and the other to deoxy sugar formation (Scheme 12.13). Pentaflates also experience sulfonyl group loss but, in addition, they undergo rearrangement to sulfurous acid esters upon irradiation (Scheme 12.14).

References

1. Jeger, O.; Schaffner, K.; Wehrli, H. *Pure Appl. Chem.* **1964,** *9,* 555.
2. Hüppi, G.; Eggart. G.; Iwasaki, S.; Wehrli, H.; Schaffner, K.; Jeger, O. *Helv. Chim. Acta* **1966,** *49,* 1986.
3. Schaffner, K. *Pure Appl. Chem.* **1968,** *16,* 75.
4. Stratenus, J. L.; Havinga, E. *Rec. Trav. Chim.* **1966,** *85,* 434.
5. Binkley, R. W.; Flechtner, T. W. In *Synthetic Organic Photochemistry;* Horspool, W. M., Ed.; Plenum Press: New York, 1984; p 375.

6. Zen, S.; Tashima, S.; Koto, S. *Bull. Chem. Soc. Jpn.* **1968**, *41*, 3025.
7. Schmid, H.; Karrer, P. *Helv. Chim. Acta* **1949**, *32*, 1371.
8. Kenner, G. W.; Murry, M. A. *J. Chem. Soc.* **1949**, 178.
9. Denney, D. B.; Goldstein, B. *J. Org. Chem.* **1956**, *21*, 479.
10. Closson, W. D.; Wriede, P.; Bank, S. *J. Am. Chem. Soc.* **1966**, *88*, 1581.
11. Pète, J.-P.; Portella, C. *Bull. Soc. Chim. Fr.* **1980**, 275.
12. Masnovi, J.; Koholic, D. J.; Berki, R. J.; Binkley, R. W. *J. Am. Chem. Soc.* **1987**, *109*, 2861.
13. Binkley, R. W. *J. Org. Chem.* **1985**, *50*, 5646.
14. Binkley, R. W.; Hehemann, D.; Koholic, D. J.; Berki, R. J.; Binkley, E. R.; Masnovi, J. *J. Carbohydr. Chem.* **1996**, *15*, 33.
15. Szarek, W. A.; Ritchie, R. G. S.; Vyas, D. M. *Carbohydr. Res.* **1978**, *62*, 89.
16. Iwakawa, M.; Martin, O. R.; Szarek, W. A. *Carbohydr. Res.* **1983**, *121*, 99.
17. Nishida, A.; Hamada, T.; Yonemitsu, O. *J. Org. Chem.* **1988**, *53*, 3386.
18. Nishida, A.; Hamada, T.; Yonemitsu, O. *Chem. Pharm. Bull.* **1990**, *38*, 2977.
19. Liu, X.-G.; Binkley, R. W.; Yeh, P. *J. Carbohydr. Chem.* **1992**, *11*, 1053.
20. Binkley, R. W.; Liu, X.-G. *J. Carbohydr. Chem.* **1992**, *11*, 183.
21. Liu, X.-G.; Binkley, R. W. *J. Carbohydr. Chem.* **1993**, *12*, 779.
22. Barford, A. D.; Foster, A. B.; Westwood, J. H. *Carbohydr. Res.*, **1970**, *13*, 189.
23. Barford, A. D.; Foster, A. B.; Westwood, J. H.; Hall, L. D.; Johnson, R. N. *Carbohydr. Res.* **1971**, *19*, 49.
24. Vegh, L.; Hardegger, E. *Helv. Chim. Acta* **1973**, *56*, 2020.
25. Boigegrain, R.-A.; Gross, B. *Carbohydr. Res.* **1975**, *41*, 135.
26. Seymour, F. R. *Carbohydr. Res.* **1974**, *34*, 65.
27. Chang, C. D.; Hullar, T. L. *Carbohydr. Res.* **1977**, *54*, 217.
28. Seymour, F. R.; Slodki, M. E.; Plattner, R. D.; Tjarks, L. W. *Carbohydr. Res.* **1976**, *46*, 189.
29. Binkley, R. W.; Koholic, D. J. *J. Org. Chem.* **1989**, *54*, 3577.
30. Szarek, W. A.; Dmytraczenko, A. *Synthesis* **1974**, 579.
31. Kishi, T.; Tsuchiya, T.; Umezawa, S. *Bull. Chem. Soc. Jpn.* **1979**, *52*, 3015.
32. Mellier, D.; Pète, J. P.; Portella, C. *Tetrahedron Lett.* **1971**, 4559.
33. Abad, A.; Mellier, D.; Pète, J. P.; Portella, C. *Tetrahedron Lett.* 1971, 4555.
34. Izawa, Y.; Kuromiya, N. *Bull. Chem. Soc. Jpn.* **1975**, *48*, 3197.
35. Mattay, J. *Angew. Chem. Int. Ed. Engl.* **1987**, *36*, 825.
36. Binkley, R. W.; Koholic, D. J. *J. Org. Chem.* **1979**, *44*, 2047.
37. Pète, J.-P.; Portella, C. *J. Chem. Res. (S)* **1979**, 20.
38. Nucci, L.; Cima, F. D.; Cavazza, M. *Tetrahedron Lett.* **1977**, 3099.
39. Mairanovsky, V. G. *Angew. Chem. Int. Ed. Engl.* **1976**, *15*, 281.
40. Yousefzadeh, P.; Mann, C. K. *Org. Chem.* **1968**, *33*, 2716.
41. Carnahan, J. C., Jr.; Closson, W. D.; Ganson, J. R.; Juckett, D. A.; Quaal, K. S. *J. Am. Chem. Soc.* **1976**, *98*, 2526.
42. Ganson, J. R.; Schulenberg, S.; Closson, W. D. *Tetrahedron Lett.* **1970**, 4397.
43. Jarrell, H. C.; Ritchie, R. G. S.; Szarek, W. A.; Jones, J. K. N. *Can. J. Chem.* **1973**, *51*, 1767.
44. Cuvigny, T.; Larchevêque, M. *J. Organomet. Chem.* **1974**, *64*, 315.
45. Hehemann, D. G.; Binkley, R. W.; Masnovi, J. *J. Org. Chem.*, **1989**, *54*, 3256.
46. Pète, J.-P.; Portella, C. *Bull. Soc. Chim. Fr.* **1985**, 195.
47. Klausener, A.; Beyer, G.; Leismann, H.; Scharf, H.-D.; Müller, E.; Runsink, J.; Görner, H. *Tetrahedron* **1989**, *45*, 4989.
48. Jortner, J.; Ottolenghi, M.; Stein, G. *J. Phys. Chem.* **1963**, *67*, 1271 and **1964**, *68*, 247.
49. Tsuchiya, T.; Nakamura, F.; Umezawa, S. *Tetrahedron Lett.* **1979**, 2805.
50. Duan, S.; Binkley, E. R.; Binkley, R. W. *J. Carbohydr. Chem.* **1995**, *14*, 1029.
51. Matsumoto, K.; Ebata, T.; Matsushita, H. *Carbohydr. Res.* **1995**, *267*, 187.
52. Hamada, T.; Nishida, A.; Matsumoto, Y.; Yonemitsu, O. *J. Am. Chem. Soc.* **1980**, *102*, 3978.
53. Hamada, T.; Nishida, A.; Yonemitsu, O. *J. Am. Chem. Soc.* **1986**, 140 (1986).

13

Sulfones and Sulfoxides

Sulfones

Background

The study of sulfone photochemistry finds its beginning with the discovery by Cava et al. that cyclic, aromatic sulfones can extrude sulfur dioxide upon photolysis to produce new, carbocyclic systems.[1] An example of this type of reaction is shown in eq 1.[1] Further examples are found in an excellent review article by Horspool in which he documents the extensive use of sulfone photolysis in synthesizing strained ring systems.[2] Acyclic sulfones also can produce sulfur dioxide upon photolysis; however, whether the sulfone is cyclic or acyclic, several conditions must be met before extrusion of sulfur dioxide will occur.

One of these conditions is that during photolysis the sulfone must absorb the incident light in order to react. Even though this is a basic requirement for photochemical reaction by direct excitation, some emphasis on this criterion is warranted here because typical aliphatic sulfones do not have significant UV absorption. The longest wavelength absorption for dimethyl sulfone, for instance, has it maximum at 180 nm.[3] Aromatic sulfones, in contrast, have absorption extending well into the UV [λ_{max} (diphenyl sulfone) 201 (ϵ 31,700), 235 (15,500), 260 (1740), 266 (2140) and 274 (1390) nm[4]]. Absorption characteristics are an important factor in the photolysis of carbohydrate sulfones because neither the sulfonyl group nor the carbohydrate portion of the molecule typically provides the necessary chromophore for UV light absorption.

Another criterion for extrusion of sulfur dioxide is that the sulfonyl radical ($\bullet SO_2R'$, Scheme 13.1), produced by fragmentation of an excited sulfone,

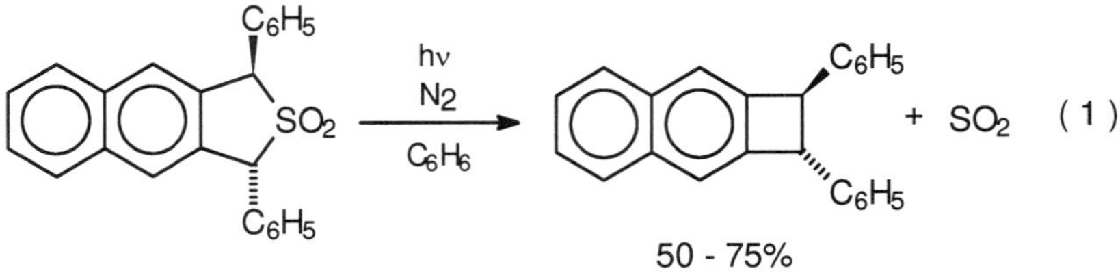

Equation 1

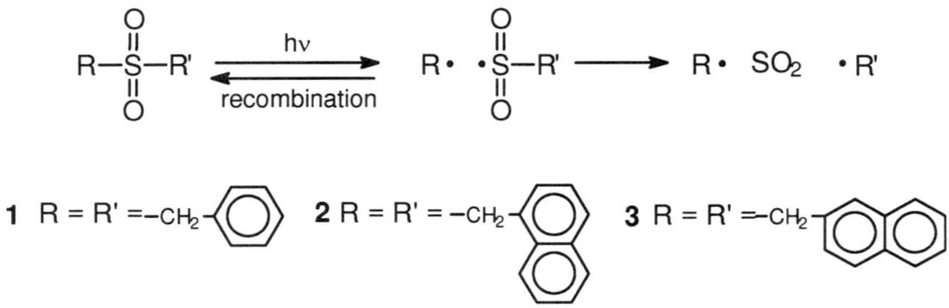

Scheme 13.1

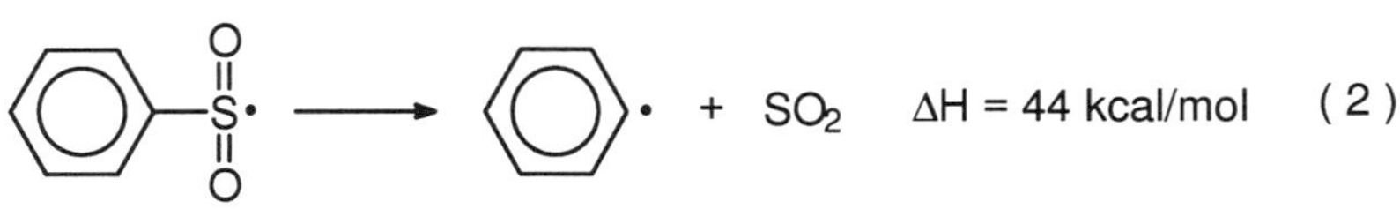

4

Equation 2

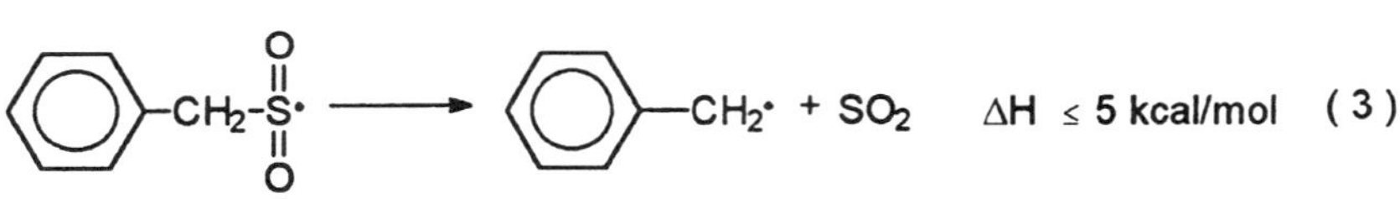

5

Equation 3

must be able to break easily its S–R′ bond. Because for sulfonyl radicals this cleavage can be quite endothermic,[5] forming a stabilized radical during bond homolysis has a significant impact on the energy of activation of reaction and, consequently, in determining whether sulfur dioxide extrusion will occur. The bond dissociation energy for the phenylsulfonyl radical (**4**, eq 2), which does not fragment readily, is estimated to be 44 kcal/mol (184 kJ/mol);[6,7] in contrast, a value no greater than 5 kcal/mol (21 kJ/mol) is estimated for the bond dissociation energy of the benzylsulfonyl radical (**5**), which fragments rapidly to form the resonance-stabilized benzyl radical (eq 3).[8] The most completely studied compounds that extrude sulfur dioxide without having the sulfonyl group as part of a ring system are the benzyl and naphthylmethyl sulfones investigated by Givens and co-workers.[9–11] These investigators determined that for the sulfonyl radicals derived from excited singlet-state reaction of the sulfones **1–3** (and several related compounds) only radical recombination (R• + •SO_2 R′ → RSO_2 R′) competed with sulfur dioxide formation (•SO_2 R′ → SO_2 + •R′).[11]

Kharasch and Khodair found that diphenyl sulfone decomposed upon photolysis in benzene to produce biphenyl; however, no sulfur dioxide was formed.[12,13] These investigators proposed the reaction mechanism shown in Scheme 13.2, and attributed failure to form sulfur dioxide to the stability of the phenylsulfonyl radical (eq 2).[6,7] The mechanism shown in Scheme 13.2 not only represents a reasonable explanation for diphenyl sulfone photoreaction, but it also serves as a beginning point for understanding the photochemistry of carbohydrate sulfones.

Photolysis of Glycosyl Phenyl Sulfones

In the first of two investigations, Collins and Whitton irradiated the α (**7**) and β (**8**) anomers of 2,3,4,6-tetra-*O*-acetyl-D-glucopyranosyl phenyl sulfone in an effort to find a new reaction for C-glycoside formation.[14] Some C-glycosides were produced but their yields were low, and too many other products were formed (eq 4) for these reactions to have any synthetic value.[14] In a second investigation, these authors extended their study to include the two anomers [α (**14**) and β (**15**)] of 2,3,4,6-tetra-*O*-acetyl-D-glucofuranosyl phenyl sulfone (eq 5).[15] The results from these photolyses were similar to those from the first study in terms of the types of products formed.

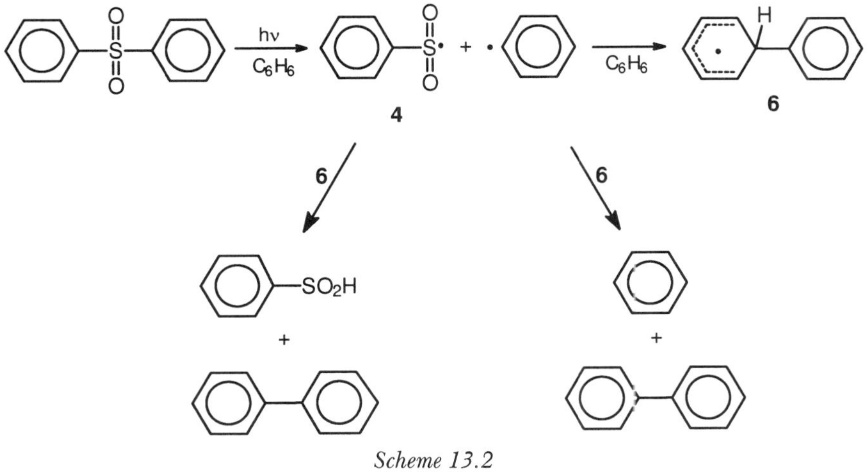

Scheme 13.2

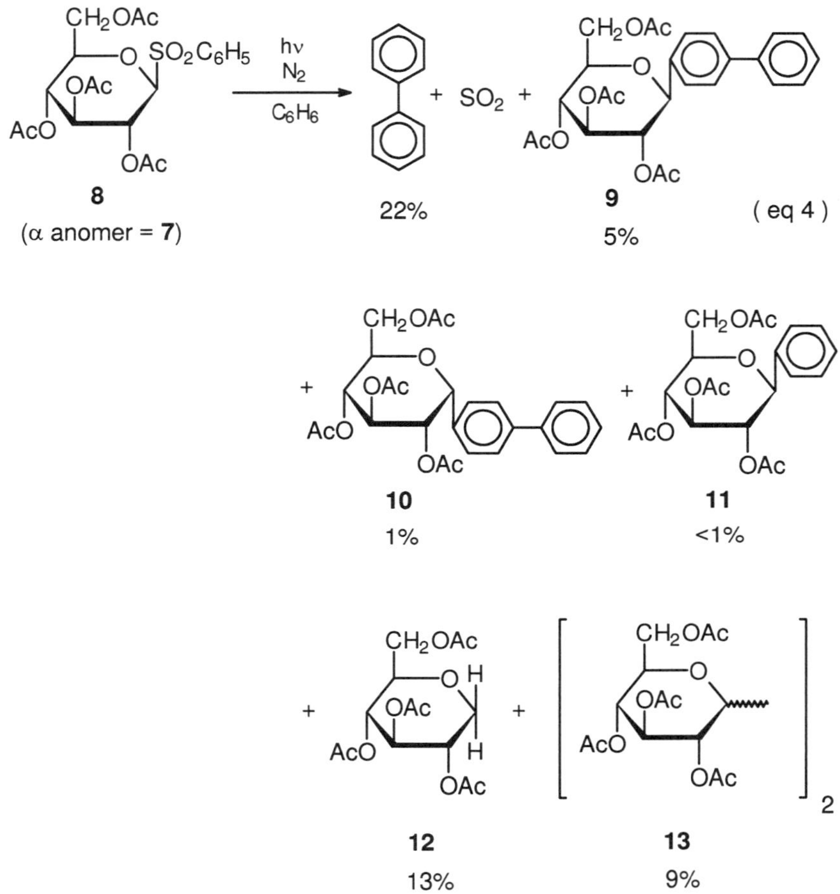

Equation 4

Reaction Mechanism

Collins and Whitton proposed free-radical mechanisms to explain product formation from photolysis of compounds **7**, **8**, **14**, and **15** (Scheme 13.3). (The mechanism shown is for reaction of compound **8**; analogous mechanisms were proposed for reaction of the other three sulfones.) The initial step in this process is the homolytic cleavage of the C_6H_5–S bond to form a phenyl radical and a sulfonyl radical (**20**). The phenyl radical adds to benzene to produce a phenylcyclohexadienyl radical (**6**), and the sulfonyl radical **20** fragments to give sulfur dioxide and the glycosyl radical **21**. Photoproducts then are formed from reactions of **6** and **21** (Scheme 13.3).

In order for the mechanism proposed by Collins and Whitton to be operative, it is necessary to assume that the sulfonyl radical **20** fragments to release

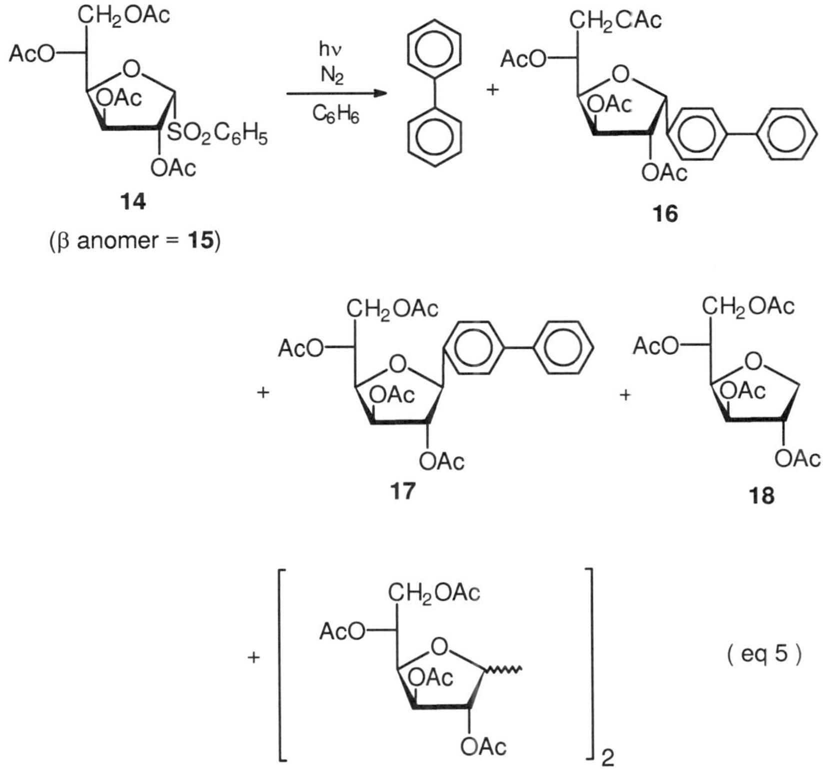

19

Equation 5

sulfur dioxide and the glycosyl radical **21** before other reactions occur. This is a reasonable possibility because the ring oxygen atom in **21** should provide stabilization for the developing radical center at C_1 and thus lower the activation energy for C_1–S bond fragmentation. It is also necessary to assume that intermediate radicals **21** and **6** are sufficiently stable that they exist in solution long enough to react with each other. (Formation of compounds **9**, **10**, and **13** certainly supports this assumption.)

Although the mechanism outlined in Scheme 13.3 explains formation of the various reaction products, some further consideration of the initial fragmentation of the excited sulfone is in order. The glycosyl radical **21** also would be formed if the excited sulfone cleaved the C_1–S bond, as shown in eq 6, rather than the C_6H_5–S bond cleavage shown as the initial step in Scheme 13.3. The reaction shown in eq 6 is arguably the major reaction pathway for the excited sulfone because this reaction produces a radical pair more stable than that from fragmenting the C_6H_5–S bond. (Givens et al. have shown that

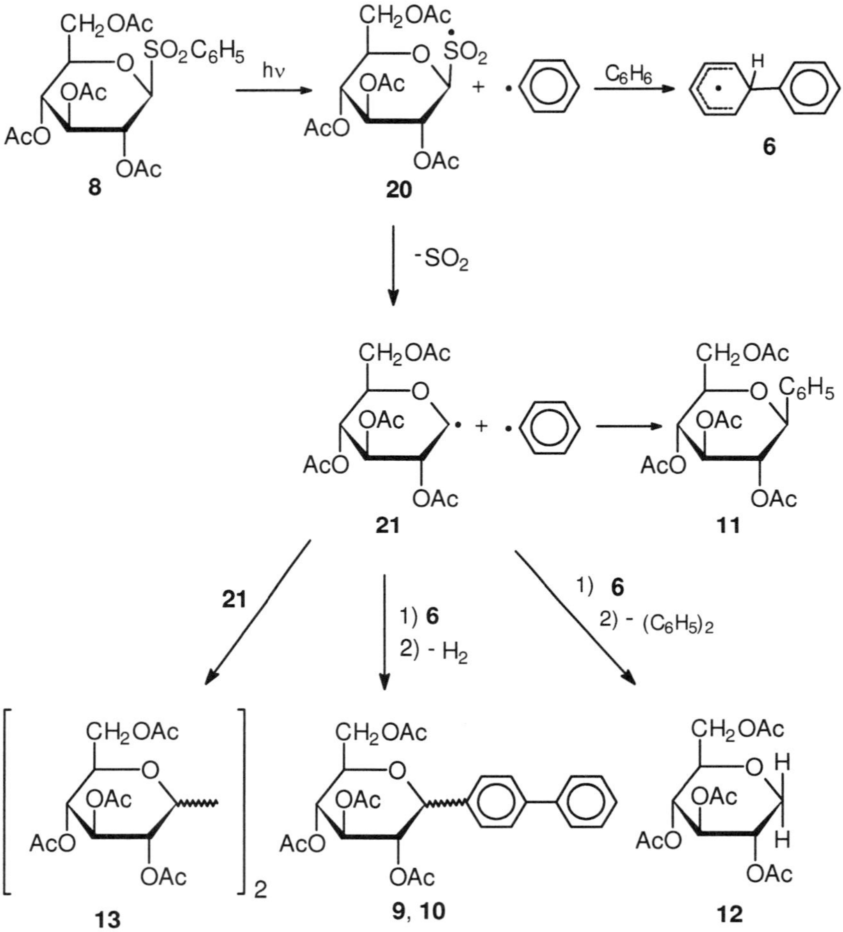

Scheme 13.3

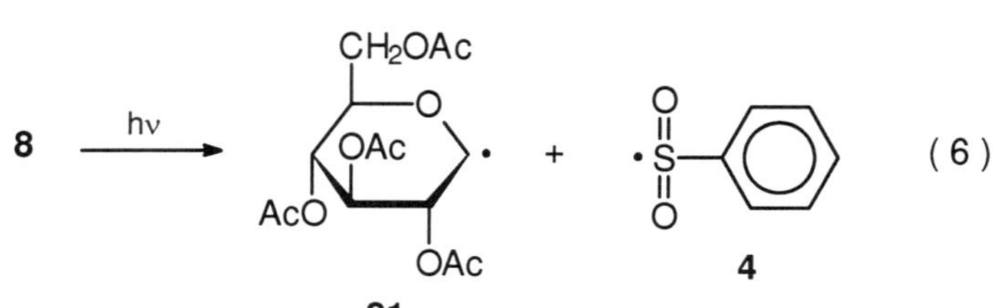

Equation 6

excited sulfones preferentially break the C–S bond that produces the more stable radical pair.[10])

Although the available data do not allow a definitive determination of which C–S bond fragmentation takes place first, some suggestive information does exist. In the photochemical reaction of diphenylsulfone,[12,13] which generates a phenyl radical from every molecule that reacts (Scheme 13.2), the yield of biphenyl is high (70–98%). From photolysis of the sulfone **8**, however, the yield of biphenyl is only 22%. This yield is consistent with relatively few phenyl radicals being formed. Preferential cleavage of the C_1–S bond as shown in eq 6 would account for the relatively low yield of biphenyl because the phenylsulfonyl radical (**4**) would not be expected to give the phenyl radical needed to produce biphenyl (eq 2).[6,7] Does the C_6H_5–S bond in the excited sulfone ever fragment? Probably, the formation of some biphenyl (and other photoproducts) is consistent with some C_6H_5–S bond cleavage. A reasonable explanation is that either of the C–S bonds in **8** can fragment upon excitation but that C_1–S bond cleavage is the major reaction pathway.

Sulfoxides

Carbohydrate Sulfoxide Photolysis

Many photochemical investigations of carbohydrates can trace their inspiration to work done on simpler organic compounds. This is not the case, however, for sulfoxides since the single paper in the carbohydrate literature describing irradiation of a sulfoxide is actually one of the earliest studies done on this class of compounds. The investigation, conducted by Horton and Jewell and published in 1966,[16] describes the photochemical conversion of 1-deoxy-1-ethylsulfinyl-D-galactitol (**22**) into galactitol (**23**) (eq 7). The authors

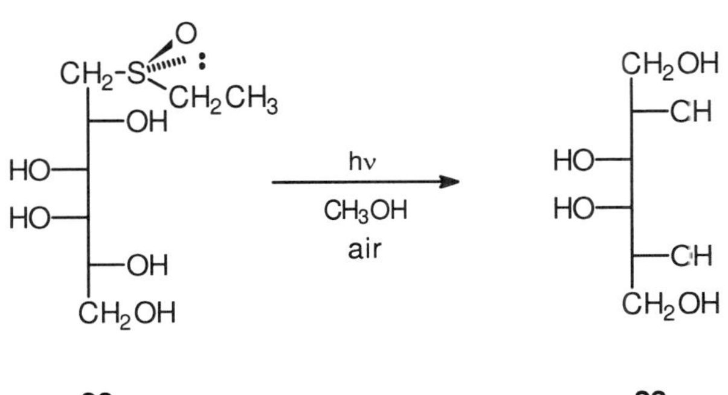

22 **23** 58% (7)

Equation 7

did not offer a mechanistic interpretation of this reaction, perhaps because little information existed in the literature at that time upon which to base a proposal. This situation has changed in the intervening years, and subsequent research has established a basis for proposing a reaction mechanism.

Reaction Mechanism

In 1968 Schultz and Schlessinger reported the first in a series of investigations on sulfoxide photolyses.[17–20] These investigators found that in addition to pyramidal inversion (eq 8),[19] which is a type of process discovered by Hammond and co-workers,[21,22] some cyclic sulfoxides also experienced rearrangement upon irradiation (Scheme 13.4).[20] In the case of the sulfoxides **24** and

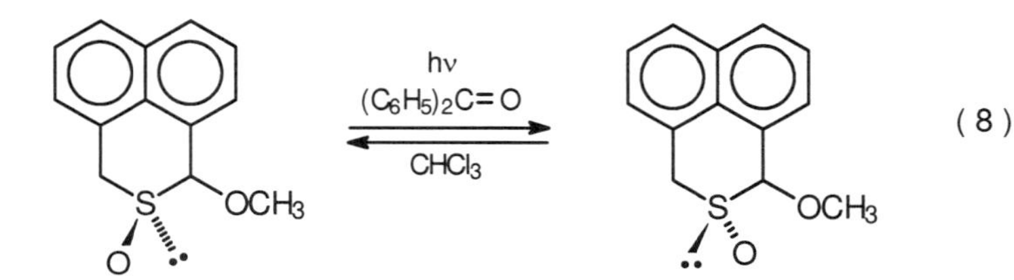

Equation 8

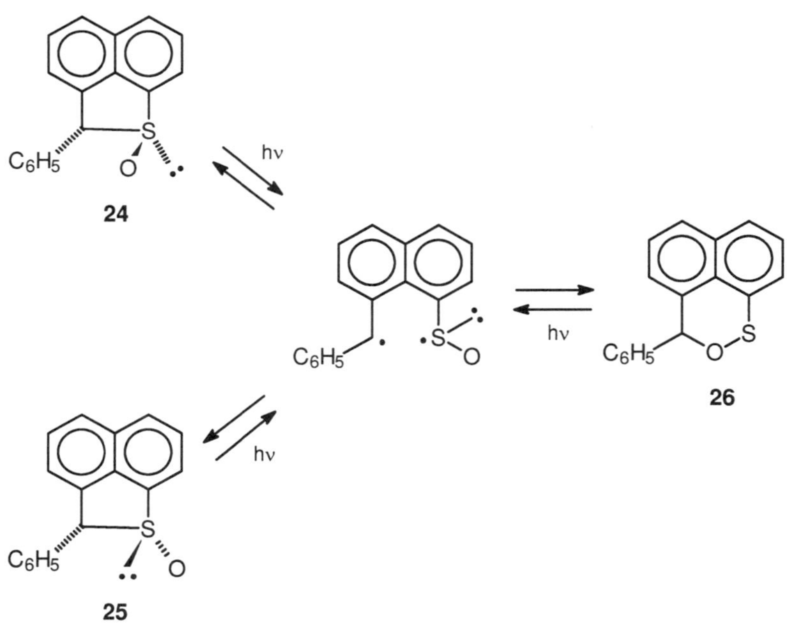

Scheme 13.4

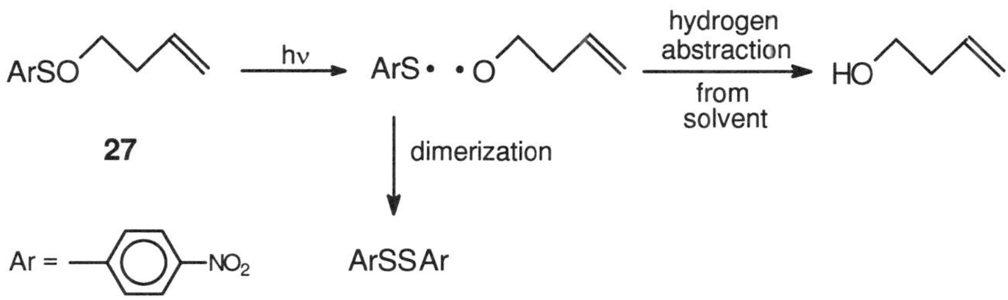

Scheme 13.5

25, Schultz and Schlessinger were able to show that the cyclic sulfenate **26** was formed in a reversible photochemical reaction.[20] (Other workers also have observed this sulfoxide to sulfenate rearrangement.[23,24]) Would this reaction still be reversible if the sulfenate were acyclic? This is a critical question with respect to the mechanism for reaction of the acyclic, carbohydrate sulfoxide **22**. The answer appears to be "probably not" if one uses as a model results from the work of Pasto and co-workers, who found that irradiation of the acyclic sulfenate **27** produced the corresponding alcohol and disulfide but no sulfoxide (Scheme 13.5).[25,26]

Armed with the findings of Schultz and Schlessinger and with those of Pasto and co-workers, one can construct a reasonable pathway for conversion of **22** into **23**. This process begins with excitation of **22*** and requires rearrangement of the excited sulfoxide to the corresponding sulfenate ester **28** (Scheme 13.6). The O–S bond in this ester then cleaves photochemically (or possibly thermally[23,24]) to produce the alkoxy radical **29**, which abstracts a hydrogen atom from the solvent to produce galactitol (**23**)

Summary

The number of photochemical investigations of carbohydrate sulfones and sulfoxides is limited; in fact, only four sulfones (**7**, **8**, **14**, and **15**), all of which are glycosyl phenyl sulfones, have been studied. Product mixtures, which are complex (eqs 4 and 5), appear to result from formation of phenylcyclohexadienyl (**6**) and glycosyl radicals. A mechanism for product formation is given in Scheme 13.3 and an alternate initial fragmentation of the excited sulfone is shown in eq 6. The photochemistry of carbohydrate sulfoxides is virtually unexplored. Only one example of photolysis of this type of compound exists in the literature. Although no mechanistic work has been done on sulfoxide photoreaction, it is possible to propose a mechanism for reaction (Scheme 13.6) based upon study of other organic sulfoxides.

*Alkyl sulfoxides, unlike alkyl sulfones, have significant UV absorption. The absorption maxima for ethyl sulfoxide are at 201 (ϵ 2130) and 214 (1210) nm.[4]

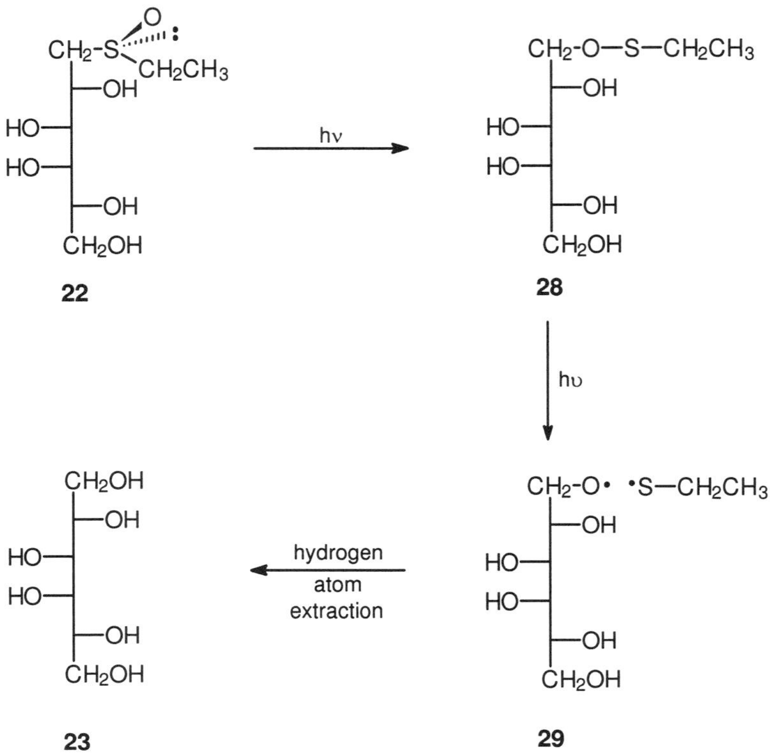

Scheme 13.6

References

1. Cava, M. P.; Schlessinger, R. H.; Van Meter, J. P. *J. Am. Chem. Soc.* **1964,** *86,* 3174.
2. Horspool, W. M. In *The Chemistry of Sulphonic Acids, Esters and their Derivatives;* Patai, S.; Rappoport, Z., Eds.; John Wiley & Sons: Chichester, NY, 1991; p 501.
3. Prochazka, M.; Paleck, M. *Coll. Czech. Chem. Commun.* **1967,** *32,* 3049.
4. Cumper, C. W. N.; Read, J. F.; Vogel, A. I. *J. Chem. Soc. (A)* **1966,** 239.
5. Bertrand, M. P. *Org. Proc. Proc. Int.* **1994,** *26,* 257.
6. Benson, S. W. *Chem. Rev.* **1978,** *78,* 23.
7. Horowitz, A. *Int. J. Chem. Kinet.* **1976,** *8,* 709.
8. Chatgilialoglu, C.; Lunazzi, L.; Ingold, K. U. *J. Org. Chem.* **1983,** *48,* 3588.
9. Givens, R. S.; Matuszewski, B. *Tetrahedron Lett.* **1978,** 861.
10. Givens, R. S.; Hrinczenko, B.; Liu, J.-S.; Matuszewski, B.; Tholen-Collison, J. *J. Am. Chem. Soc.* **1984,** *106,* 1779.
11. Gould, I. R.; Tung, C.; Turro, N. J.; Givens, R. S.; Matuszewski, B. *J. Am. Chem. Soc.* **1984,** *106,* 1789.
12. Kharasch, N.; Khodair, A. I. A. *J. Chem. Soc., Chem. Commun.* **1967,** 98.
13. Khodair, A. I.; Nakabayashi, T.; Kharasch, N. *Int. J. Sulfur Chem.* **1973,** *8,* 37.
14. Collins, P. M.; Whitton, B. R. *J. Chem. Soc., Perkin Trans. 1,* **1974,** 1069.
15. Collins, P. M.; Whitton, B. R. *Carbohydr. Res.* **1974,** *36,* 293.

16. Horton, D.; Jewell, J. S. *J. Org. Chem.* **1966,** *31,* 509.
17. Schultz, A. G.; DeBoer, C. D.; Schlessinger, R. H. *J. Am. Chem. Soc.* **1968,** *90,* 5314.
18. Schultz, A. G.; Schlessinger, R. H. *J. Chem. Soc., Chem. Commun.* **1969,** 1483.
19. Schultz; A. G.; Schlessinger, R. H. *J. Chem. Soc., Chem. Commun.* **1970,** 1051 and 1294.
20. Schultz, A. G.; Schlessinger, R. H. *Tetrahedron Lett.* **1973,** 3605.
21. Cooke, R. S.; Hammond, G. S. *J. Am. Chem. Soc.* **1968,** *90,* 2958.
22. Mislow, K.; Axelrod, M.; Rayner, D. R.; Gotthardt, H.; Coyne, L. M.; and G. S. Hammond, B. S. *J. Am. Chem. Soc.* **1965,** *87,* 4958.
23. Still, I. W. J.; Chauhan, M. S.; and Thomas, M. T. *Tetrahedron Lett.* **1973,** 1311.
24. Kowalewski, R.; Margaretha, P. *Helv. Chim. Acta* **1993,** *76,* 1251.
25. Pasto, D. J.; L'Hermine, G. *J. Org. Chem.* **1990,** *55,* 5815.
26. Pasto, D. J.; Cottard, F. *Tetrahedron Lett.* **1994,** *35,* 4303.

14

Thio Sugars and Their Derivatives

Introduction

Thio sugars are defined as "carbohydrate derivatives in which one or more of the oxygen atoms bonded directly to the carbon skeleton have been replaced by sulfur."[1] This bonding arrangement determines much of the photochemistry of these compounds because many photochemical reactions of thio sugars and their derivatives involve C–S bond homolysis. Examples of this type of reaction are the conversion of sulfur-containing nucleosides into cyclonucleosides, the photoreduction of sugar dialkyl dithioacetals, and the formation of *C*-glycosides from *S*-nucleosides. Other photochemical reactions of thio sugars are ionic in nature and depend upon the ability of a sulfur atom to function as a donor in an electron-transfer reaction. Examples of this type of process include glycoside formation from 1-thioglycosides and deoxy sugar synthesis from thioethers. Together the reactions of thio sugars and their derivatives include some of the earliest photolyses of carbohydrates as well as some of the most recently discovered reactions.

Direct excitation of thio sugars presents no experimental difficulty in excited state formation because compounds containing sulfide linkages easily absorb UV radiation. Diethyl sulfide, for example, has an absorption maximum at 202 nm ($\in$ 2090) with a molar absorptivity falling below 500 only at wavelengths longer than 220 nm. Diphenyl sulfide has UV maxima at 201 ($\in$ 38,100), 250 (11,300), and 275 (5630) nm.[2]

Formation of Cyclonucleosides

Ueda and co-workers conducted a systematic study of the photolysis of nucleosides in which phenylthio groups (and related, sulfur-containing substituents) were attached to various positions in these molecules (Tables 14.1–14.4).[3–10] Irradiation of these compounds typically produced cyclonucleosides and, in some cases, photoreduction products (i.e., compounds in which the phenylthio group had been replaced by a hydrogen atom). These researchers proposed a free-radical mechanism (Scheme 14.1) to account for the observed

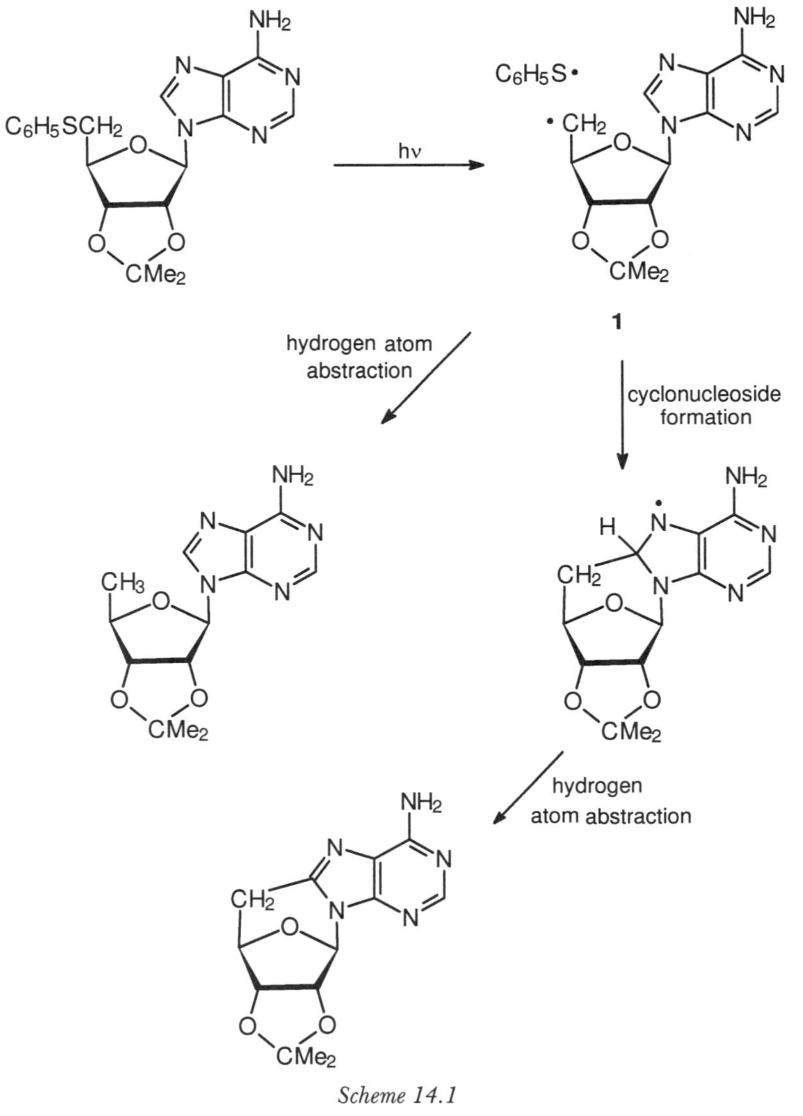

Scheme 14.1

reactions.[3] According to this mechanism, the course of reaction is determined by a competition between intramolecular (cyclonucleoside formation) and intermolecular (hydrogen atom abstraction) processes originating from the intermediate **1**. Any changes in molecular structure that affect this competition necessarily influence the balance between these two pathways. One such change is the length of the new bridging chain between the sugar and nitrogenous base portions of the molecule. In those cases where the bridge is a one- or two-carbon atom unit, cyclonucleoside formation generally is observed, but if the potential bridge is greater than two carbon atoms in length, photoreduction takes place (Table 14.1).

Trimethyl phosphite is added to photolysis mixtures during cyclonucleoside formation in order to optimize product yield. If this reagent is omitted, yields are reduced and considerable unidentified material is formed. Ueda et al. proposed that the beneficial effect of trimethyl phosphite on product formation is due to its ability to trap phenylthiyl radicals.[3] This is a reasonable explanation because phenylthiyl radicals, which are produced in the initial step shown in Scheme 14.1, are capable of reactions such as addition to π-systems, coupling with molecular oxygen, and dimerization. Any of these processes would increase the complexity of the reaction mixture. In support of their proposal for the role of trimethyl phosphite, Ueda et al. cite the work of Walling and Rabinowitz,[13] who determined that thiyl radicals are removed from solution by reacting with triethyl phosphite in the manner shown in Scheme 14.2.

In addition to studying compounds with sulfur-containing substituents at the 5′ position (Table 14.1), the Ueda group also investigated nucleosides with a phenylthiomethyl (or related) group placed at either the 2′ or 3′ position and found that photolysis also produced cyclonucleosides (Tables 14.2 and 14.3). (The proposed mechanism for these reactions is analogous to that shown in Scheme 14.1.) Even irradiating a compound with the phenylthio group at the 8 position in the adenine ring led to formation of a cyclonucleoside (Table 14.4). The mechanism for this process is similar to that for irradiation of other phenylthio compounds, but it is made slightly more complex by the need for internal hydrogen atom abstraction prior to cyclization (Scheme 14.3).

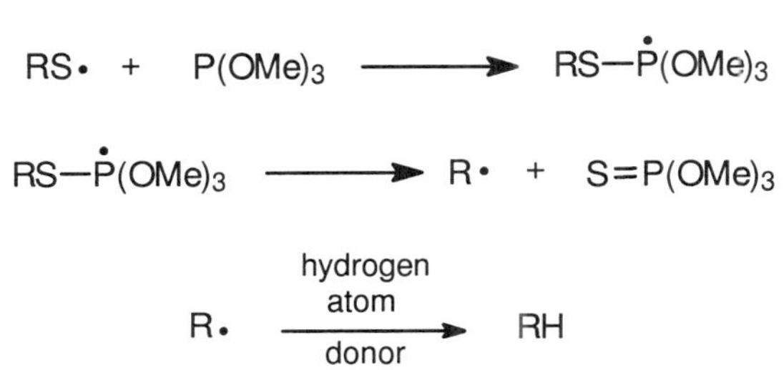

Scheme 14.2

Table 14.1. Photolysis of Nucleosides with Sulfur-Containing 5′ Substituents[a]

Reactant	*Product and Percent Yield*	*Ref.*
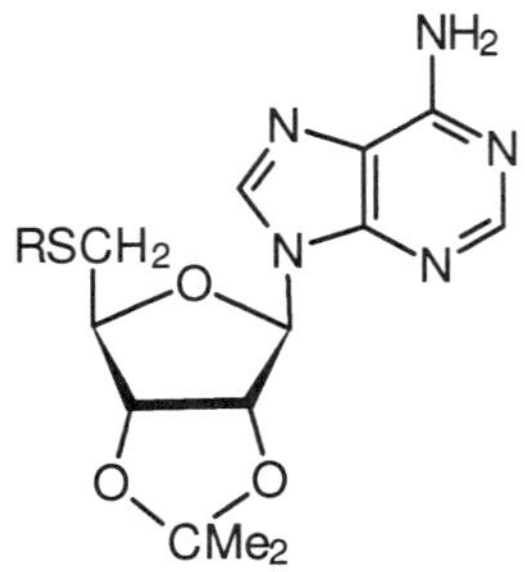	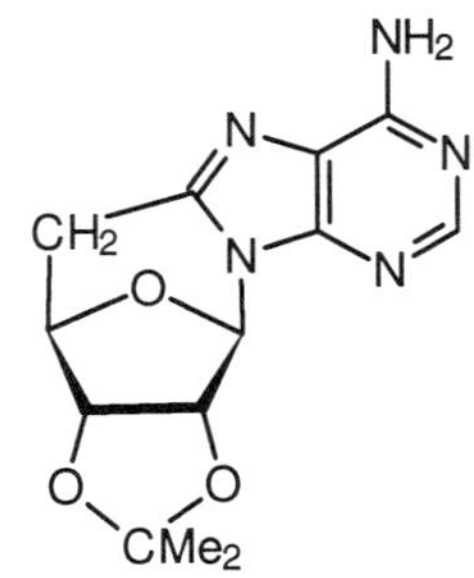	
R = C_6H_5-	66%	3, 4
R = $C_6H_5CH_2$-	44%	3
R = $CH_2{=}CHCH_2$-	not	3
R = $CH_3CH_2CH_2$-	reported	3
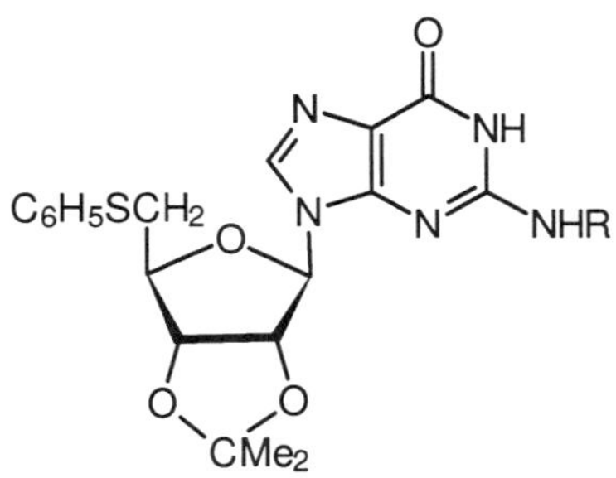	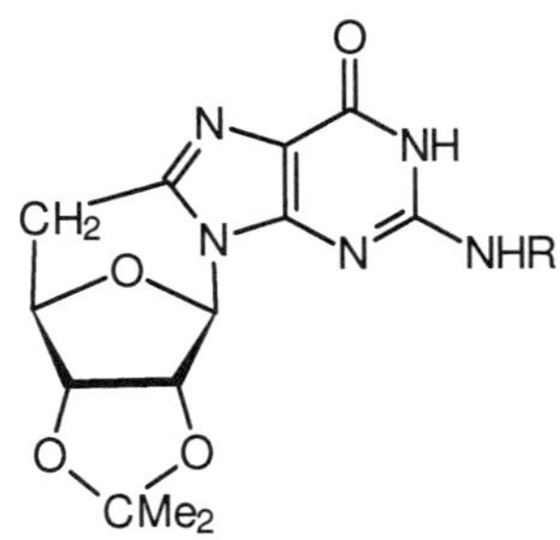	5
R = Bz	47%	
R = H	89%	
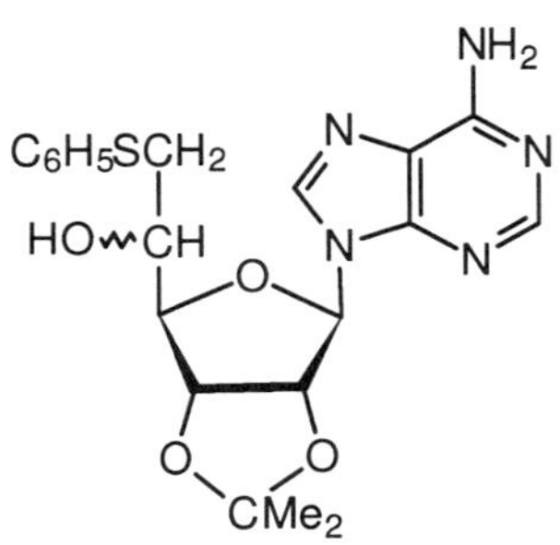	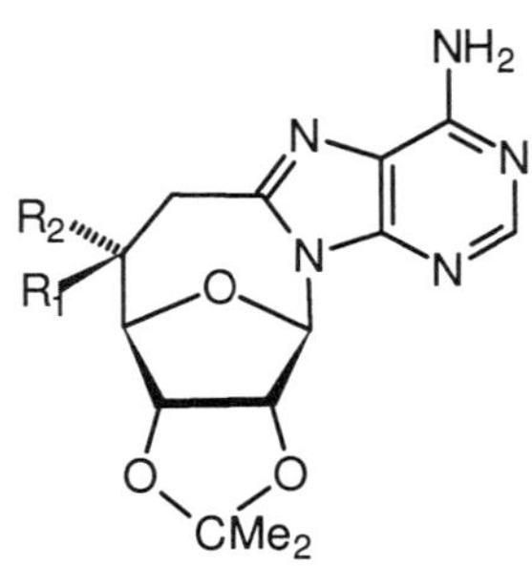	6[b]
R_2 = H, R_1 = OH	20%	
R_2 = OH, R_1 = H	17%	

Continued on next page

Table 14.1. Photolysis of Nucleosides with Sulfur-Containing 5′ Substituents[a] —*Continued*

Reactant	*Product and Percent Yield*	*Ref.*
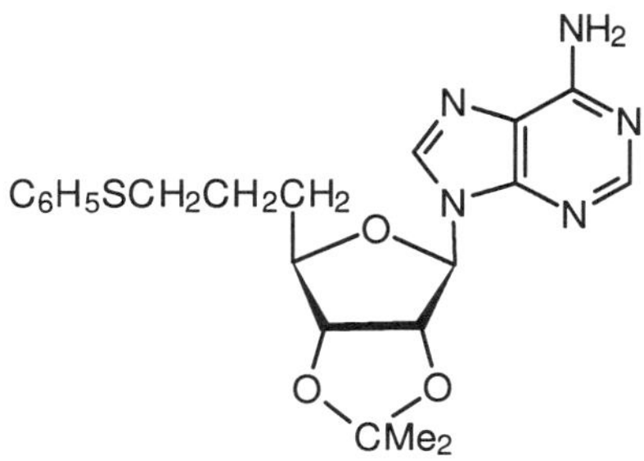	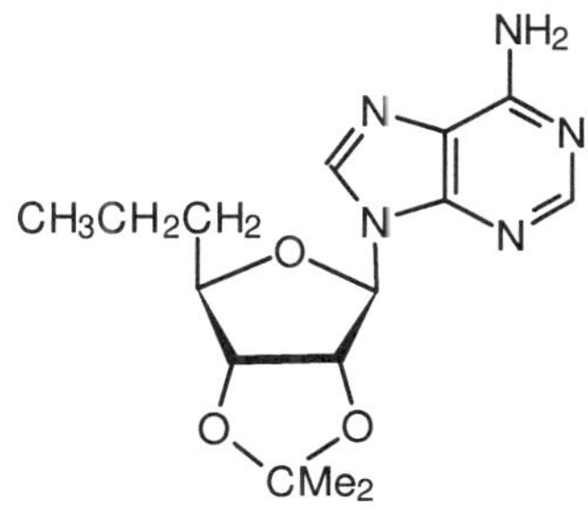 67%	6

[a]Irradiations were conducted in acetonitrile under an inert atmosphere with added trimethyl phosphite.

[b]An additional product was formed in which the phenylthio group in the reactant was replaced by a hydrogen atom.

Table 14.2. Photolysis of Nucleosides with Sulfur-Containing 2′ Substituents[a]

Reactant	*Product and Percent Yield*	*Ref.*
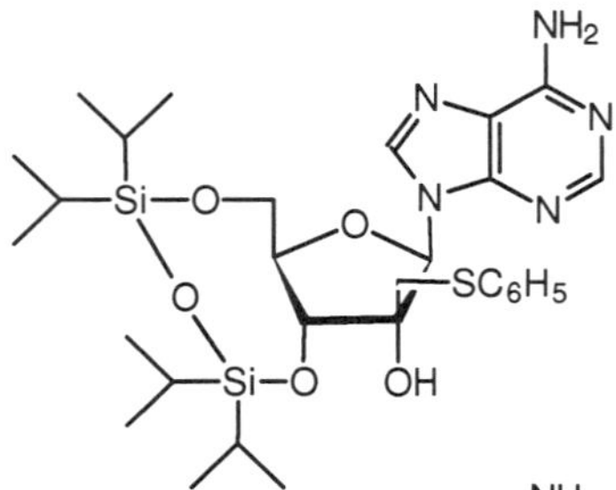	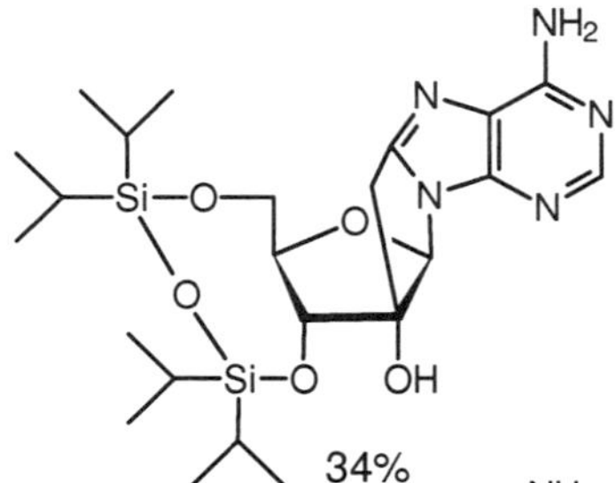 34%	7
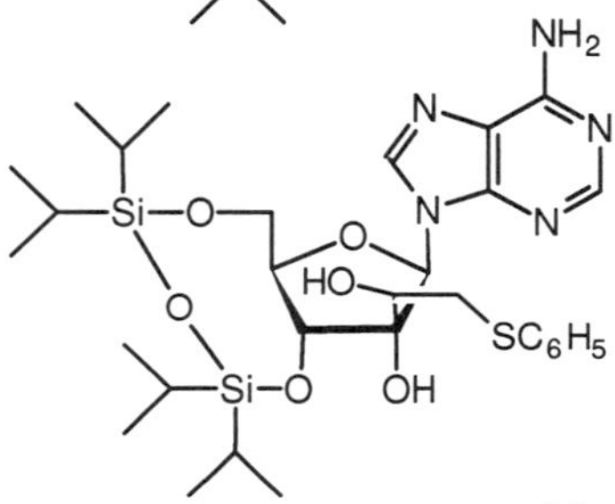	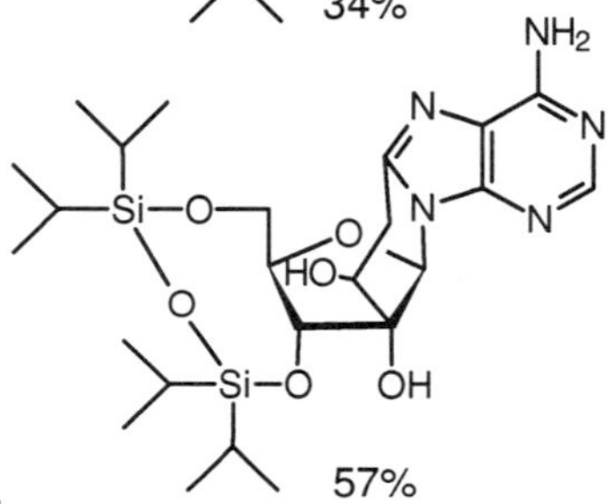 57%	8
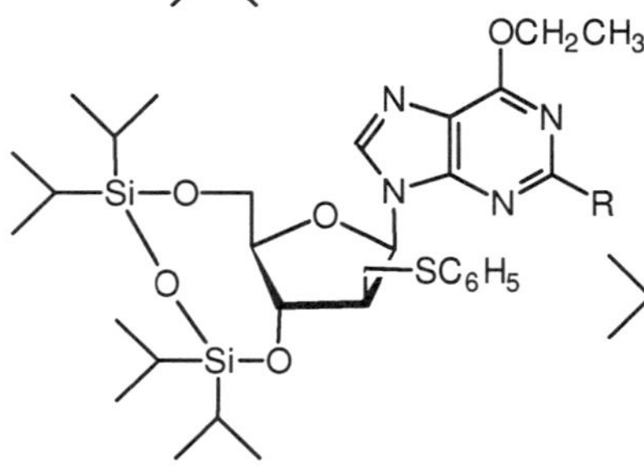 R = NHAc	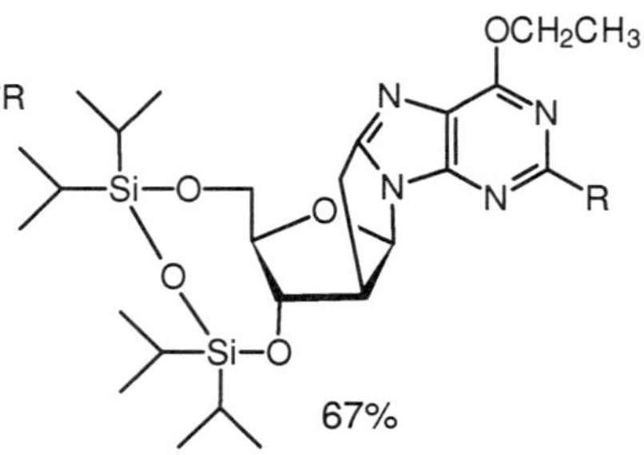 67%	8
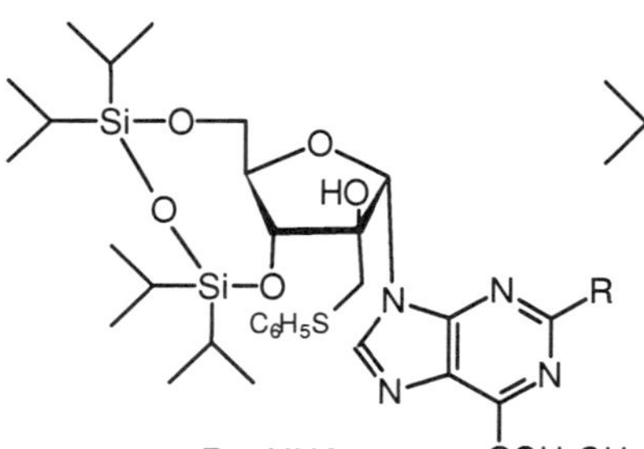 R = NHAc	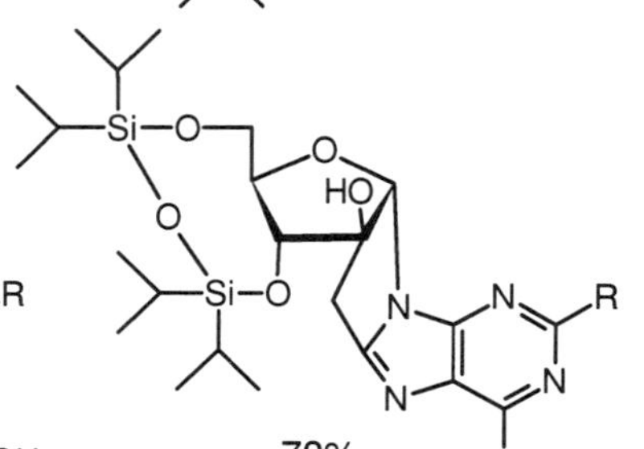 73%	8

[a]Irradiations were conducted in acetonitrile under an inert atmosphere with trimethyl phosphite.

Table 14.3. Photolysis of Nucleosides with Sulfur-Containing 3′ Substituents[a]

Reactant	*Product*	*Percent Yield*	*Ref.*
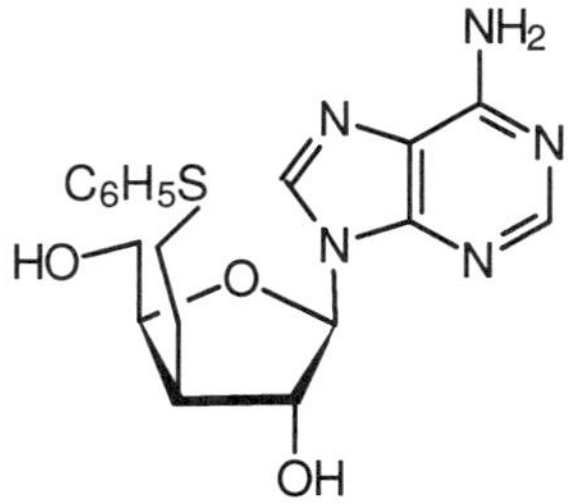	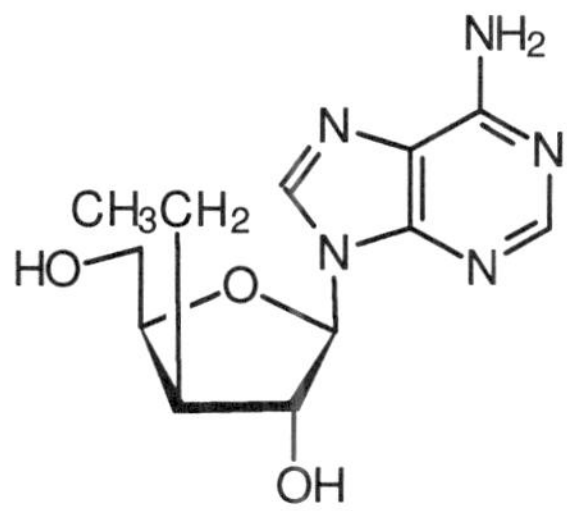	65	9
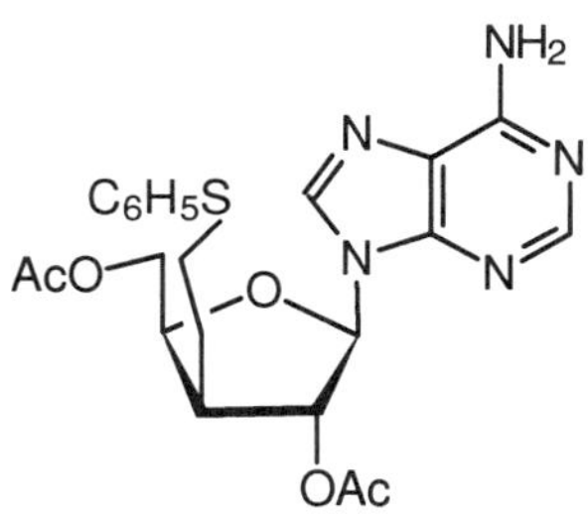	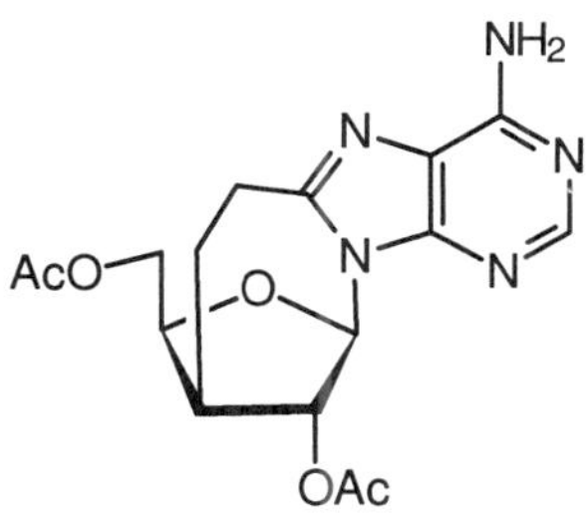	40	9

[a]Irradiations were conducted in acetonitrile under an inert atmosphere with trimethyl phosphite present.

Table 14.4. Photolysis of Nucleosides with Sulfur-Containing 8 Substituents[a]

Reactant	*Product*	*Percent Yield*	*Ref.*
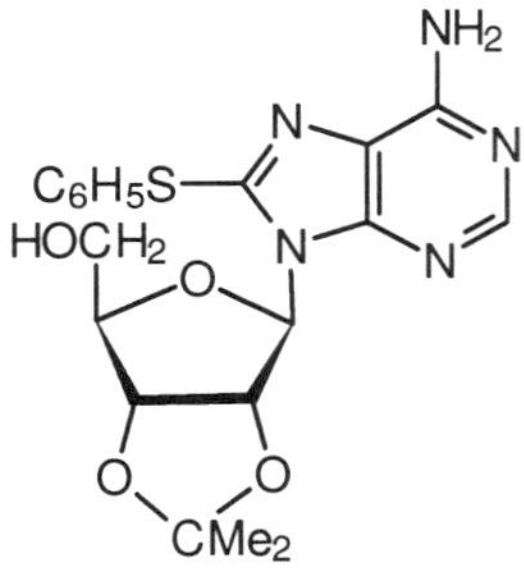	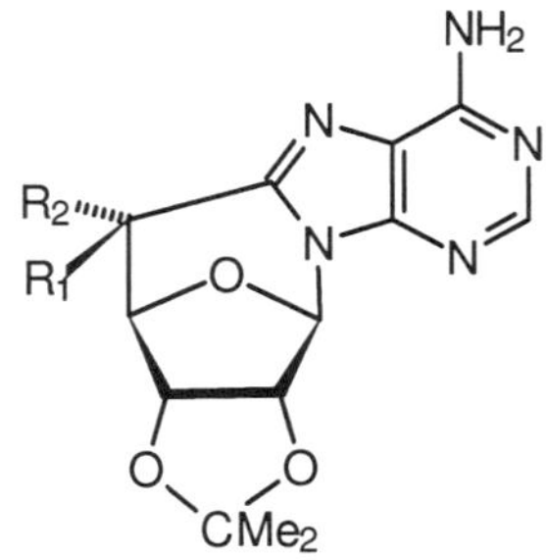		10
	R_1 = OH, R_2 = H	13	
	R_1 = H, R_2 = OH	8	
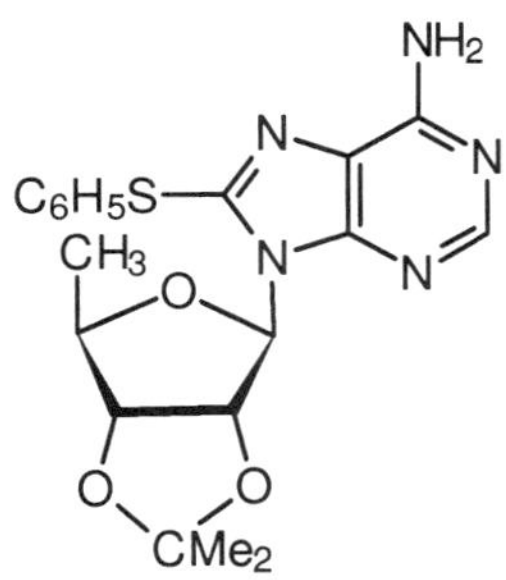	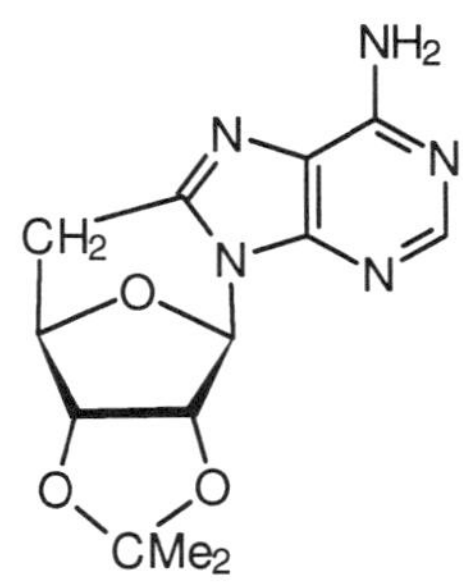	75	11, 12[b]

[a]Irradiations were conducted in acetonitrile under an inert atmosphere with trimethyl phosphite present.

[b]Irradiation was conducted in acetonitrile under an inert atmosphere with *t*-butyl peroxide present.

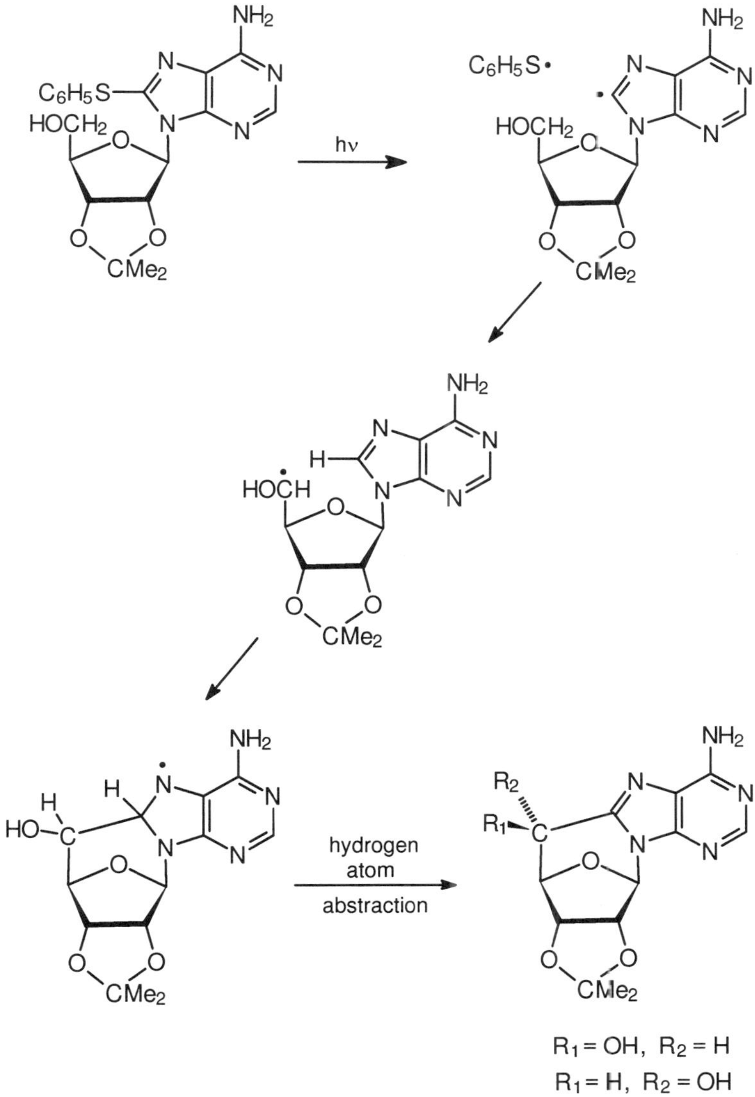

Scheme 14.3

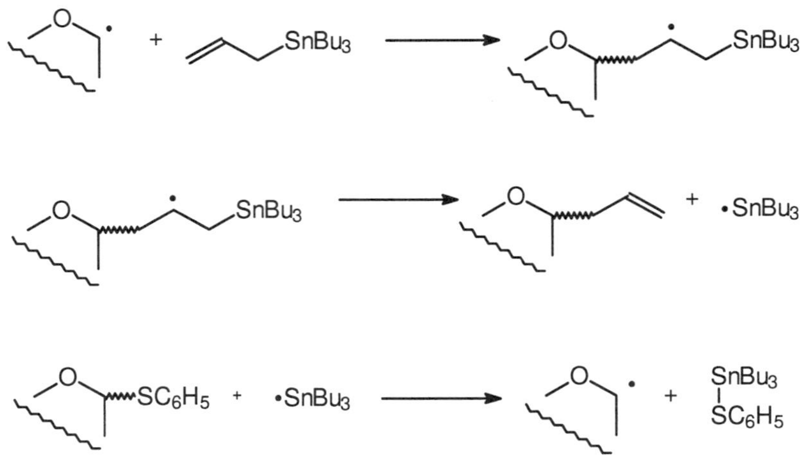

Scheme 14.4

*Substitution Reactions**

Substitution by an Allyl Group: The Keck Reaction

Photochemically based substitution reactions involving thio sugars can occur in several different ways.[14–23] Keck et al. found that a phenylthio glycoside can participate in a photochemically initiated chain reaction in which the phenylthio substituent is replaced by an allyl group to produce a *C*-glycoside (the Keck reaction).[14–16] Earlier Keck and Yates had found this reaction to be a free-radical process for which they proposed the mechanism shown in Scheme 14.4.[17] Introduction of allyl groups in this fashion is quite selective in that it does not take place for simple thioethers. The reason for this selectivity is thought to be that the stabilizing effect of the ring oxygen atom on the developing radical center at C-1 (Scheme 14.4, third equation) is essential for C–S bond cleavage.[14]

Substitution by a Hydrogen Atom

β-Glycoside Synthesis: The Kahne Reaction

In a process that has mechanistic similarity to the Keck reaction, Kahne and co-workers discovered a new method for the synthesis of β-glycosides.[24,25] This new reaction depends upon the replacement of a methylthio group by a hydrogen atom in a chain reaction involving tri-*n*-butylstannane as a source of

*We will continue in this chapter the policy adopted in Chapter 11 of discussing photochemically initiated, chain reactions of carbohydrates but not including tables of these reactions.

both hydrogen atoms and tri-*n*-butylstannyl radicals (Scheme 14.5). The radicals that initiate this reaction are produced by photolysis of azobis(isobutyronitrile) (AIBN). In a striking example of the potential of this reaction, Kahne et al. converted the α-glycoside **2** into the β-glycoside **3** (eq 1). The importance of this reaction can be appreciated when one considers the problems (discussed in Chapter 15) that often are encountered in β-glycoside synthesis. (In reference 24 Kahne presents an excellent analysis of the factors that contribute to determining the stereochemistry at C-1 in glycoside synthesis from glycosyl radicals.)

Sugar Diethyl Dithioacetal Reduction

In one of the pioneering studies in carbohydrate photochemistry, Horton and Jewell irradiated D-galactose diethyl dithioacetal (**4**) in methanol (in the presence of air) and found that one of the ethylthio groups was replaced by a hydrogen atom to form 1-*S*-ethyl-1-thiogalactitol (**6**) (Scheme 14.6).[26] These investigators also determined that, upon irradiation, **6** experienced replacement of the second ethylthio group with a hydrogen atom to form L-fucitol (**8**). To account for the observed substitution reactions these authors proposed the intermediacy of radicals **5** and **7** (Scheme 14.6). In a similar study Matsuura et al. irradiated four per-*O*-acetylated diethyl dithioacetals under nitrogen in *t*-butyl alcohol to form the corresponding per-*O*-acetylated ethyl-1-thioalditols (eq 2).[27] Conducting these reactions in an inert atmosphere undoubtedly improved product yields by preventing the reaction of radical intermediates with oxygen.

$$CH_3SR + \bullet SnBu_3 \longrightarrow CH_3SSnBu_3 + R\bullet$$

$$R\bullet + HSnBu_3 \longrightarrow RH + \bullet SnBu_3$$

R = carbohydrate residue

Scheme 14.5

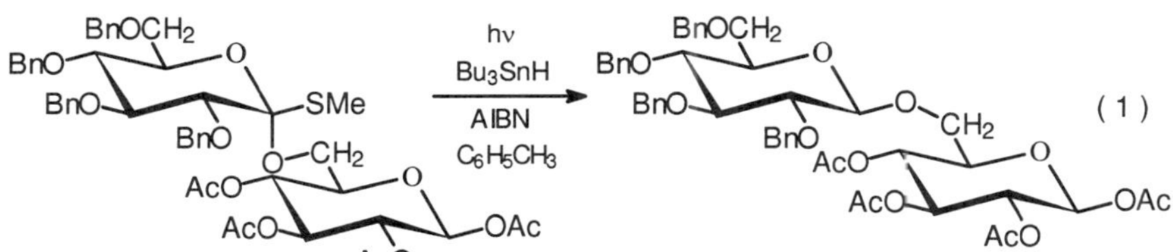

2

3

75% yield

β:α > 10:1

Equation 1

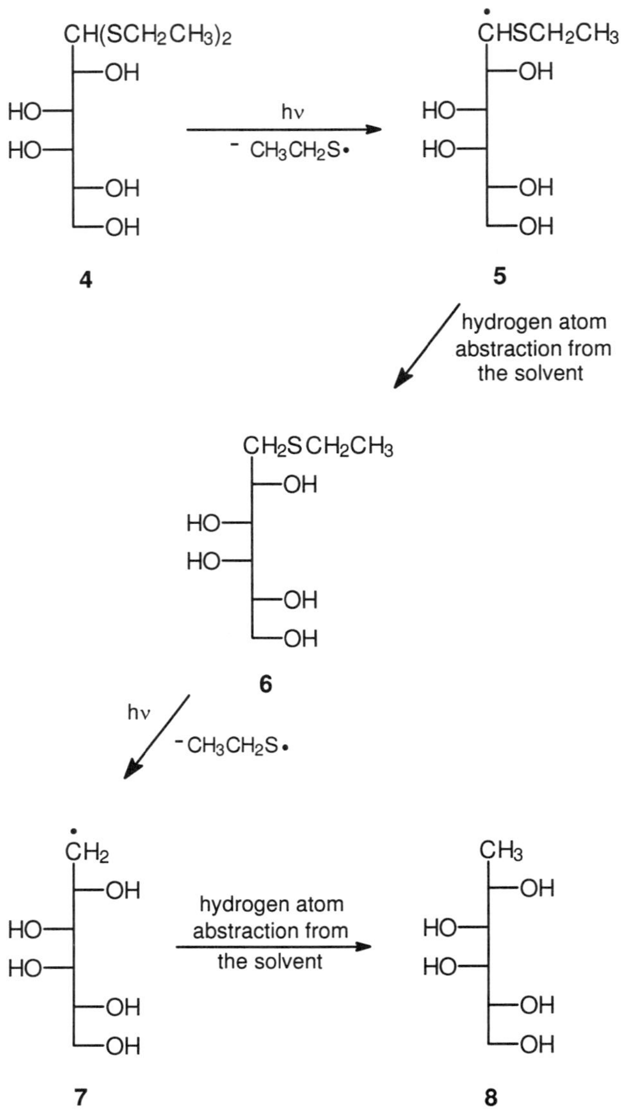
CH(SCH2CH3)2
OH
HO
HO
OH
OH
4
hν
- CH3CH2S•
CHSCH2CH3
5
hydrogen atom
abstraction from
the solvent
CH2SCH2CH3
6
hν
-CH3CH2S•
CH2
7
CH3
8

Scheme 14.6

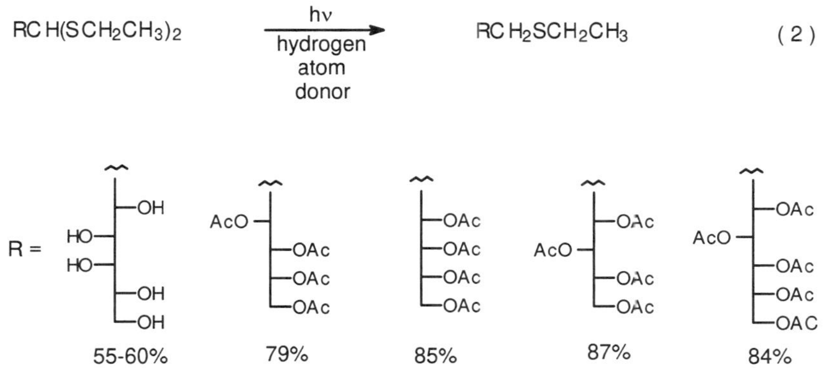

Equation 2

Glycoside Synthesis

A substitution process quite different from the Keck and Kahne reactions takes place upon irradiation of a phenylthio glycoside in the presence of an electron acceptor. Griffin et al. studied such a reaction as a possible method for oligosaccharide synthesis.[28] These researchers found that photolysis of phenyl 1-thio-β-D-glucopyranoside (**9**) in acetonitrile–methanol containing 1,4-dicyanonaphthalene (DCN) produced the methyl glycosides **10** and **11**. They proposed a mechanism for this reaction (Scheme 14.7) in which the glycosyl cation **13** was formed by transfer of an electron from **9** to DCN to generate the radical cation **12**, an intermediate in the formation of **13**. The cation **13** then reacted with methanol to form a mixture of the methyl glycosides **10** and **11**. Complicating the reaction was the unexpected appearance of the photoproducts D-glucono-1,4-lactone (**15**) and D-gluconic acid (**16**) (Chart 14.1). Griffin et al. did not investigate the formation of compounds **15** and **16**, but they did garner further support for the intermediacy of the cation **13** by finding that reaction in the absence of methanol produced 1,6-anhydro-β-D-glucopyranose (**14**).

Furuta et al. have synthesized glycosides, including a disaccharide, by irradiation of a 1-selenophenyl glycoside under electron-transfer conditions in the presence of various alcohols (eq 3).[29] The selenoglysoside in eq 3 is only one of those that have been photolyzed, but the others appear to react by photochemically initiated, chain reactions.[30–32]

Rearrangement Reactions

S-Nucleoside Photolyses: *C*-Nucleoside Formation

Photolysis of compounds in which an arylthio group is attached to an anomeric carbon atom—but other reagents such as electron acceptors or organotin com-

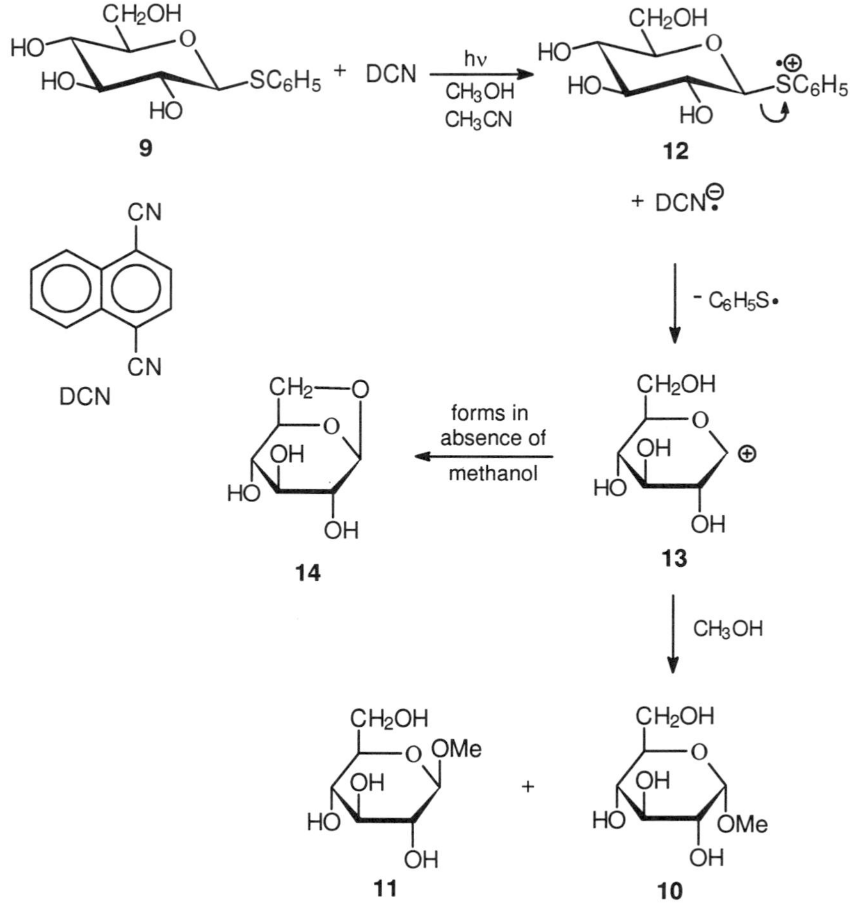

Scheme 14.7

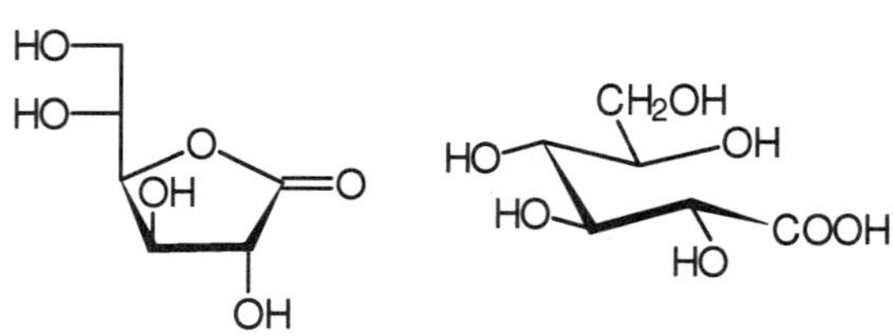

15 16

Chart 14.1. D-Glucono-1,4-lactone (15) and D-Gluconic Acid (16)

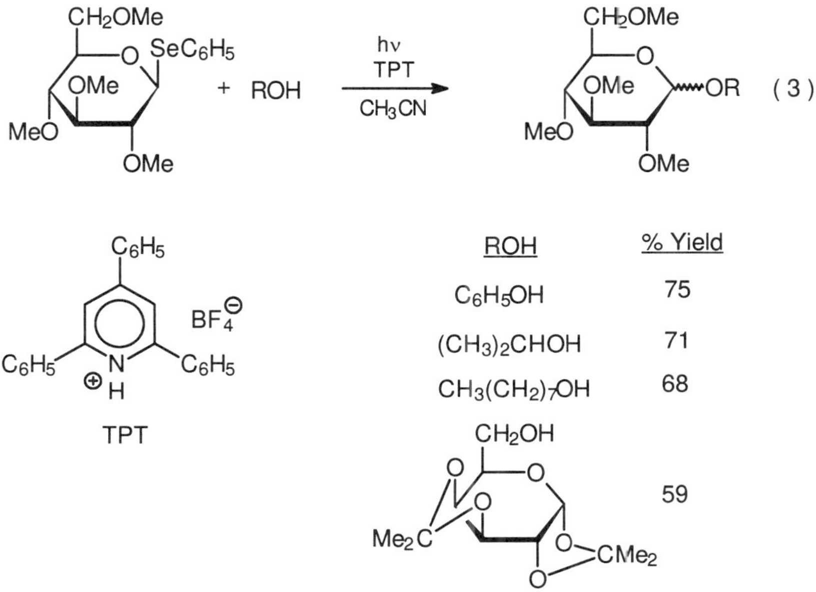

Equation 3

pounds are absent—results in C–S bond cleavage and the formation of rearrangement products that include *C*-nucleosides. Fourrey and Jouin irradiated the *S*-nucleoside **17** and isolated products formed from C_1–S bond homolysis followed by radical recombination at ring positions *ortho* and *para* to the sulfur atom (i.e., all positions with significant radical character) (Scheme 14.8).[34] Irradiations of similar *S*-nucleosides gave the same type of products, although not all the possible isomers from recombination were observed (eqs 4 and 5).[35]

Substituted Cyclitol Photolysis: Deoxy Sugar Formation

Photoinduced electron transfer between benzophenone and thiophenyl substituted compounds represents a novel process for synthesis of deoxy sugars.[36] Benzophenone (or an intermediate produced by its photolysis) is involved at each stage of reaction. This process begins when excited benzophenone accepts an electron from the sulfur atom in an appropriately substituted cyclitol, such as **18**, to form the benzophenone radical anion and a radical cation (e.g., **19**) derived from the cyclitol (Scheme 14.9).[37] The benzophenone radical anion then functions as a base during the ring-opening of **19** to give the aldehyde **20**. Disproportionation involving **20** and the benzhydrol radical regenerates benzophenone and produces the dideoxy sugar **21**. (Added thiophenol can replace the benzhydrol radical as a hydrogen atom source.) Gravel and co-workers discovered this reaction and have provided the mechanistic interpretation of this

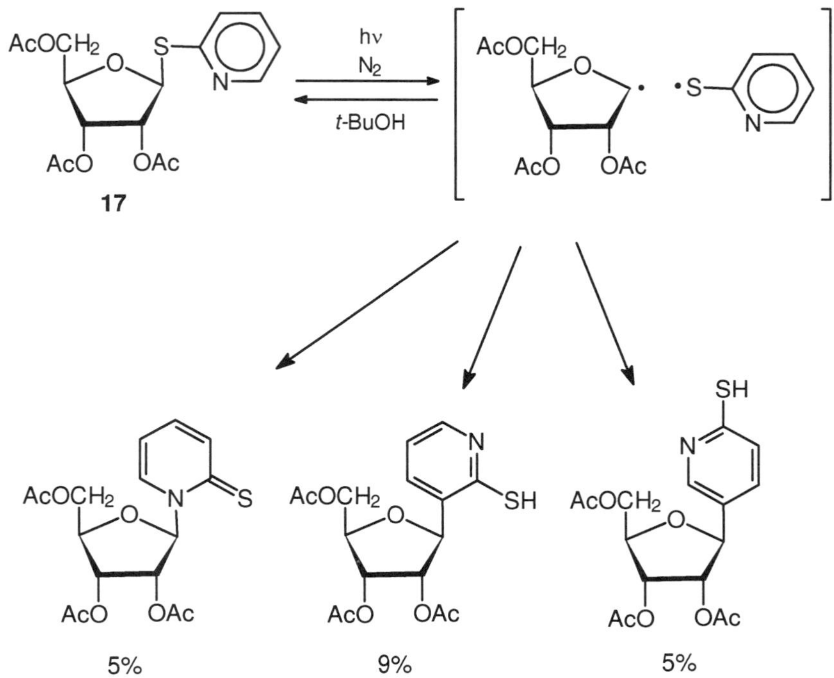

Scheme 14.8

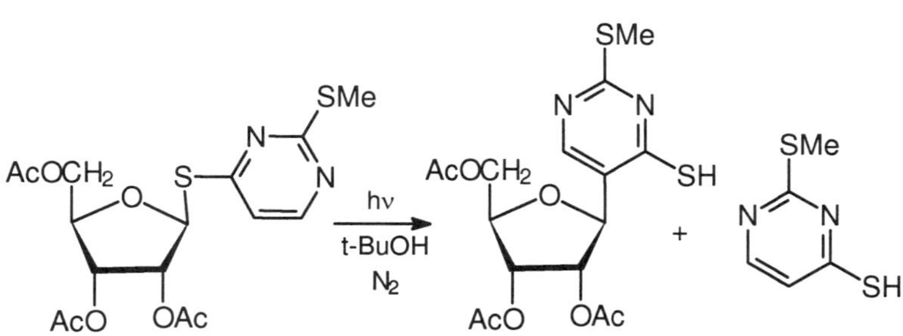

(4)

Equation 4

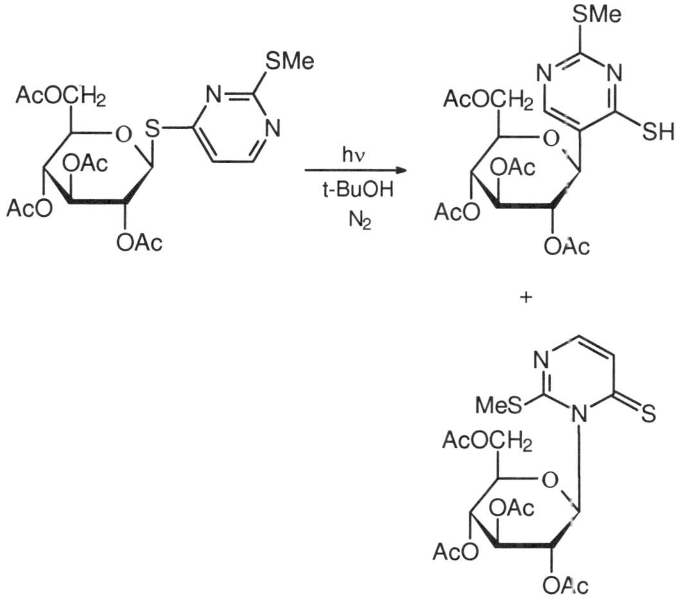

(5)

Equation 5

process outlined in Scheme 14.9.[36,37] The various deoxy sugars that these researchers have synthesized in this fashion are shown in Table 14.5.

Irradiation of Selenium-Containing Mono- and Oligonucleotides

Giese et al. found that irradiation of selenium-containing nucleotides and oligonucleotides (Table 14.6) generates the same radicals that are formed from Norrish Type I reactions of analogous compounds (Scheme 2.10, Table 2.4, Chapter 2). These investigators demonstrated that photolysis of selenium-containing oligonucleotides leads to cleavage of the nucleotide chain.[41]

Summary

Two types of processes dominate photolyses of thio sugars. The first of these consists of radical reactions brought about by C–S bond homolysis, and the second includes electron-transfer reactions in which the sulfur atom serves as an electron source. The synthetically most useful radical processes are chain reactions that result in either allyl group introduction (the Keck reaction,

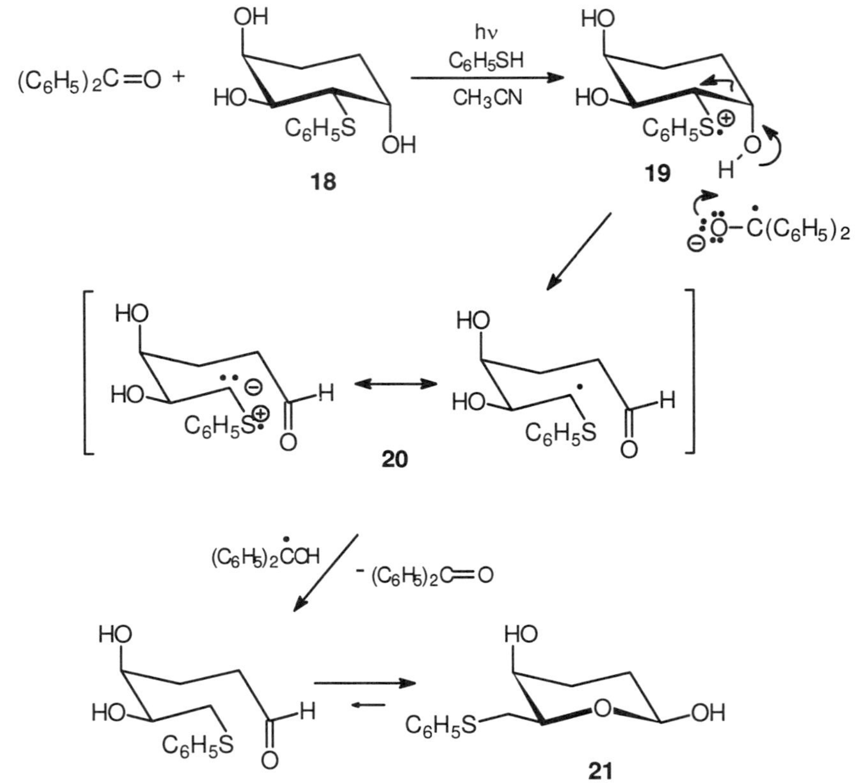

Scheme 14.9

Scheme 14.4) or β-glycoside synthesis (the Kahne reaction, Scheme 14.5); also, a nonchain cycloaddition process that converts nucleosides in cyclonucleosides has been used extensively (Scheme 14.1, Tables 14.1–14.4). Electron-transfer reactions are less frequently encountered in the photolysis of thio sugars. The most intriguing electron-transfer process is one that transforms substituted cyclitols into deoxy sugars (Scheme 14.9, Table 14.5).

References

1. Horton, D.; Wander J. D. In *The Carbohydrates;* Pigman, W.; Horton, D., Eds.; Academic Press: New York; 1980, Vol 1B, p 799.
2. Cumper, C. W. N.; Read, J. F.; Vogel, A. I. *J. Chem. Soc. (A)* **1966**, 239.
3. Matsuda, A.; Muneyama, K.; Nishida, T.; Sato, T.; Ueda, T. *Nucleic Acids Res.* **1976**, *3*, 3349.
4. Matsuda, A.; Tezuka, M.; Ueda, T. *Nucleic Acids Res.* **1976**, s13.

Table 14.5. Photolysis of Substituted Cyclitols

Reactant	*Product and Percent Yield*

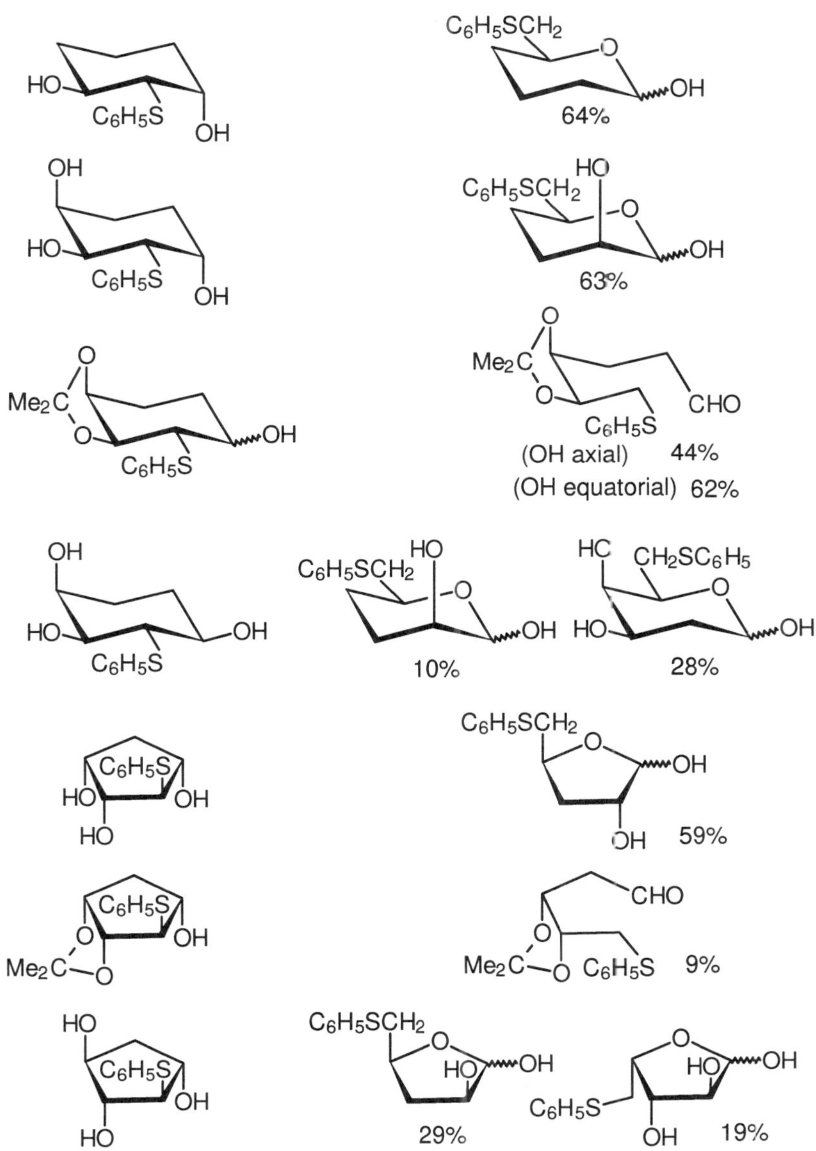

[a]Information in this table comes from reference 36.

Table 14.6. Selenonucleoside Photolysis

Reactant	*Product*	*Percent Yield*	*Ref.*
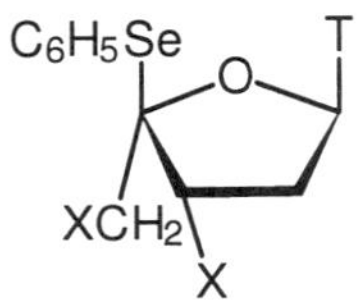 + MeOH + Bu_3SnH	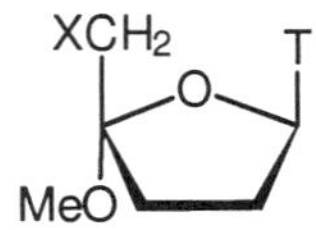	28	38
	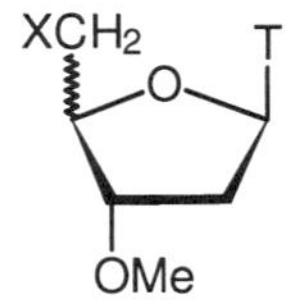	34	
X = $(EtO)_2PO_2$	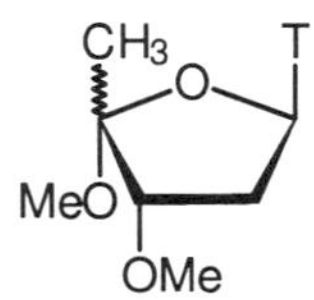	2	
	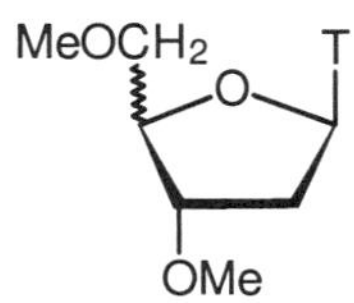	27	
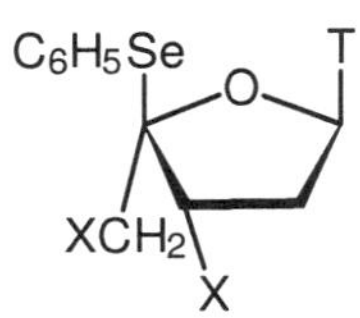 + $CH_2{=}CHCH_2OH$ + Bu_3SnH	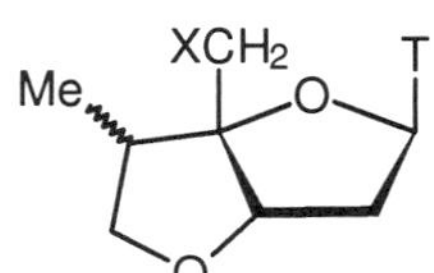 +	30	38
X = $(EtO)_2PO_2$	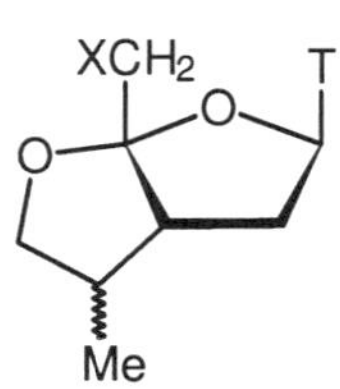	10	
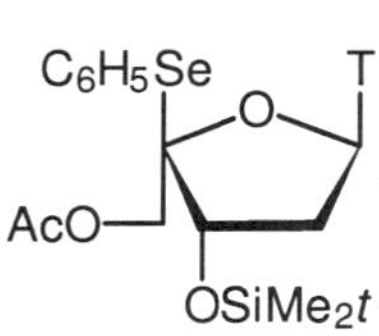 + C_6H_5SH	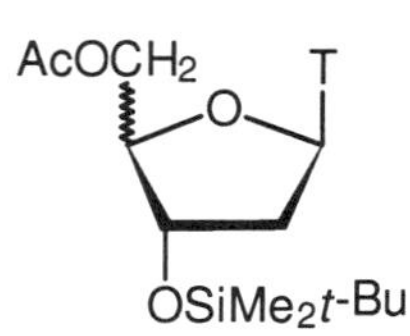	70-75	39

Continued on next page

Table 14.6. Selenonucleoside Photolysis—*Continued*

Reactant	*Product*	*Percent Yield*	*Ref.*
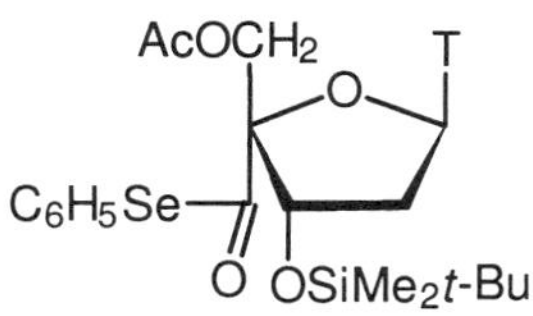 + C_6H_5SH	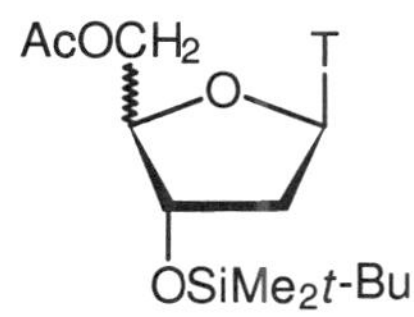	70-75	39
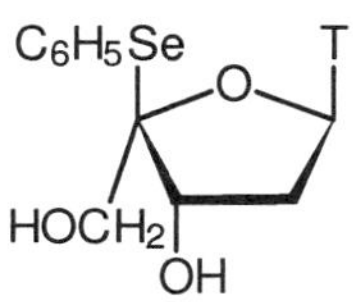 + O_2	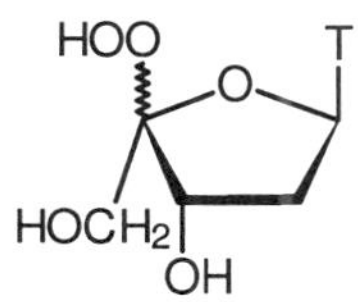	51-74	40
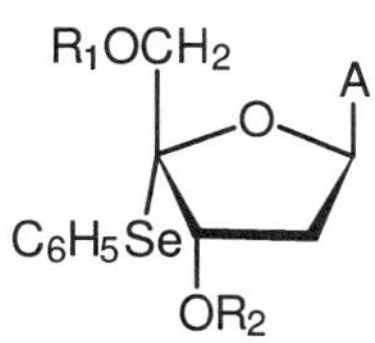	R_1OH + R_2OH	90	41

R_1 =

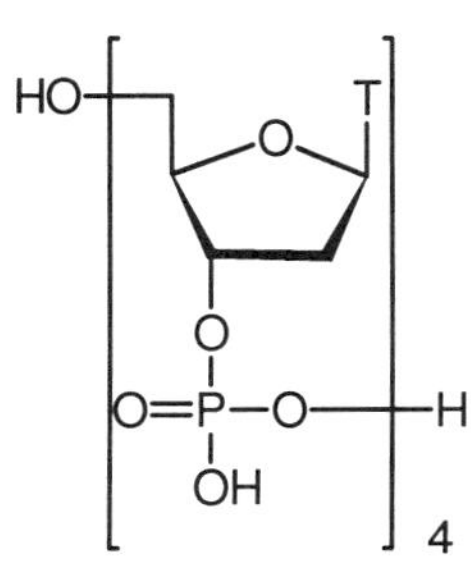

R_2 =

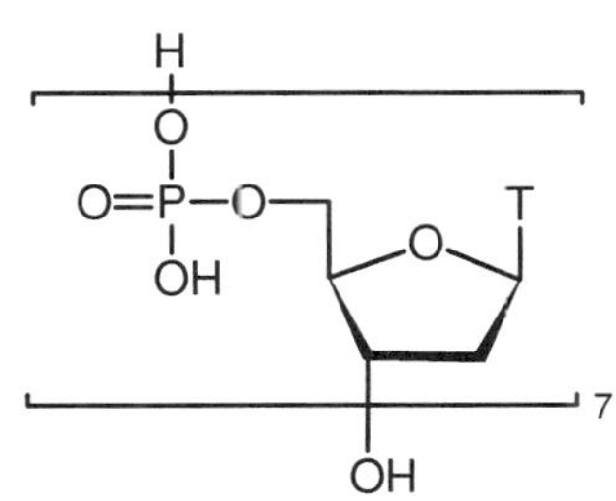

5. Matsuda, A.; Tezuka, M.; Niizuma, K.; Sugiyama, E.; Ueda, T. *Tetrahedron* **1978**, *34*, 2633.
6. Matsuda, A.; Ueda, T. *Chem. Pharm. Bull.* **1986**, *34*, 1573.
7. Usui, H.; Ueda, T. *Chem. Pharm. Bull.* **1986**, *34*, 15.
8. Usui, H.; Ueda, T. *Chem. Pharm. Bull.* **1986**, *34*, 1518.
9. Usui, H.; Matsuda, A.; Ueda, T. *Chem. Pharm. Bull.* **1986**, *34*, 1961.
10. Matsuda, A.; Tezuka, M.; Ueda, T. *Tetrahedron* **1978**, *34*, 2449.
11. Gani, D.; Johnson, A. W.; Lappert, M. F. *J. Chem. Soc., Chem. Commun.* **1980**, 1244.
12. Gani, D.; Johnson, A. W.; Lappert, M. F. *J. Chem. Soc., Perkin Trans. 1* **1980**, 3065.
13. Walling, C.; Rabinowitz, R. *J. Am. Chem. Soc.* **1957**, *79*, 5326.
14. Keck, G. E.; Enholm, E. J.; Kachensky, D. F. *Tetrahedron Lett.* **1984**, *25*, 1867.
15. Keck, G. E.; Kachensky, D. F.; Enholm, E. J. *J. Org. Chem.* **1984**, *49*, 1464.
16. Keck, G. E.; Kachensky, D. F.; Enholm, E. J. *J. Org. Chem.* **1985**, *50*, 4317.
17. Keck, G. E.; Yates, J. B. *J. Am. Chem. Soc.* **1982**, *104*, 5829.
18. Marco-Contelles, J. L.; Fernández, C.; Gómez, A.; Martín-León, N. *Tetrahedron Lett.* **1990**, *31*, 1467.
19. Wåglund, T.; Claesson, A. *Acta Chem. Scand.* **1992**, *46*, 73.
20. Fiandor, J.; Tam, S. Y. *Tetrahedron Lett.* **1990**, *31*, 597.
21. De Mesmaeker, A.; Libreton, J.; Hoffmann, P.; Freier, S. M. *Synlett* **1993**, 677.
22. Grøtli, M.; Undheim, K. *Acta Chem. Scand.* **1995**, *49*, 217.
23. Keck, G. E.; Kachensky, D. F. *J. Org. Chem.* **1986**, *51*, 2487.
24. Kahne, D.; Yang, D.; Lim, J. J.; Miller, R.; Paguaga, E. *J. Am. Chem. Soc.* **1988**, *110*, 8716.
25. Kahne, D.; Yang, D.; Lim, J. J.; Miller, R.; Paguaga, E. *Chemtracts: Org. Chem.* **1988**, *2*, 193.
26. Horton, D.; Jewell, J. S. *J. Org. Chem.* **1966**, *31*, 509.
27. Matsuura, K.; Araki, Y.; Ishido, Y. *Bull. Chem. Soc. Jpn.* **1973**, *46*, 2261.
28. Griffin, G. W.; Bandara, N. C.; Clarke, M. A.; Tsang, W.-S.; Garegg, P. J.; Oscarson, S.; Silwanis, B. A. *Heterocycles* **1990** *30*, 939.
29. Fourrey, J.-L.; Henry, G.; Jouin, P. *Tetrahedron Lett.* **1980**, *21*, 455.
30. Giese, B.; Rückert, B.; Gröninger, S.; Muhn, R.; Linder, H. J. *Liebigs Ann. Chem.* **1988**, 997.
31. De Mesmaeker, A. Waldner, A.; Hoffmann, P.; Hug, P.; Winkler, T. *Synlett* **1992**, 285.
32. Furuta, T.; Takeuchi, K.; Iwamura, M. *J. Chem. Soc., Chem. Commun.* **1966**, 157.
33. Haraguchi, K.; Tanaka, H.; Saito, S.; Kinoshima, S.; Hosoe, M.; Kanmuri, K; Yamaguchi, K.; Miyasaka, T. *Tetrahedron* **1996**, *52*, 9469.
34. Fourrey, J.-L.; Jouin, P. *J. Org. Chem.* **1979**, *44*, 1892.
35. Fourrey, J.-L.; Henry, G.; Jouin, P. *J. Am. Chem. Soc.* **1977**, *99*, 6753.
36. Gravel, D.; Farmer, L.; R. Denis, C.; Schultz, E. *Tetrahedron Lett.* **1994**, *35*, 8981.
37. Gravel, D.; Farmer, L.; Ayotte, C. *Tetrahedron Lett.* **1990**, *31*, 63.
38. Giese, B.; Beyrich-Graf, X.; Burger, J.; Kesselheim, C.; Senn, M.; Schäfer, T. *Angew. Chem. Int. Ed. Engl.* **1993**, *32*, 1742.
39. Giese, B.; Erdmann, P.; Giraud, L.; Göbel, T.; Petretta, M.; Schäfer, T.; von Raumer, M. *Tetrahedron Lett.* **1994**, *35*, 2683.
40. Giese, B.; Beyrich-Graf, X.; Erdmann, P.; Giraud, L.; Imwinkelried, P.; Müller, S. N.; Schwitter, U. *J. Am. Chem. Soc.* **1995**, *117*, 6146.
41. Giese, B.; Dussy, A.; Elie, C.; Erdmann, P.; Schwitter, U. *Angew. Chem. Int. Ed. Engl.* **1994**, *33*, 1861.

15

Thiocarbonyl Compounds

Introduction

Thiocarbonates, thiocarbamates, xanthates, and xanthate derivatives are among the thiocarbonyl-containing carbohydrates that are known to be photochemically reactive. A common thread that runs through the photochemistry of these compounds is the ease with which they generate radicals by homolytic cleavage of carbon–sulfur and sulfur–sulfur bonds. Although radical formation is the most frequent result from excitation of thiocarbonyl compounds, ionic reactions also occur when photolyses are conducted in the presence of appropriate, electron-transfer agents.

Esters of *N*-hydroxyl-2-thiopyridone are yet another type of photolabile, thiocarbonyl compound; however, these esters occupy a unique position among sulfur-containing carbohydrates because *O*-acyl-*N*-hydroxy-2-thiopyridones are starting materials for photochemically initiated chain reactions that generate carbohydrate radicals. The addition and substitution reactions of these radicals have been studied in considerable detail and found to have important synthetic applications.

The lowest energy excited states for thiocarbonyl compounds typically are considerably lower in energy than those for the corresponding carbonyl compounds. [The longest wavelength absorption for heptan-4-thione has its maximum at 503 nm (ϵ = 10) while that for 4-heptanone is at 278 nm (ϵ = 15).[1]] The energy gap between the two lowest excited singlet states, S_1 (n,π*) and S_2 (π,π*) in a typical thioketone, is sufficiently large that fluorescence as well as reaction can take place from the upper excited singlet state (S_2); consequently, for many thiocarbonyl compounds different reactivity is possible from excitation with UV light (S_2 formed) as opposed to visible light (S_1 formed). Excited state reactivity is further influenced by the difference in energy between the lowest singlet and triplet excited states (n,π*) because this energy difference is small enough that S_1 is converted rapidly into T_1. The practical outcome of this arrangement of energy levels is that, unlike most compounds, thiocarbonyl compounds often react from S_2 or T_1 states (rather than the common pattern of reacting from S_1 or T_1 states).[1]

*Reactions of Photochemically Generated Carbon Radicals**

In 1983 Barton et al. reported that carbon radicals could be generated easily by photochemically (and thermally) initiated reactions of *O*-acyl-*N*-hydroxyl-2-thiopyridone derivatives (**1**, also referred to as thiohydroxamic acid derivatives) (eq 1).[2] They further determined that hydrogen atom abstraction by these carbon radicals took place readily in the presence of effective donors (e.g., *t*-butyl mercaptan) and proposed the chain reaction shown in Scheme 15.1 to explain product formation. Later experiments demonstrated that the quantum yield for disappearance of starting material could be as high as 60, depending on the reaction conditions, and that a variety of visible-light sources, including sunlight, could initiate reaction.[3,4] [The observation by Alam et al. that photolysis of **1** produced an absorption at 490 nm, attributed to the 2-thiopyridyl radical, further supported the proposed mechanism (Scheme 15.1).[5]] In analyzing this process, Barton attributed its success in generating carbon radicals to (a) a carbonyl for thiocarbonyl exchange, (b) the increased aromaticity in converting a 2-thiopyridone structure into a substituted pyridine, and (c) an increase in entropy as a result of reaction.[6]

Addition Reactions

Alkenes

In their extensive investigation of the reactions of carbon radicals generated from *O*-acyl-*N*-hydroxy-2-thiopyridones, Barton et al. found that in the presence of reactive alkenes, addition to a double bond (Scheme 15.2) is a synthetically useful reaction.[7] This reaction has since been used extensively by Barton[7–12,19–21,23] and others.[13–18,22] In a specific demonstration of the value of *O*-acyl-*N*-hydroxy-2-thiopyridone photolysis in synthesis, Barton, Géro, and co-workers reported that photoinitiated reaction between compounds **2** and **3** (eq 2) was the critical step in a synthesis of the nucleoside antibiotic sinefungin.[11,12] In another and different type of illustration of the synthetic potential of this reaction, Garner and co-workers showed that the carbohydrate portion

*We will continue in this chapter the policy adopted in earlier chapters of discussing photochemically initiated, chain reactions of carbohydrates but not including tables of these reactions.

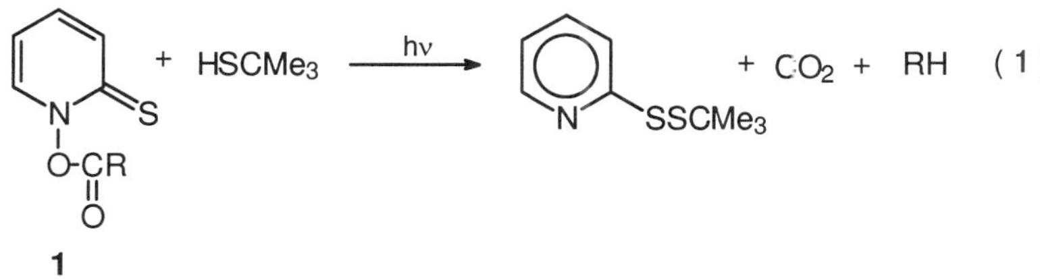

Equation 1

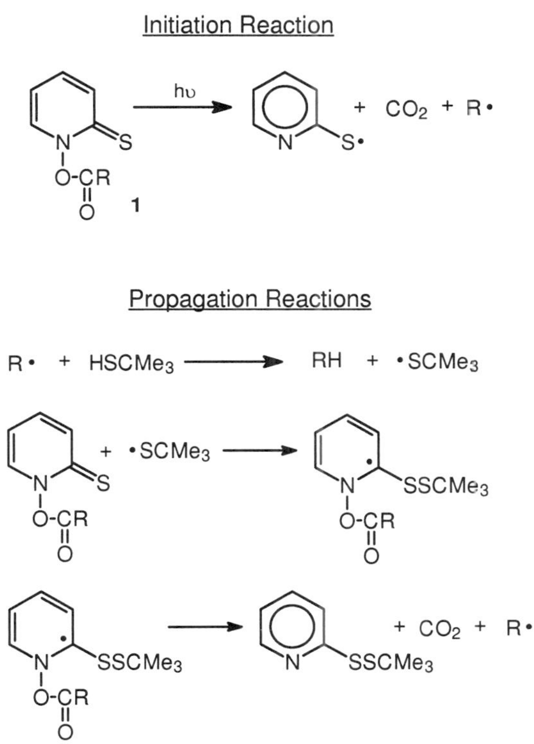

Scheme 15.1

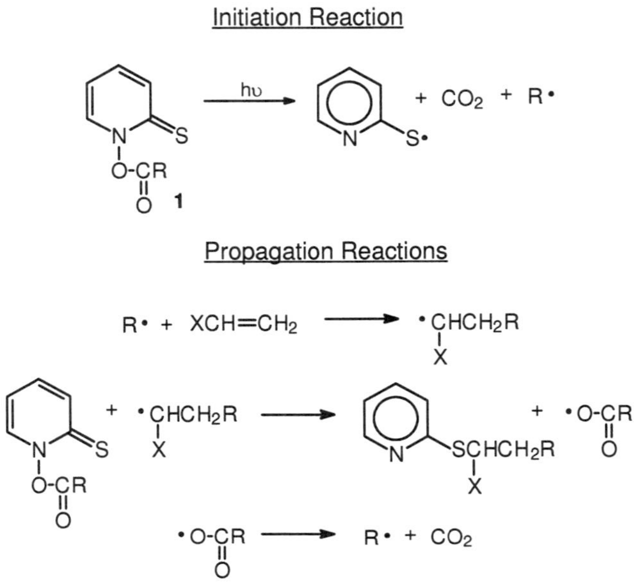

X = electron-withdrawing group

Scheme 15.2

of a radical can dramatically influence the stereoselectivity of its reaction with an alkene (Scheme 15.3).[17,18]

Protonated Heterocyclic Compounds

Further research on *N*-hydroxy-2-thiopyridone esters has shown that addition reactions also occur between carbohydrate radicals and protonated heteroaromatic compounds.[25] Togo, Yokoyama, and co-workers extended this work to include carbohydrates (eq 3),[15] a modification that results in formation of *C*-nucleosides.[13–16,26]

Radical Exchange

In 1990 Barton and Ramesh introduced a process for formation of carbon radicals that they termed "radical exchange."[27] Radical exchange begins with the introduction of tellurium into a molecule by a displacement reaction with the nucleophile anisyl telluride (p-$CH_3OC_6H_4Te^-$) (Scheme 15.4). This tellurium-containing compound then reacts with a methyl radical, generated from photolysis of *O*-acetyl-*N*-hydroxy-2-thiopyridone, to give the corresponding carbohydrate radical (Scheme 15.5). The methyl radical exchanges with the carbohydrate radical because the methyl radical forms a stronger bond to tellurium.

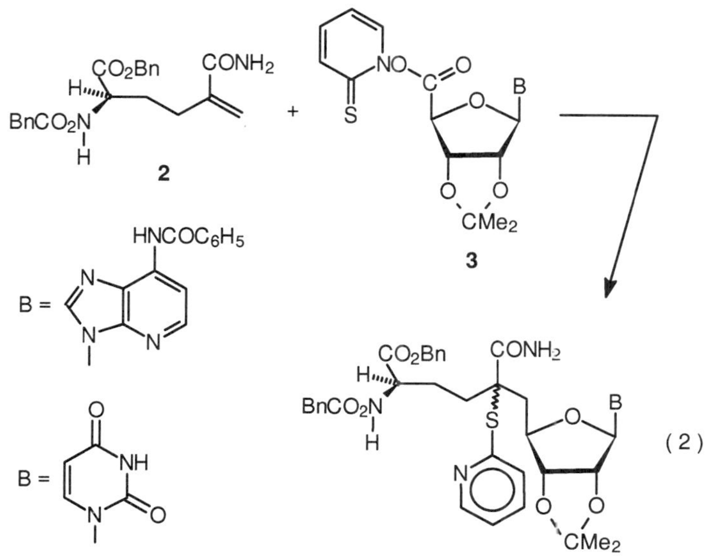

Equation 2

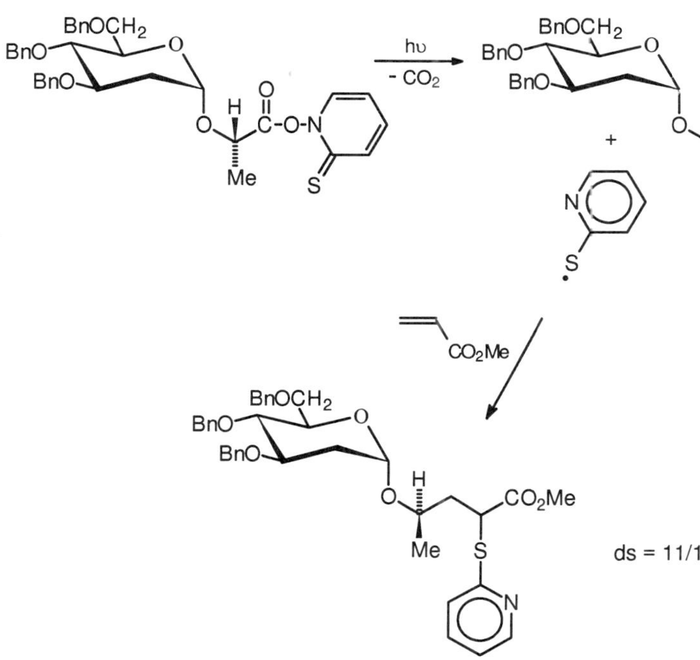

Scheme 15.3

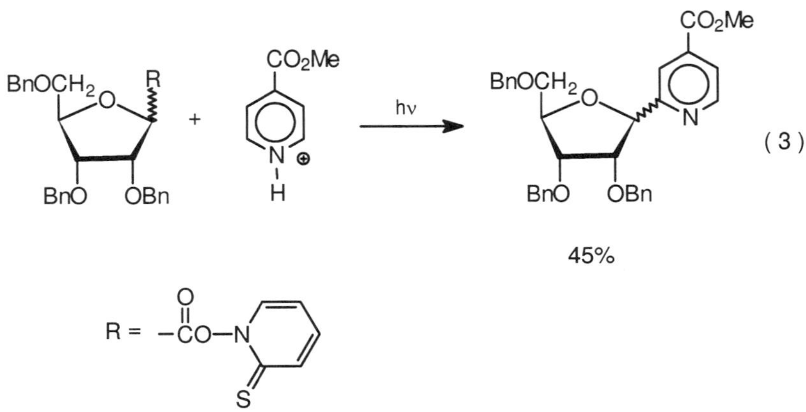

Equation 3

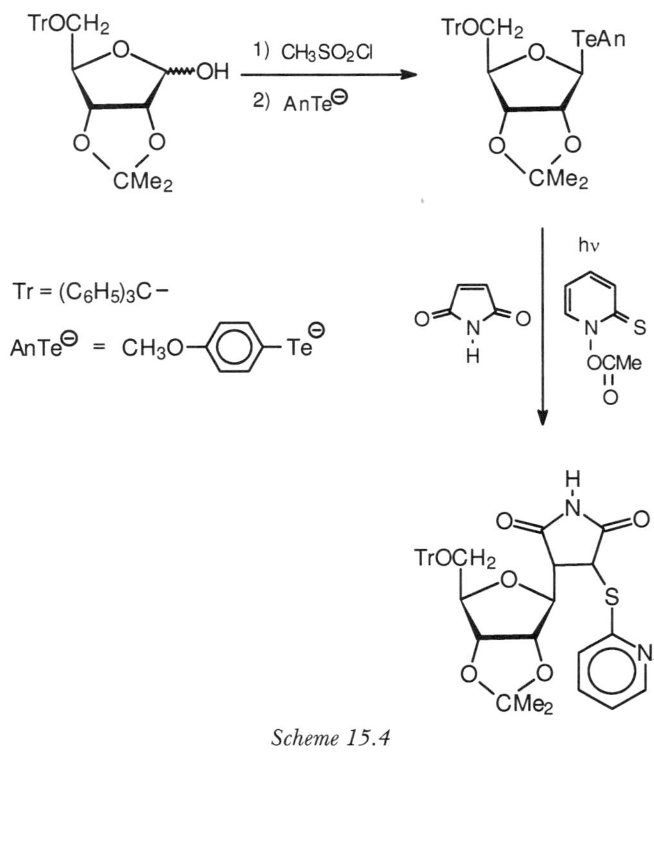

Scheme 15.4

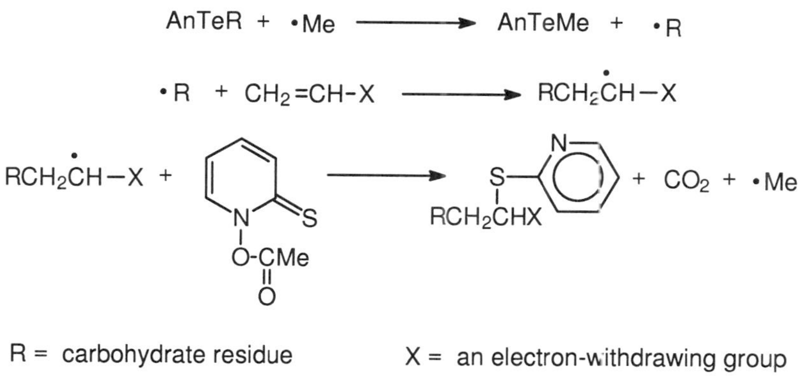

Scheme 15.5

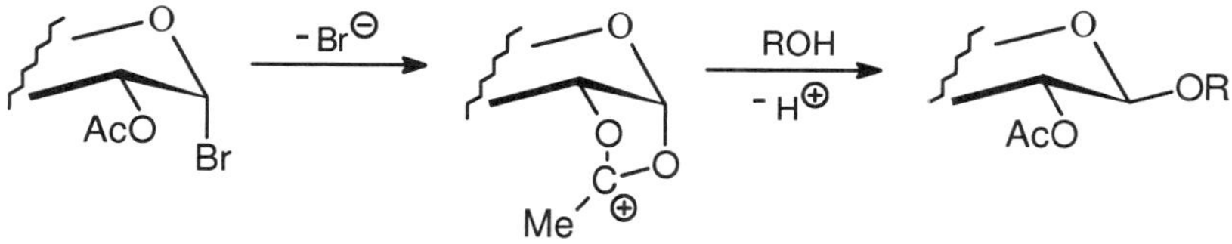

Scheme 15.6

β-Glycoside Synthesis

An interesting application of 2-thiopyridone ester photochemistry is found in the synthesis of β-glycosides.[28–37] Classical methods for β-glycoside formation depend upon anchimeric assistance by a C-2 group to control the stereochemistry at C-1 during glycoside formation (e.g., Scheme 15.6). Without this group participation, the thermodynamically more stable compound, usually the α-anomer, is normally the major product. Since participation by a C-2 group is not possible in 2-deoxy sugars, other methods are needed for efficient β-glycoside synthesis of these compounds.

Crich and Ritchie discovered an improved method for β-glycoside synthesis when they determined that photolysis of appropriate esters of *N*-hydroxy-2-thiopyridone in the presence of thiols preferentially produced β-glycosides.[28] The basic mechanism for this process had been proposed in an earlier publication and is outlined in Scheme 15.1.[1] Although this proposal describes the general reaction mechanism, it does not address the question of stereoselectivity in glycoside formation. The critical step determining stereochemistry at C-1 in this process is the transfer of a hydrogen atom to an intermediate 1-glycosyl radical. If this transfer occurs preferentially to the α-face of the radical, a β-glycoside is formed (Scheme 15.7). Why should hydrogen atom transfer to the α-face of the radical be more probable? Crich and Lim proposed that a conformational equilibrium exists between anomeric radicals **4** and **5** and that this equilibrium favors **4**.[31] In support of this proposal Somsák and Ferrier noted that the greater stabilization of an axial σ-radical (e.g., **4**), when compared to its equatorial conformer (e.g., **5**), can be attributed to better interaction between the axial lone pair of electrons on the ring-oxygen atom and the half-filled orbital at C-1 in the σ-radical.[38] If this explanation is correct, it follows that a hydrogen atom donor is more likely to encounter **4** than **5**, and if these two conformers have comparable reactivity with the donor, the β-glycoside will be formed in preference.

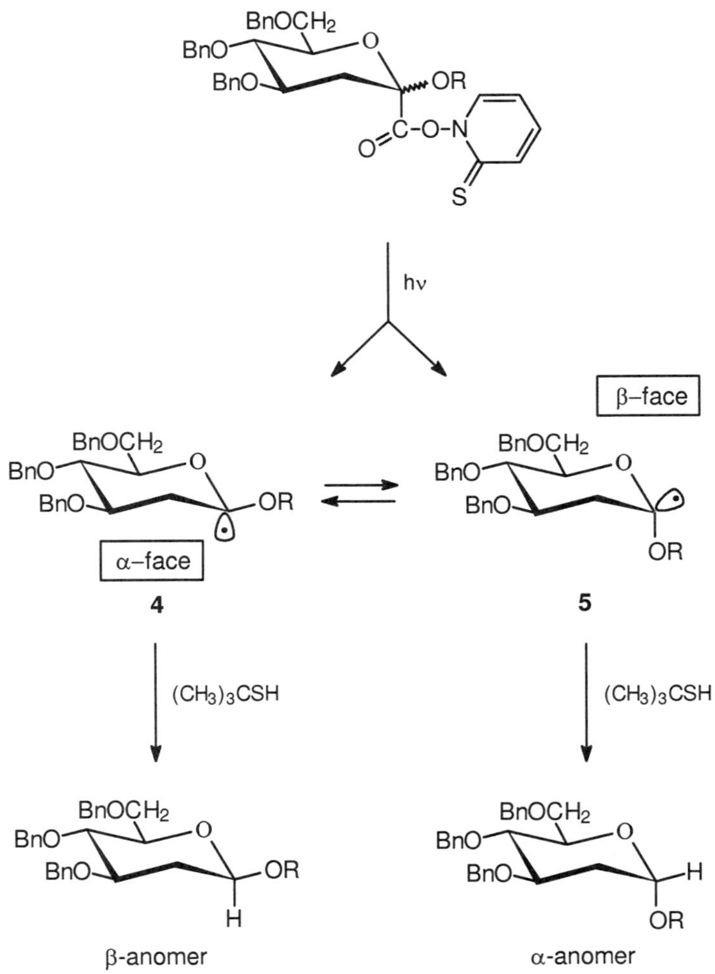

Scheme 15.7

The situation with respect to conformational equilibria of 1-glycosyl radicals becomes more complicated when ESR studies are taken into consideration. Experiments show that the major conformer of an anomeric radical can be either a boat, chair, or half-chair, depending upon the structure of the parent system.[39–45] As noted in the previous paragraph, preferred conformation is a vital factor in determining product stereochemistry when a 2-deoxy (or other) 1-glycosyl radical abstracts a hydrogen atom from a donor. The abstraction necessary for glycoside formation may occur with approximately equal ease from different conformations. If the favored radical conformer is a half-chair or flattened boat (Figure 15.1), a strong preference for hydrogen atom transfer to a particular face of the radical is unlikely. Rychnovsky et al. have performed ab initio calculations on anomeric radicals in an effort to determine the basis for preferential β-glycoside formation.[46] Their findings support the proposal that stereoselectivity in reactions of anomeric radicals is determined primarily by conformational preference and that a pyramidal, axial radical (e.g., **4**) is generally more stable than its equatorial counterpart (e.g., **5**) or any other conformer; hence, hydrogen atom abstraction favors β-glycoside formation. Furthermore, Rychnovsky et al. determined that introduction of a powerful, electron-withdrawing group at the 2-position “flattens the radical center and makes the boat conformation ... more accessible”.[46] This latter finding, which also was reached by Giese from considering frontier molecular orbital interactions,[44] seems to explain the ESR result that some 1-glycosyl radicals with strong electron-withdrawing substituents at C-2 can adopt a boat or half-chair conformation while radicals without electron-withdrawing substituents (or with less powerful ones) exist in chair conformers and exhibit stereoselectivity in reaction consistent with axial anomeric radicals being the major reactive intermediates. (The stability of anomeric radicals also is discussed in Chapters 3 and 5.)

Formation of 2-Thiopyridine Derivatives

In the absence of a compound that reacts rapidly with carbon radicals (e.g., a mercaptan or an electron-deficient alkene), irradiation of *O*-acyl-*N*-hydroxy-2-

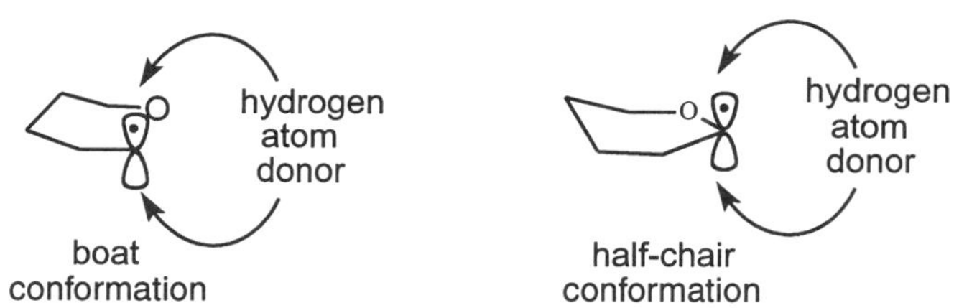

Figure 15.1. Boat and Half-Chair Conformations of 1-Glycosyl Radicals

thiopyridones produces 2-thiopyridine derivatives (Scheme 15.8). At present this reaction does not have important synthetic application in carbohydrate chemistry, although several examples of its use have been reported.[47,48] Formation of 2-thiopyridine derivatives is often a side reaction when carbon radicals are generated for other purposes (e.g., addition reactions.).[15]

Introduction of Bromine

Photolysis of *O*-acyl-*N*-hydroxy-2-thiopyridones also can be used to synthesize halogenated carbohydrates.[49,50] Halogenation is conducted in the presence of $BrCCl_3$, which serves as a source of bromine atoms (Scheme 15.9).

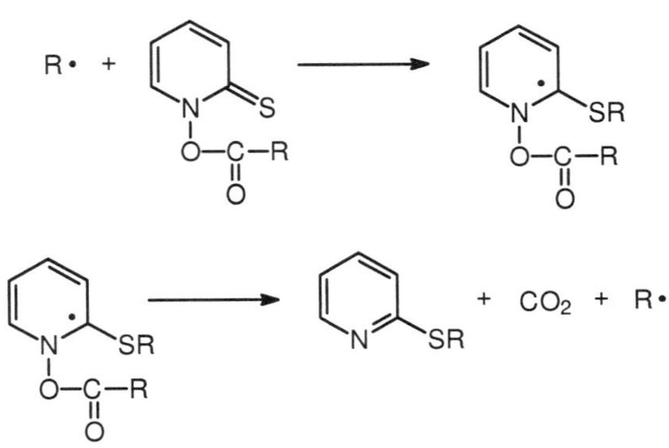

R = carbohydrate residue

Scheme 15.8

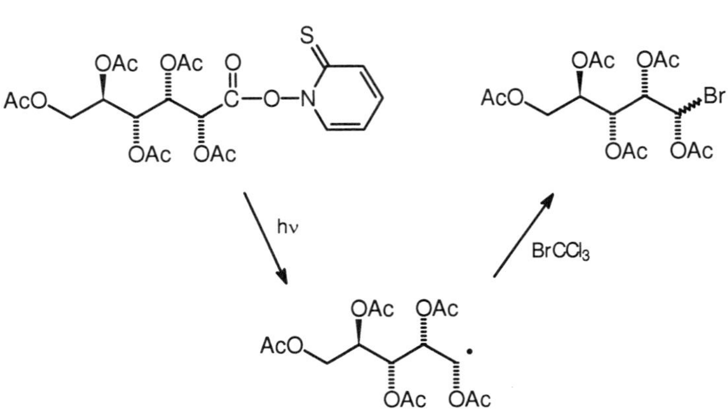

Scheme 15.9

Electron-Transfer Reactions

Deoxy Sugar Synthesis

Kishi et al. tested the photochemical reactivity of a group of thiocarbonyl-containing compounds (Table 15.1) in the presence of over a dozen solvents and found that little photochemical reaction occurred except in aqueous hexamethylphosphoramide (HMPA).[51] In this solvent, replacement of the sulfur-containing substituent with a hydrogen atom was a common occurrence that usually was accompanied by photohydrolysis (Table 15.1). The reason that photoreaction in HMPA was different from that in other solvents was that HMPA functions as a donor in electron-transfer reactions. (The electron-donating ability of HMPA is discussed in Chapter 6.)

Rearrangement Reactions

Photoinduced, electron-transfer reactions also take place when an appropriate electron acceptor is added to photolysis mixtures containing thiocarbonyl

Table 15.1. Photochemical Reactions of Thiocarbonyl Compounds with HMPA

Reactant	*Product and Percent Yield*	
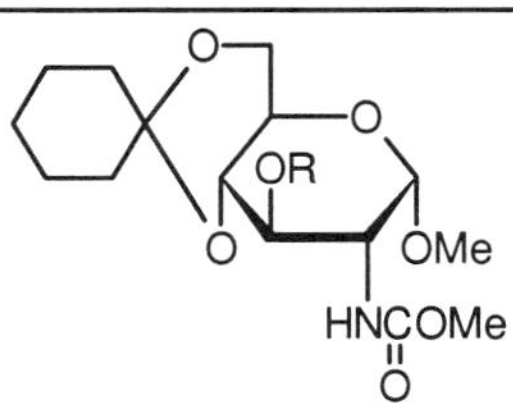	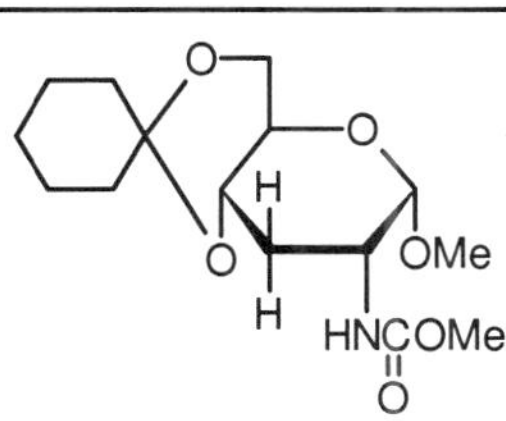	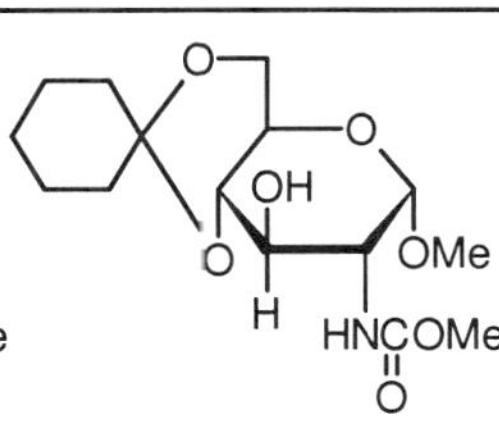
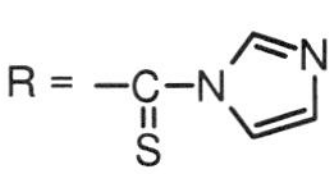	54%	17%
R = $-C(=S)SCH_3$	57%	12%
R = $-C(=S)N(CH_3)_2$	72%	10%
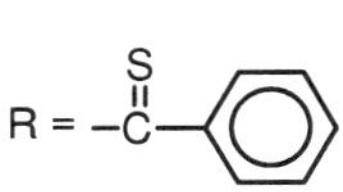	none	30%

[a]These data are from reference 51.

compounds. Barton et al. adopted this approach for reactions of xanthates, thiocarbonates, and related compounds by adding tris(4-bromophenyl)aminium hexachloroantimonate to photolysis mixtures.[52] As might be expected in these reactions, the radical cations that form sometimes exhibit reactivity characteristic of cations (Scheme 15.10) and sometimes radicals (Scheme 15.11). In one case (eq 4) Barton et al. observed a reaction for which they proposed both ionic and radical reactivity.[52]

Homolytic Carbon–Sulfur and Sulfur–Sulfur Bond-Cleavage Reactions

Deoxy Sugar Synthesis from Sugar Dimethylthiocarbamates

In addition to electron-transfer reaction in the presence of HMPA, another method for synthesis of deoxy sugars involves photolysis of sugar dimethylthiocarbamates in methanol. Horton and co-workers have shown that deoxy sugar formation occurs from the photolysis of these compounds; however, the major reaction pathway in each case leads to regeneration of the sugars from which the dimethylthiocarbamates were synthesized (Table 15.2).[53,54] Even though these investigators noted that it was possible to improve the yield of deoxy sugars by recycling the alcohols produced (ROH, Table 15.2), the low yields from this reaction reduce its attractiveness as a synthetic procedure. Several possible mechanisms for this reaction have been proposed,[54,55] but with-

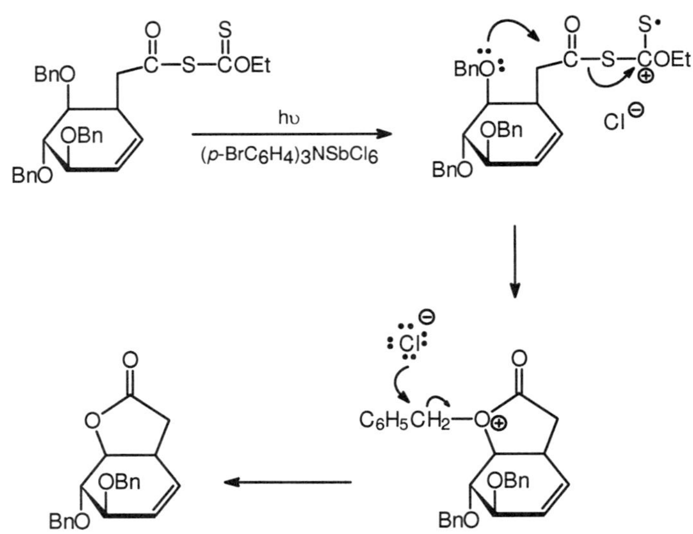

Scheme 15.10

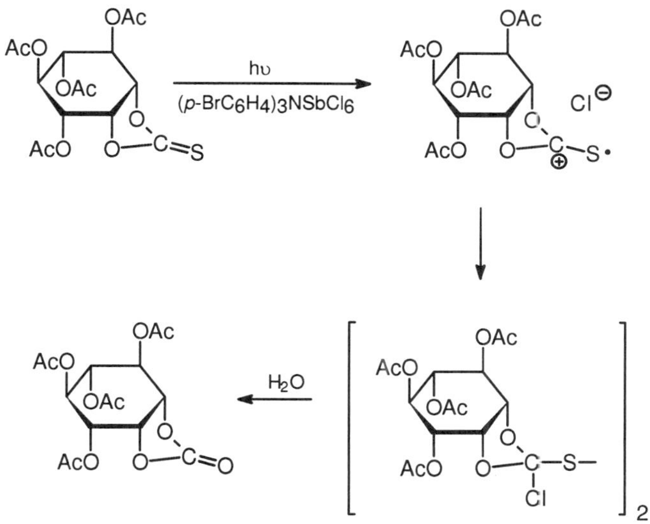

Scheme 15.11

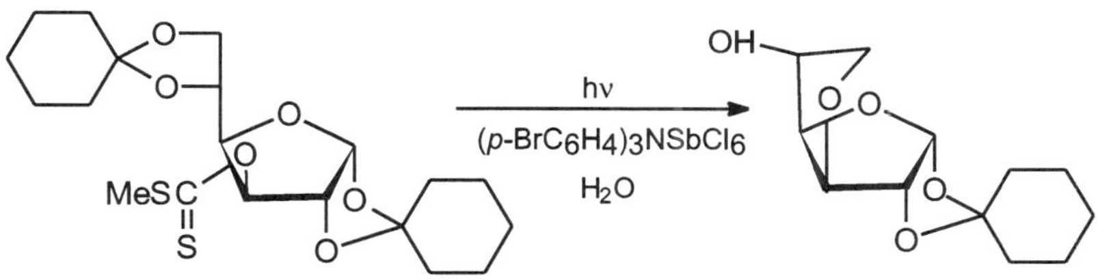

(4)

Equation 4

out further study, it is not possible to choose among the alternatives. The better deoxy sugar yield obtained by photolysis of a dimethylthiocarbamate in the presence of HMPA (Table 15.1) suggests that, in general, dimethylthiocarbamate photolyses should be conducted in the presence of an electron donor.

Reactions of Sulfur-Containing Anhydrides

Several sulfur-containing anhydrides have been investigated and found to eliminate carbon oxysulfide (COS) upon irradiation.[56,57] Two of these reactions involve the photolysis of the "xanthic anhydrides" **6** and **7**, and the others occur upon irradiation of the oxidatively coupled xanthates **8–10** (Table 15.3). Product yields given in Table 15.3 are for reactions conducted in dilute solutions. In concentrated solutions the reaction mixtures become more complex.[57] One report of xanthate photolysis appears to involve hydrolytic removal of isopropylidene groups catalyzed by photochemically generated acid.[58]

Table 15.2. Photochemcial Reactions of Dimethylthiocarbamates

Reactant	*Product and Percent Yield*		*Ref.*
$R{-}OC(=S)NMe_2$	RH +	ROH	53, 54
R = 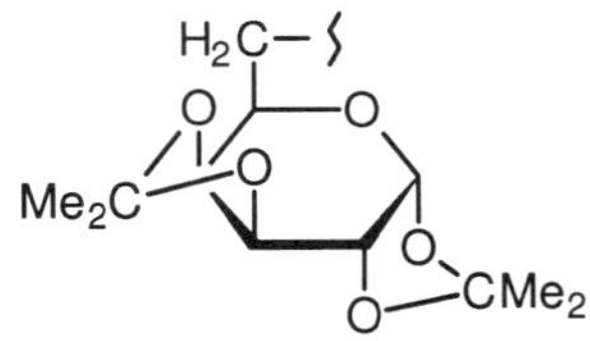	15%	36%	
R = 	19%	17%	
R = 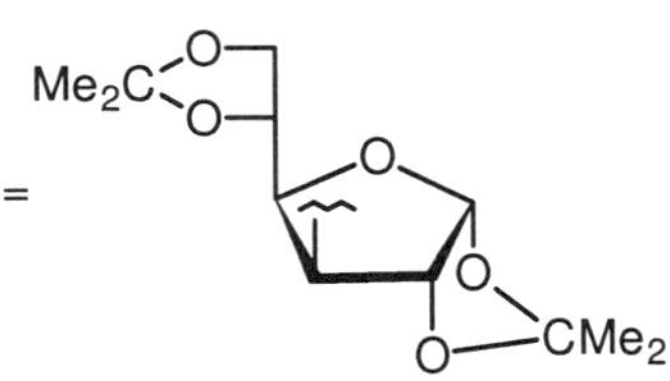	17-23%	26-32%	
R = 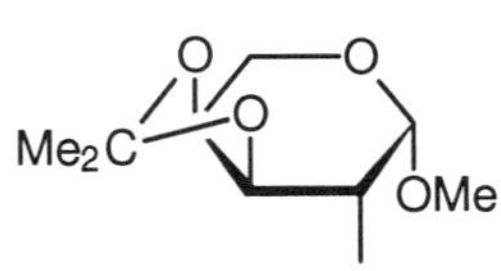	11%	36%	

Table 15.3. Photochemical Reactions of Sulfur-Containing Anhydrides

Reactant	*Product*	*Percent Yield*	*Ref.*
ROC(=S)SC(=S)OR	RSC(=S)OR		56
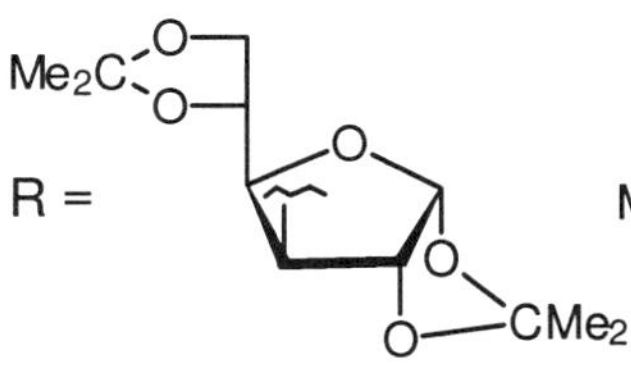 **6** 70%	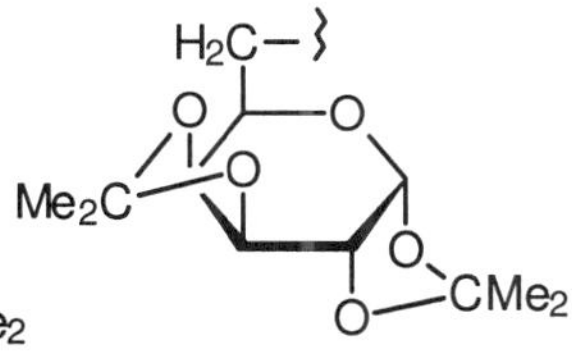 **7** 81%		
ROC(=S)SSC(=S)OR	RSC(=S)OR		57
	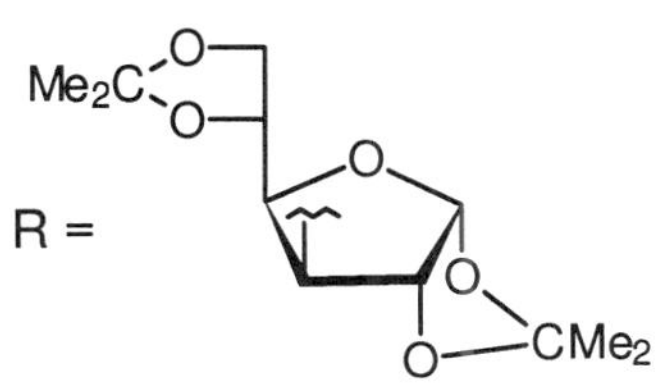 **8**	34	
	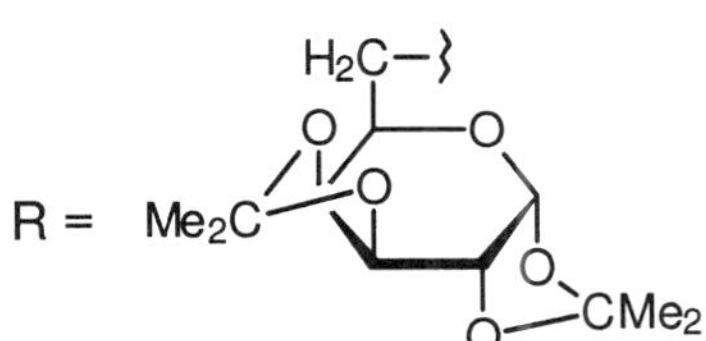 **9**	74	
	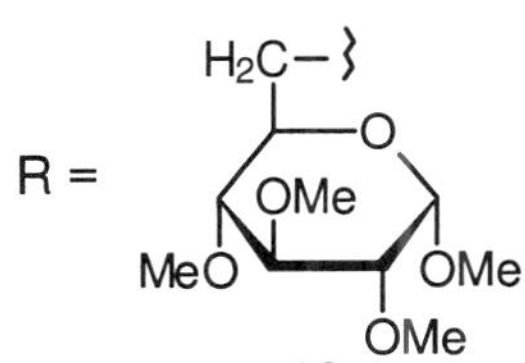 **10**	47	

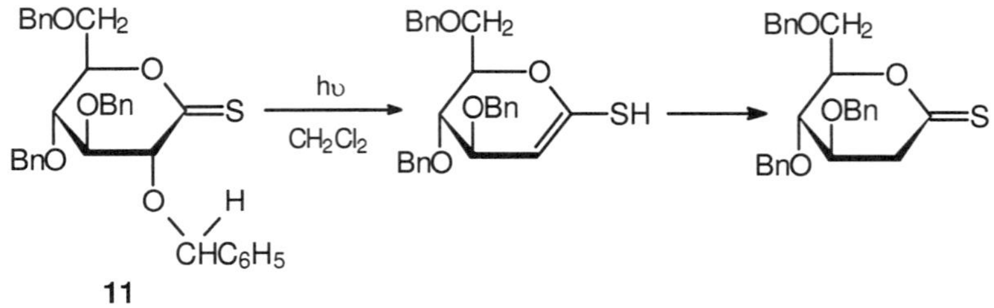

Scheme 15.12

Norrish Type II Reaction

Although the Norrish Type II reaction holds a prominent position in the photochemistry of carbonyl-containing carbohydrates (Chapter 3), it is not a common reaction for thiocarbonyl compounds. Hürzeler et al. proposed that Type II reaction occurs upon irradiation of compound **11** (Scheme 15.12),[59] but this reaction represents the only example of a thiocarbonyl-containing carbohydrate that appears to undergo Type II reaction.

Summary

The photolysis of an *O*-acyl-*N*-hydroxy-2-thiopyridone derivative (**1**) initiates a chain reaction that generates carbon radicals (Scheme 15.1) that add to a variety of electron-deficient alkenes (Scheme 15.2) and protonated heteroaromatic compounds. Reaction of anomeric radicals generated from 2-thiopyridone photolysis provides a pathway for forming β-glycosidic linkages in compounds for which this process is difficult (Scheme 15.7).

Several types of thiocarbonyl-containing carbohydrates, including thiocarbonates, thiocarbamates, xanthates, and xanthate derivatives, have been the subject of photochemical investigation (Tables 15.1–15.3). A common feature of the photochemistry of these compounds is the ease with which carbon–sulfur and sulfur–sulfur bonds fragment homolytically.

References

1. Coyle, J. D. *Tetrahedron* **1985**, *41*, 5393.
2. Barton, D. H. R.; Crich, D.; Motherwell, W. B. *J. Chem. Soc., Chem. Commun.* **1983**, 939.
3. Barton, D. H. R.; Blundell, P.; Jaszberenyi, J. C. *J. Am. Chem. Soc.* **1991** *113*, 6937.
4. Barton, D. H. R.; Jaszberenyi, J. C.; Tang, D. *Tetrahedron Lett.* **1993** *34*, 3381.
5. Alam, M. M.; Watanabe, A.; Ito, O. *J. Org. Chem.*, **1995** *60*, 3440.
6. Barton, D. H. R.; Crich, D.; Kretzschmar, G. *J. Chem. Soc., Perkin Trans. 1* **1986**, 39.
7. Barton, D. H. R.; Gateau-Olesker, A.; Géro, S. D.; Lacher, B.; Tachdjian, C.; Zard, S. Z. *J. Chem. Soc., Chem. Commun.* **1987**, 1790.

8. Barton, D. H. R.; Géro, S. D.; Quiclet-Sire, B.; Samadi, M. *J. Chem. Soc., Chem. Commun.* **1988**, 1372.
9. Barton, D. H. R.; Géro, S. D.; Quiclet-Sire, B.; Samadi, M.; Ozbalik, N.; Sarma, J. C.; Ramesh, M. *Pure Appl. Chem.* **1988**, *60,* 1549.
10. Barton, D. H. R.; Géro, S. D.; Quiclet-Sire, B.; Samadi, M. *Tetrahedron: Asymmetry* **1994**, *5,* 2123.
11. Barton, D. H. R.; Géro, S. D.; Quiclet-Sire, B.; Samadi, M. *J. Chem. Soc., Perkin Trans. 1* **1991**, 981.
12. Barton, D. H. R.; Géro, S. D.; Lawrence, F.; Robert-Gero, M.; Quiclet-Sire, B.; Samadi, M. *J. Med. Chem.* **1992**, *35,* 63.
13. Togo, H.; Fujii, M.; Ishigami, S.; Yokoyama, M. *Nucleic Acids Symp Ser.* **1991**, *25,* 89.
14. Togo, H.; Fujii, M.; Ikuma, T.; Yokoyama, M. *Tetrahedron Lett.* **1991**, *32,* 3377.
15. Togo, H.; Ishigami, S.; Fujii, M.; Ikuma, T.; Yokoyama, M. *J. Chem. Soc., Perkin Trans. 1* **1994**, 2931.
16. Togo, H.; Ishigami, S.; Yokoyama, M. *Chem. Lett.* **1992**, 1673.
17. Garner, P. P.; Cox, P. B.; Klippenstein, S. J. *J. Am. Chem. Soc.* **1995**, *117,* 4183.
18. Garner, P.; Leslie, R.; Anderson, J. T. *J. Org. Chem.* **1996**, *61,* 6754.
19. Barton, D. H. R.; Camara, J.; Cheng, X.; Géro, S. D.; Jaszberenyi, J. C.; and Quiclet-Sire, B. *Tetrahedron* **1992**, *48,* 9261.
20. Barton, D. H. R.; Géro, S. D.; Quiclet-Sire, B.; Samadi, M. *J. Chem. Soc., Chem. Commum.* **1989**, 1000.
21. Barton, D. H. R.; Géro, S. D.; Quiclet-Sire, B.; Samadi, M. *Tetrahedron Lett.* **1989**, *30,* 4969.
22. Wnuk, S. F.; Robins, M. J. *J. Am. Chem. Soc.* **1996**, *118,* 2519.
23. Barton, D. H. R.; Jaszberenyi, J. C.; Liu, W.; Shinada, T. *Tetrahedron* **1996**, *52,* 2717.
24. Ziegler, F. E.; Harran, P. G. *Synlett* **1995**, 493.
25. Barton, D. H. R.; Garcia, B.; Togo, H.; Zard, S. Z. *Tetrahedron Lett.* **1986**, *27,* 1327.
26. Togo, H.; Aoki, M.; Yokoyama, M. *Tetrahedron Lett.* **1991**, *32,* 6559.
27. Barton, D. H. R.; Ramesh, M. *J. Am. Chem. Soc.* **1990**, *112,* 891.
28. Crich, D.; Ritchie, T. J. *J. Chem. Soc., Chem. Commun.* **1988**, 1461.
29. Crich, D.; Ritchie, T. J. *J. Chem. Soc., Perkin Trans. 1* **1990**, 945.
30. Crich, D.; Lim, L. B. L. *Tetrahedron Lett.* **1990**, *31,* 1897.
31. Crich, D.; Lim, L. B. L. *J. Chem. Soc., Perkin Trans. 1* **1991**, 2205.
32. Crich, D.; Lim, L. B. L. *J. Chem. Soc., Perkin Trans. 1* **1991**, 2209.
33. Crich, D.; Lim, L. B. L. *Tetrahedron Lett.* **1991**, *32,* 2565.
34. Sugai, T.; Shen, G.-J.; Ichikawa, Y.; Wong, C.-H. *J. Am. Chem. Soc.* **1993**, *115,* 413.
35. Crich, D.; Ritchie, T. J. *Carbohydr. Res.* **1989**, *190,* C3.
36. Crich, D.; Hermann, F. *Tetrahedron Lett.* **1993**, *34,* 3385.
37. Crich, D.; Hwang, J.-T.; Yuan, H. *J. Org. Chem.* **1996**, *61,* 6189.
38. Somsák, L.; Ferrier, R. L. *Adv. Carbohydr. Chem. Biochem.* **1991**, *49,* 37.
39. Dupuis, J.; Giese, B.; Rüegge, D.; Fischer, H.; Korth, H.-G.; Sustmann, R. *Angew. Chem. Int. Ed. Engl.* **1984**, *23,* 896.
40. Korth, H.-G.; Sustmann, R.; Dupuis, J.; Giese, B. *J. Chem. Soc., Perkin Trans. 2* **1986**, 1453.
41. Korth, H.-G.; Sustmann, R.; Groeninger, K. S.; Witzel, T.; Giese, B. *J. Chem. Soc., Perkin Trans. 2* **1986**, 1461.
42. Korth, H.-G.; Sustmann, R.; Groeninger, K. S.; Leisung, M.; Giese, B. *J. Org. Chem.* **1988**, *53,* 4364.
43. Korth, H.-G.; Sustmann, R.; Giese, B.; Rueckert, B.; Groeninger, K. S. *Chem. Ber.* **1990**, *123,* 1891.
44. Giese, B. *Angew. Chem. Int. Ed. Engl.* **1989**, *28,* 969.
45. H.-G.; Praly, J.-P.; Somsák, L.; R. Sustmann, R. *Chem. Ber.* **1990**, *123,* 1155.
46. Rychnovsky, S. D.; Powers, J. P.; LePage, T. J. *J. Am. Chem. Soc.* **1992**, *114,* 8375.

47. Saksena, A. K.; Ganguly, A. K.; Girijavallabhan, V. M.; Pike, R. E.; Chen, Y.-T.; Puar, M. S. *Tetrahedron Lett.* **1992**, *33,* 7721.
48. Barton, D. H. R.; Chern, C.-Y.; Jaszberenyi, J. C.; Shinada, T. *Tetrahedron Lett.* **1993**, *34,* 6505.
49. Giese, B.; Carboni, B.; Göbel, T.; Muhn, R.; Wetterich, F. *Tetrahedron Lett.* **1992**, *33,* 2673.
50. Wendeborn, S.; Wolf, R. M.; De Mesmaeker, A. *Tetrahedron Lett.* **1995**, *36,* 6879.
51. Kishi, T.; Tsuchiya, T.; Umezawa, S. *Bull. Chem. Soc. Jpn.* **1979**, *52,* 3015.
52. Barton, D. H. R.; Dalko P. I.; Géro, S. D. *Tetrahedron Lett.*, **1992**, *33,* 1883.
53. Bell, R. H.; Horton, D.; Williams, D. M. *J. Chem. Soc., Chem. Commun.* **1968**, 323.
54. Bell, R. H.; Horton, D.; Williams, D. M.; Winter-Mihaly, E. *Carbohydr. Res.* **1977**, *58,* 109.
55. Binkley, R. W. *Adv. Carbohydr. Chem. Biochem.* **1981**, *38,* 105.
56. Forbes, J.; Zard, S. Z. *Tetrahedron* **1993**, *49,* 8257.
57. Stout, E. I.; Doane, W. M.; Russell, C. R.; Jones, L. B. *J. Org. Chem.* **1975**, *40,* 1331.
58. Descotes, G.; Faure, A.; Kryczka, B.; Bouchu, M. N. *Bull. Acad. Pol. Sci., Ser. Sci.Chim.* **1979**, *27,* 173.
59. Hürzeler, M.; Bernet, B.; Vasella, A. *Helv. Chim. Acta* **1993**, *76,* 995.

16

Nitro Compounds

Introduction

Information on the photochemistry of carbohydrates protected by *o*-nitrobenzyl and related groups first appeared in the literature in the early 1920's. At that time Tanasescu and co-workers began publishing a series of papers on the photochemistry of *o*-nitrobenzylidene-protected sugars.[1–13] Identification of the products from these reactions proved to be difficult because photolysis generated complex mixtures of nitroso compounds, materials that were experimentally difficult to handle and characterize. These early workers lacked the analytical techniques and instrumentation needed to properly separate and identify most nitroso compounds.[14,15] Furthermore, appropriate chemical reagents and procedures to convert these photoproducts into more easily characterizable materials were unavailable or unknown to these investigators.

Many of the difficulties surrounding the photochemical reactions of compounds containing *o*-nitrobenzyl and related groups were overcome by discoveries made in the 1970's. Zehavi et al. recognized the value of the *o*-nitrobenzyl group as a photoremovable protecting group (eq 1) and solved the problem created by formation of nitroso compounds by developing techniques for separating them chromatographically from other photoproducts.[16,17] In another approach to product purification, Zehavi and Patchornik immobilized the *o*-nitrobenzyl group by attaching it to a polymer support.[18] In this way the nitroso compounds formed by photolysis were eliminated by filtering the solution after irradiation to remove the polymer (eq 2).

Physical removal of nitroso compounds from photochemical reaction mixtures is not a possibility following irradiation of *o*-nitrobenzylidene-protected sugars because the *o*-nitrosobenzoyl and carbohydrate portions of the photoproduct are part of the same molecule (eq 3). A procedure for changing the

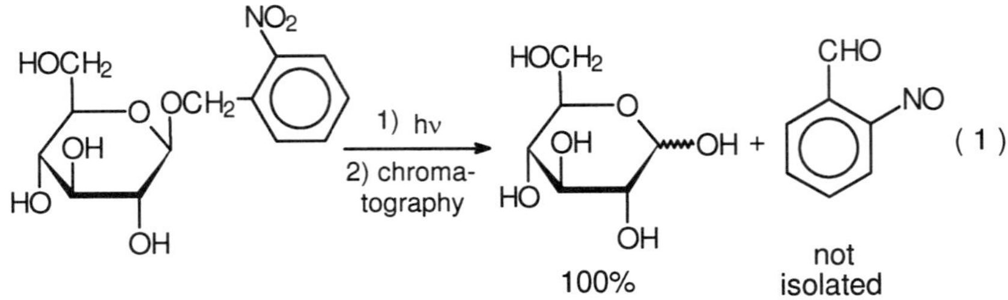

Equation 1

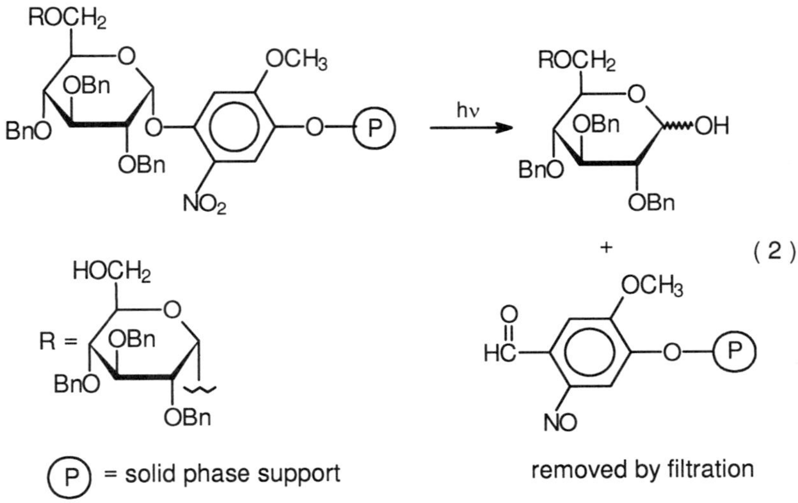

Equation 2

nitroso group was discovered by Collins and co-workers, who found that nitroso groups could be easily and quantitatively oxidized to nitro groups by reaction with trifluoroperacetic acid (eq 3).[14,15] Since the *o*-nitrobenzoates thus produced did not present the characterization problems associated with their *o*-nitroso counterparts, product yields and structures (after oxidation) could be determined by standard procedures.

o-*Nitrobenzyl Ethers*

Characteristics of *o*-Nitrobenzyl Protection

A primary synthetic requirement for protecting hydroxyl groups as *o*-nitrobenzyl ethers is that the *o*-nitrobenzyl group be stable under a variety of reaction conditions, including the presence of acids and bases.[19–21] An equally impor-

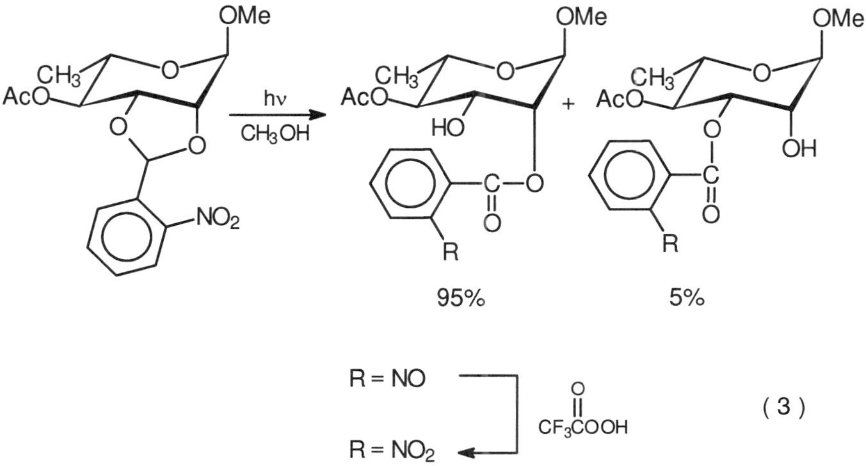

(3)

Equation 3

tant criterion for any protecting group is that deprotection occurs under mild conditions. The *o*-nitrobenzyl group also satisfies this requirement because it is removed at room temperature in neutral solution with radiation of sufficiently long wavelength ($\lambda > 320$ nm) that light is not absorbed by other common chromophores. For *o*-nitrobenzyl-protected compounds this means that competing photoreactions generally do not take place during deprotection.

Mechanism of Photoreaction of *o*-Nitrobenzyl Ethers

A proposed mechanism for the photochemical reaction of an *o*-nitrobenzyl ether is pictured in Scheme 16.1.[22–25] According to this proposal, the ether **1** undergoes intramolecular hydrogen abstraction to generate the *o*-quinoid intermediate **6**. Completion of the deprotection process occurs when the anion **7** cyclizes to **8**, which then cleaves to give *o*-nitrosobenzaldehyde and an alkoxide ion.

Hydrogen abstraction by an *o*-nitrobenzyl ether can occur from its lowest (n,π*) singlet (**2**) or triplet (**3**) state.[25] If reaction is from the singlet state (**2**), compound **6** can form by a concerted [1,5]-hydrogen shift (Scheme 16.1). It is also possible that singlet state reaction is stepwise and involves the singlet diradical **4**, but as Gravel et al. have pointed out, the energy difference between the singlet diradical **4** and the *o*-quinoid intermediate **6** is "likely to be small and may even be insignificant, so that it may be more appropriate to consider the reaction from the singlet excited state as a concerted [1,5]-sigmatropic shift."[22]

Support for the mechanism pictured in Scheme 16.1 comes from several sources. One of these is that aromatic nitro compounds are known to abstract hydrogen atoms; consequently, the formation of intermediates **4–6** has ample

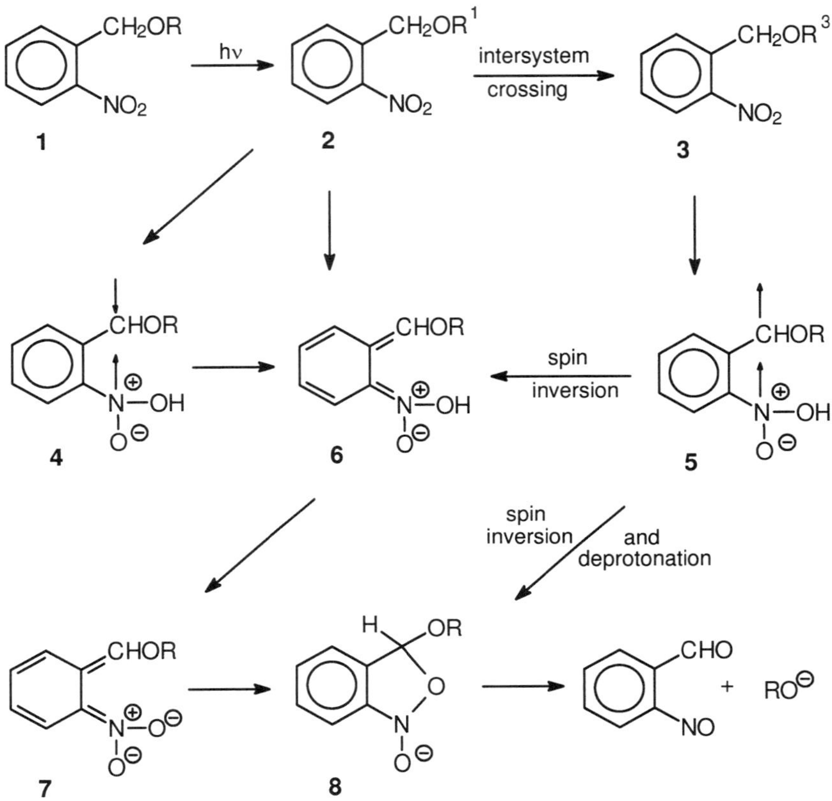

Scheme 16.1

precedent.[26,27] Another piece of supporting information is that the *o*-quinoid species **6** has been observed directly in flash photolysis experiments. Also, ^{18}O labeling has provided the basis for determining that the oxygen atom found in the carbonyl group of the *o*-nitrosobenzaldehyde photoproduct originally was part of the nitro group.[28] Since flash photolysis experiments have been particularly informative in determining the mechanism of photoreaction of *o*-nitrobenzyl compounds, the results from these experiments deserve further comment.

From study of a collection of acetals of *o*-nitrobenzaldehyde and *o*-nitrobenzyl ethers, Yip et al. arrived at several conclusions regarding the mechanism of photochemical reaction.[25] These workers first determined that flash photolysis of the acetal **9** (Chart 16.1) in tetrahydrofuran produced a transient with an absorption maximum at ca. 435 nm. Since the structurally similar acetal **10**, which has no hydrogen atom appropriately placed for abstraction, did

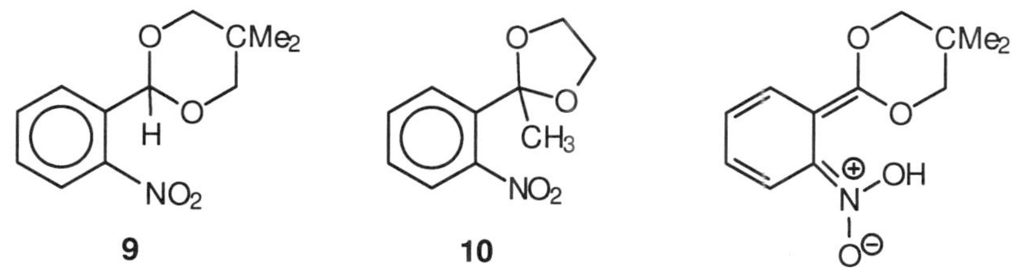

9 10 11

Chart 16.1. Acetals of o-*Nitrobenzaldhyde*

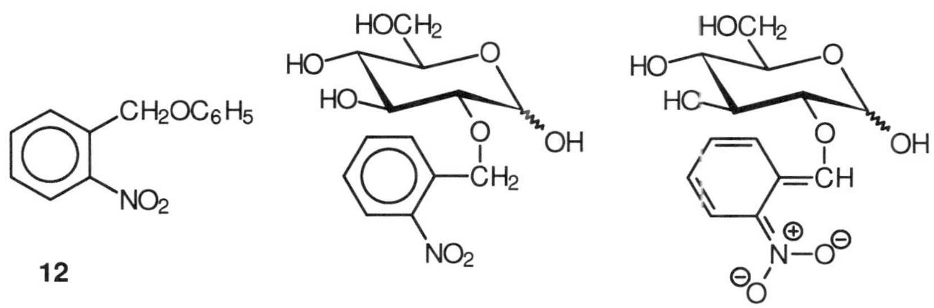

12 13 14

Chart 16.2. o-*Nitrobenzyl Ethers*

not produce such a transient, Yip et al. concluded that a benzylic hydrogen atom on a carbon atom *ortho* to the nitro group was necessary for transient formation. Based on this observation and the knowledge that a similar absorption is produced by the *o*-quinoid intermediate from excitation of *o*-nitrotoluene, these investigators assigned the structure **11** to the transient produced from photolysis of **9**. Flash photolysis of the *o*-nitrobenzyl ether **12** (Chart 16.2) produced a similar, 435-nm absorbing transient that also was assigned an *o*-quinoid structure.

McCray and Trentham noted that the intermediate **6** is not observed during flash photolyses conducted in methanol or water because proton transfer to the medium rapidly forms the anion **7**.[24] Under these conditions **6** is not observed, but **7** is. For example, photolysis of 2-*O*-(*o*-nitrobenzyl)-D-glucose (**13**) in aqueous solution produces a transient with an absorption maximum at 406 nm assigned to the *o*-quinoid anion **14** (Chart 16.2).[24,29]

A question about the mechanism shown in Scheme 16.1 that remained unanswered for many years concerned whether the diradical **5** could cyclize without passing through the *o*-quinoid structure **6**. Although **6** is firmly established in the photochemistry of *o*-nitrobenzyl compounds, Gravel et al. showed that when formation of **6** is not structurally possible, the triplet diradical **5** experiences spin inversion and cyclization to give **8** (Scheme 16.1).[22]

Photochemical Reactions of *o*-Nitrobenzyl-Protected Compounds

Types of Compounds Deprotected by Photolysis

The range of compounds that have been protected by *o*-nitrobenzyl groups and then deprotected photochemically is impressive. It includes mono- and oligosaccharides (Tables 16.1 and 16.2), nucleosides (Table 16.3), oligonucleotides (Table 16.4), deoxyamino sugars (Table 16.5), and all of these types of compounds attached to solid phase supports (Tables 16.6 and 16.7). In general, deprotection occurs in good yield, and irradiation times with standard equipment are usually short (typically a few hours).

o-Nitrobenzyl Protection in Oligoribonucleotide Synthesis

A common application of *o*-nitrobenzyl protection occurs during the synthesis of oligoribonucleotides. *o*-Nitrobenzyl groups are particularly well suited for this task because, as Hayes et al. have observed, "the nitrobenzyl ether linkage is resistant to both acid and base, and therefore satisfies an important requirement of oligoribonucleotide synthesis by remaining intact throughout all of the commonly used construction and deprotection steps, until its final cleavage by irradiation with long-wavelength UV light. Moreover, the nitrobenzyl group shows no tendency to migrate from the 2′- to the 3′-hydroxyl."[43]

One problem that does arise during photodeprotection of pyrimidine-containing oligoribonucleotides is that a [2 + 2] cycloaddition between pyrimidine rings competes with *o*-nitrobenzyl group reaction. This cycloaddition occurs because pyrimidine rings are capable of absorbing some of the incident UV light. When Greenberg and Gilmore irradiated compound **15** (Chart 16.3), an eicosamer of polythymidine (T_{20}) linked to a polymer support, they found that 7% of the contiguous thymidines were transformed into thymine•thymine cycloaddition products (**16**).[63] With further experimentation, Verkatesan and Greenberg were able to reduce this amount of cycloadditon product to 3% by altering the structure of the group linking the oligoribonucleotide with the polymer support (to give **17**) and by selecting a light source that emitted UV radiation captured more effectively by the *o*-nitrobenzyl group.[55] Greenberg and co-workers have had considerable success with reducing side reactions and increasing oligonucleotide yields by optimizing the structure of the photolabile linking group.[55,62]

Product yields for photochemical removal of *o*-nitrobenzyl protection from, at least some, synthetic oligoribonucleotides is pH-dependent.[43] In one instance, photolysis at pH 3.5 produces the deprotected oligonucleotide ApApApC in nearly quantitative yield, but at pH 7.5 there is a 35–40% yield of unidentified material. Although the reason for this pH dependence is not known, this phenomenon also has been observed during photolysis of *o*-nitrobenzyl-protected amino acids.[68–70] In these cases low yields in neutral

Table 16.1. Sugars Protected as *o*-Nitrobenzyl Ethers

Reactant	*Product*	*Percent Yield*	*Ref.*
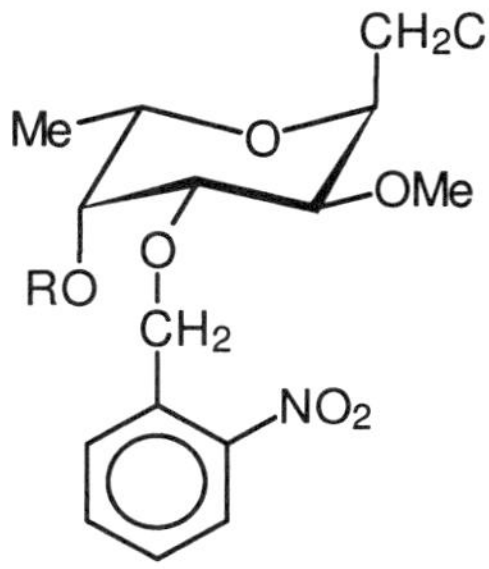	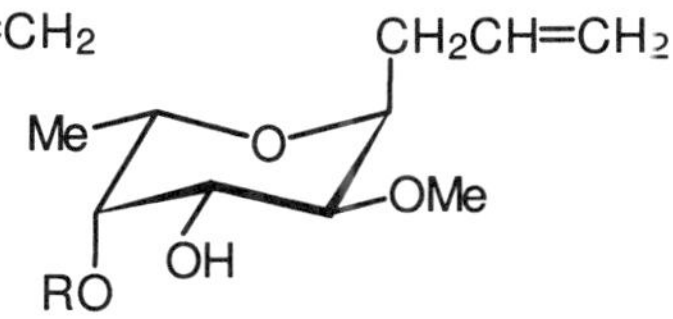	75	30
R = 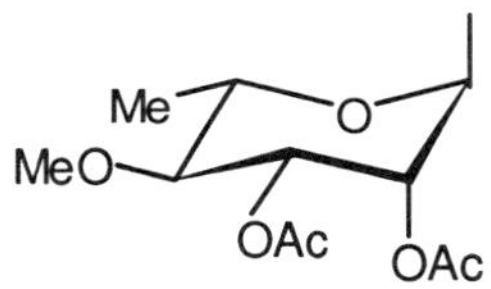			
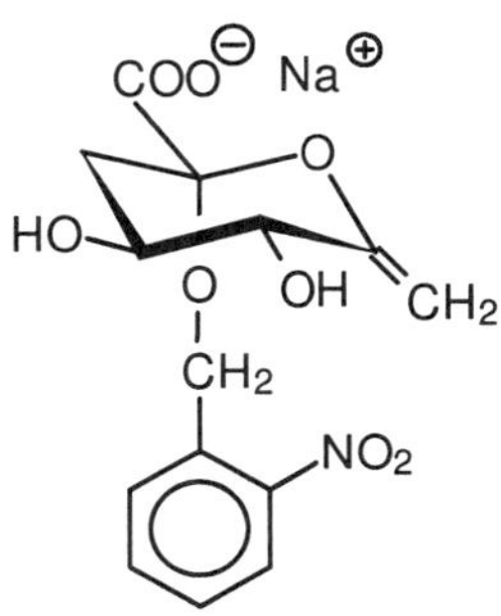	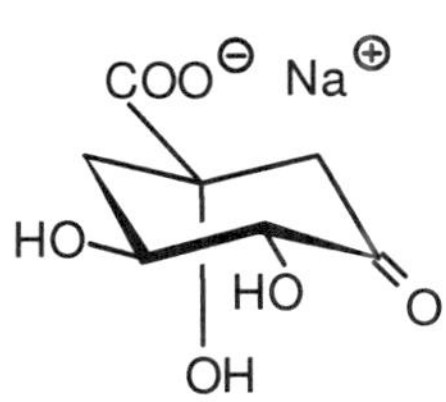	100	31
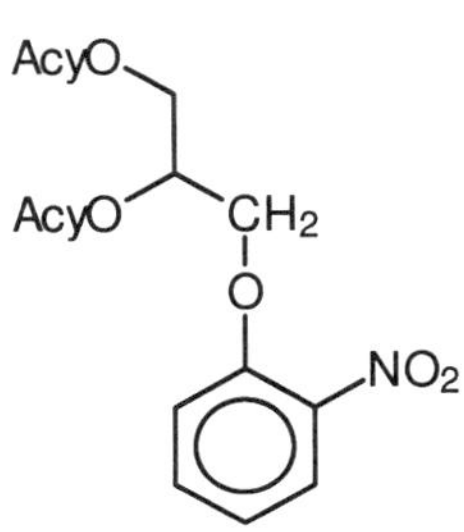	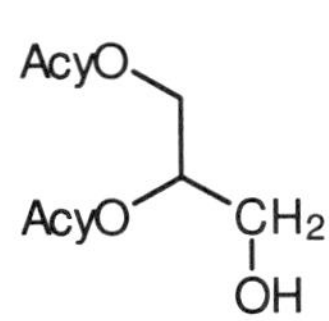	50-60	32

Acy = 9 -, 10 -, 11 -, and 12 - thiasteryl groups

Continued on next page

Table 16.1. Sugars Protected as *o*-Nitrobenzyl Ethers—*Continued*

Reactant	*Product*	*Percent Yield*	*Ref.*
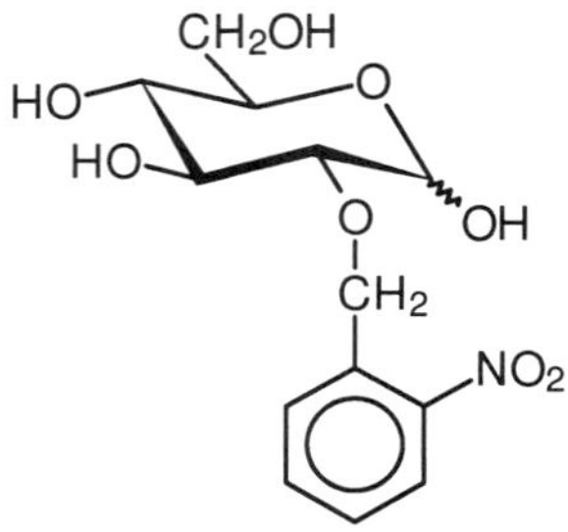	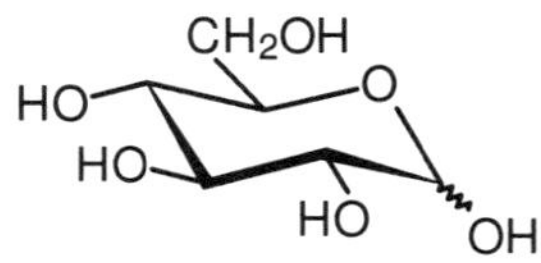	100	23
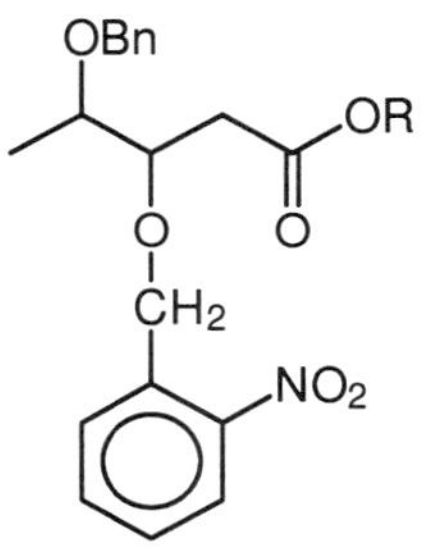	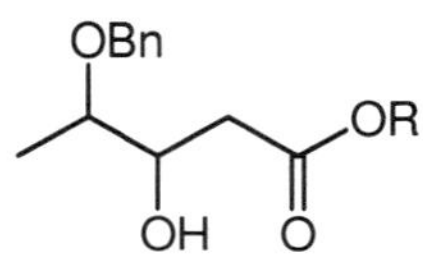	74	33

R = 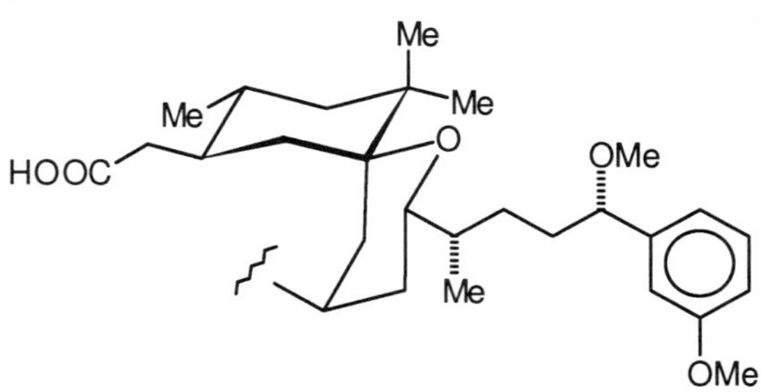

Table 16.2. Photochemical Reactions of *o*-Nitrobenzyl Glycosides

Reactant	*Product*	*Percent Yield*	*Ref.*
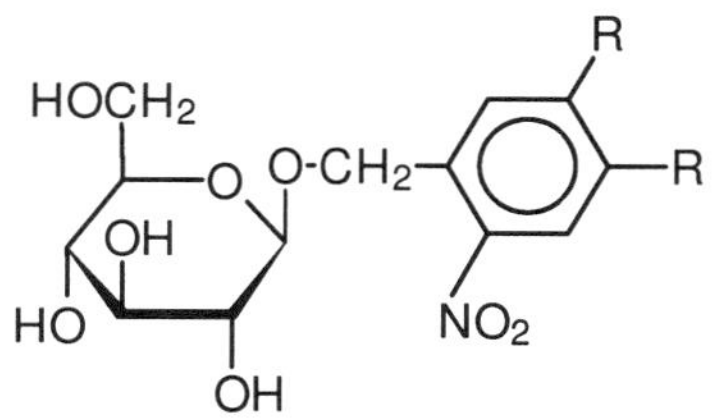	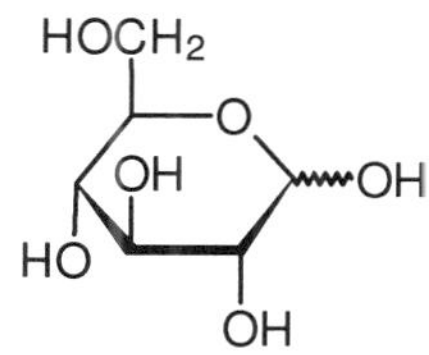		16
R = OMe		100	
R = H		100	
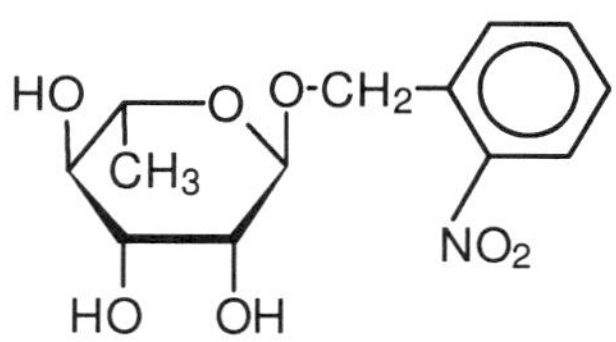	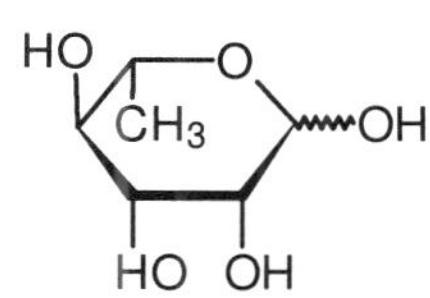	100	17
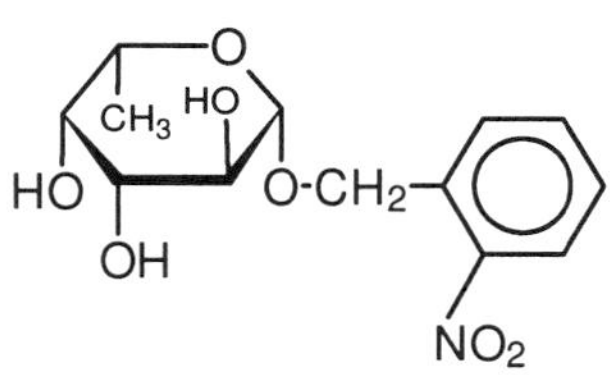	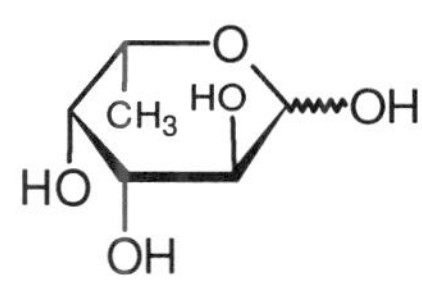	100	17
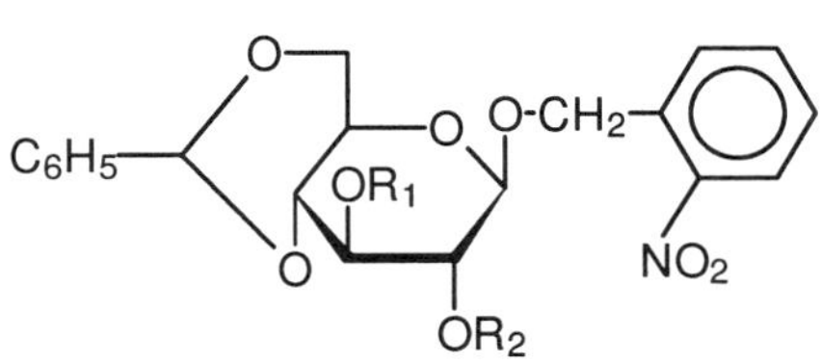	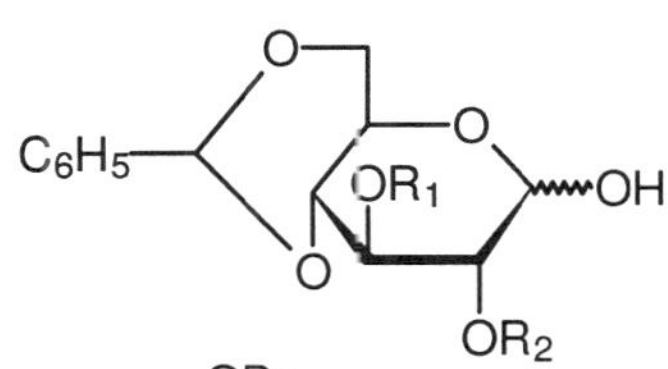		
$R_1 = R_2 =$ 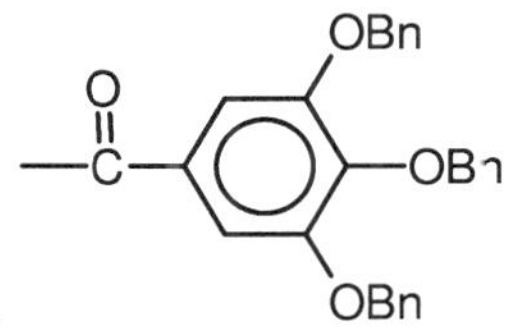		66	37
$R_1, R_2 =$ 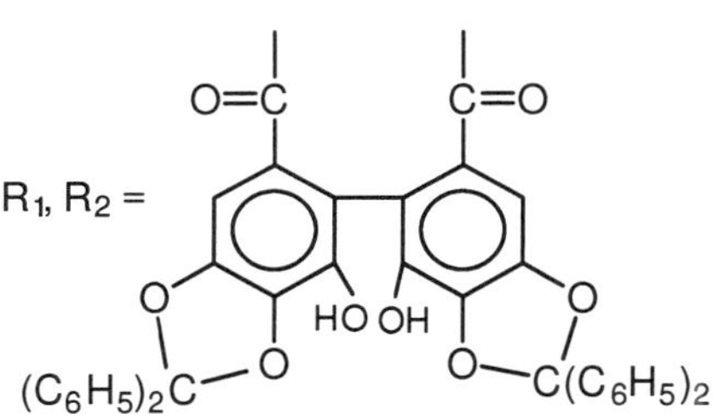		52	37

Continued on next page

Table 16.2. Photochemical Reactions of *o*-Nitrobenzyl Glycosides—*Continued*

Reactant	*Product*	*Percent Yield*	*Ref.*
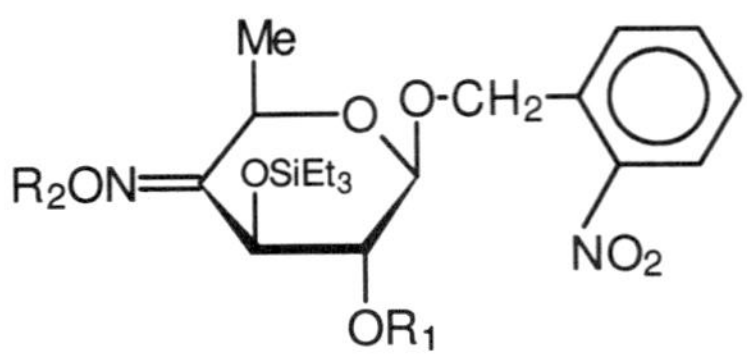	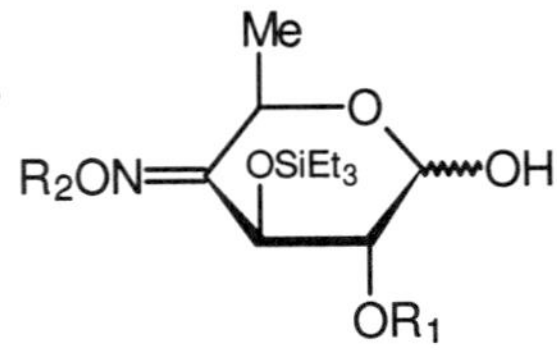		
R_1 = 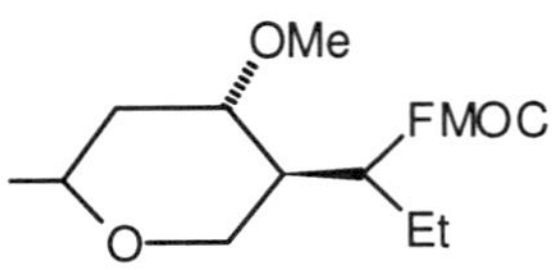			
$R_2 = C_6H_5CH_2$		95	34
R_2 = 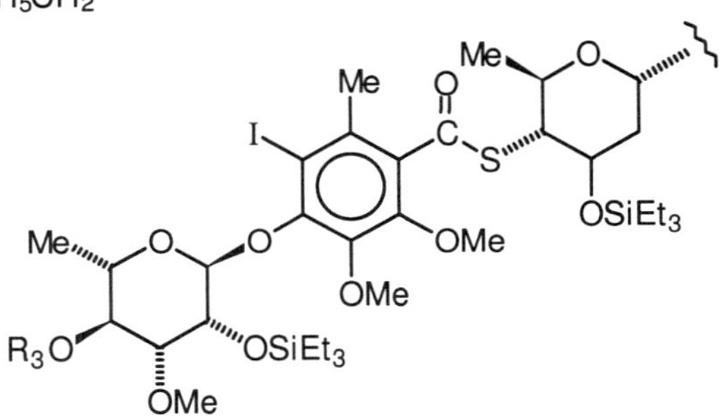			
$R_3 = Et_3Si$		95	35
R_3 = The remainder of the head-to-tail dimer of the calicheamicin oligosaccaride (a molecule with ten rings)		70[a]	36

[a]See reference 36 for the complete structure.

Table 16.3. Photochemical Reactions of 2-*O*-(*o*-Nitrobenzyl)-Protected Nucleosides

Reactant	*Product*	*Percent Yield*	*Ref.*
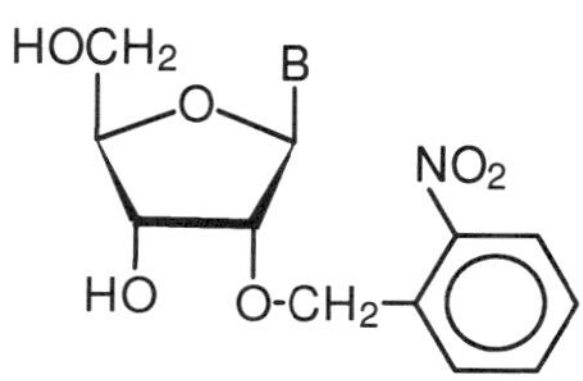	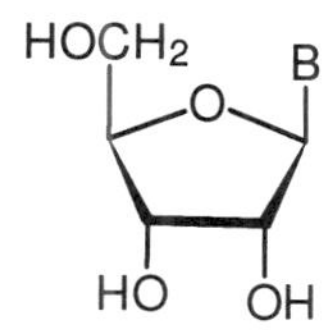		
B = 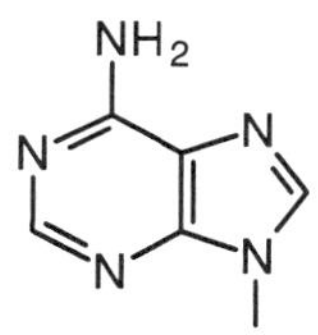 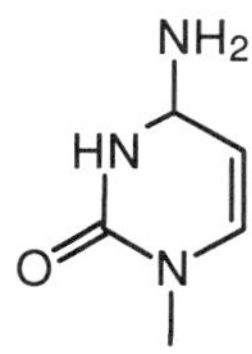	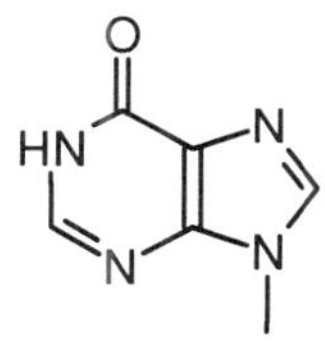	100	38
B = 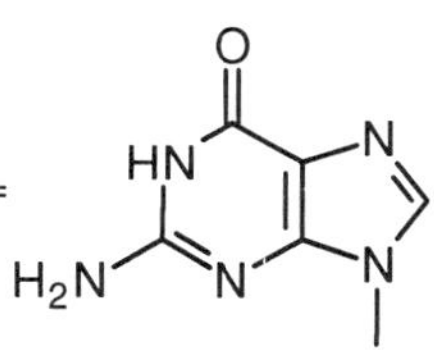		a	39
B = 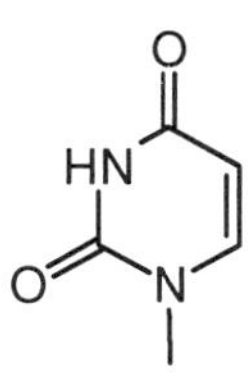		100	38
		a	40
		93	41

[a]The yield was not reported.

Table 16.4. Photochemical Reactions of 2′-*O*-(*o*-Nitrobenzyl)-Protected Polynucleotides[a]

Reactant	*Product*	*Percent Yield*	*Ref.*
U(nB)pU	UpU	94-95	41
U(nB)pA	UpA	94-95	41
A(nB)pU	ApU	b	43
CpCpA(nB)	CpCpA	95	42
GpG(nB)p	GpGP	c	45
CpU(nB)p	CpUp	c	45
A(nB)(S)pA	Ap(S)A	d	44
A(nB)pA(nB)pA(nB)pC	ApApApC	b	43
ApApA(nB)p	ApApAp	c	45
ApGpC(nB)p	ApGpCp	c	45
e	CCUGUCACCGC	27	46
e	GGACGAAACAGG	19	46
e	GCGGUCUGAUGAGUCC	13	46
e	CCUGUCmACCGC	29	46
e	GCGGYCACAGG	22	46
e	GGACGA	12	46
e	GAGUCC	24	46
e	UACUAAC	28	47
e	GUAUGUUAAUA	17	47
e	AGCAGCCUGG	c	43, 48
e	AUCGAAA	30	49

Continued on next page

Table 16.4. Photochemical Reactions of 2′-*O*-(*o*-Nitrobenzyl)-Protected Polynucleotides[a]—*Continued*

Reactant	*Product*	*Percent Yield*	*Ref.*
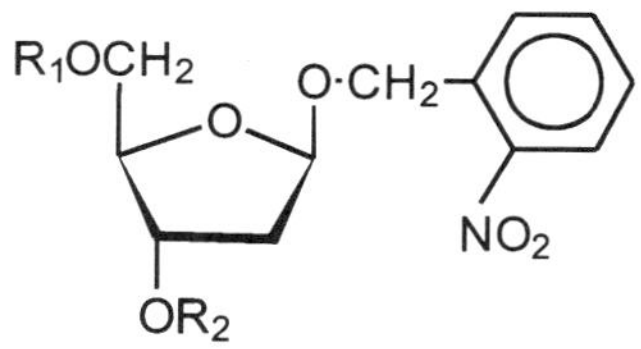	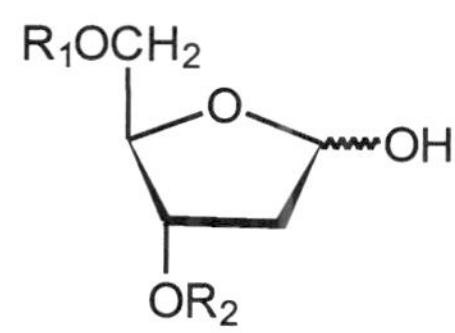		
R_1 = 5'-T_{10} - R_2 = T_{20}-3'		c	50
R_1 = CTT R_2 = TCC		100[e]	51
R_1 = AACGTGAGTGCCGTG R_2 = GTGCCGTGACTGCA		100[e]	51

[a]Standard shorthand representations for nucleotides and oligonucleotides are used.[67] (nB) = *o*-nitrobenzyl group.

[b]Product yields were better in acidic solution.

[c]The yield was not reported.

[d]One of the oxygen atoms attached to phosphorous has been replaced by a sulfur atom.

[e]The reactants were the corresponding ribooligonucleotides with each nucleotide containing a 2'-*O*-(*o*-nitrobenzyl) group.

[f]Reaction was conducted in 0.1 M aqueous ammonium formate.

Table 16.5. Photochemical Reactions of N-*o*-Nitrobenzyl-Protected Carbohydrates

Reactant	*Product*	*Percent Yield*	*Ref.*
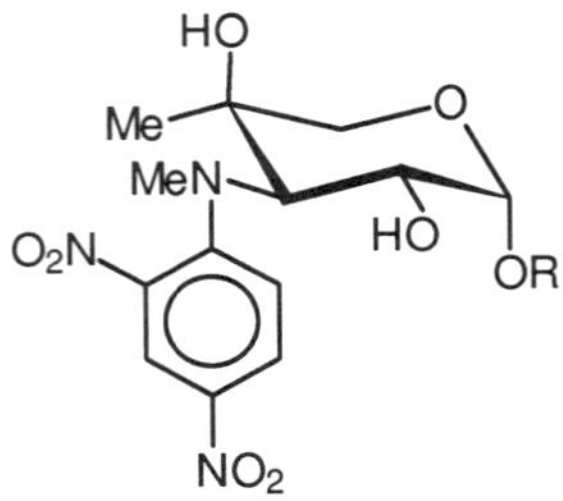 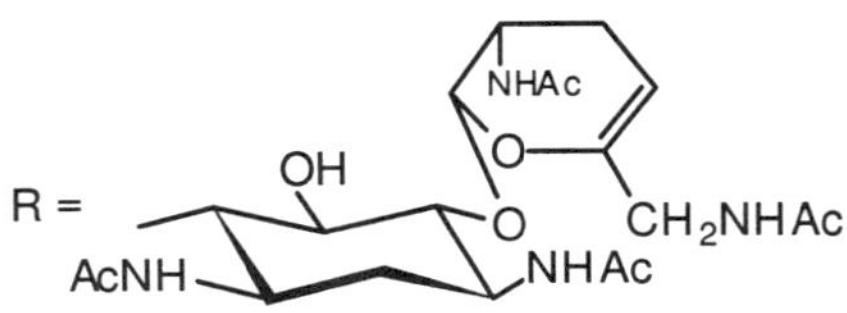	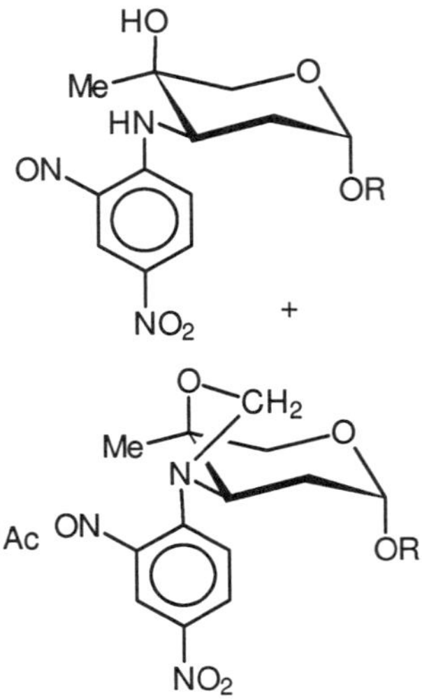		52
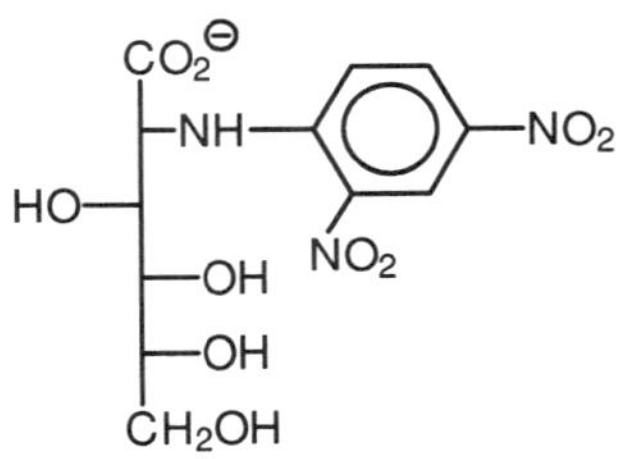	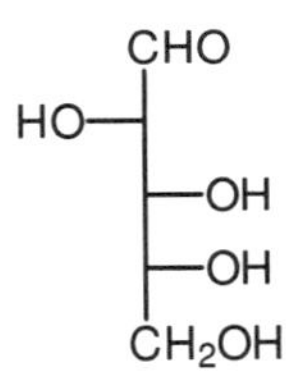	52	53
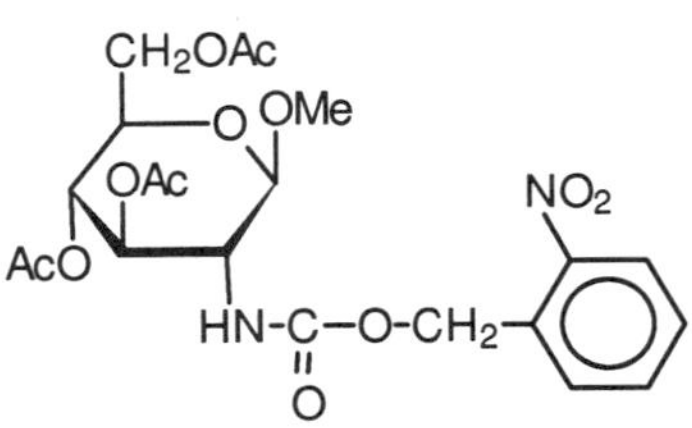	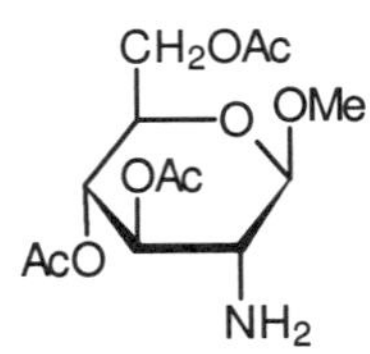	50	54
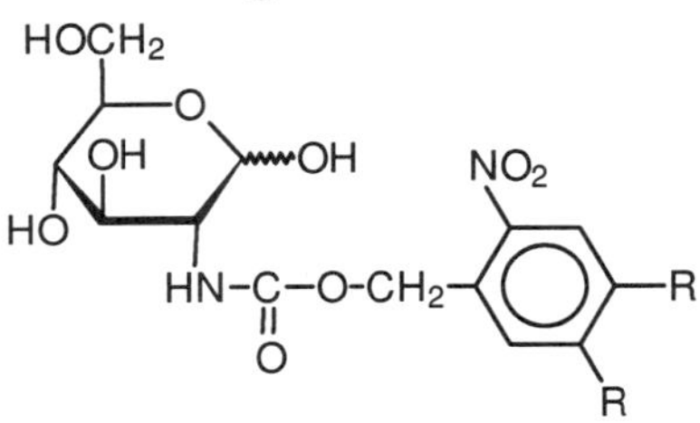	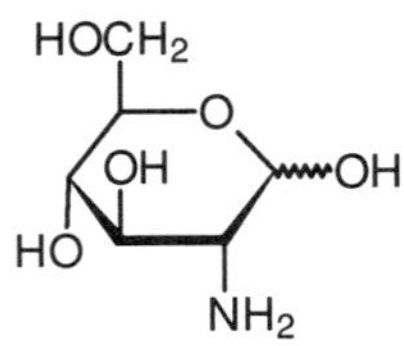 R = H R = OMe	 95 70	47

Table 16.6. Sugars Released from Solid Phase Supports by Photolysis[a]

Photolabile Linker	*Sugar Released*	*Percent Yield*	*Ref.*
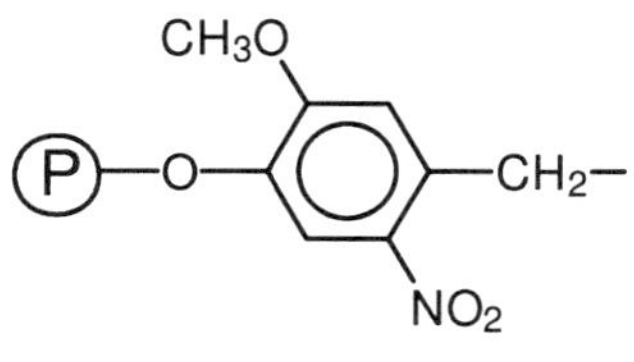	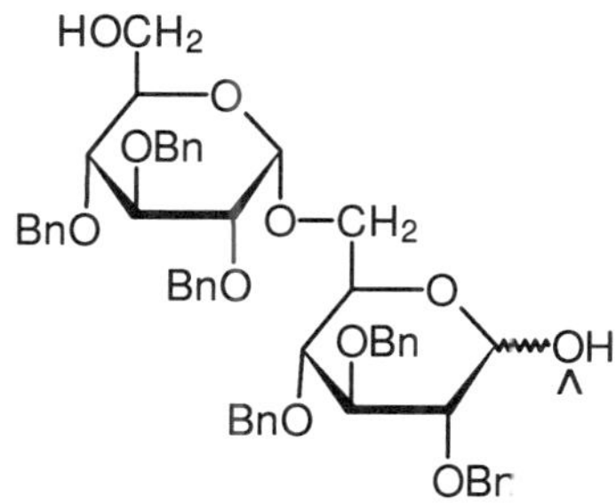	11	18
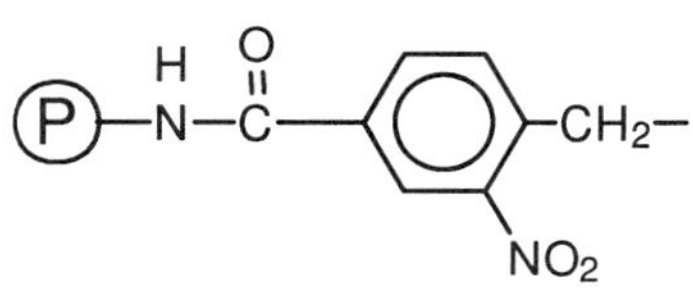	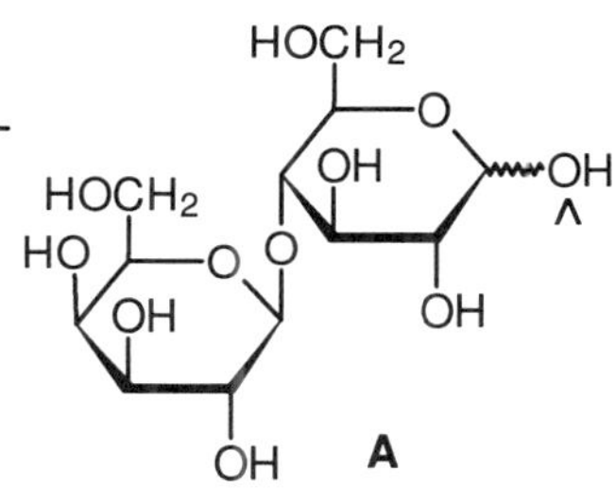 **A**	88[b]	59
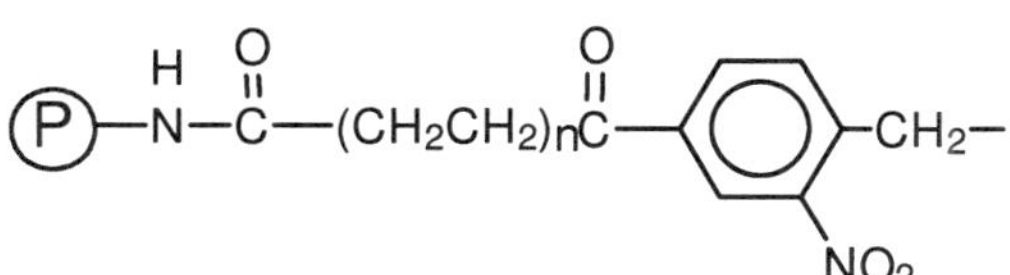 n = 3 or 4	**A**	51	60
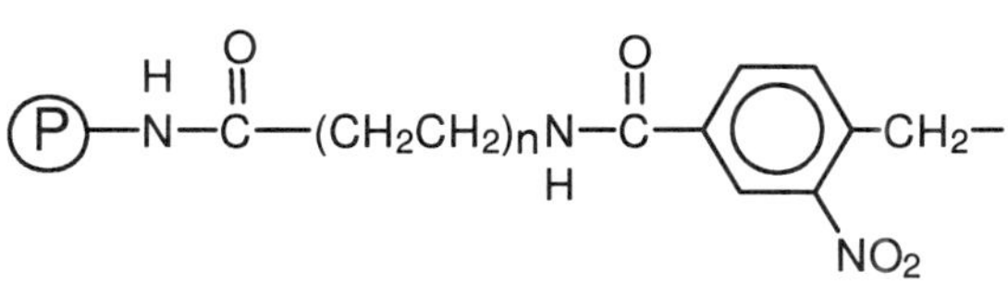	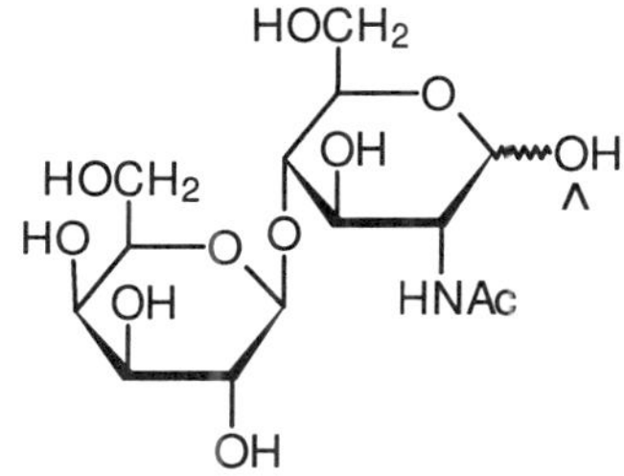	2-30	61

Continued on next page

Table 16.6. Sugars Released from Solid Phase Supports by Photolysis[a]—*Continued*

Photolabile Linker	*Sugar Released*	*Percent Yield*	*Ref.*
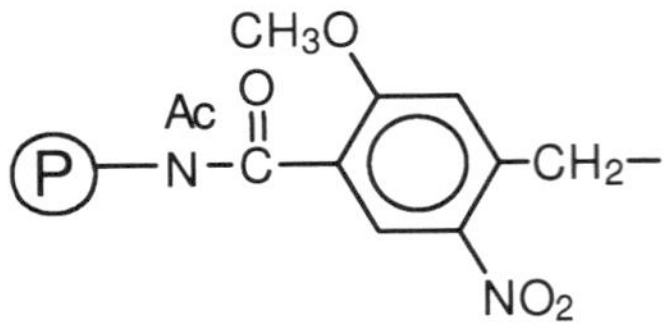	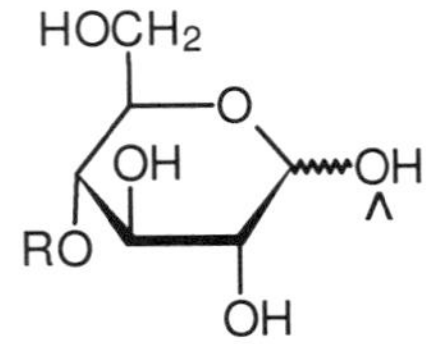	d	
R = β-D-Gal*p*-(1→4)		100	56, 57
R = β-D-Gal*p*-(1→4) -β-D-Glc*p*-(1→4)		100	56, 57
R = [β-D-Glc*pN*Ac-(1→4)]n-β-D-Glc*p*-(1→ 4) n = 1, 2		c	56
R = [α-D-Glc*pN*Ac-(1→4)]n n = 1, 2		c	57, 58
R = (α-D-Glc*p*)m-[α-D-Glc*pN*Ac-(1→4)]2 m≥1		65	57

[a]The "^" indicates the position of attachment of the carbohydrate to the solid phase support.
[b]A water soluble polymer was used.
[c]The yield was not reported.
[d]The linker without the *N*-acetyl group was used in some reactions.

Table 16.7. Oligonucleotides Released from Solid Phase Supports by Photolysis[a]

Photolabile Linker	*Oligonucleotide Released*	*Percent Yield*	*Ref.*
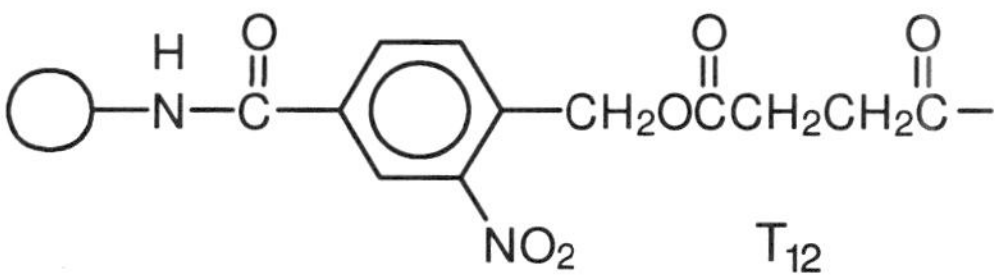	T_{12}	83-87	63
	T_{20}	70-78	63
		70-92[b]	62
		86	64
	5'-d(TACGCAATCCTAGATTAAT)	70-87[b]	62
		82	
		72-76	63
	5'-d(CATAIGCATGTAGCTGAT)	—	65
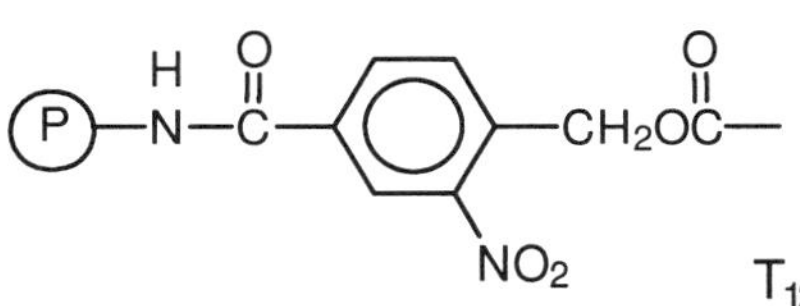	T_{12}	32-56	63
	5'-d(CATACTTAACTT)	59	63
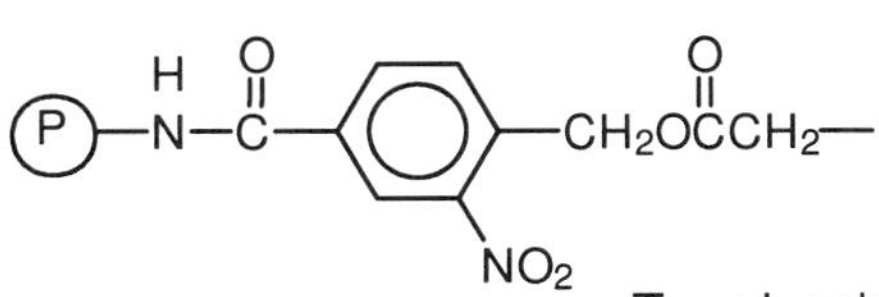	T_{12}-glycolate	79	62
	T_{20}-glycolate	62	65
	5'-d(TACGCAATGCTAGATCTAAT)-glycolate	45-57	65

[a]The reported yields are relative to the conventional ammonolysis yields for releasing oligonucleotides from the solid phase support.
[b]Four different photolabile linkers were studied. They differed in the number of methylene groups (1-4) between the two carbonyl groups in the photolabile linker.

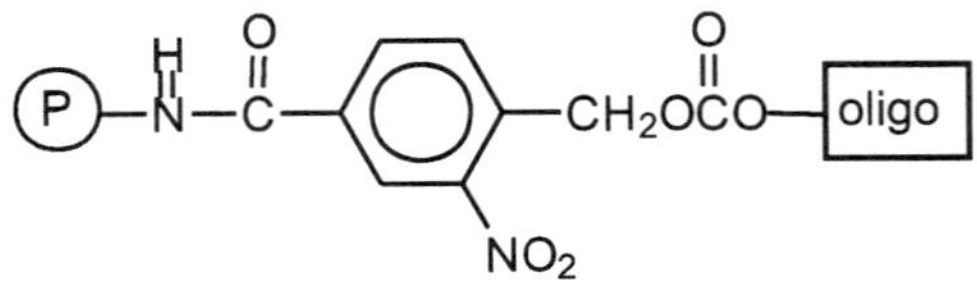

15

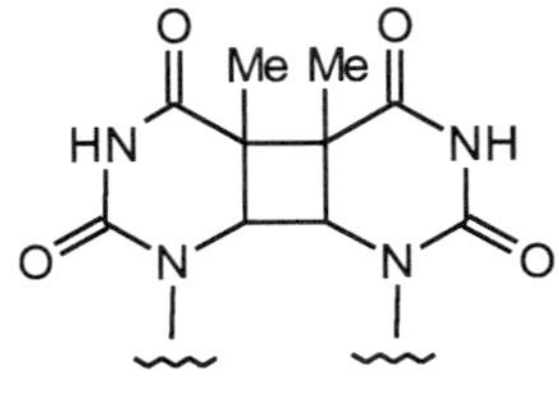

16

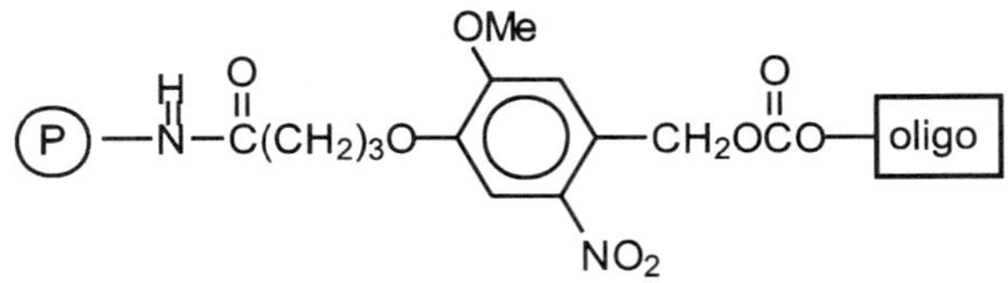

17

(P) = solid phase support   oligo = oligoribonucleotide

Chart 16.3. Polymer-Supported o*-Nitrobenzyl Ethers*

solution were attributed to reaction of the liberated amino acid with the *o*-nitrosobenzaldehyde produced by photolysis.[68] Yields of deprotected compound greatly improved when photolyses were conducted in acidic solution, where reactive nucleophilic centers in the amino acids were protonated.

There is another problem created by the presence of *o*-nitrosobenzaldehyde in photochemical reaction mixtures. Under typical reaction conditions *o*-nitrosobenzaldehyde is converted in a bimolecular reaction into an azobenzene derivative that acts as an internal light filter to slow the photochemical reaction.[69,70] Whenever it is possible to have the *o*-nitrosobenzoyl group remain attached to a polymer support after photolysis, azobenzene formation is effectively eliminated.

o-*Nitrobenzylidene-Protected Sugars*

Following the early investigations by Tanasescu and co-workers,[1–13] the Collins group undertook an extensive study of the photolysis of *o*-nitrobenzylidene-protected sugars. Photochemical reaction offered, in effect, a means for partially removing benzylidene groups without the acid catalysis that normally is required for this reaction.[14,15] As mentioned in the introductory section to this chapter, Collins recognized that the nitroso compounds formed by photolysis of *o*-nitrobenzylidene-protected sugars could be oxidized easily to the corresponding nitro compounds. This finding was a key to the success of these investigations because oxidation converted the photoproducts into separable, characterizable compounds. Collins and co-workers have used this process of photolysis and oxidation to partially deprotect a number of *o*-nitrobenzylidene-containing sugars (Table 16.8).

Although introducing *o*-nitrobenzylidene group protection into a molecule also introduces a new chiral center, the temporary increase in the number of stereoisomers caused by forming this new ring system does not create a lasting problem. Mixtures of these stereoisomers can be irradiated as easily and effectively as pure compounds, and photochemical opening of the newly formed 1,3-dioxane ring eliminates the temporarily created chiral center.[74] Ring opening does produce a mixture of regioisomers that can require chromatographic separation for purification. In many instances, however, ring opening is highly regioselective (eq 3), and purification of the major isomer can be achieved by crystallization.

In analogy to the reaction of an *o*-nitrobenzyl ether, photolysis of an *o*-nitrobenzylidene acetal is assumed to involve formation of an *o*-quinoid intermediate (**19**, Scheme 16.2).[14,15] According to the mechanism proposed for this reaction, cyclization of deprotonated **19** then leads to **20**, which ring opens to generate the photoproducts **21** and **22**. The major product in this reaction (**21**) is the regioisomer with the axial, acyloxy group. This type of regioselectivity is consistently observed in photochemical, ring-opening reactions of *o*-nitrobenzylidene-protected sugars (Table 16.8); in fact, forming a regioisomer with an acyloxy group in an axial position occurs in general for ring-opening reactions of ortho-esters and related compounds (such as **20**).[77–79]

m- *and* p-*Nitrophenyl Glycosides*

Two research groups have studied the photochemistry of *m*- and *p*-nitrophenyl glycosides. Baugh et al. found that irradiation of *m*- or *p*-nitrophenyl glycosides in methanol resulted in radical anion formation, but products from these reactions were not isolated.[80] Yamane et al. reported that photolysis of a *p*-nitrophenyl glycoside of a uronic acid produced low yields of deprotected sugar acids.[81]

Sugar Nitrates

Upon photolysis in methanol, sugar nitrates produce the corresponding hydroxy compounds in excellent yields (Table 16.9).[82–84] Binkley and Koholic proposed that this reaction occurred by homolytic cleavage of an oxygen–nitrogen bond followed by hydrogen abstraction from the solvent (Scheme 16.3).[82,83] Replacement of the nitro group by a hydrogen atom is typical of this

Table 16.8. Photochemical Reactions of Compounds with an *o*-Nitrobenzylidene Group[a]

Reactant	*Product*	*Percent Yield*	*Ref.*
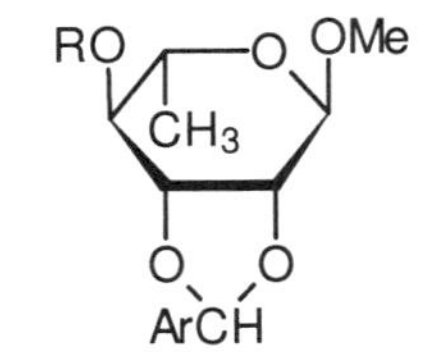	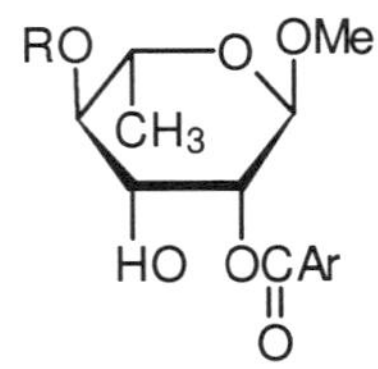		
Ar = 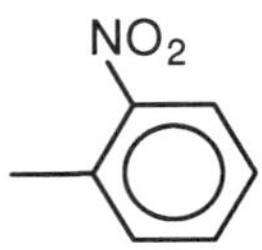	R = Ac	88 95[b]	14 15
	R = 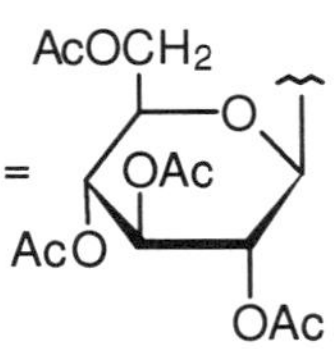	c 60[d] 62[e]	72 75 70
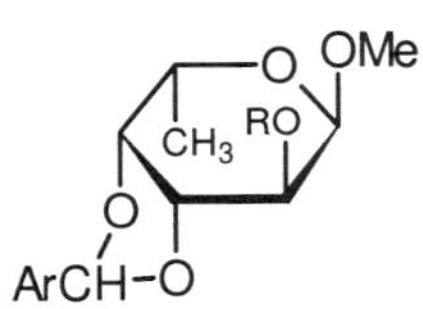	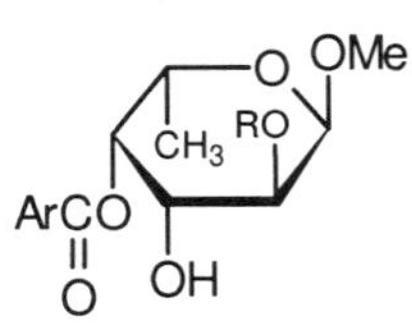		
	R = Ac	86	14
	R = 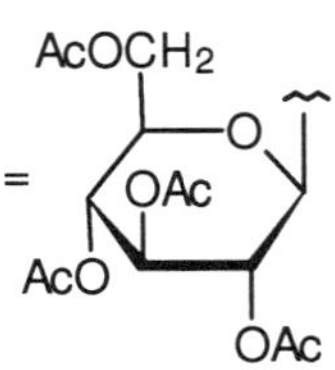	95	72, 73
	R = 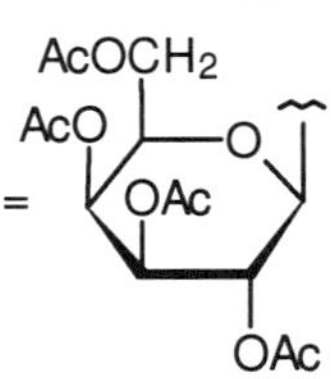	c 93[f]	72 73

Continued on next page

Table 16.8. Photochemical Reactions of Compounds with an *o*-Nitrobenzylidene Group[a]—*Continued*

Reactant	*Product*	*Percent Yield*	*Ref.*
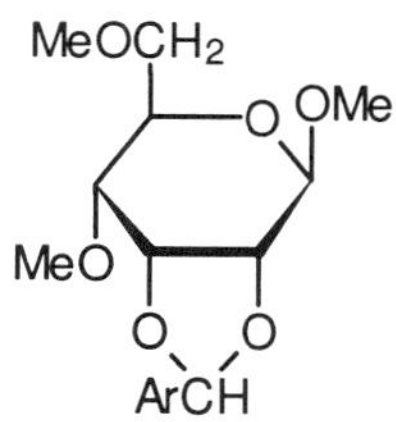	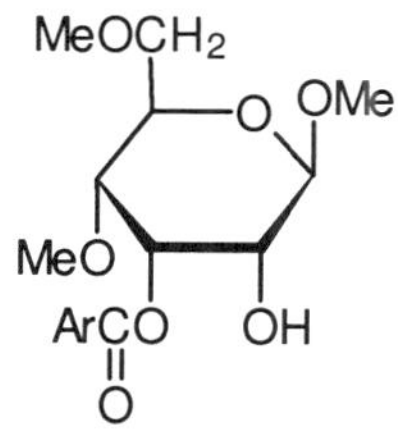	93	14
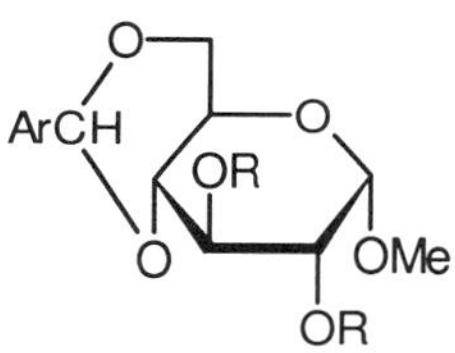	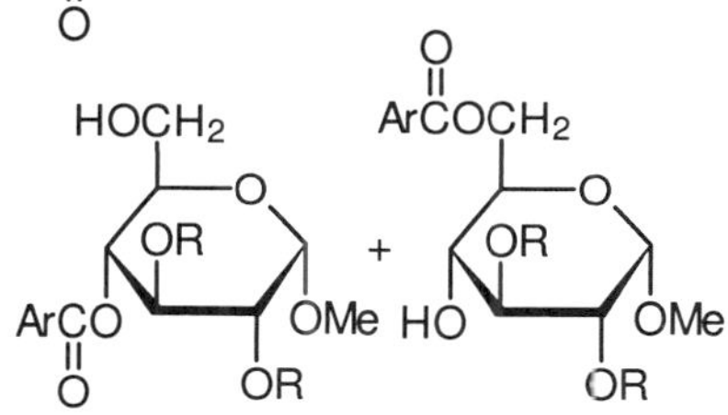		
R = Ac	27% (first product); 63%, 48% (second product)		14 71
R = H	c; c		14, 6
R = CH_3	c; c		14
R = Ts	40%; c		71
	c; c		14, 6
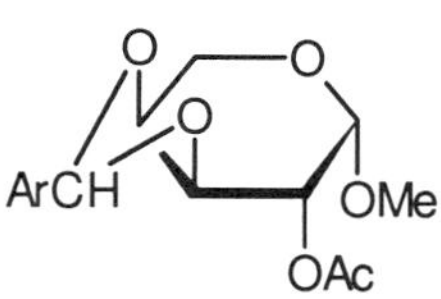	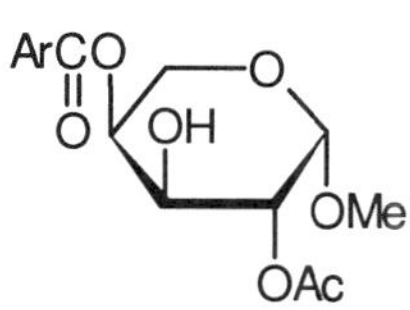	c 95[b]	14 71
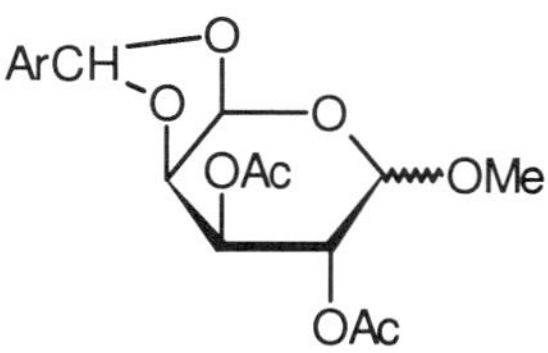	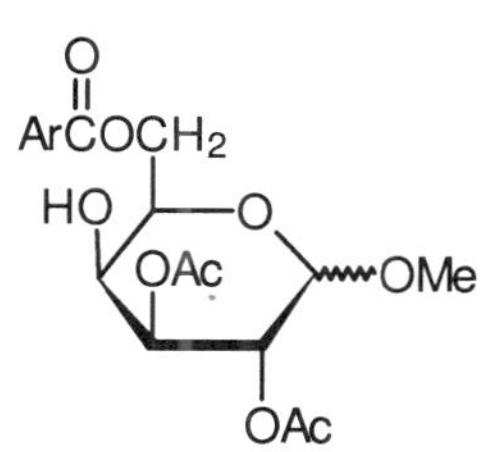	c 34[g] 53[h]	14 71 71
	+ 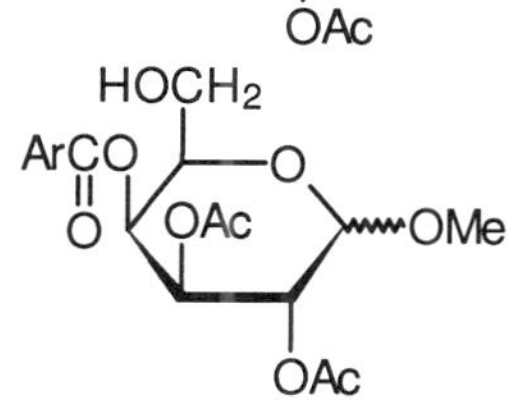	c 66[g] 28[h]	14

Continued on next page

Table 16.8. Photochemical Reactions of Compounds with an *o*-Nitrobenzylidene Group[a]—*Continued*

Reactant	*Product*	*Percent Yield*	*Ref.*
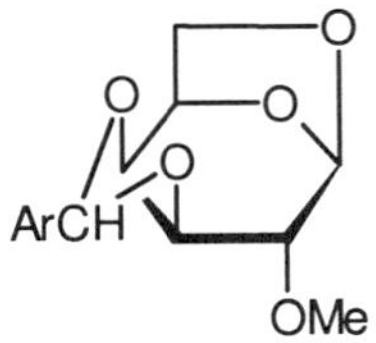	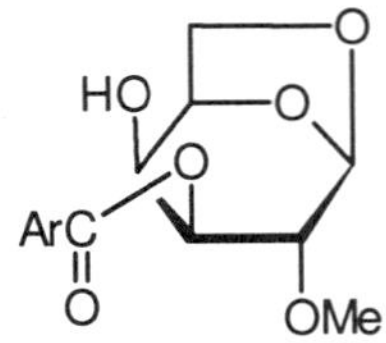	89	15
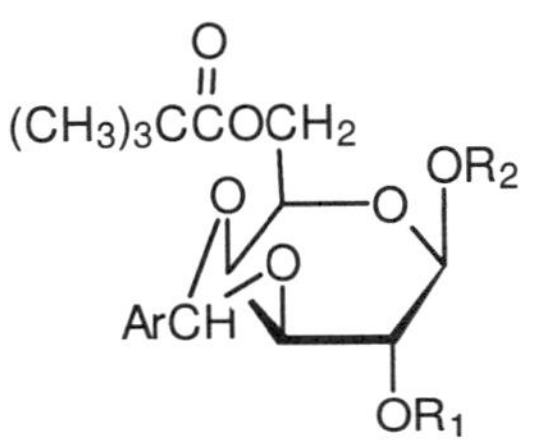	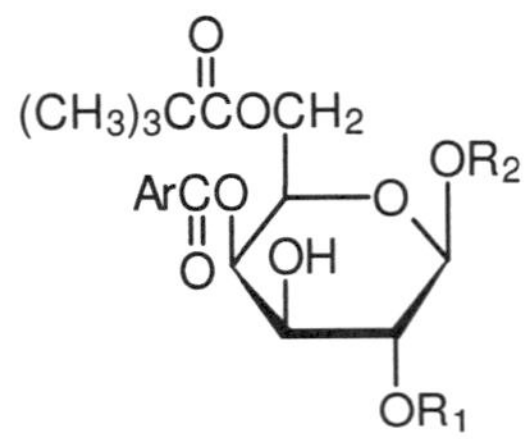	73	74

R_1 = 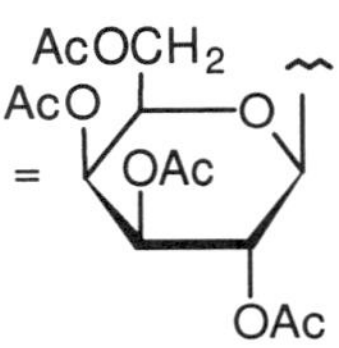

$R_2 = (CH_2)_3CO_2Me$

[a]Ar = *o*-nitrobenzyl group. The nitroso group in each photoproduct has been oxidized to a nitro group by treatment with trifluoroperacetic acid.
[b]5% of the 3-*O*-(*o*-nitrobenzoyl) compound was formed.
[c]The yield was not reported.
[d]A 45% yield of the 3-*O*-(*o*-nitrobenzoyl) compound was formed.
[e]A 12% yield of the 3-*O*-(*o*-nitrobenzoyl) compound was formed.
[f]A small amount of the 3-*O*-(*o*-nitrobenzoyl) compound was formed.
[g]The β-anomer was irradiated.
[h]The α-anomer was irradiated.

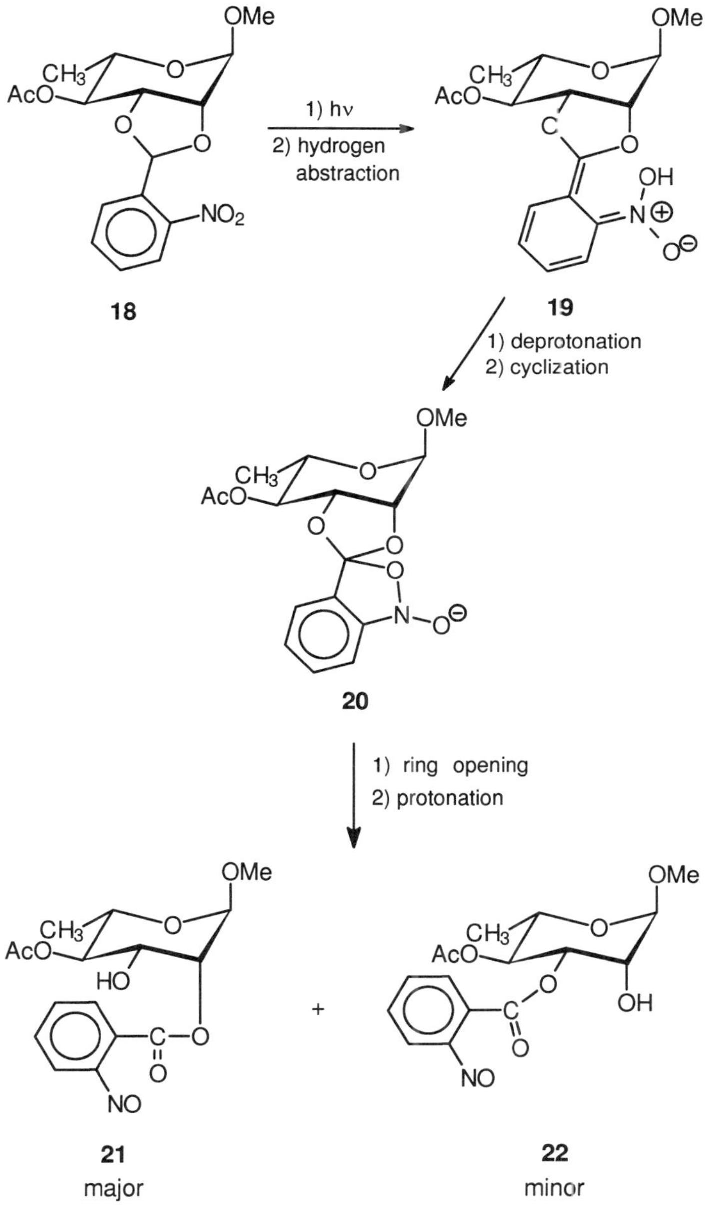
OMe
CH3
AcO
O
1) hν
2) hydrogen abstraction
NO2
18
19
OH
N⊕
O⊖
1) deprotonation
2) cyclization
20
1) ring opening
2) protonation
HO
OH
NO
21 major
+
22 minor

Scheme 16.2

Table 16.9. Photochemical Reactions of Sugar Nitrates[a]

Reactant	*Product*	*Percent Yield*	*Ref.*
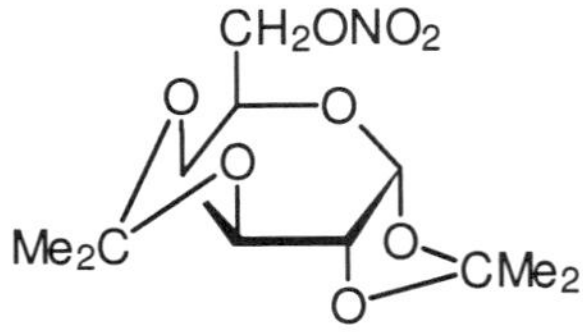	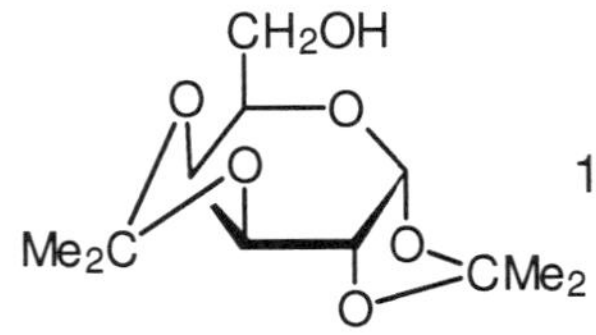	100	81, 82
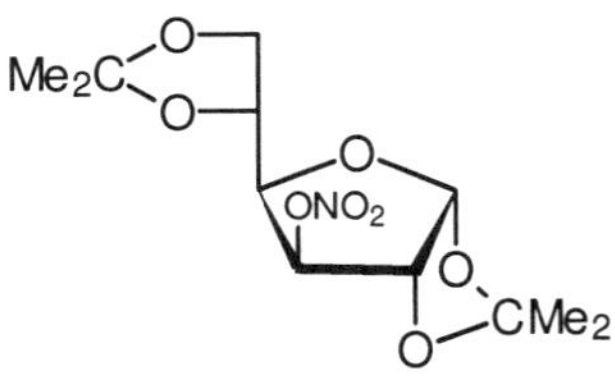	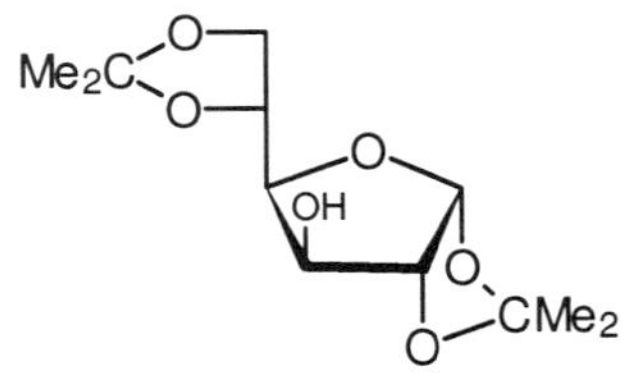	92	81, 82
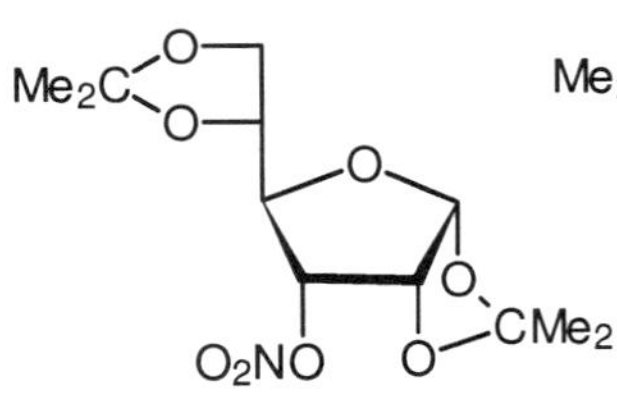	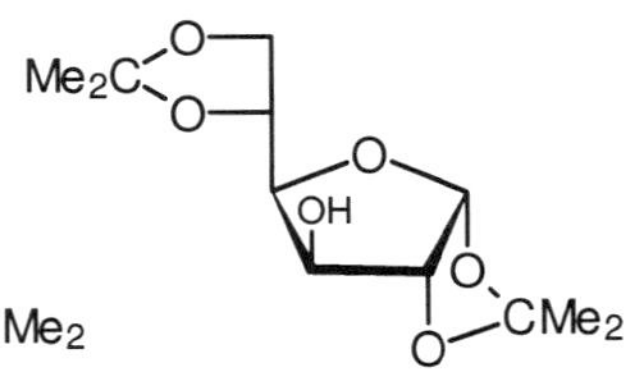	100	81, 82
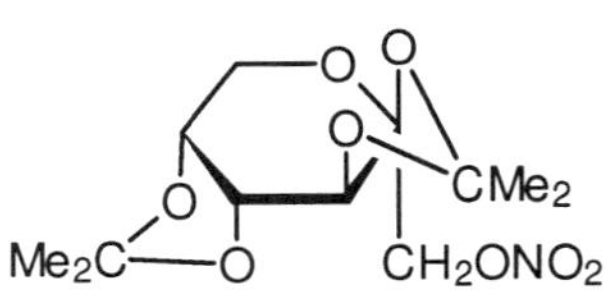	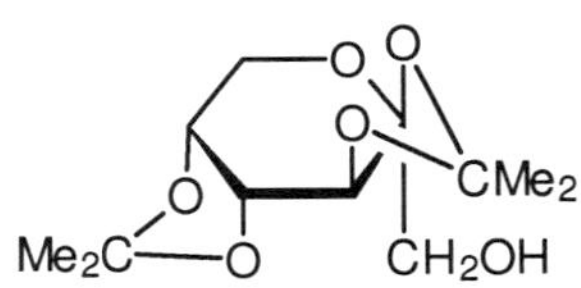	97	82
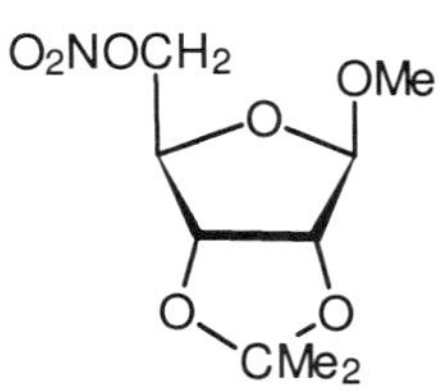	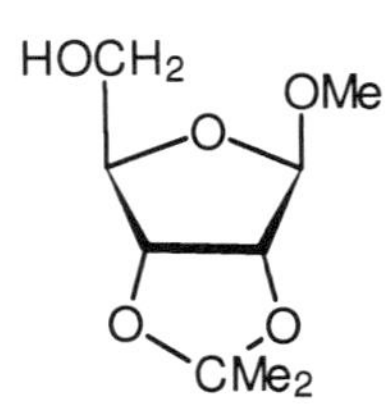	96	82
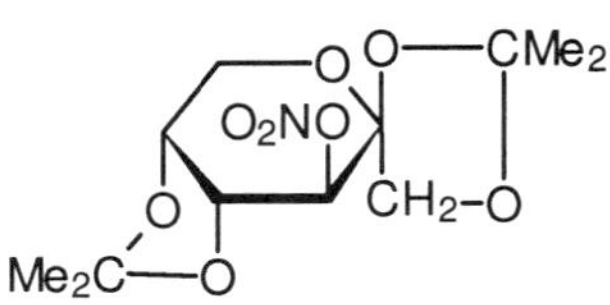	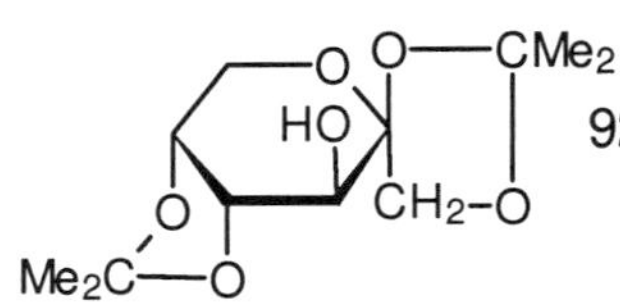	92	82

Continued on next page

Table 16.9. Photochemical Reactions of Sugar Nitrates[a]—*Continued*

Reactant	*Product*	*Percent Yield*	*Ref.*
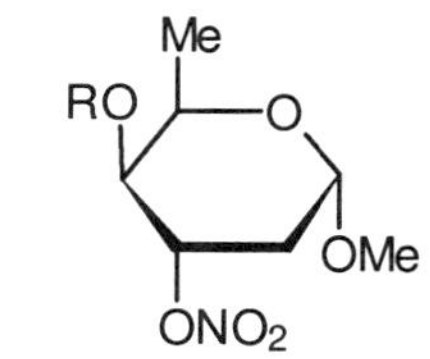	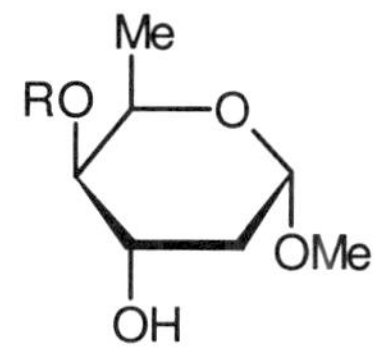		83
R = Bz		90	
R = H		b	
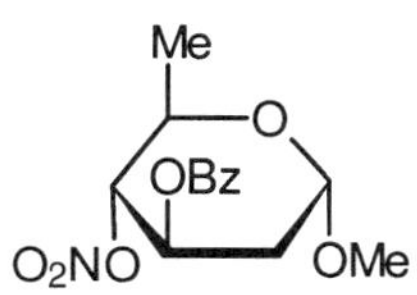	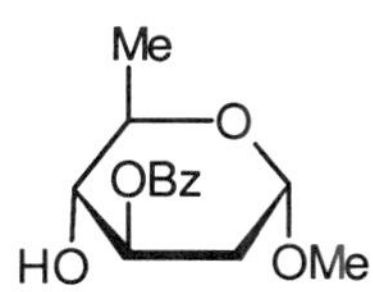	b	83

[a]Methanol was the reaction solvent.
[b]The yield was not reported.

$$RONO_2 \xrightarrow[-\ NO_2]{h\nu} RO\bullet \xrightarrow[\text{from the solvent}]{\text{hydrogen abstraction}} ROH$$

Scheme 16.3

reaction, but rearrangement takes place when the alkoxy radical formed by photolysis is highly strained (see Scheme 12.6 in Chapter 12).

Nitro Alkenes

Photolysis of non-aromatic nitro compounds is a rare event in carbohydrate chemistry. The only investigation of this type of compound is one conducted by Szarek, Jones and co-workers, who found that the expected photochemical *cis–trans* isomerization of the nitro alkene **23** was accompanied by a rearrangement that produced the α,β-unsaturated ketones **25** and **26** (eq 4).[85,86] Although no mechanistic studies were conducted on this reaction, a possible sequence of events is outlined in Scheme 16.4.

Summary

Among the several types of carbohydrates containing nitro groups that have been photolyzed, the most important are compounds protected by *o*-nitrobenzyl groups. This protection is readily removed by photolysis (Tables 16.1–

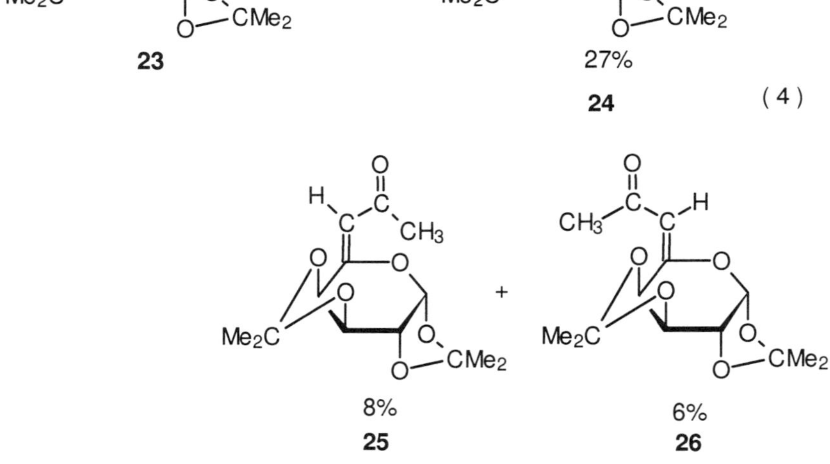

23 24 (4)

25 26

Equation 4

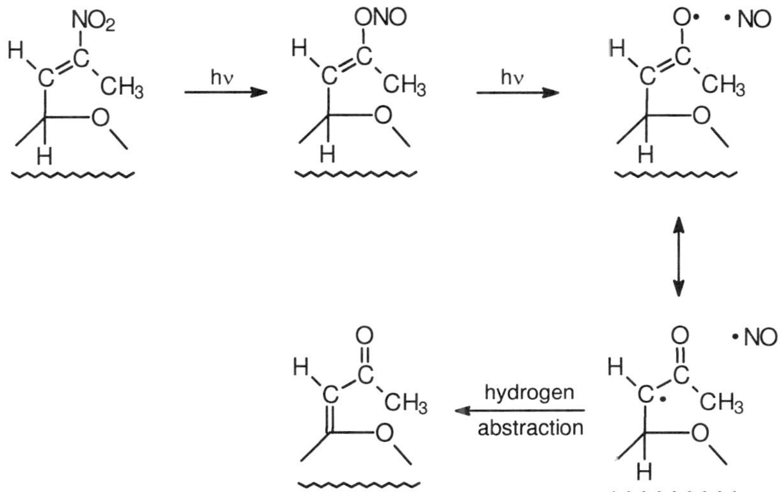

Scheme 16.4

16.7). Deprotection begins with hydrogen abstraction by an excited *o*-nitrobenzyl group, a process that produces an unstable *o*-quinoid photoproduct. Spontaneous (at room temperature) reaction of this photoproduct completes the deprotection by eliminating *o*-nitrosobenzaldehyde (Scheme 16.1).

Other nitro compounds that have been irradiated include nitrates and *o*-nitrobenzylidene acetals. Photolysis of *o*-nitrobenzylidene acetals converts them into *o*-nitrosobenzoates, compounds that are easily oxidized to *o*-nitrobenzoates (eq 3). Nitrated carbohydrates are converted into the corresponding hydroxy compounds by photolysis (Scheme 16.3).

References

1. Tanasecu, I. *Bul. Soc. Stünte Cluj* **1924**, *2,* 111.
2. Tanasecu, I.; Tanasecu, H. *Bul. Soc. Stünte Cluj* **1925** *2,* 369.
3. Tanasecu, I.; Macovski, E. *Bull. Soc. Chim.* **1930,** *47,* 457.
4. Tanasecu, I.; Macovski, E. *Bull. Soc. Chim.* **1930**, *47,* 86.
5. Tanasecu, I.; Macovski, E. *Bull. Soc. Chim.* **1933**, *53,* 1097.
6. Tanasecu, I.; Craciunescu, E. *Bull. Soc. Chim.* **1936**, *3,* 581.
7. Tanasecu, I.; Craciunescu, E. *Bull. Soc. Chim.* **1936**, *3,* 1511.
8. Tanasecu, I.; Craciunescu, E. *Bull. Soc. Chim.* **1936**, *3,* 1517.
9. Tanasecu, I.; Iliescu, I. *Bull. Soc. Chim.* **1938**, *5,* 1446.
10. Tanasecu, I.; Ionescu, M. *Bull. Soc. Chim.* **1940**, *7,* 77.
11. Tanasecu, I.; Ionescu, M. *Bull. Soc. Chim.* **1940**, *7,* 84.
12. Tanasecu, I.; Ionescu, M. *Bull. Soc. Chim.* **1940**, *7,* 90.
13. Tanasecu, I.; Otea, A. *Stud. Univ. Babes-Bolyai, Ser. Chem.* **1972**, *17,* 113.
14. Collins, P. M.; Oparaeche, N. N. *J. Chem. Soc., Chem. Commun.* **1972**, 532.

15. Collins, P. M.; Oparaeche, N. N. *J. Chem. Soc., Perkin Trans. 1* **1975**, 1695.
16. Zehavi, U.; Amit, B.; Patchornik, A. *J. Org. Chem.* **1972**, *37*, 2281.
17. Zehavi, U.; Patchornik, A. *J. Org. Chem.* **1972**, *37*, 2285.
18. Zehavi, U.; Patchornik, A. *J. Am. Chem. Soc.* **1973**, *95*, 5673.
19. Amarnath, V.; Broom, A. D. *Chem. Rev.* **1977**, *77*, 183.
20. Zehavi, U. *Adv. Carbohydr. Chem. Biochem.* **1988**, *88*, 179.
21. Binkley, R. W.; Flechtner, T. W. In *Synthetic Organic Photochemistry;* Horspool, W. M., Ed.; Plenum: New York, 1984; p 375.
22. Gravel, D.; Giasson, R.; Blanchet, D.; Yip, R. W.; Sharma, D. K. *Can. J. Chem.* **1991**, *69*, 1193.
23. Corrie, J. E. T. *J. Chem. Soc., Perkin Trans. 1* **1993**, 2161.
24. McCray, J. A.; Trentham, D. R. *Annu. Rev. Biophys. Biophys. Chem.* **1989**, *18*, 239 .
25. Yip, R. W.; Sharma, D. K.; Giasson, R.; Gravel, D. *J. Phys. Chem.* **1985**, *89*, 5328.
26. Dessauer, R.; Paris, J. P. In *Advances in Photochemistry;* Noyes, W. A.; Hammond, G. S.; Pitts, J. N., Eds.; Interscience: New York, 1963; p 275.
27. Gilbert, A.; Baggott, J. *Essentials of Molecular Photochemistry*; CRC: Boca Raton, FL, 1991; p 437.
28. Amit, B.; Ben-Efraim, D. A.; Patchornik, A. *J. Chem. Soc., Perkin Trans. 1* **1976**, 57.
29. Corrie, J. E. T.; Katayama, Y.; Reid, G. P.; Anson, M.; Trentham, D. R. *Philos. Trans. R. Soc. London, Ser. A* **1992**, *340*, 233.
30. Aspinall, G. O.; Crane, A. M.; Gammon, D. W.; Ibrahim, I. H.; Khare, N. K.; Chatterjee, D.; Rivoire, B.; Brennan, P. J. *Carbohydr. Res.* **1991**, *216*, 337.
31. Bartlett, P. A.; Satake K., *J. Am. Chem. Soc.* **1988**, *110*, 1628.
32. Li, R.; Pascal, Jr., R. A. *J. Org. Chem.* **1993**, *58*, 1952.
33. Ireland, R. E.; Thaisrivongs, S.; Dussault, P. H. *J. Am. Chem. Soc.* **1988**, *110*, 5768.
34. Nicolaou, K. C.; Schreiner, E. P.; Stahl, W. *Angew. Chem. Int. Ed. Engl.* **1991**, *30*, 585.
35. Nicolaou, K. C.; Schreiner, E. P.; Iwabuchi, Y.; Suzuki, T. *Angew. Chem. Int. Ed. Engl.* **1992**, *31*, 340.
36. Nicolaou, K. C.; Ajito, K.; Komatsu, H.; Smith, B. M.; Bertinato, P.; Gomez-Paloma, L. *J. Chem. Soc., Chem. Commun.* **1996**, 1495.
37. Feldman, K. S.; Sambandam, A. *J. Org. Chem.* **1995**, *60*, 8171.
38. Bartholomew, D. G.; Broom, A. D. *J. Chem. Soc., Chem. Commun.* **1975**, 38.
39. Ohtsuka, E.; Tanaka, S.; Ikehara, M. *Synthesis* **1977**, 453.
40. Ohtsuka, E.; Tanaka, S.; Ikehara, M. *Nucleic Acid Chem.* **1978**, *1*, 401.
41. Ohtsuka, E.; Tanaka, S.; Ikehara, M. *Nucleic Acids Res.* **1974**, *1*, 1351.
42. Ohtsuka, E.; Tanaka, S.; Ikehara, M. *Chem. Pharm. Bull.* **1977**, *25*, 949.
43. Hayes, J. A.; Brunden, M. J.; Gilham, P. T.; Gough, G. R. *Tetrahedron Lett.* **1985**, *26*, 2407.
44. Marlier, J. F.; Benkovic, S. J. *Tetrahedron Lett.* **1980**, *21*, 1121.
45. Ohtsuka, E.; Tanaka, T.; Tanaka, S.; Ikehara, M. *J. Am. Chem. Soc.* **1987**, *100*, 4580.
46. Odai, O.; Kodama, H.; Hiroaki, H.; Sakata, T.; Tanaka, T.; Uesugi, S. *Nucleic Acids Res.* **1990**, *18*, 5955.
47. Tanaka, T.; Tamatsukuri, S.; Ikehara, M. *Nucleic Acids Res.* **1986**, *14*, 6265.
48. Ohtsuka, E.; Tanaka, T.; Ikehara, M. *J. Am. Chem. Soc.* **1979**, *101*, 6409.
49. Ohtsuka, E.; Tanaka, S.; Ikehara, M. *J. Am. Chem. Soc.* **1978**, *100*, 8210.
50. Horn, T.; Downing, K.; Gee, Y.; Urdea, M. S. *Nucleosides Nucleotides* **1991**, *10*, 299.
51. P'eoc'h, D.; Meyer, A.; Imbach, J. L.; Rayner, B. *Tetrahedron Lett.* **1991**, *32*, 207.
52. Nagabhushan, T. L; Wright, J. J.; Cooper, A. B.; Turner, W. N.; Miller, G. H. *J. Anitbiot.* **1978**, *31*, 43.
53. El Ashmawy, A. E.; Horton, D.; Philips, K. D. *Carbohydr. Res.* **1969**, *9*, 350.
54. Amit, B.; Zehavi, U.; Patchornik, A. *J. Org. Chem.* **1974**, *39*, 192.
55. Venkatesan, H.; Greenberg, M. M. *J. Org. Chem.* **1996**, *61*, 525.
56. Zehavi, U.; Sadeh, S.; Herchman, M. *Carbohydr. Res.* **1983**, *124*, 23.
57. Zehavi, U.; Herchman, M. *Carbohydr. Res.* **1986**, *156*, 371.

58. Zehavi, U.; Herchman, M. Köpper, S. *Carbohydr. Res.* **1992**, *228*, 255.
59. Zehavi, U.; Herchman, M. *Carbohydr. Res.* **1984**, *128*, 160.
60. Köpper, S. *Carbohydr. Res.* **1994**, *265*, 161.
61. Wiemann, T.; Taubken, N.; Zehavi, U. *Carbohydr. Res.* **1994**, *257*, C1.
62. Yoo, D. J.; Greenberg, M. M. *J. Org. Chem.* **1995**, *60*, 3358.
63. Greenberg, M. M.; Gilmore, J. L. *J. Org. Chem.* **1994**, *59*, 746.
64. Greenberg, M. M. *Tetrahedron Lett.* **1993**, *34*, 251.
65. Matray, T. J.; Greenberg, M. M. *J. Am. Chem. Soc.* **1994**, *116*, 6931
66. Greenberg, M. M. *Tetrahedron* **1995**, *51*, 29.
67. Lehninger, A. L. *Biochemistry;* Worth: New York, 1975; p 322.
68. Patchornik, A.; Amit, B.; Woodward, R. B. *J. Am. Chem. Soc.* **1970**, *92*, 6333.
69. Amit, B.; Zehavi, U.; Patchornik, A. *Israel J. Chem.* **1974**, *12*, 103.
70. Ajayaghosh, A.; Pillai, V. N. R. *Tetrahedron Lett.* **1995**, *36*, 777.
71. Baines, F. C.; Collins, P. M.; Farnia, F. *Carbohydr. Res.* **1985**, *136*, 27.
72. Collins, P. M.; Oparaeche, N. N.; Munasinghe, V. R. N. *J. Chem. Soc., Perkin Trans. 1* **1975**, 1700.
73. Collins, P. M.; Munasinghe, V. R. N. *J. Chem. Soc., Chem. Commun.* **1981**, 362.
74. Collins, P. M.; Munasinghe, V. R. N. *J. Chem. Soc., Perkin Trans. 1* **1983**, 921.
75. Collins, P. M.; Eder, H. *J. Chem. Soc., Perkin Trans. 1* **1983**, 927.
76. Collins, P. M.; Munasinghe, V. R. N. *J. Chem. Soc., Perkin Trans. 1* **1983**, 1879.
77. King, J. R.; Allbutt, A. D. *Can. J. Chem.* **1969**, *47*, 1455.
78. King, J. R.; Allbutt, A. D. *Can. J. Chem.* **1970**, *48*, 1754.
79. Deslongchamps, P.; Atlani, P.; Frehel, D.; Malaval, A. *Can. J. Chem.* **1972**, *50*, 2405.
80. Bauch, P. J.; Kershaw, K.; Phillips, G. O. *Carbohydr. Res.* **1972**, *22*, 233.
81. Yamane, Y.; Miyazaki, M.; Sakai, K.; Ajiki, C.; Chang, L. Z.; Yoshimoto, K. *Chem. Pharm. Bull.* **1969**, *17*, 1183.
82. Binkley, R. W.; Koholic, D. J. *J. Org. Chem.* **1979**, *44*, 2047.
83. Binkley, R. W.; Koholic, D. J. *J. Carbohydr. Chem.* **1984**, *3*, 85.
84. Binkley, R. W.; Abadulaziz, M. A. *J. Org. Chem.* **1987**, *52*, 4713.
85. Howarth, G. B.; Lance, D. G.; Szarek, W. A.; Jones, J. K. N. *J. Chem. Soc., Chem. Commun.* **1968**, 1349.
86. Howarth, G. B.; Lance, D. G.; Szarek, W. A.; Jones, J. K. N. *Can. J. Chem.* **1969**, *47*, 81.

17

Azo, Diazo, Azido, and Related Compounds

Formation of Divalent Intermediates

Azo and diazo compounds, along with salts of tosylhydrazones, form a highly photochemically reactive group of unsaturated, nitrogen-containing compounds. Although all of these compounds are capable of several different types of reaction, their principal photochemical process, where carbohydrates are concerned, is generation of an intermediate with a divalent carbon atom (i.e., a carbene, Scheme 17.1). This favored formation of a carbene is partially attributable to the basic reactivity of these compounds; that is, all are capable of carbene formation. Carbene formation also is favored by the particular compounds investigated because these compounds form carbenes in preference to other types of reaction. For example, the majority of research conducted on azo compounds has been on reactions of the diazirine **1**, a compound that readily loses nitrogen upon photolysis to generate the zwitterionic carbene **2** (Scheme 17.2).[1] If, however, a structurally different type of azo compound, such as one containing a pyrazoline ring, is irradiated, loss of nitrogen produces a 1,3-diradical (Scheme 17.3).[2]

Although the basic reactivity of a carbene is determined by its divalent carbon atom, the multiplicity of the carbene also has an effect upon product formation.[3] Reactions of a singlet carbene are those of a zwitterionic intermediate, and reactions of a triplet carbene are those of a species with two radical centers on the same carbon atom (Figure 17.1). (Although different hybridizations occur in other carbenes, the divalent carbon atom in the carbene shown in Figure 17.1 is pictured as being sp^2 hybridized because this type of hybridization has been invoked in explaining the reactions of the zwitterionic carbene **2**.[1])

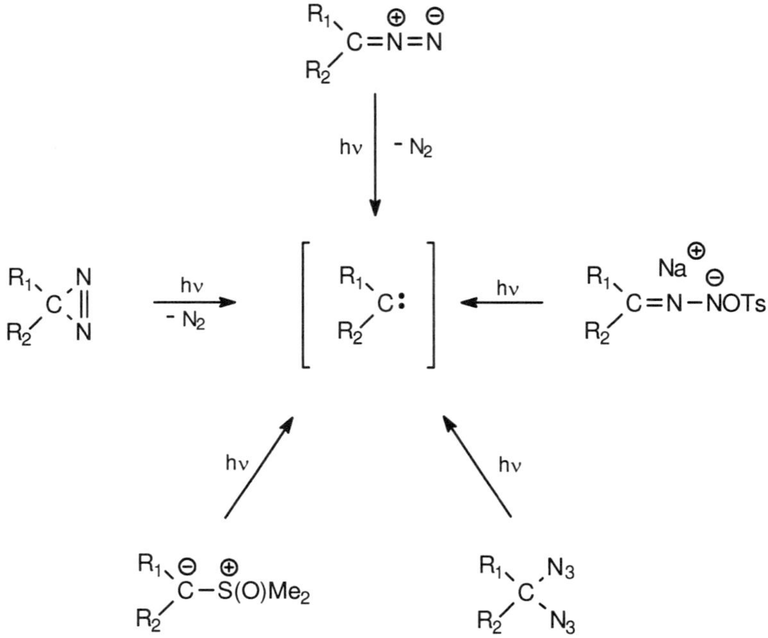

Scheme 17.1

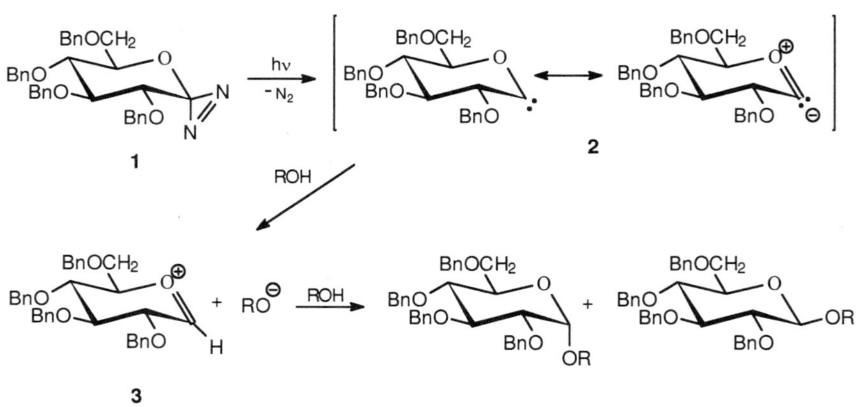

thermal reaction R = $(CF_3)_2CHOH$	17%	60%
thermal reaction R = $(CH_3)_2CHOH$	20%	20%
photochemical reaction R = $(CH_3)_2CHOH$ (- 70 °C)	36%	36%

Scheme 17.2

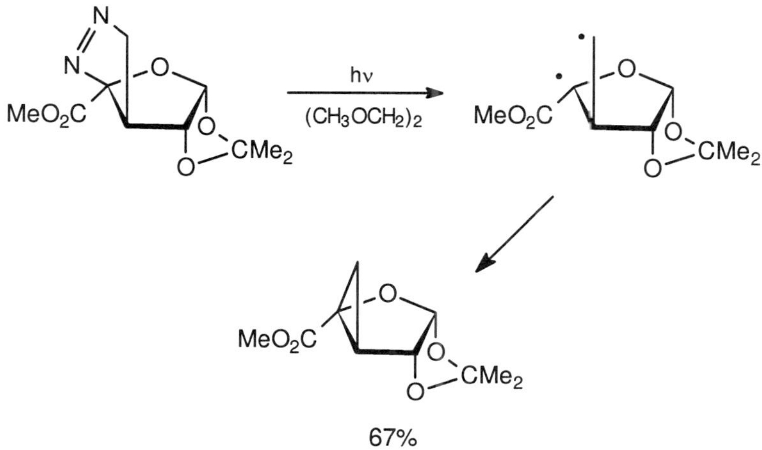

Scheme 17.3

Figure 17.1 Carbene singlet and triplet states

*Azo Compounds**

Diazirines

Vasella and co-workers have studied the photochemical and thermal reactions of glycosylidene diazirines as starting materials in glycoside synthesis.[1,5–14] They have proposed a process in which compound **1** (and related diazirines) loses nitrogen to generate the glycosylidene carbene **2**, an intermediate that (a) adds to unsaturated compounds to give products with three-membered rings (eq 1)[5] and (b) reacts with alcohols to form glycosides (Scheme 17.2).

Since the sp^2 hybrid orbital on the divalent carbon atom in **2** (Scheme 17.2) contains an electron pair, C-1 is a basic, nucleophilic center. Briner and Vasella have argued that the reaction of an alcohol with this carbene (**2**) occurs in two steps.[1,7] The first of these is protonation to give the 1-glycosyl cation, and the second is glycoside formation by reaction of this cation with an alcohol (Scheme 17.2). Alcohols fall into two categories when forming glycosides by

Aliphatic azo compounds have characteristic (n,π) absorption in the region from 320–380 nm.[4]

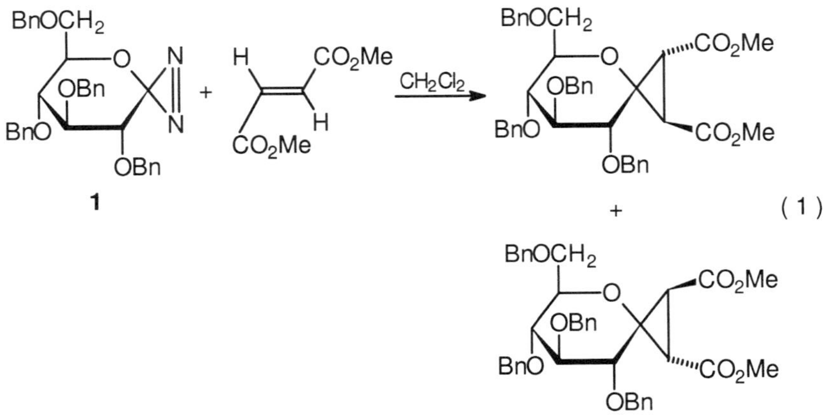

Equation 1

reaction with **2**. A strongly acidic (weakly nucleophilic) alcohol such as $(CF_3)_2CHOH$ reacts stereoselectively to give a 77% yield of anomeric glycosides (α:β = 22:78), but a weakly acidic (more nucleophilic) alcohol such as $(CH_3)_2CHOH$ reacts nonstereoselectively to give a lower glycoside yield (39%, α:β = 50:50). Both of these reactions occur at room temperature without irradiation. Briner and Vasella attributed the stereoselectivity associated with reaction of weakly nucleophilic alcohols to participation by the neighboring 2-*O*-benzyloxy group in directing the incoming alcohol to the β-face of the glycosyl cation (**2**).[1,7] In a more nucleophilic solvent the 2-*O*-benzyl group cannot compete with the solvent in coordinating with the cationic center and, consequently, the stereoselectivity of the reaction disappears.

Photolyses of **1** generally are conducted well below room temperature where product yields from reaction of the glycosylidene carbene **2** with less acidic alcohols are good. [The yield from reaction of **2** with $(CH_3)_2CHOH$ at –70 °C is 71% (Scheme 17.2).] Even under these conditions, however, reaction remains nonstereoselective (α:β = 50:50). If a nucleophilic solvent such as tetrahydrofuran (THF) is introduced into the photochemical reaction mixture, stereoselectivity dramatically increases (α:β = 8:92). Vasella has argued that the high yield of β-anomer in the presence of THF is the result of formation of the tetrahydrofuranylium ion **4**, an intermediate that preferentially forms as an α-anomer from which THF is displaced by an alcohol to give a ß-glycoside (eq 2).[1]

Vasella and co-workers studied the photochemical and thermal reactions of **1** with a number of diols[8,9] and triols[10,11] and observed that regioselective reaction occurs when there is a relatively large difference in acidity among the hydroxyl groups.[1,8–11] When differences exist, a more acidic hydroxyl group transfers a proton to the basic center (C-1) of the zwitterionic carbene (**2**) more rapidly than a less acidic hydroxyl group. Regioselective reaction results

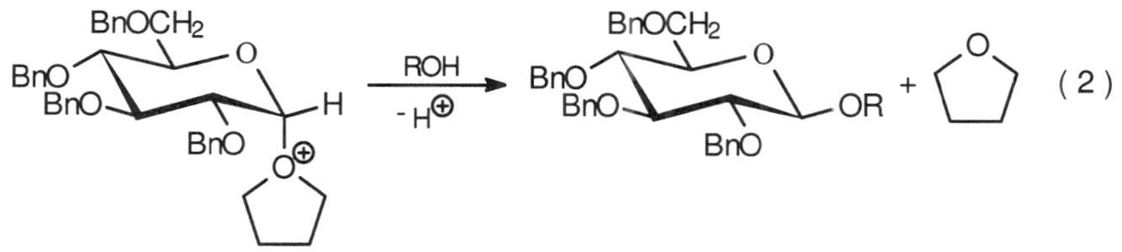

4

Equation 2

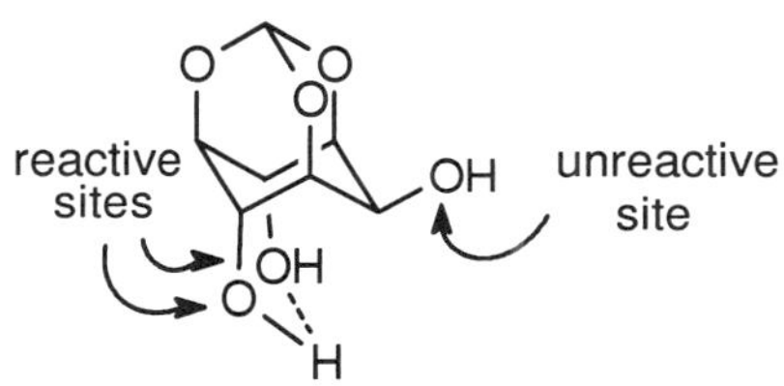

5

Figure 17.2 Importance of hydrogen bonding to hydroxyl group reactivity

because proton transfer is the first step in glycoside formation. Since differences in acidity correlate with differences in intramolecular hydrogen bonding, Vasella has been able to relate regioselectivity to the strength of internal hydrogen bonding.[1] For example, he observed that the strongest intramolecular hydrogen bonding in compound **5** existed between 1,3-diaxial hydroxyl groups and that, consequently, reaction occurred exclusively at an axial hydroxyl group (Figure 17.2).

With one exception, all of the reports of diazirine photolyses have involved formation of glycosylidene carbenes. This exception comes from the work of Lehmann and Petry, who studied diazirines as possible photoaffinity labels.[15] These investigators found that each diazirine photolyzed gave a divalent intermediate that experienced a hydrogen migration to form a carbonyl compound. An example of this type of reaction is shown in Scheme 17.4.

Pyrazolines

Carbohydrates in which the azo group is part of a pyrazoline ring eliminate molecular nitrogen upon photolysis to produce diradicals that cyclize to give three-membered rings. Bonjouklian and Ganem have taken advantage of this type of reaction in the early stages of macrolide antibiotic synthesis by photochemically generating the cyclopropane intermediate shown in Scheme 17.3.[2] This type of reaction has been used to introduce cyclopropane rings into other carbohydrates (Table 17.1).

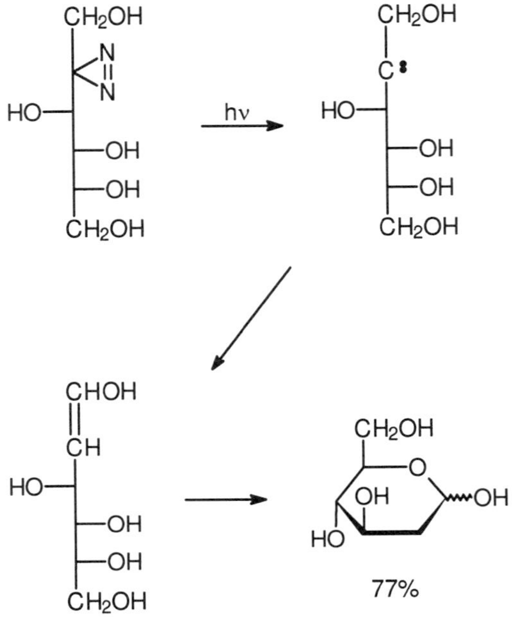

Scheme 17.4

Tosylhydrazones

Mangholz and Vasella have irradiated salts of tosyl hydrazones as an alternative procedure to the reaction of diazirines for generating zwitterionic carbenes.[21] The uses of this reaction in carbohydrate chemistry all involve forming a glycosylidene carbene (Scheme 17.1).[21–23] Although salts of tosylhydrazones are easier to prepare than diazirines, they give lower product yields upon photolysis. At least part of the reason for lower yields is that the *p*-toluenesulfinate ion liberated during carbene formation reacts with the carbene to form a glycosyl sulfone.[1,23]

Sulfonium Ylides

Kunieda and Witkop found that irradiation of sulfonium ylides produces divalent intermediates.[24,25] The reported examples of this reaction involve pyrimidine nucleoside sulfonium ylides that experience carbene insertion into a C–H (or O–H) bond to form cyclonucleosides. An example of this type of reaction is shown in eq 3.

Table 17.1. Photochemical Reactions of Cyclic Azo Compounds

Reactant	*Product*	*Percent Yield*	*Ref.*
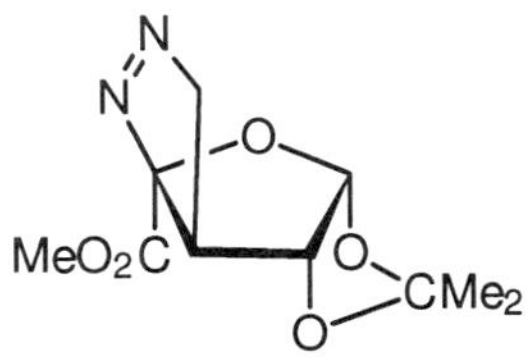	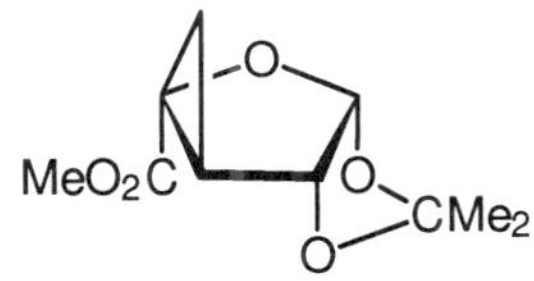	67	2
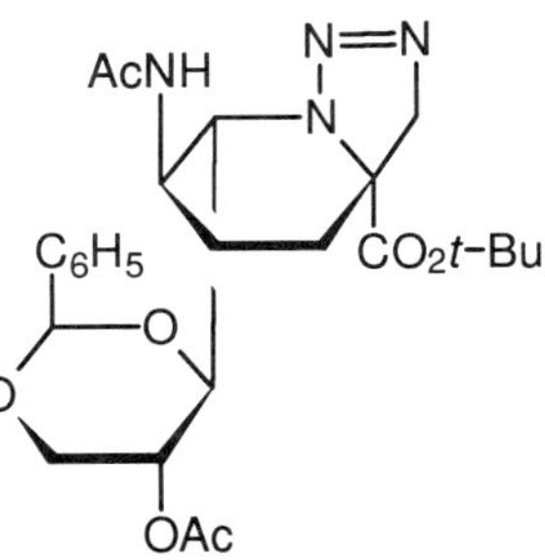	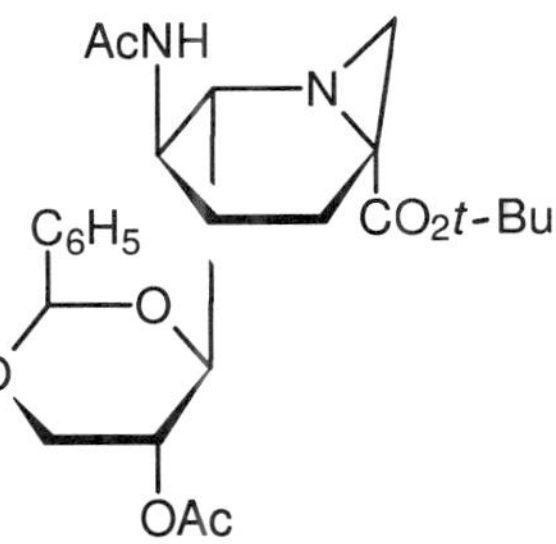	40[a]	16
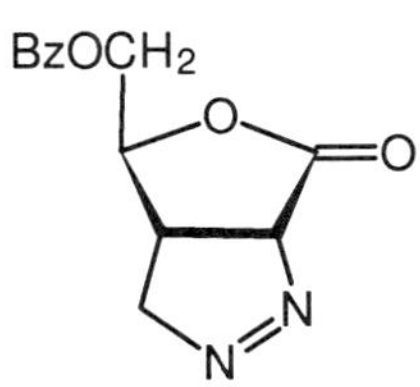	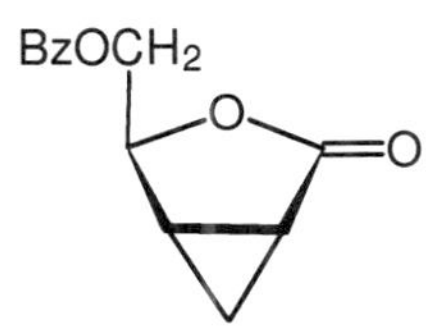	83	17
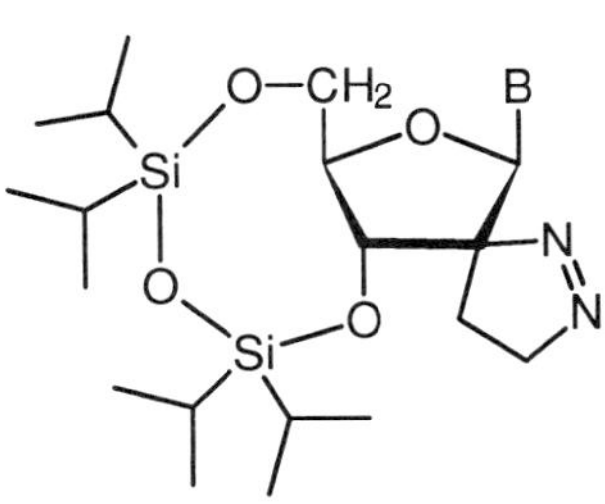	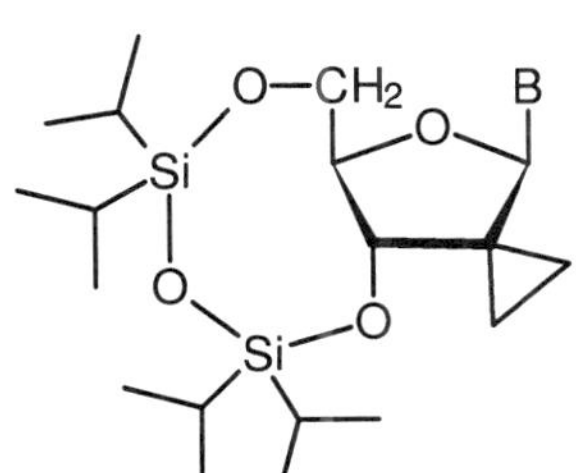	92[a]	18

B = 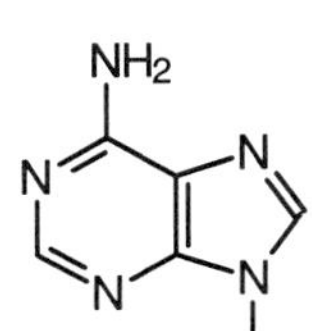

Continued on next page

Table 17.1. Photochemical Reactions of Cyclic Azo Compounds—*Continued*

Reactant	*Product*	*Percent Yield*	*Ref.*
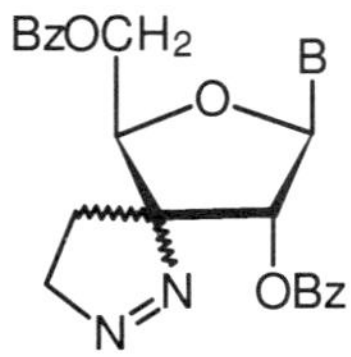	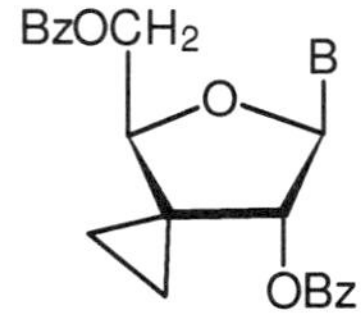	[a]	19
B = 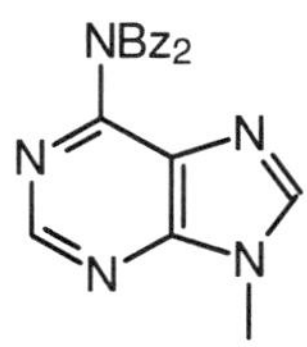 90%	B = 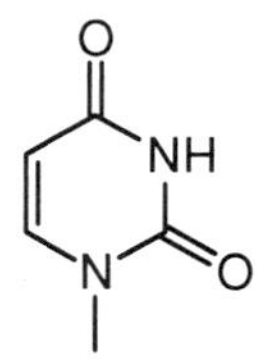 80%		
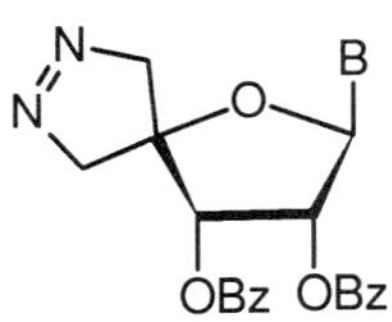	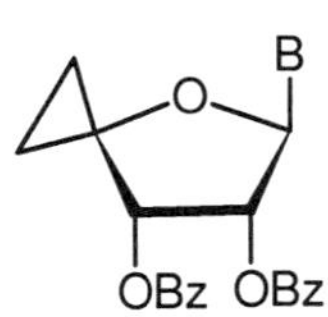	[a]	19
B = 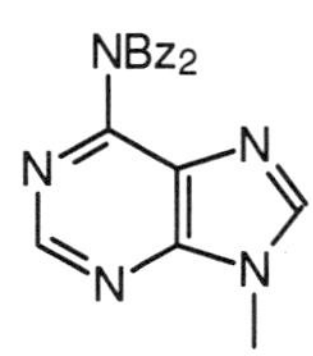	45%		
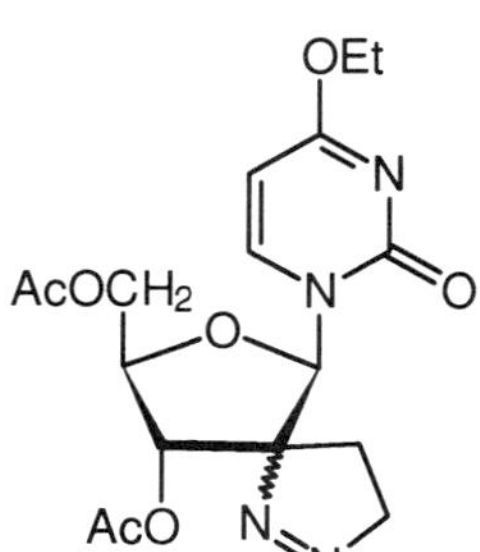	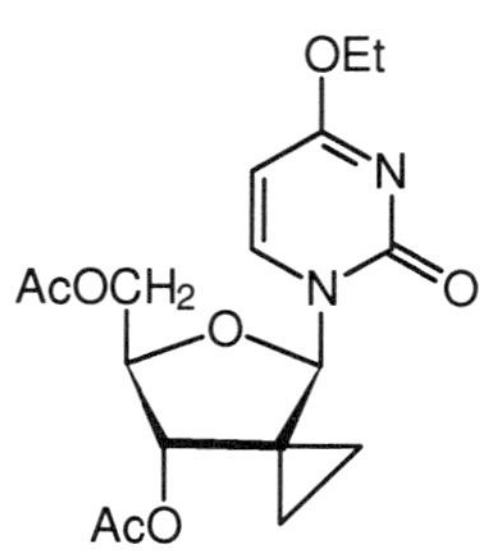	65	20

[a]Benzophenone was added as a sensitizer.

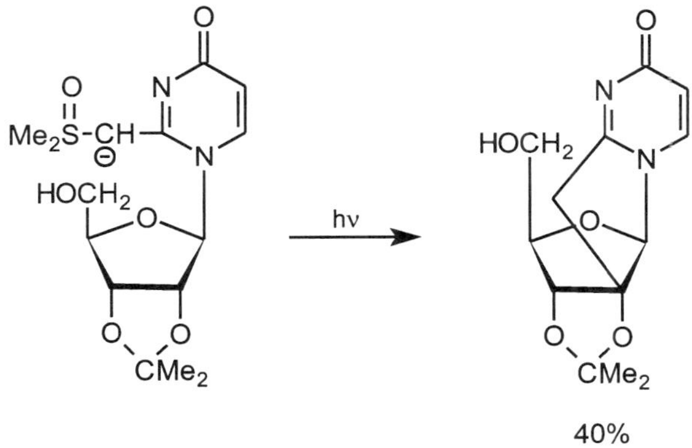

(3)

Equation 3

Diazoketones

Photolysis of an α-diazoketone generates a carbene next to an carbonyl group. This type of carbene typically undergoes intramolecular rearrangement to form a ketene (the Wolff rearrangement).[26,27] If the reaction solvent is an alcohol, the ketene combines with the solvent to form the corresponding ester (Scheme 17.5). Since the photochemical Wolff rearrangement is an intramolecular process, it can produce a stable, ground-state product directly only from a singlet ketocarbene. Singlet carbenes are implicated in the reaction of

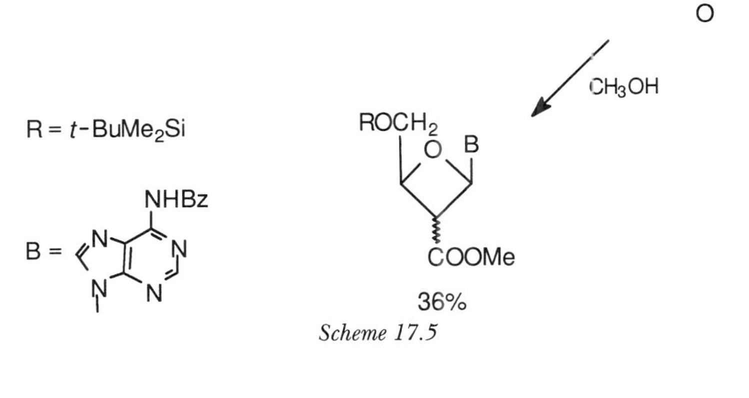

Scheme 17.5

most carbohydrate diazoketones because the Wolff rearrangement is a common process (Table 17.2). Triplet-sensitized decomposition of diazoketones generates carbenes that do not undergo Wolff rearrangement; rather, they experience typical radical reactions, such as hydrogen abstraction.[28] Although this type of reaction is rare in carbohydrate photochemistry, an example is shown in eq 4.[32]

Azides

Photolysis of an azide also generates a divalent intermediate; however, in this case, the divalent atom is nitrogen (a nitrene). The most common reaction for a nitrene, where carbohydrates are concerned, is hydrogen migration to form an imine (Scheme 17.6). This imine can be hydrolyzed to the corresponding aldehyde or ketone under mild conditions, such as treatment with aqueous acetic acid (eq 5).[34] The imine is rarely isolated not only because it is easily hydrolyzed but also because, at least in some cases, the imine does not have a monomeric structure.[35] The variety of carbohydrate azides that have been converted into carbonyl compounds includes cyclic and acyclic structures and ranges from simple sugars (Table 17.3) to nucleosides (Table 17.4) and polymeric carbohydrates (eq 6).[53–58] The vast majority of compounds studied have

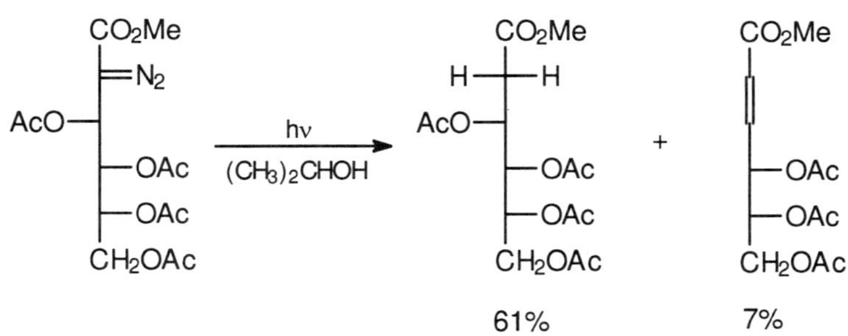

(4)

Equation 4

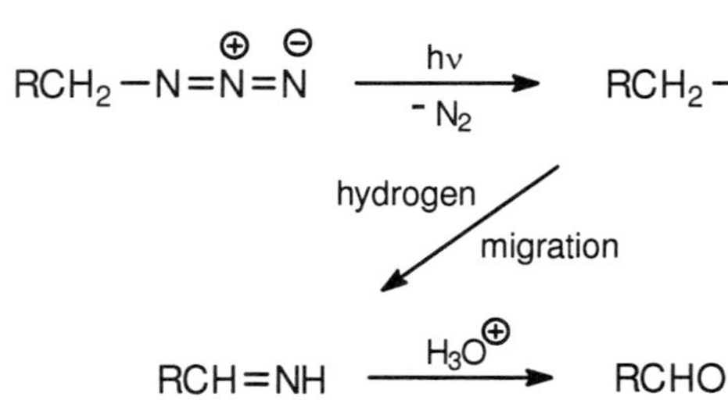

Scheme 17.6

Table 17.2. Photochemical Reactions of Diazo Compounds: The Wolff Rearrangement

Reactant	*Product and Percent Yield*		*Ref.*
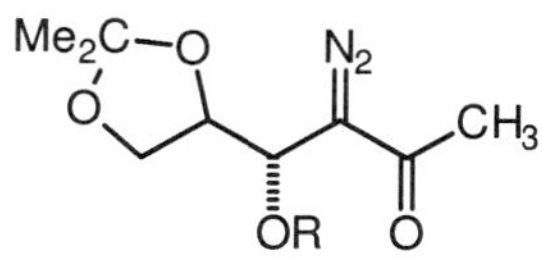 + CH_3OH	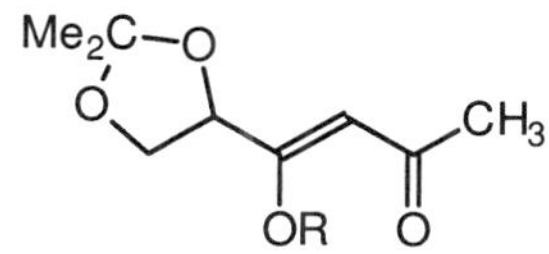 **A** + 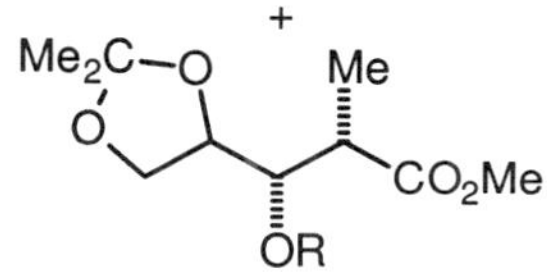 **B**		29
	A	B	
R = H	100	-	
R = Ac	15	85	
R = $C(O)NHC_6H_5$	35	65	
R = Me	70	30	
R = Me_3Si	95	5	
R = t–$BuMe_2Si$	45	55	
R = t–$Bu(C_6H_5)_2Si$	55	45	
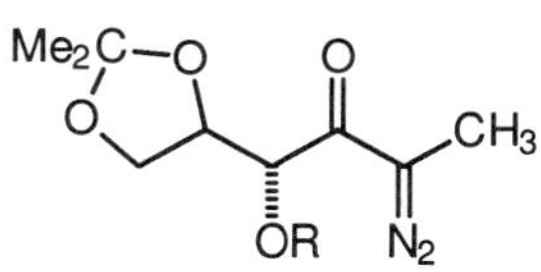 + CH_3OH	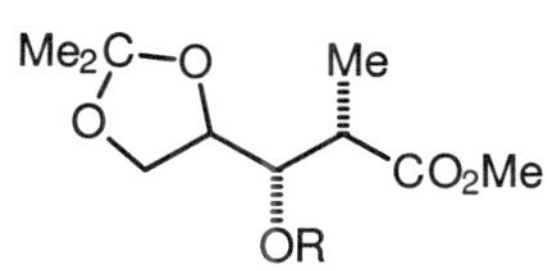 **C** + 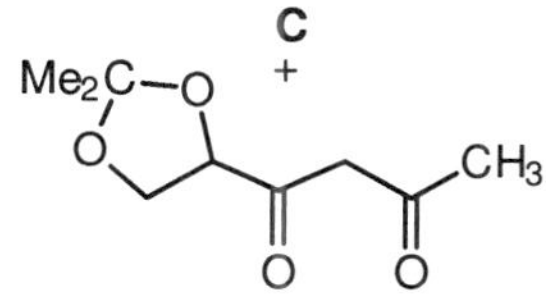 **D**		30
	C	D	
R = t–$BuMe_2Si$	90	-	
R = H	55a	45	

Continued on next page

Table 17.2. Photochemical Reactions of Diazo Compounds: The Wolff Rearrangement—*Continued*

Reactant	*Product and Percent Yield*	*Ref.*
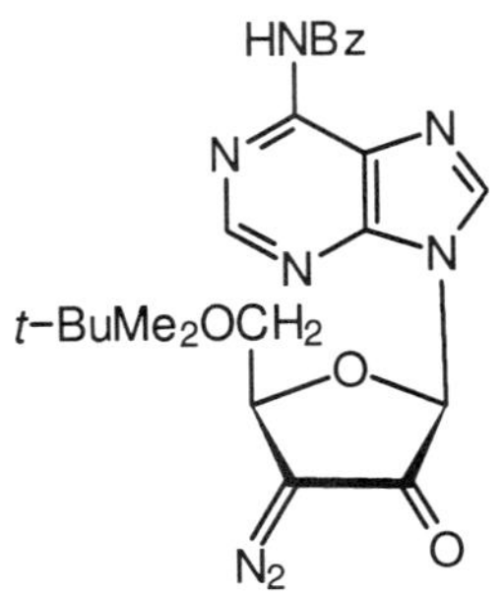	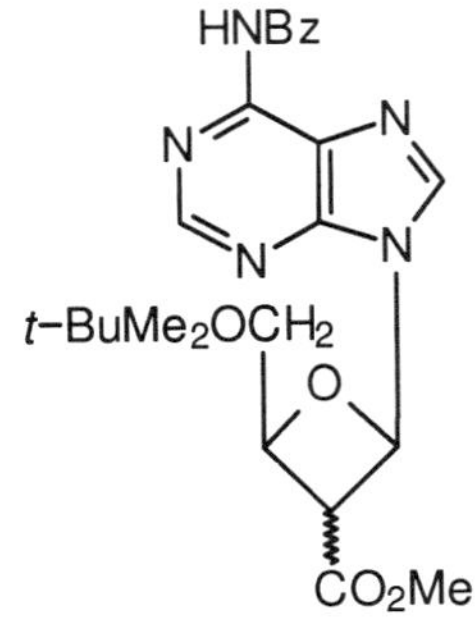 36%[c]	31
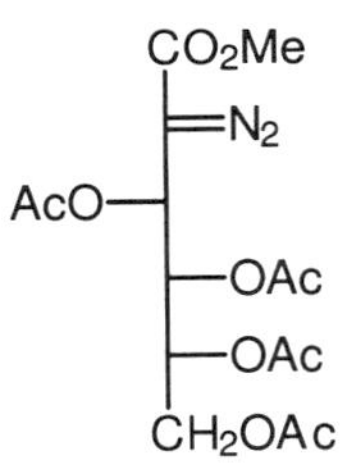	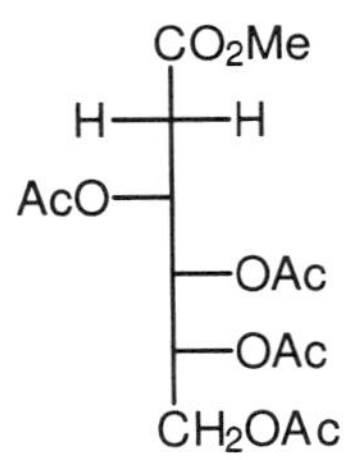 61%[b] 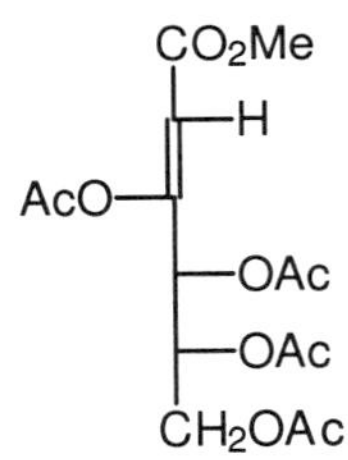 7%[b] major[c]	32
$RCH_2-C(=N_2)CO_2Et$ R = 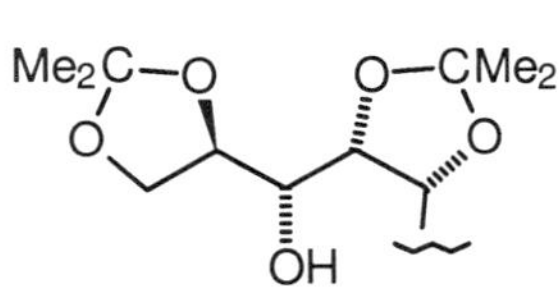	$RCH{=}CHCO_2Et$	33

[a]A mixture of C-2 epimers was formed.
[b]The reaction was conducted in 2-propanol.
[c]The reaction was conducted in methanol.

Table 17.3. Photochemical Reactions of Azides[a]

Reactant	*Product*	*Percent Yield*	*Ref.*
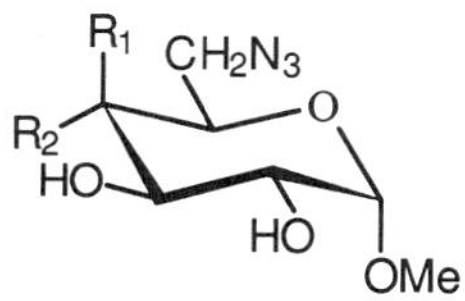	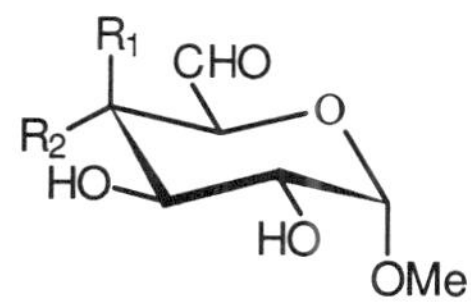		
R_1 = H, R_2 = OH		90^b	34
R_1 = OH, R_2 = H		100^b	
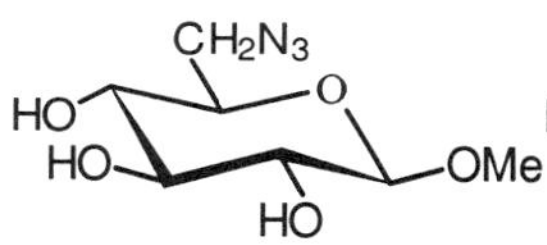	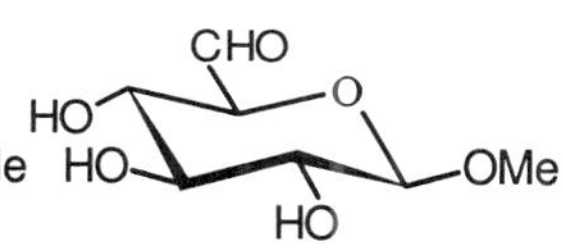	75-80^b	34
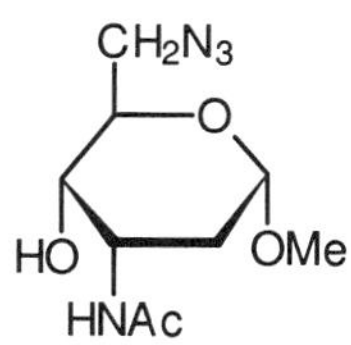	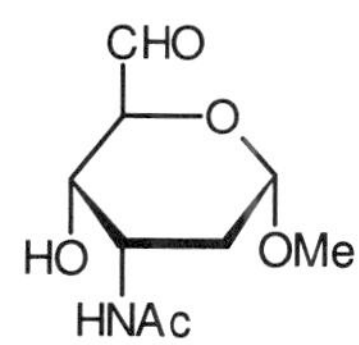	83	35
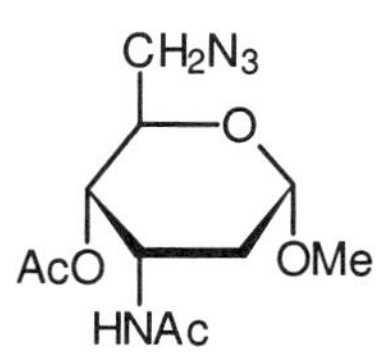	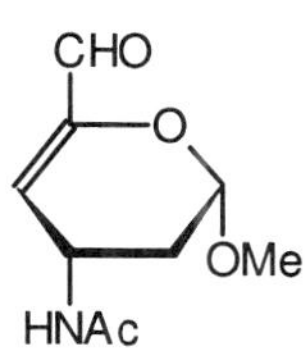	57	35
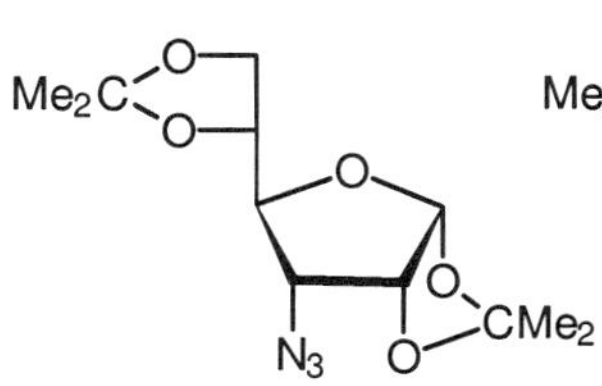	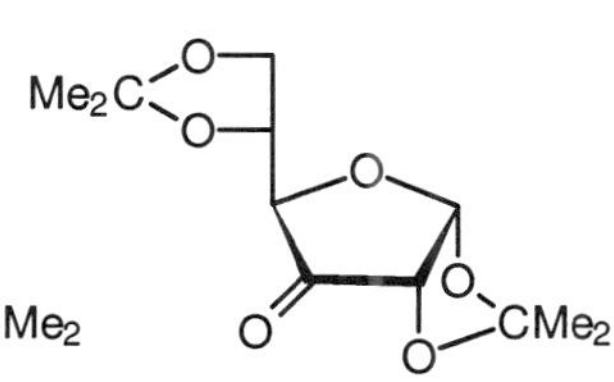	34	36
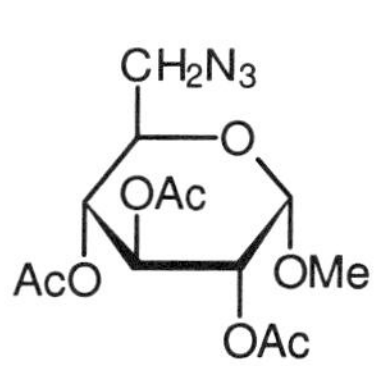	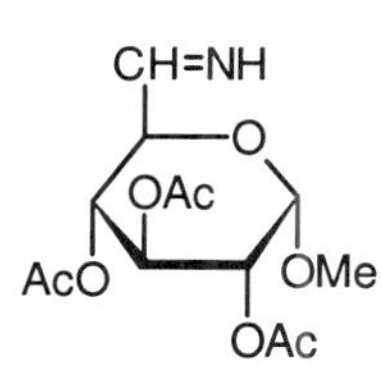	33 50	37 38

Continued on next page

Table 17.3. Photochemical Reactions of Azides[a]—*Continued*

Reactant	*Product*	*Percent Yield*	*Ref.*
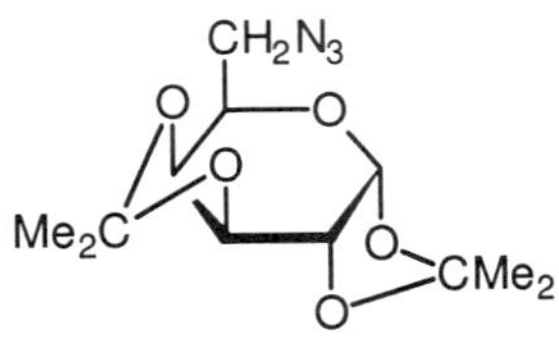	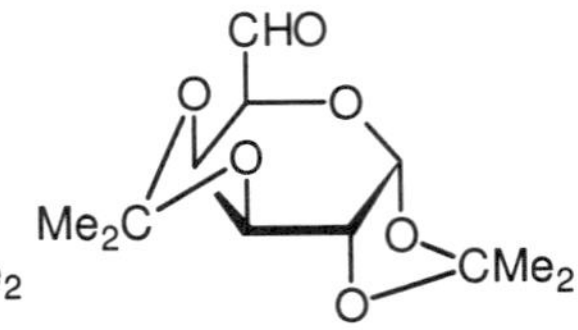	39[c]	39
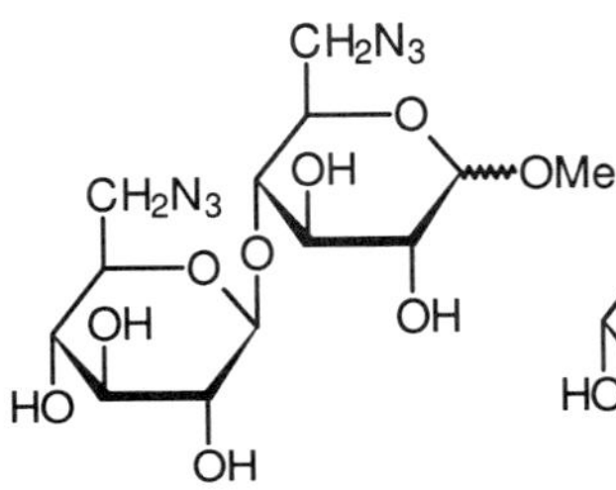	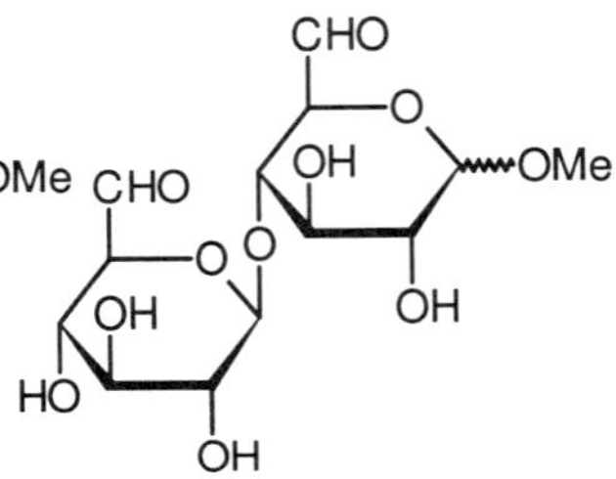	–	40
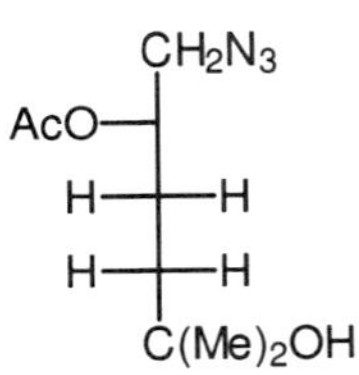	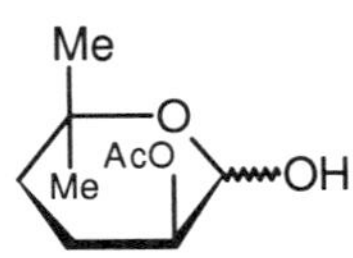	51	41
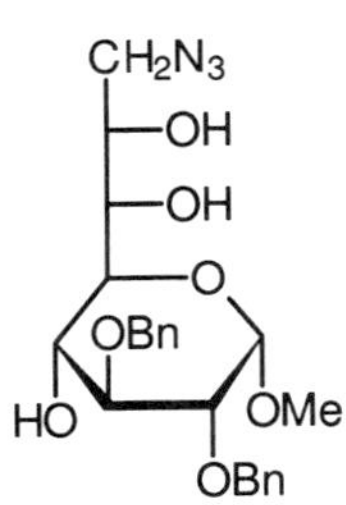	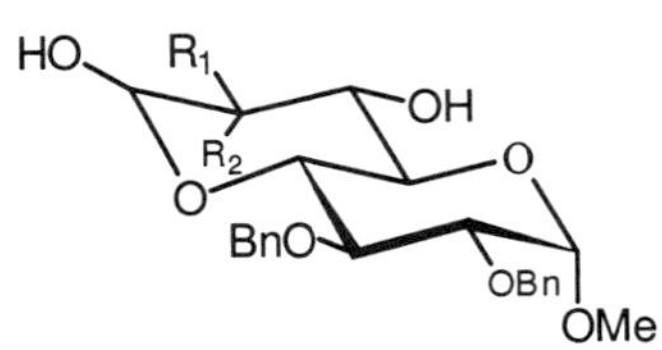		
	R_1 = H, R_2 = OH	33	42
	R_1 = OH, R_2 = H	57	43
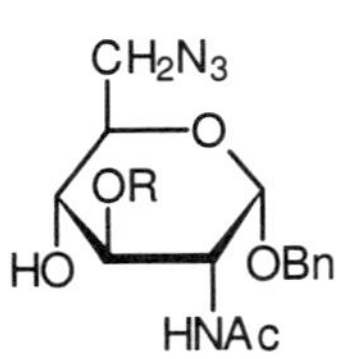	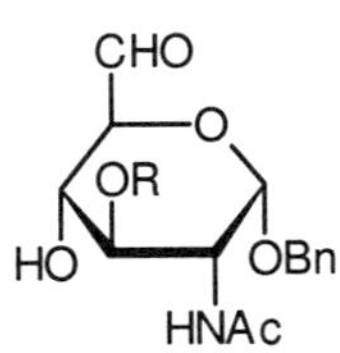	76	44

R = CH_3CHCO-L-Ala-D-IsoGln($OCH_2C_6H_5$)

Continued on next page

Table 17.3. Photochemical Reactions of Azides[a]—*Continued*

Reactant	*Product*	*Percent Yield*	*Ref.*
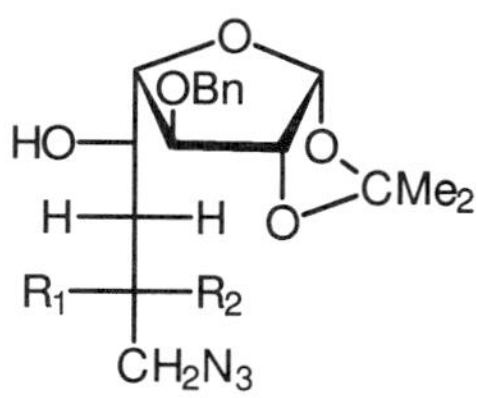	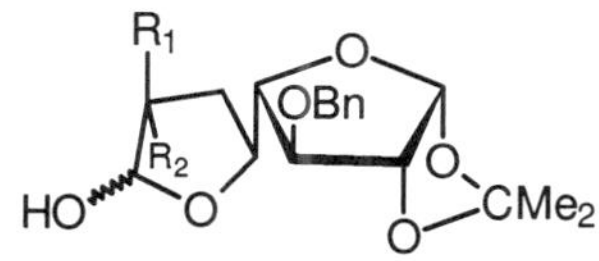		
	R_1 = OH, R_2 = H	16	45
	R_1 = H, R_2 = OH	25	45
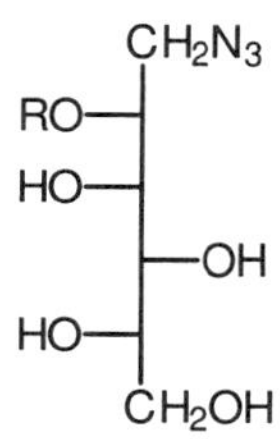 R = 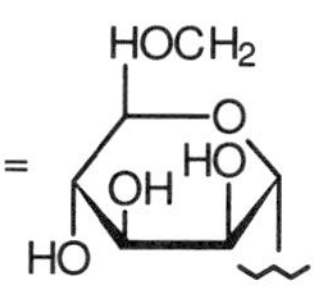	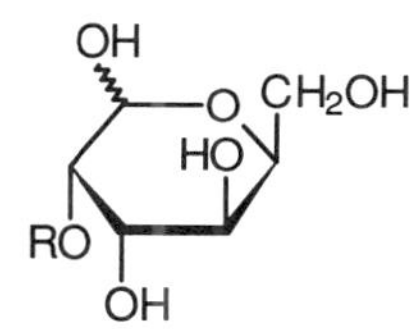	48	46
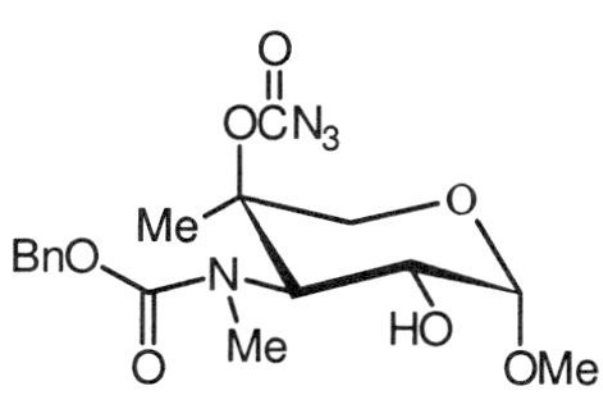	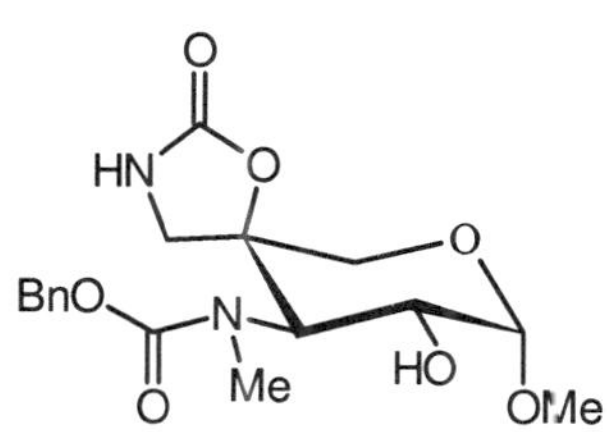	12	47
	+ 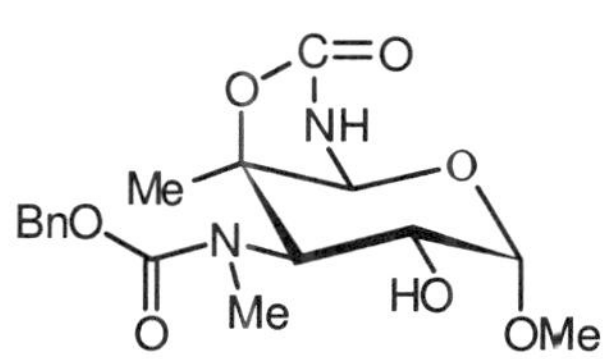	23	
	+ 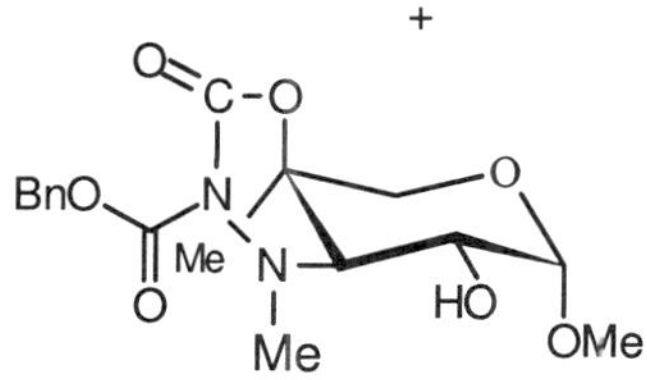	22	

Continued on next page

Table 17.3. Photochemical Reactions of Azides[a]—*Continued*

Reactant	*Product*	*Percent Yield*	*Ref.*
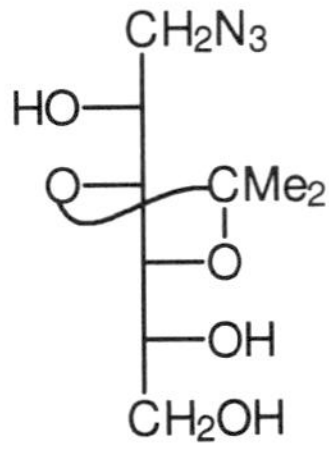	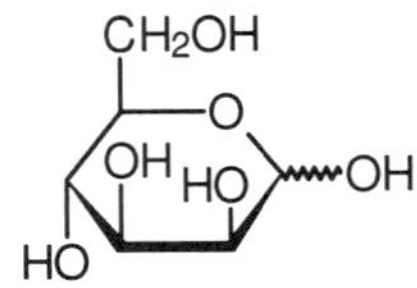	55[d]	48
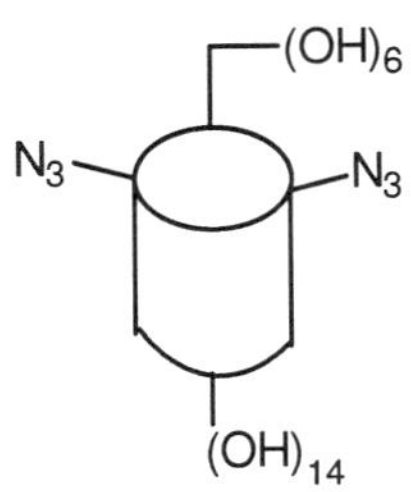 6A,6D-diazo-6A,6D-dideoxy-β-cyclodextrin	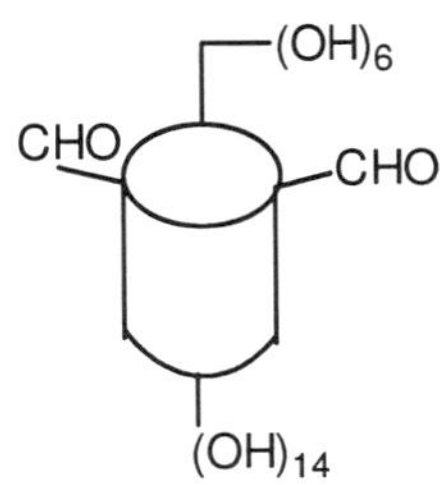	61[e]	49
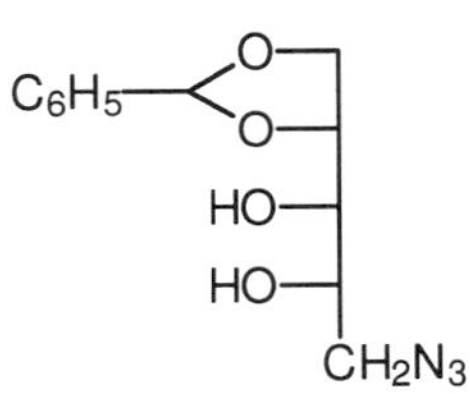	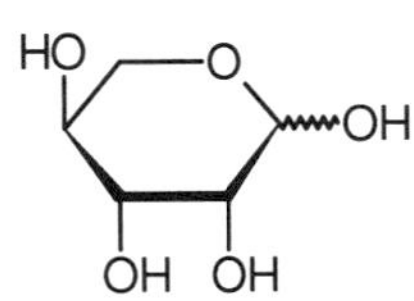	82[d]	50
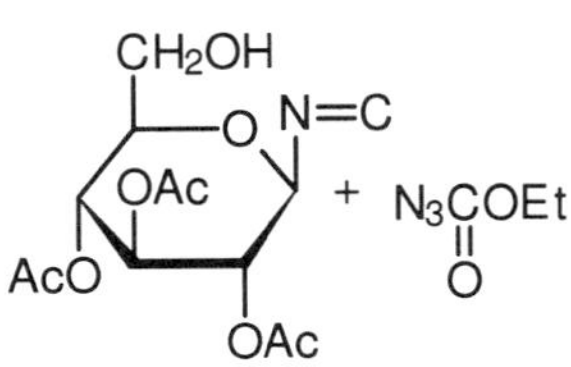	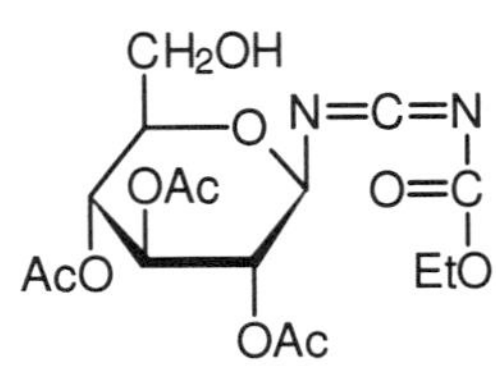	75	51

Continued on next page

Table 17.3. Photochemical Reactions of Azides[a]—*Continued*

Reactant	*Product*	*Percent Yield*	*Ref.*
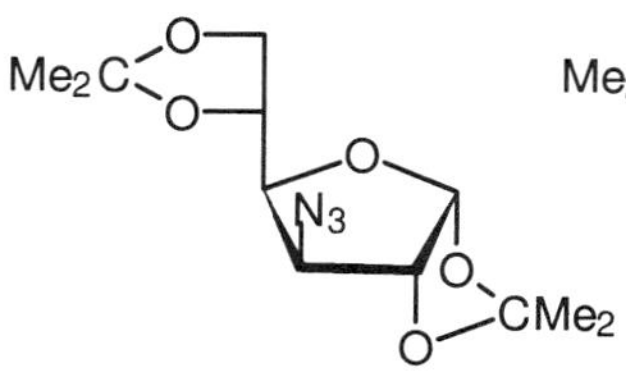	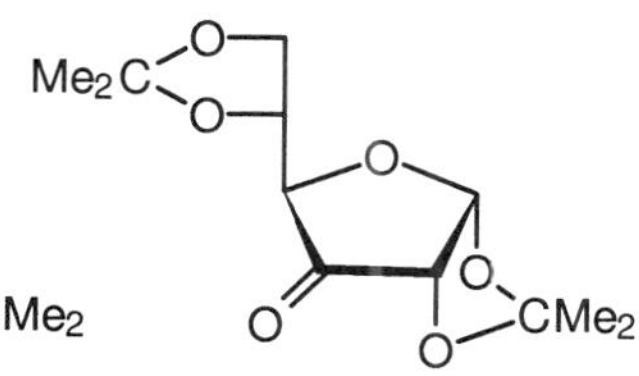	15	52

[a]Unless an imine photoproduct is shown, the product was isolated after imine hydrolysis.

[b]The product was isolated as a hydrate.

[c]Reaction of an α-cyclodextrin derative also was reported.

[d]The acetal was hydrolyzed during product isolation.

[e]The product was isolated as the dicarboxylic acid. The methylated cyclodextrin gave a 68% product yield.

Table 17.4. Photochemical Reactions of Nucleoside Azides

Reactant	*Product*	*Percent Yield*	*Ref.*
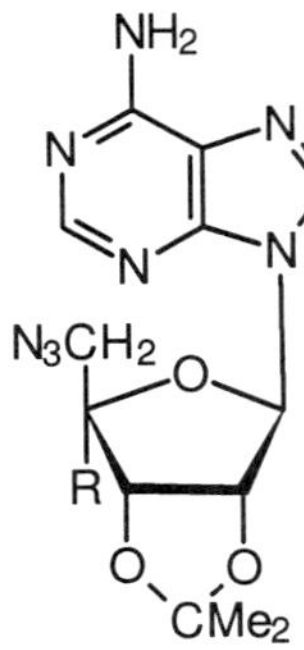	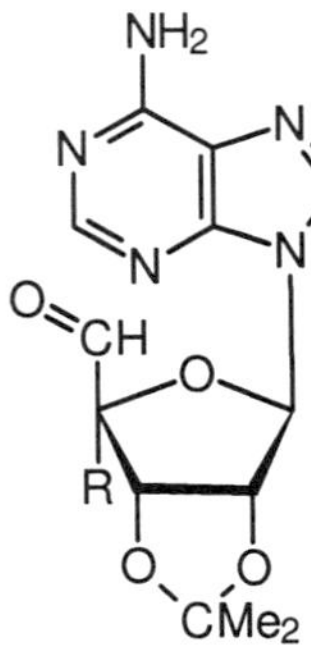		
R = H		54(58)[a]	59
R = F		b	60
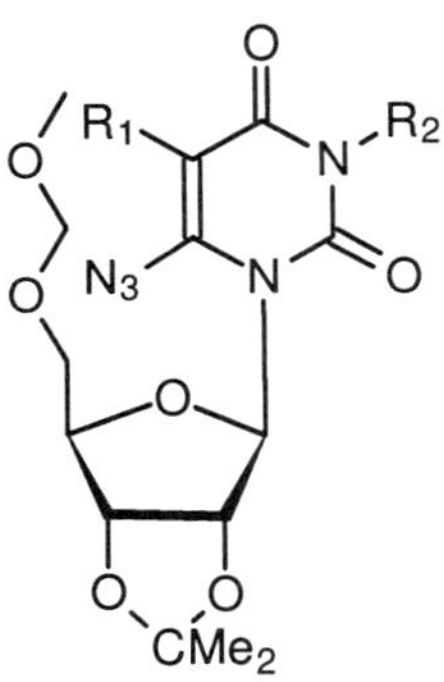	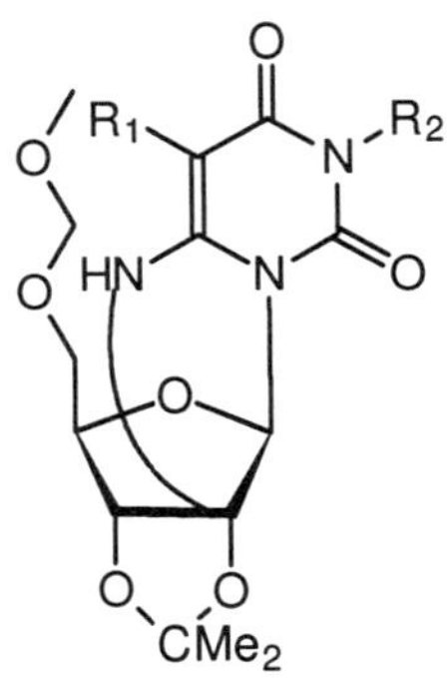		61
$R_1 = R_2 = H$		79	
$R_1 = H, R_2 = Me$		60	
$R_1 = Me, R_2 = H$		86	

[a]The yield was 58% when the 2',3'-*O*-benzylidene compound was irradiated.

[b]The yield was not reported.

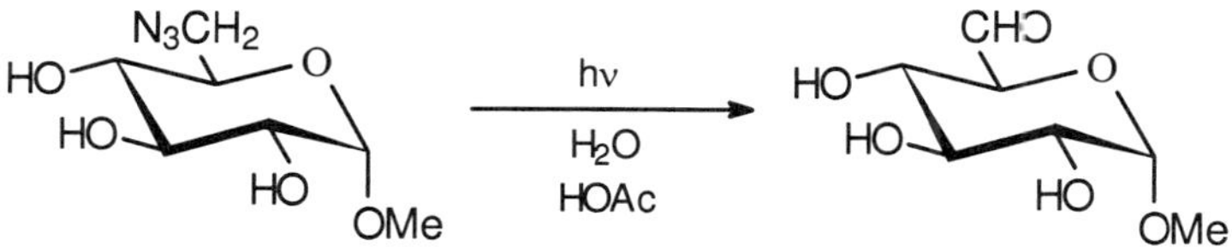

(5)

Equation 5

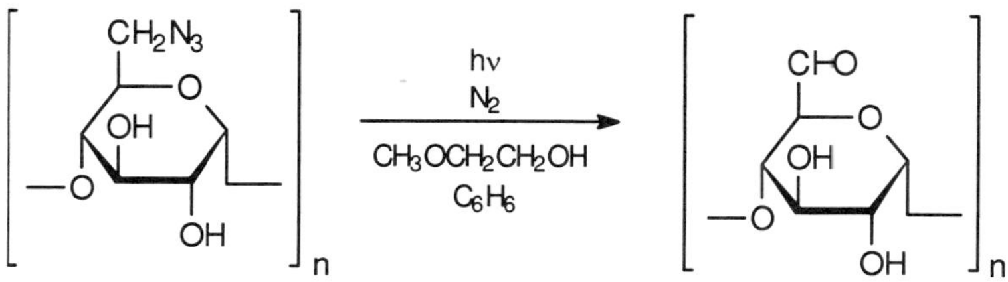

(6)

Equation 6

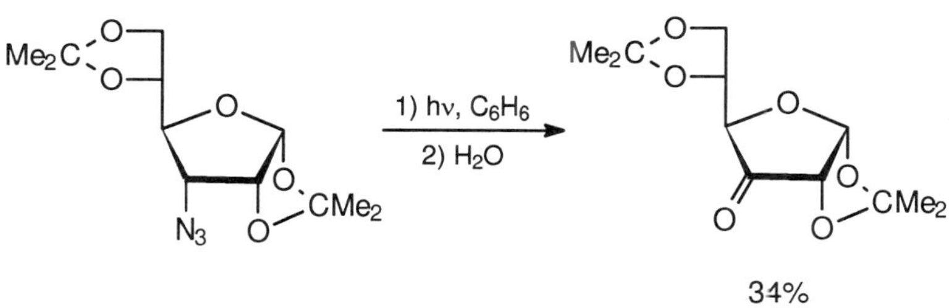

(7)

Equation 7

been primary azides, although there are examples of irradiation of secondary azido compounds (eq 7).[36]

Having an azido group attached to an anomeric carbon atom opens additional reaction pathways (Table 17.5). Szarek and co-workers found that photolysis of the glycosyl azide **6** produced an aldose with one less carbon atom than the number present in the starting material.[62] They suggested that the iminolactone **7** was an intermediate in aldose formation (Scheme 17.7). These investigators also determined that some glycosyl azides (e.g., **8**) formed unstable photoproducts that reverted to the starting material upon standing (Scheme 17.8).[62,63] They proposed the structure **9** for the unstable photoproduct generated from **8**, but were unable to determine why structurally similar compounds, such as **6** and **8**, should exhibit such a difference in reactivity. Further study did show that a necessary, but not sufficient, condition for formation of an unstable photoproduct such as **9** (Scheme 17.7) was the presence of an unprotected hydroxyl group on C-2.

When an anomeric azide does not have a hydrogen atom attached to C-1, hydrogen migration is replaced by other migration reactions; that is, one of the groups attached to the anomeric carbon atom migrates to the divalent

Table 17.5. Photochemical Reactions of Anomeric Azides

Reactant	*Product and Percent Yield*	*Ref.*
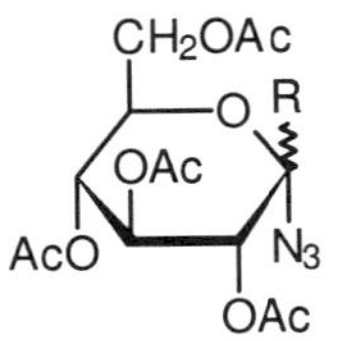	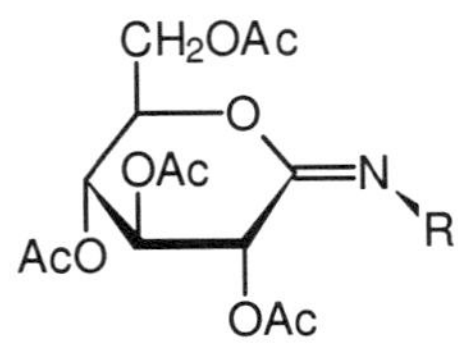 **A** + 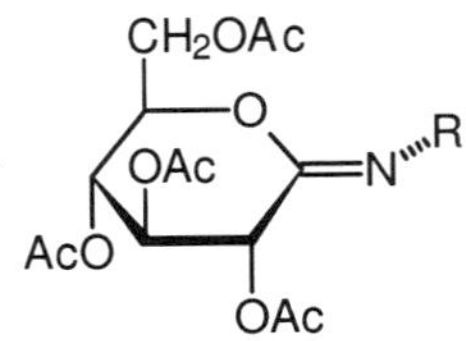 **B** + 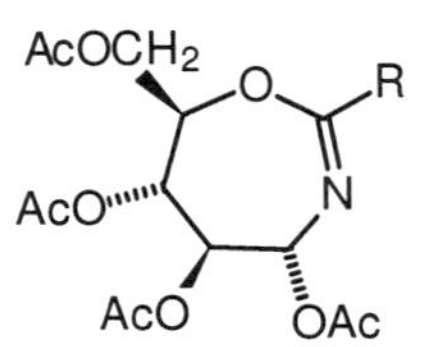 **C** + 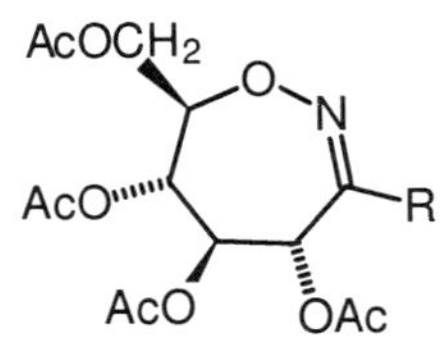 **D**	64, 65

	A	B	C	D
R = α-OMe	10%-15%	30%-45%	0%-35%	0%-5%
R = β-OMe	0%-5%	5%-15%	0%-10%	50%-70%
R = α-CN	-	-	50%	-

Reactant	*Product and Percent Yield*	*Ref.*
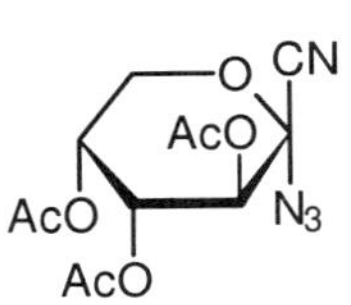	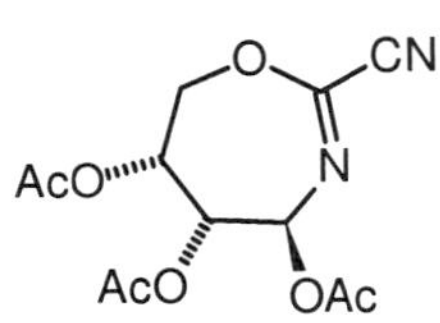 58%	65
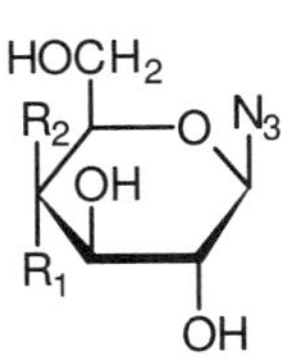	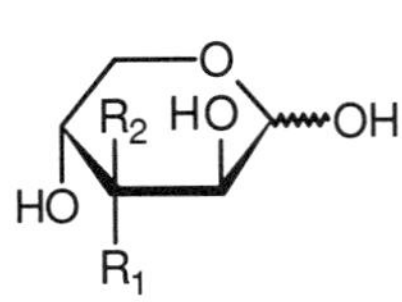	62, 63
R_1 = OH, R_2 = H	47%	
R_1 = H, R_2 = OH	60%	

Continued on next page

Table 17.5. Photochemical Reactions of Anomeric Azides—*Continued*

Reactant	*Product and Percent Yield*	*Ref.*
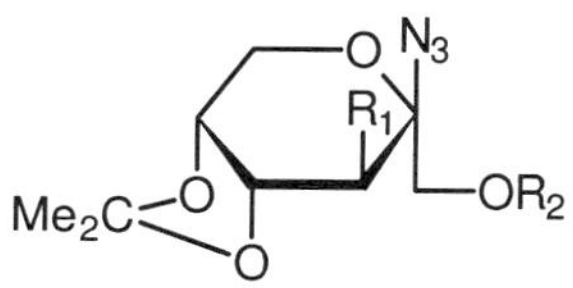	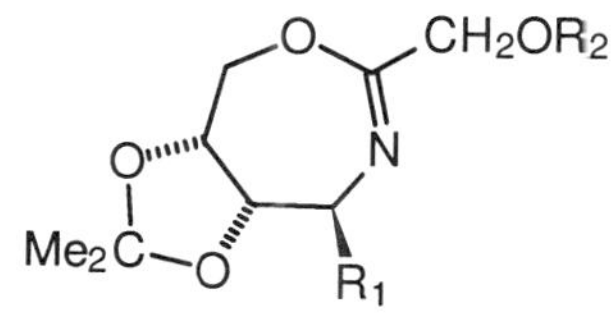 **E** + 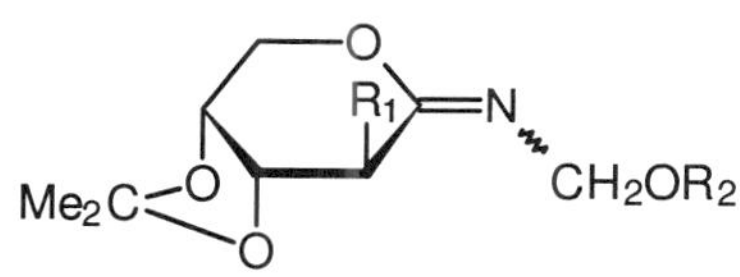 **F**	66

R₁	**R₂**	**E**	**F**
OMe	Ac (α)	72%	8%
OMe	Ac (β)	59%	22%
OMe	Ms	35%	6%
H	Ac (α)	50%	10%
H	Ac (β)	50%	-

Reactant	*Product and Percent Yield*	*Ref.*
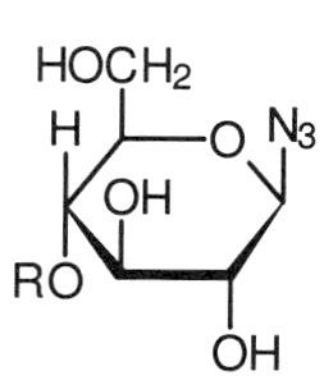	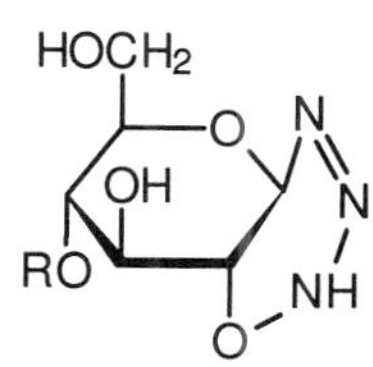	62, 63[a]

R = 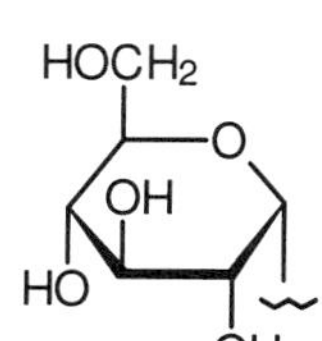

Continued on next page

Table 17.5. Photochemical Reactions of Anomeric Azides—*Continued*

Reactant	*Product and Percent Yield*	*Ref.*
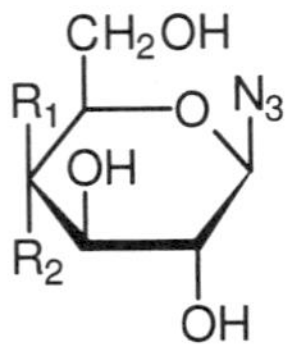	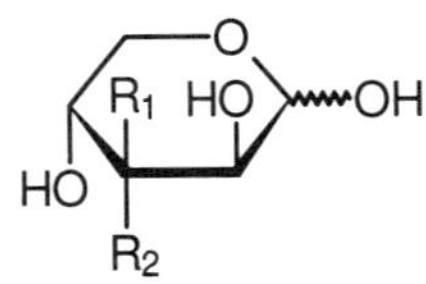	
R_1 = OH, R_2 = H	65%	62, 63
R_1 = H, R_2 = 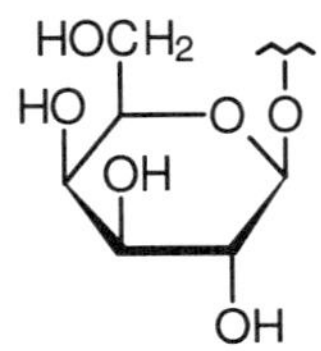	51%	63
R_1 = H, R_2 = 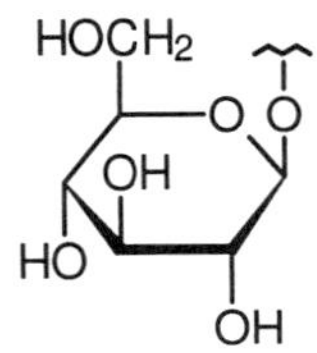	53%	63

[a]Two other anomeric azides appeared to give a similar type of product upon photolysis but the product structures were not established.

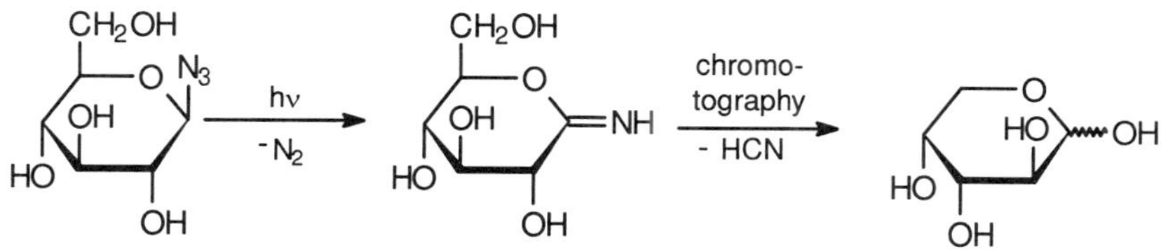

6 7

Scheme 17.7

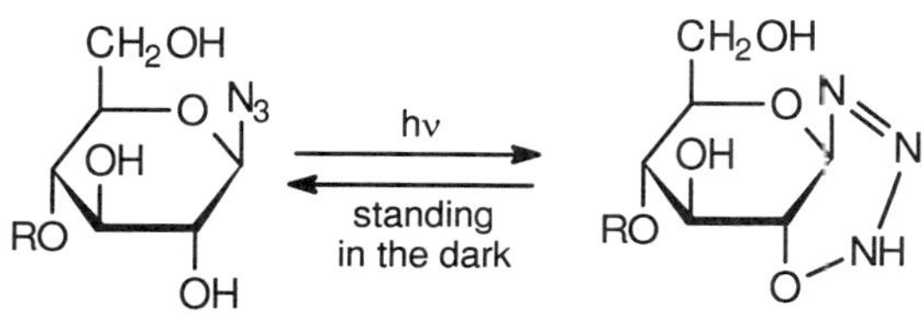

8 9

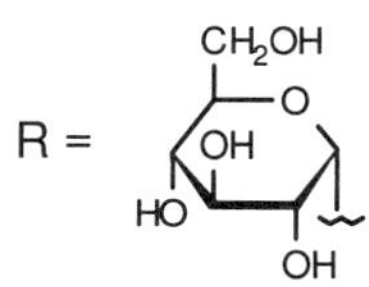

Scheme 17.8

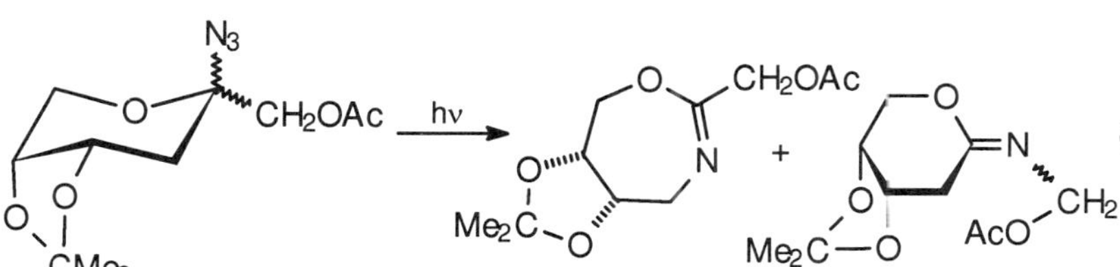

(8)

50% 10%

Equation 8

nitrogen (eq 8).[64–66] By studying different types of anomeric azides (Table 17.5), Praly et al. were able to establish the migratory aptitudes H>>CH_3> C_6H_5>>OR.[66]

In still another type of reaction for forming divalent intermediates, Descotes and co-workers found that carbenes could be generated by photolysis of anomeric diazides (Scheme 17.1).[67,68] Although initial experiments indicated that this reaction would be an effective method for carbene formation, later work showed that it was not.[67–72] Early papers[67,68] had reported high yields of cyclopropanes from photolyses of anomeric diazides in the presence of acrylonitrile (Table 17.6), but further experimentation revealed that the major reaction pathway was rearrangement to form sugar tetrazoles.[69–72]

Table 17.6. Photochemical Reactions of Anomeric Diazides

Reactant	*Product and Percent Yield*	*Ref.*
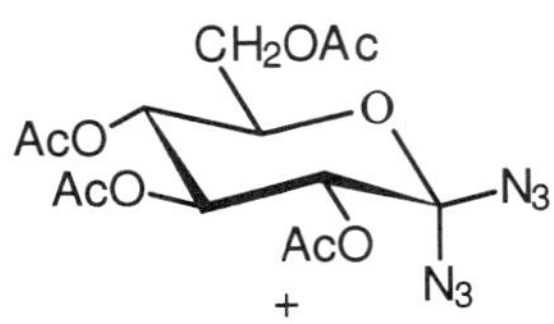 + CH_2=CHCN	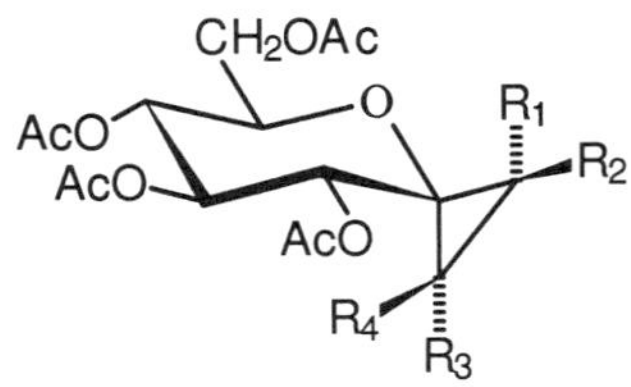	67, 68
	$R_1 = R_2 = R_4 = H, R_3 = CN$ 25%	
	$R_1 = R_3 = R_4 = H, R_2 = CN$ 12%	
	$R_1 = R_2 = R_3 = H, R_4 = CN$ 15%	
	$R_2 = R_3 = R_4 = H, R_1 = CN$ 13%	
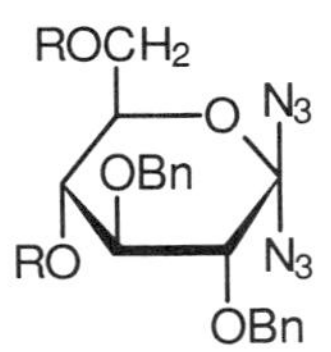	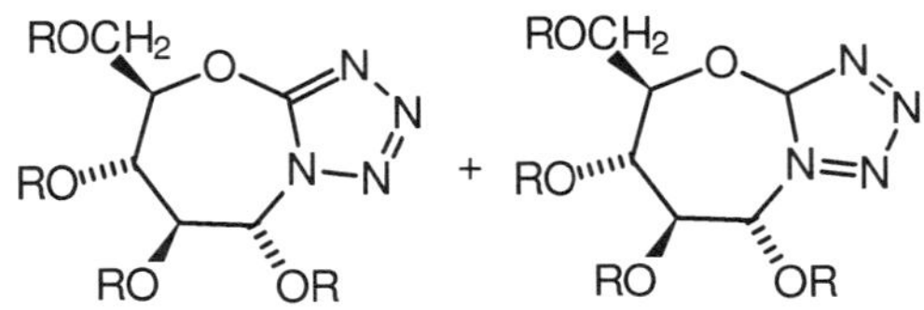 **A** + **B**	
R = Ac	**A** 46%, **B** 10%	69
R = Bn	**A** 51-55%, **B** 20%	69, 70, 71
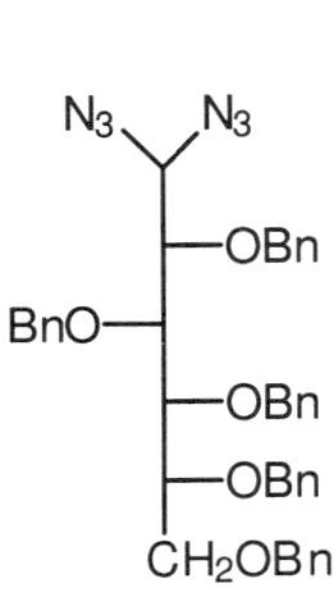	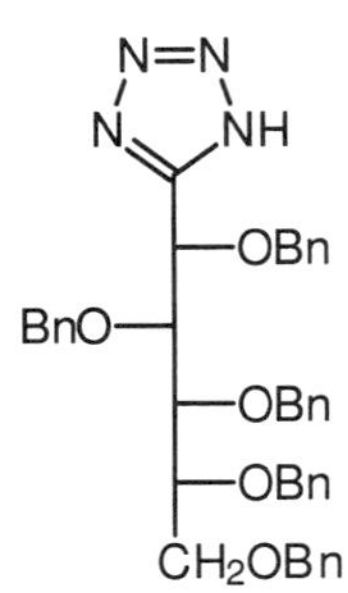 29% 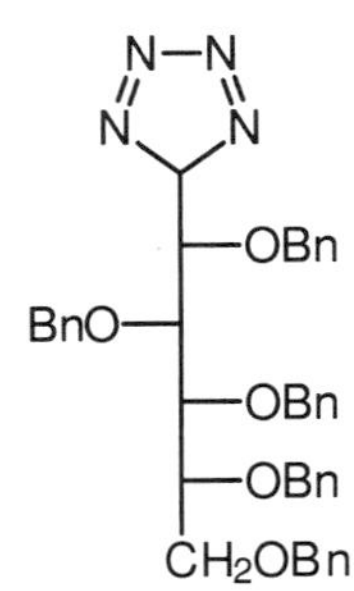 22%	71, 72

A mechanism for tetrazole formation is outlined in Scheme 17.9.[69–72] Since the same reaction is observed both upon direct irradiation and upon sensitization by acetophenone, Yokoyama et al. concluded that reaction takes place from a triplet state of the diazide.[72] The azido nitrene **10,** which forms upon loss of molecular nitrogen, is a precursor to the two ring-expanded intermediates **11** and **12**. The details of the reaction as it progresses from **11** and **12** to the tetrazole photoproducts have not yet been determined.

Summary

Photolysis of diazirines, salts of tosylhydrazones, and diazoketones all lead to the formation of carbenes (Scheme 17.1). In carbohydrate chemistry the primary interest in generating these carbenes is to use them as intermediates in glycoside synthesis. Nitrenes, formed by azide photolysis, are valuable because they undergo a 1,2-hydrogen atom shift to produce the corresponding imines,

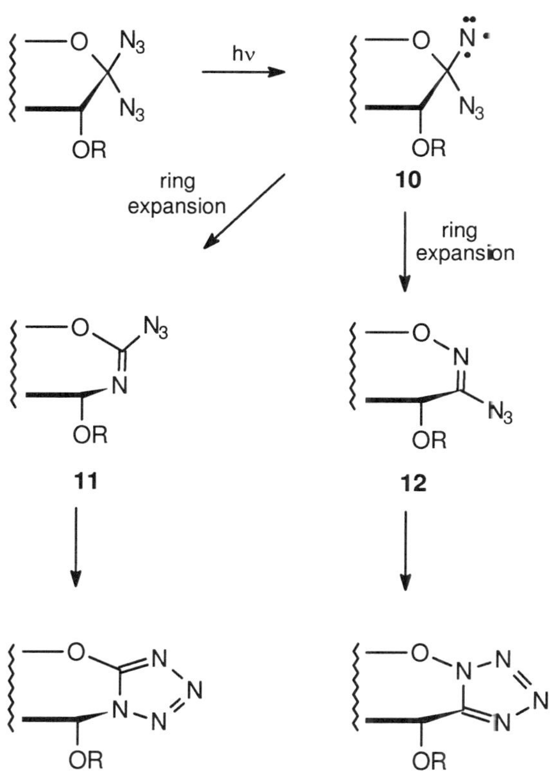

Scheme 17.9

which hydrolyze easily to carbonyl compounds (Table 17.3). If hydrogen migration during azide photolysis is structurally prohibited, other groups will migrate (Table 17.5).

References

1. Vasella, A. *Pure Appl. Chem.* **1993**, *65*, 731.
2. Bonjouklian, R.; Ganem, B. *Carbohydr. Res.* **1979**, *76*, 245.
3. Turro, N. J. *Modern Molecular Photochemistry;* Benjamin/Cummings Publishing: Menlo Park, CA, 1978; p 552.
4. Dürr, H.; Ruge, B. *Top. Curr. Chem.* **1976**, *66*, 53.
5. Vasella, A.;Waldraff, C. A. A. *Helv. Chim. Acta* **1991**, *74*, 585.
6. Vasella, A.; Briner, K.; Soundararajan, N.; Platz, M. S. *J. Org. Chem.* **1991**, *56*, 4741.
7. Briner, K.; Vasella, A. *Helv. Chim. Acta* **1992**, *75*, 621.
8. Muddasani, P. R.; Bozó, E.; Bernet, B.; Vasella, A. *Helv. Chim. Acta* **1994**, *77*, 257.
9. Muddasani, P. R.; Bernet, B.; Vasella, A. *Helv. Chim. Acta* **1994**, *77*, 334.
10. Uhlmann, P.; Vasella, A. *Helv. Chim. Acta* **1994**, *77*, 1175.
11. Bozó, E.; Vasella, A. *Helv. Chim. Acta* **1994**, *77*, 745.
12. Vasella, A.; Dhar, P.; Witzig, C. *Helv. Chim. Acta* **1993**, *76*, 1767.
13. Vasella, A.; Witzig, C. *Helv. Chim. Acta* **1995**, *78*, 1971.
14. Vasella, A. *Pure Appl. Chem.* **1991**, *63*, 507.
15. Lehmann, J.; Petry, S. *Carbohydr. Res.* **1993**, *239*, 133.
16. Bernet, B.; Murty, A. R. C. B.; Vasella, A. *Helv. Chim. Acta* **1990**, *73*, 940.
17. Beard, A. R.; Butler, P. I.; Mann, J.; Parlett, N. K. *Carbohydr. Res.* **1990**, *205*, 87.
18. Samano, V.; Robins, M. J. *J. Am. Chem. Soc.* **1992**, *114*, 4007.
19. Samano, V.; Robins, M. J. *Tetrahedron Lett.* **1994**, *35*, 3445.
20. Czernecki, S.; Mulard, L.; Valéry, J.-M.; Commerçon, A. *Can. J. Chem.* **1993**, *71*, 413.
21. Mangholz, S. E.; Vasella, A. *Helv. Chim. Acta* **1991**, *74*, 2100.
22. Somsák, L.; Praly, J.-P.; Descotes, G. *Synlett* **1992**, 119.
23. Mangholz, S. E.; Vasella, A. *Helv. Chim. Acta* **1995**, *78*, 1020.
24. Kunieda, T.; Witkop, B. *J. Am. Chem. Soc.* **1969**, *71*, 7752.
25. Kunieda, T.; Witkop, B. *J. Am. Chem. Soc.* **1971**, *93*, 3487.
26. Meier, H.; Zeller, K. *Angew. Chem. Int. Ed. Eng.* **1975**, *14*, 32.
27. Wentrup, C. *Top. Curr. Chem.* **1976**, *62*, 173.
28. Turro, N. J. *Modern Molecular Photochemistry;* Benjamin/Cummings Publishing: Menlo Park, CA, 1978; pp 554-555.
29. López-Herrera, F. J.; Sarabia-García, F. *Tetrahedron. Lett.* **1993**, *34*, 3467.
30. López-Herrera, F. J.; Sarabia-García, F. *Tetrahedron. Lett.* **1994**, *35*, 2929.
31. Norbeck, D. W.; Kramer, J. B. *J. Am. Chem. Soc.* **1988**, *110*, 7217.
32. Horton, D.; Philips, K. D. *Carbohydr. Res.* **1972**, *22*, 151.
33. Sarabia-García, F.; López-Herrera, F. J.; Pino-González, M. S. *Tetrahedron Lett.* **1994**, *35*, 6709.
34. Singh, S.; Nambiear, S.; Porter, R. A.; Sander, T. L.; Taylor, K. G.; Doyle, R. J. *J. Org. Chem.* **1989**, *54*, 2300.
35. Horton, D.; Weckerle, W.; Winter, B. *Carbohydr. Res.* **1979**, *70*, 59.
36. Whistler, R. L.; Doner, L. W. *J. Org. Chem.* **1970**, *35*, 3562.
37. Whistler, R. L.; Anisuzzaman, A. K. M. *J. Org. Chem.* **1969**, *34*, 3823.
38. Horton, D.; Leutzow, A. E.; Wease, J. C. *Carbohydr. Res.* **1968**, *8*, 366.
39. Gibson, A. R.; Melton, L. D.; Slessor, K. N. *Can. J. Chem.* **1974**, *52*, 3905.
40. Beving, H. F. G.; Leutzow, A. E.; Theander, O. *Carbohydr. Res.* **1975**, *41*, 105.

41. Szarek, W. A.; Vyas, D. M.; Chen, L.-Y. *Carbohydr. Res.* **1977**, *53*, C1.
42. Jarrell, H. C.; Szarek, W. A. *Can. J. Chem.* **1978**, *56*, 144.
43. Jarrell, H. C.; Szarek, W. A. *Can. J. Chem.* **1979**, *57*, 924.
44. Durette, P. L.; Rosegay, A.; Walsh, M. A. R.; Shen, T. Y. *Tetrahedron Lett.* **1979**, 291.
45. Jarrell, H. C.; Szarek, W. A. *Carbohydr. Res.* **1978**, *67*, 43.
46. Tsuchiya, T.; Miyake, T.; Kageyama, S.; Umezawa, S.; Umezawa, H.; Takita, T. *Tetrahedron Lett.* **1981**, *22*, 1413.
47. Wright, J. J. K.; Albarella, J. A.; Lee, P. *J. Org. Chem.* **1982**, *47*, 523.
48. Petrušová, M.; Fedoronko, M; Petruš, L. *Chem. Papers* **1986**, *40*, 655.
49. Kuroda, Y.; Kobayashi, O.; Suzuki, Y.; Ogoshi, H. *Tetrahedron Lett.* **1989**, *30*, 7225.
50. Petrušová, M.; Matulová, M.; Fedoronko, M; Petruš, L. *Synthesis* **1991**, 209.
51. Kozlowska-Gramsz, E.; Descotes, G. *Tetrahedron Lett.* **1982**, *23*, 1585.
52. Clode, D. M.; Horton, D. *Carbohydr. Res.* **1970**, *14*, 405.
53. Clode, D. M.; Horton, D.; Meshreki, M. H.; Shoji, H. *J. Chem. Soc., Chem. Commun.* **1969**, 694.
54. Clode, D. M.; Horton, D. *Carbohydr. Res.* **1971**, *17*, 365.
55. Horton, D.; Luetzow, A. E.; Theander, O. *Carbohydr. Res.* **1973**, *27*, 268.
56. Clode, D. M.; Horton. D. *Carbohydr. Res.* **1970**, *12*, 477.
57. Clode, D. M.; Horton, D. *Carbohydr. Res.* **1971**, *19*, 329.
58. Horton, D.; Luetzow, A. E.; Theander, O. *Carbohydr. Res.* **1973**, *26*, 1.
59. Baker, D. C.; Horton, D. *Carbohydr. Res.* **1972**, *21*, 393.
60. Jenkins, I. D.; Verheyden, J. P. H.; Moffatt, J. G. *J. Am. Chem. Soc.* **1971**, *93*, 4323.
61. Miyasaka, T.; Tanaka, H.; Satoh, K.; Imahashi, M.; Yamaguchi, K.; Iitaka, Y. *J. Heterocyclic Chem.* **1987**, *24*, 873.
62. Plenkiewicz, J.; Hay, G. W.; Szarek, W. A. *Can. J. Chem.* **1974**, *52*, 183.
63. Szarek, W. A.; Achmatowicz, Jr., O.; Plenkiewicz, J.; Radatus, B. K. *Tetrahedron* **1978**, *34*, 1427.
64. Di Stéfano, C.; Descotes, G.; Praly, J.-P. *Tetrahedron Lett.* **1994**, *35*, 93.
65. Praly, J.-P.; Di Stéfano, C.; Descotes, G.; Faure, R; Somsák, L.; Eperjesi, I. *Tetrahedron Lett.* **1995**, *36*, 3329.
66. Praly, J.-P.; Bonnevie, C.; Huag, P.; Descotes, G. *Tetrahedron* **1996**, *52*, 9057.
67. Praly, J.-P.; El Kharraf, Z.; Descotes, G. *Tetrahedron Lett.* **1990**, *31*, 4441.
68. Descotes, G.; El Kharraf, Z.; Faure, R.; Fenet, B.; Praly, J.-P. *J. Carbohydr. Chem.* **1991**, *10*, 959.
69. Praly, J.-P.; Di Stéfano, C.; Descotes, G.; Faure, R. *Tetrahedron Lett.* **1994**, *35*, 89.
70. Yokoyama, M.; Matsushita, M.; Hirano, S.; Togo, H. *Tetrahedron Lett.* **1993**, *34*, 5097.
71. Yokoyama, M.; Hirano, S.; Hachiya, T.; Togo, H. *Chem. Lett.* **1994**, 1107.
72. Yokoyama, M.; Hirano, S.; Matsushita, M.; Hachiya, T.; Kobayashi, N.; Kubo, M.; Togo, H.; Seki, H. *J. Chem. Soc., Perkin Trans. 1* **1995**, 1747.

18

Amines and Amine Derivatives

Introduction

A wide variety of amines and amine derivatives have been irradiated in the hope of discovering new and useful synthetic reactions; unfortunately, few of these investigation have uncovered photochemical processes that have a clear synthetic application. Two reactions that have been used in synthesis are the formation of new ring systems by imide photolysis,[1–3] and photodemethylation of anthracycline glycosides.[4] Most photolyses of amine derivatives, however, are best described as exploratory studies.

The diversity in structure of amine derivatives insures that there is not a single mechanistic pattern for their reactions. Some of these compounds undergo photochemical electron transfer, and others, in particular those with carbonyl groups, abstract hydrogen atoms in a manner characteristic of aldehydes and ketones. Perhaps the most common reaction of amine derivatives, however, is photofragmentation of a bond to nitrogen leading to radical intermediates.

Amides

Amides are often synthesized to protect amino groups in aminodeoxy sugars. When this protection involves formation of an amide of phenylacetic acid, the amino sugar can be regenerated by photolysis (Scheme 18.1).[5] This reaction is made possible by the stability of the benzyl (**2**) and acyl (**3**) radical intermediates generated by an α-cleavage reaction similar to that observed during photolysis of aldehydes and ketones (Chapter 2). The radical **3** becomes an aminodeoxy sugar by decarbonylation followed by hydrogen abstraction (Scheme

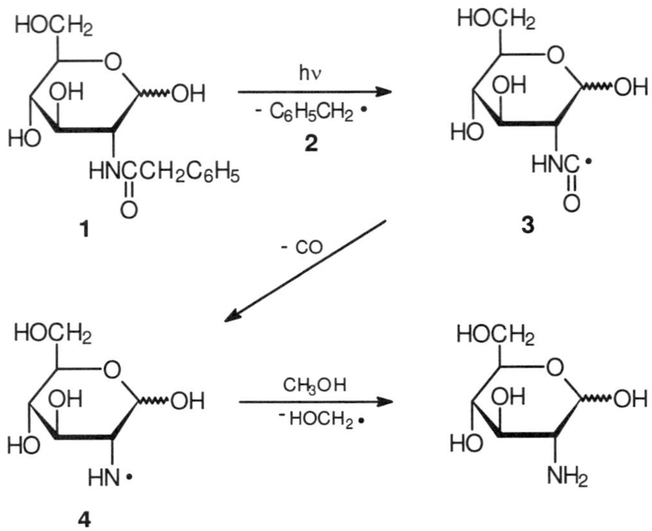

Scheme 18.1

18.1). In a related reaction, oxidative deprotection of amides of phenylglyoxylic acid generates imines that hydrolyze readily to the corresponding carbonyl compounds in low overall yield (Table 18.1).[6]

Imides

Sowa and Thiem discovered that glycosylimides abstract appropriately placed hydrogen atoms internally to generate new ring systems and create ring-expanded photoproducts (Scheme 18.2).[1–3] These investigators found that in such reactions, ring conformation is the decisive factor in determining which of these two reaction pathways will be followed. For the conformation with an equatorial succinimido group (**5e**), 1,6-abstraction produces the ketoamide **7** by way of the intermediate hydroxyamide **6**. Production of **6** is analogous to the cyclobutanol formation (Yang reaction) that typically accompanies a Norrish Type II reaction (Chapter 3). When the succinimido group is axial (**5a**), 1,7-hydrogen abstraction leads to the tricyclic photoproduct **9** (Scheme 18.2).

Since substantial changes in conformational equilibria are brought about by small changes (a few kilocalories per mole) in the relative energy of the conformers, an alteration in structure, such as replacing a protecting group, establishes a new conformational equilibrium between **5a** and **5e** and, in so doing, changes the relative amounts of photoproducts formed.[1] As can be seen from inspection of Table 18.2, photoproduct yields are dependent upon the identity of the protecting groups.

Table 18.1. Photochemical Reactions of Aromatic Amides

Reactant	*Product*	*Percent Yield*	*Ref.*
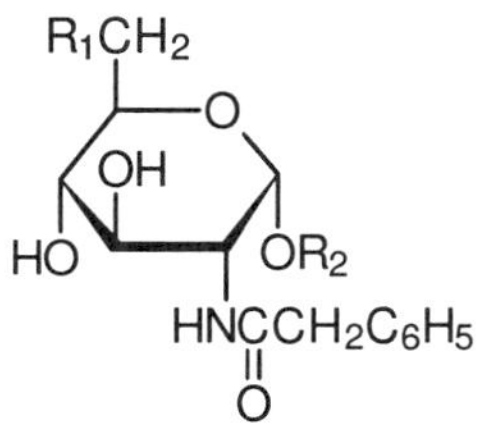	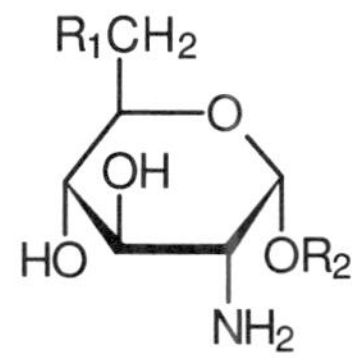		5
R_1 = OH, R_2 = H		73	
R_1 = Br, R_2 = Me		63	
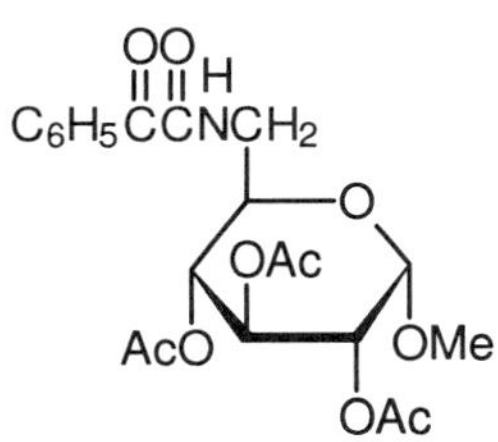	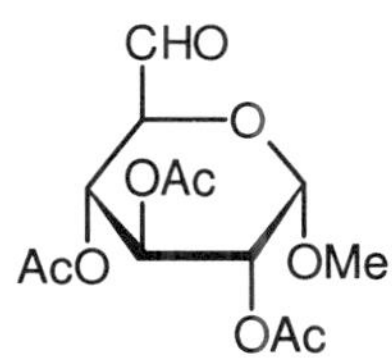	3	6
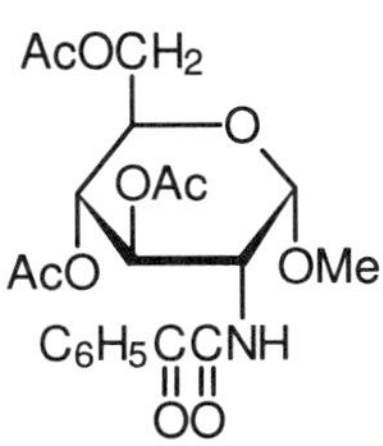	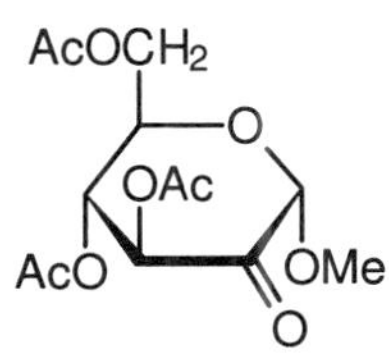	11	6

Table 18.2. Photochemical Reactions of Imides

Reactant	*Product and Percent Yield*		*Ref.*
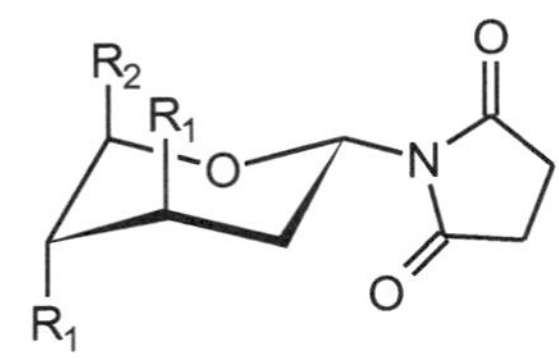	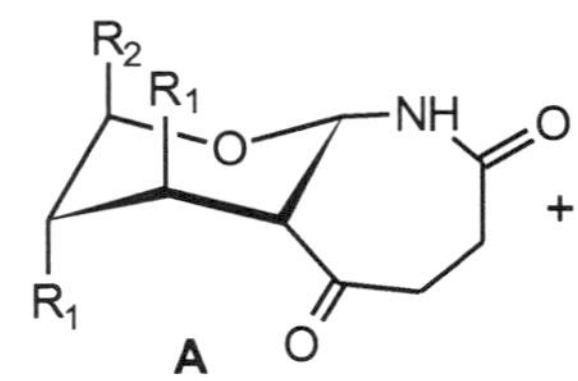 A +	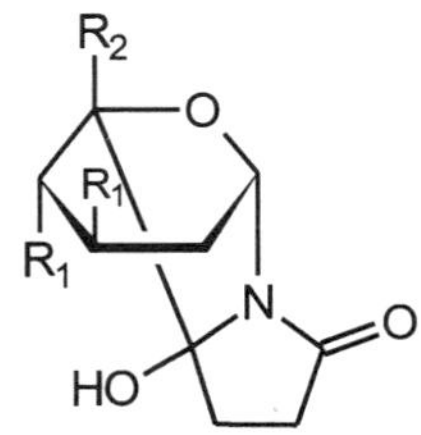 B	
	A	B	
R_1 = OSi*t*-BuMe$_2$, R_2 = H	52%	b	1, 2
R_1 = OSi*t*-BuMe$_2$ R_2 = CH$_2$OSi*t*-BuMe$_2$	21%	62%	1, 2[a]
R_1 = OSiMe$_3$, R_2 = CH$_2$OSiMe$_3$	b	83%	1
R_1 = OMe, R_2 = CH$_2$OSiMe$_3$	11%	56%	2[a]
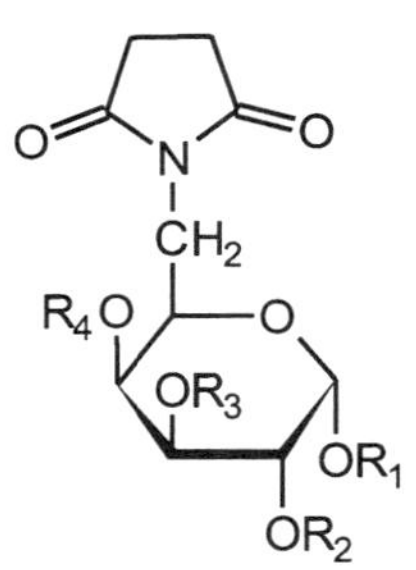	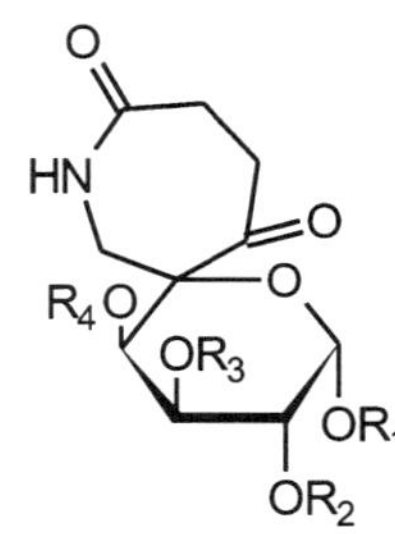		3
$R_1 = R_2, R_3 = R_4$ = CMe$_2$	36%		
$R_1 = R_2 = R_3 = R_4$ = SiMe$_3$	49%		
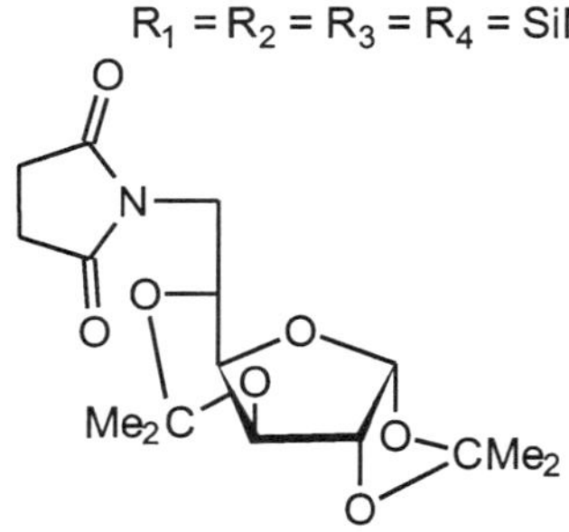	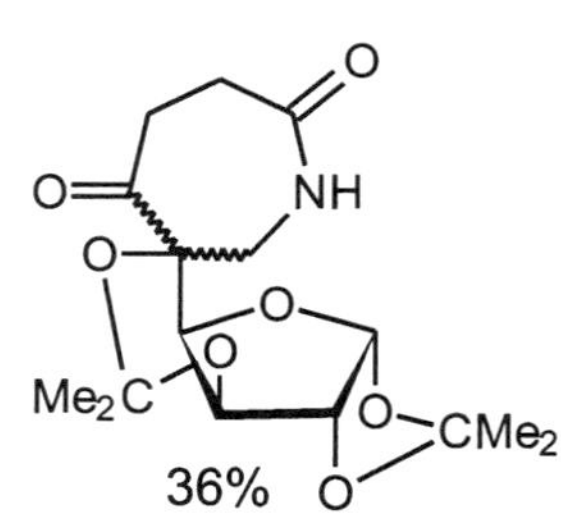 36%		3

[a]Prolonged irradiation produced secondary photoproducts.
[b]No product was detected.

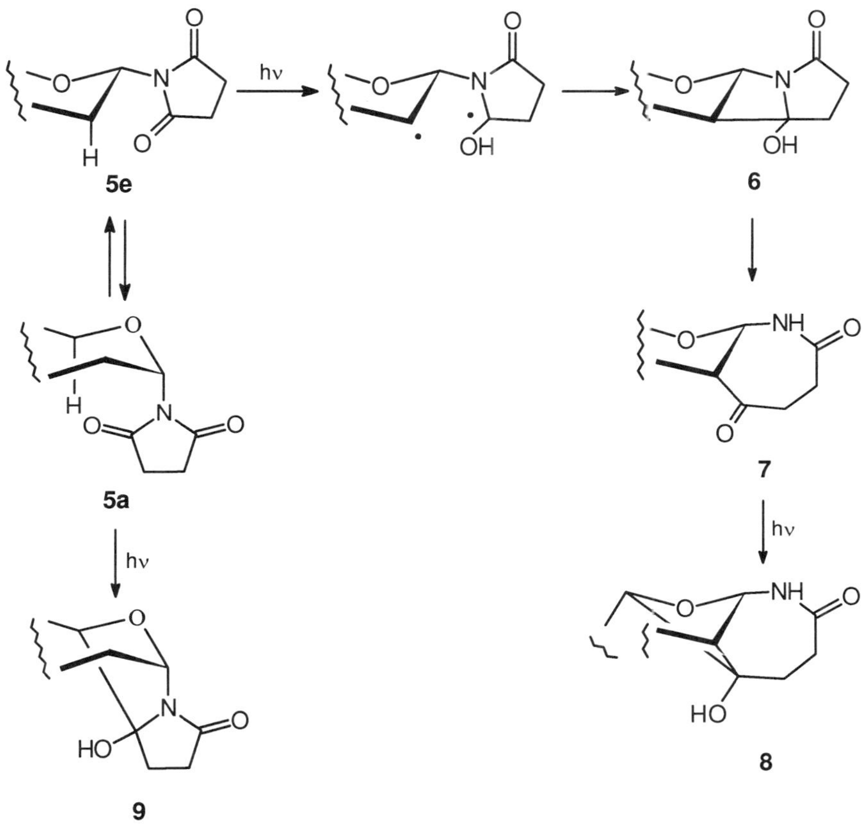

Scheme 18.2

N,N-*Dimethylamines*

Irradiation of an anthracycline glycoside that has an *N,N*-dimethylamino group in the carbohydrate portion of the molecule (e.g., **10**) represents a pathway for removal of one of the *N*-methyl groups (eq 1). More than a dozen analogs of **10**, each of which has a different aglycon, undergo photochemical demethylation in yields ranging from 13–86%.[4,7,8] Additional analogs of **10** in which the monosaccharide residue has been replaced by a trisaccharide unit experience the same reaction.[9,10] The first *N*-methyl group in **10** is removed with ease, but although the second methyl group also can be removed by photolysis, reaction is much slower and lower in yield.[7,8] Even though a mechanism for this reaction has not been proposed, it is known that a halogenated solvent (usually chloroform) is necessary for reaction and that a hydroxylic solvent (methanol) is always included. The necessity for a halogenated solvent suggests that the reaction

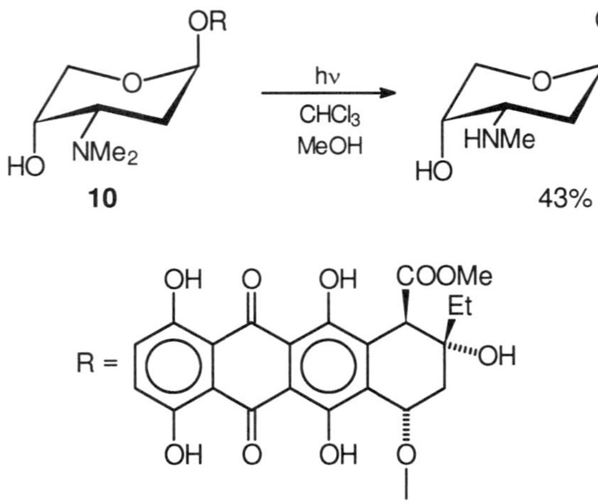

(1)

Equation 1

involves a halogen atom at some stage, and the need for methanol indicates that solvolytic decomposition of an intermediate imine takes place.

Photocyclization of Nucleosides

If a nucleoside has a free 5′-hydroxyl group, cyclization can result from irradiation in the presence of an electron acceptor. Maki and co-workers discovered a number of examples of this type of reaction (Table 18.3).[12–15] Since photocyclization takes place only in the presence of an electron acceptor, these investigators proposed the electron-transfer mechanism shown in Scheme 18.3.

Hydroxylamines

Photolysis of hydroxylamines leads to formation of imino lactones (Table 18.4). Binkley and Binkley proposed that the first step in this process is homolytic cleavage of the nitrogen–oxygen bond to give a radical pair.[16,17] Disproportionation then produces an iminolactone. While some of these lactone photoproducts are stable enough to be isolated, others spontaneously lose hydrogen cyanide upon standing to form a sugar with one less carbon atom than the number found in the starting material (Scheme 18.4)

Azines, Hydrazines, and Oximes

Several photochemical reactions of compounds related to hydroxylamines have been reported (Table 18.5). These reactions are similar to those of

Table 18.3. Photocyclization of Nucleosides by Electron Transfer

Reactant	*Product*	*Percent Yield*	*Ref.*
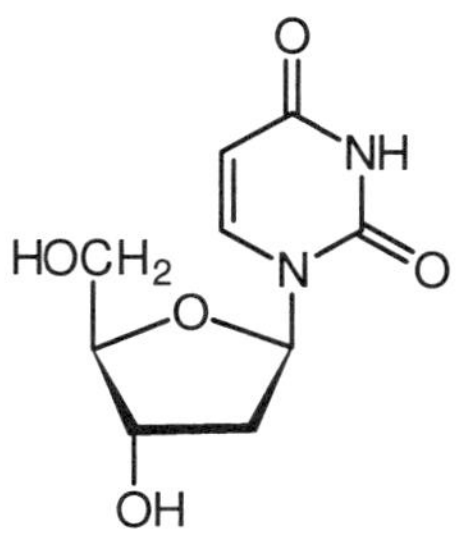	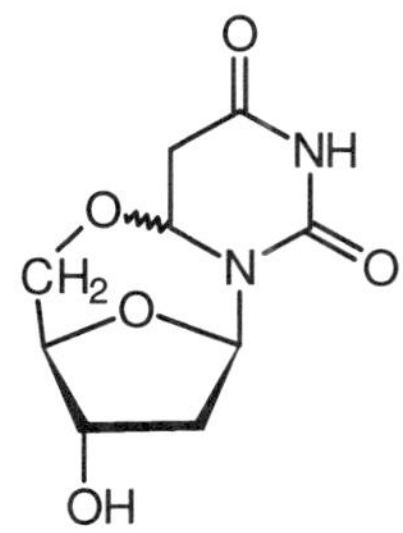	9[b]	11
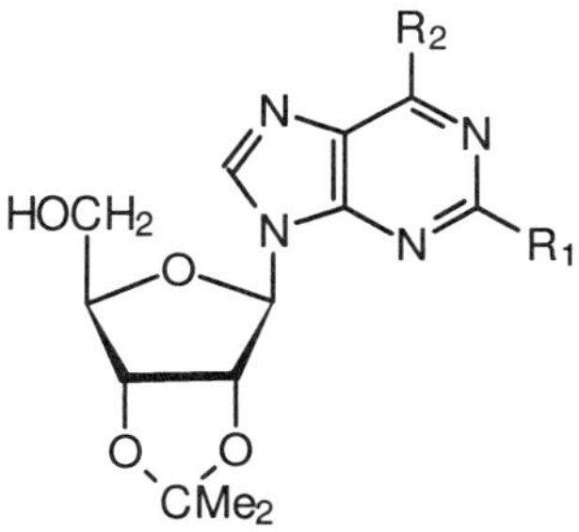	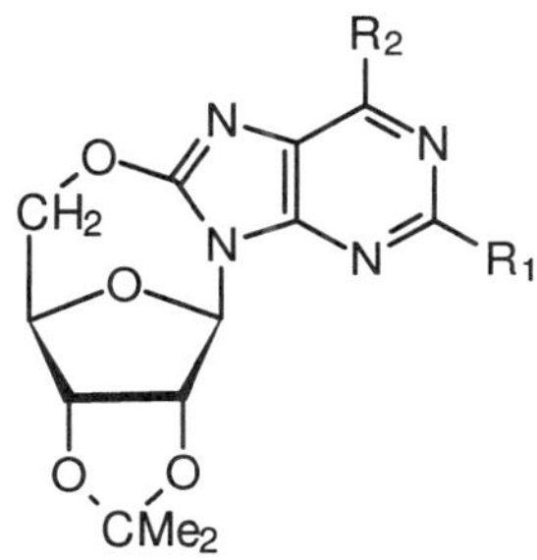		
R_1 = $NHCOC_6H_5$, R_2 = OH		70	12
R_1 = H, R_2 = $NHCOC_6H_5$		18	12, 13
		5	14, 15
R_1 = H, R_2 = NH_2		20	12, 14, 15
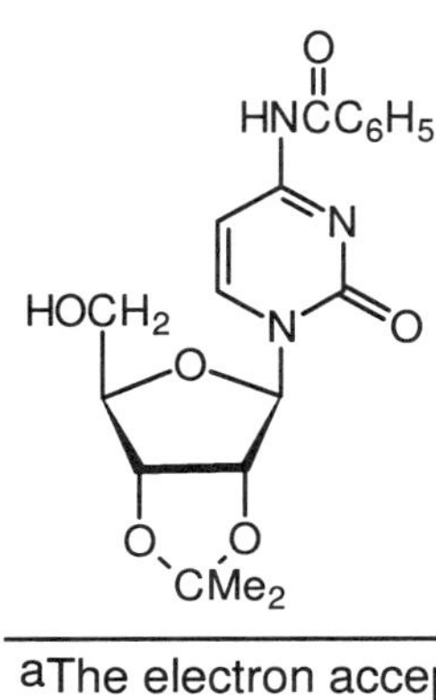	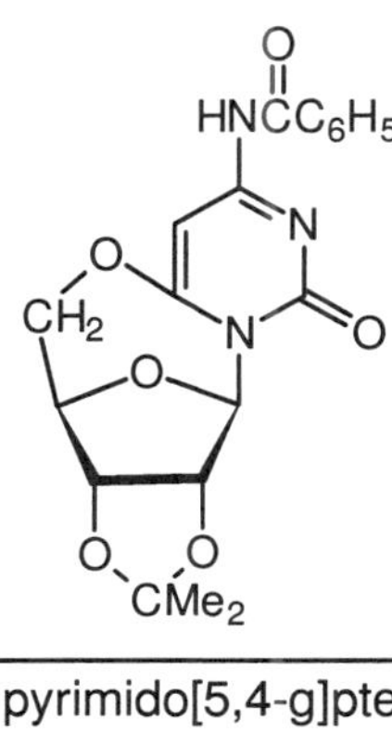	7	12

[a]The electron acceptor was pyrimido[5,4-g]pteridinetetraone *N*-oxide.
[b]The major products were photohydrates.

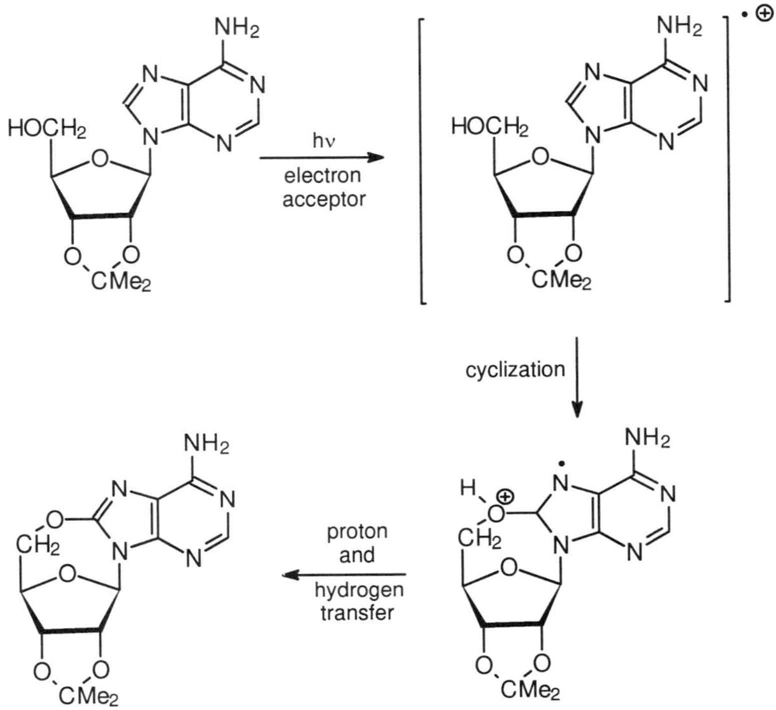

Scheme 18.3

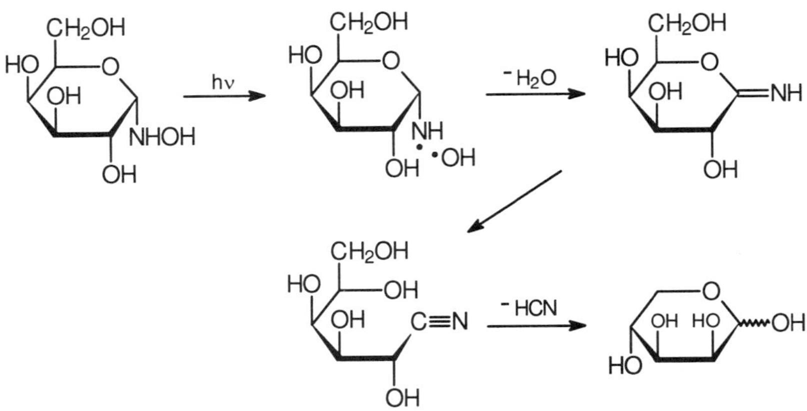

Scheme 18.4

Table 18.4. Photochemical Reactions of Hydroxyl Amines

Reactant	*Product*	*Percent Yield*	*Ref.*
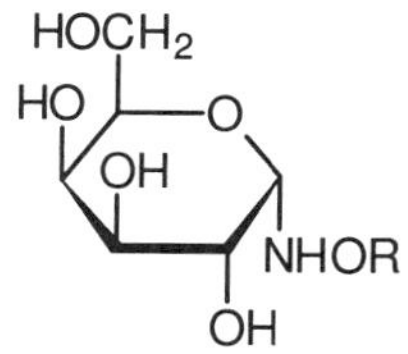	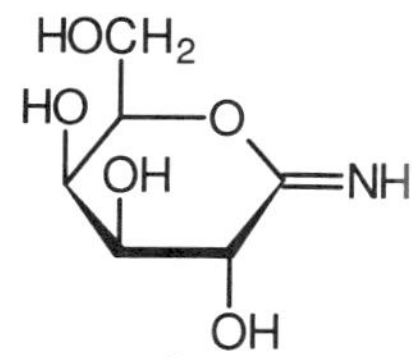 +	81	16, 17
R = H R = CH_3	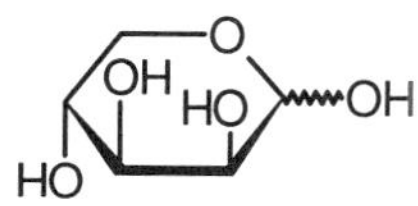	14	
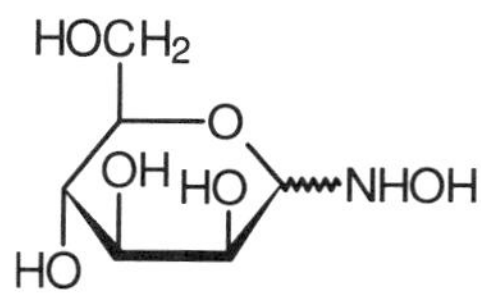	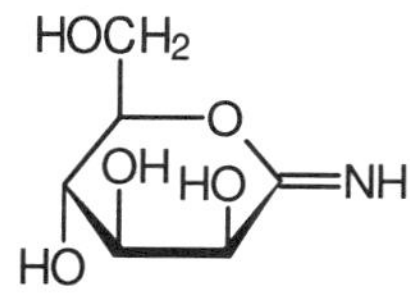	33	17
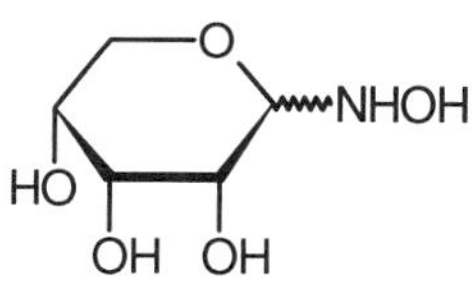	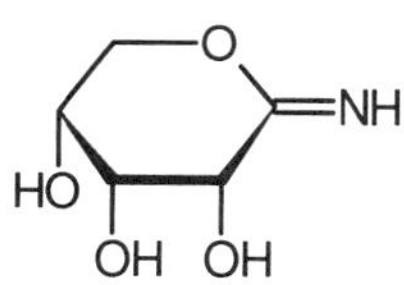	23	17
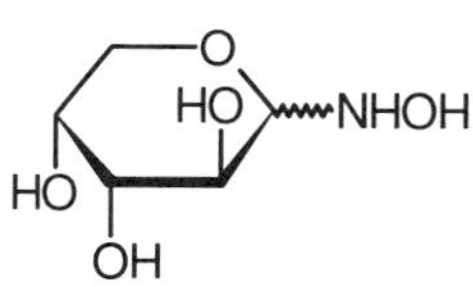	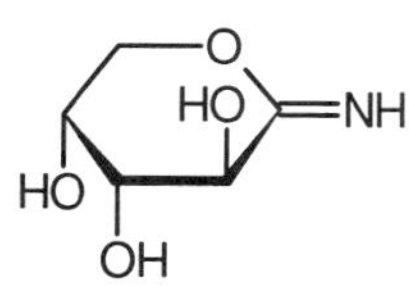	28	17
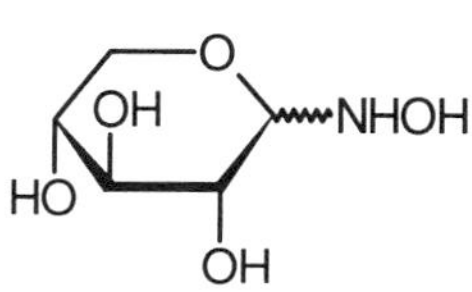	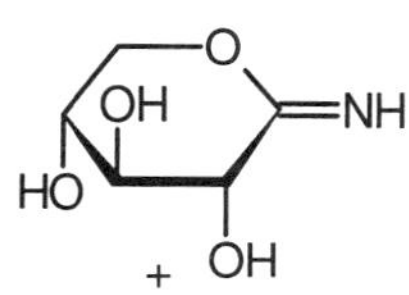 +	15	17
	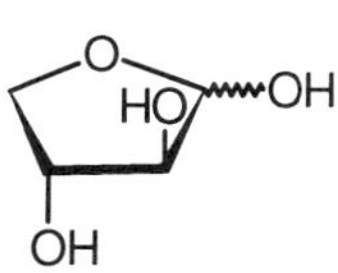	18	

Table 18.5. Photochemical Reactions of Azines, Hydrazines, Oximes, and Osazones

Reactant	*Product*	*Percent Yield*	*Ref.*
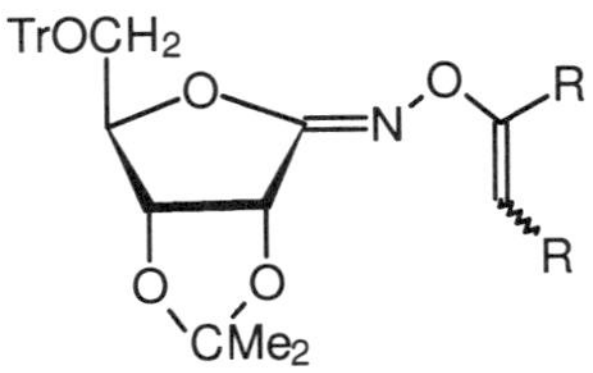 R = CO_2Me	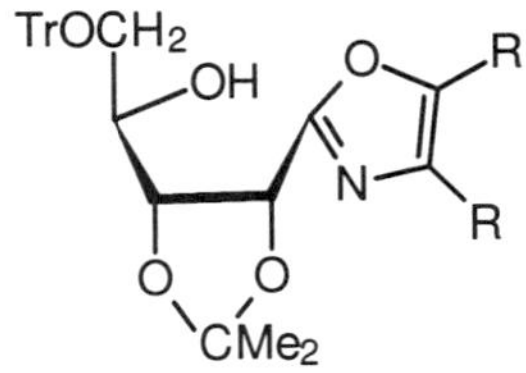	35	18
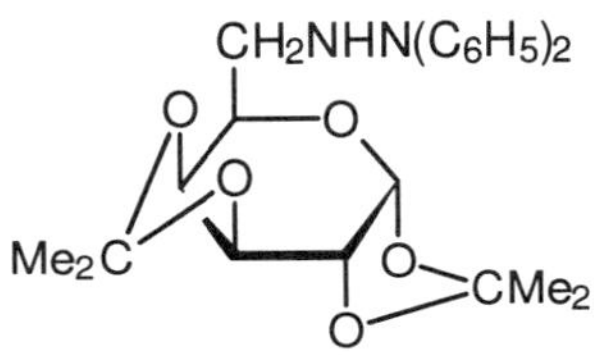	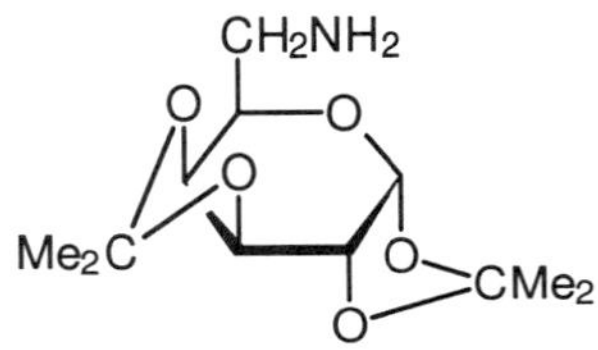	27	19
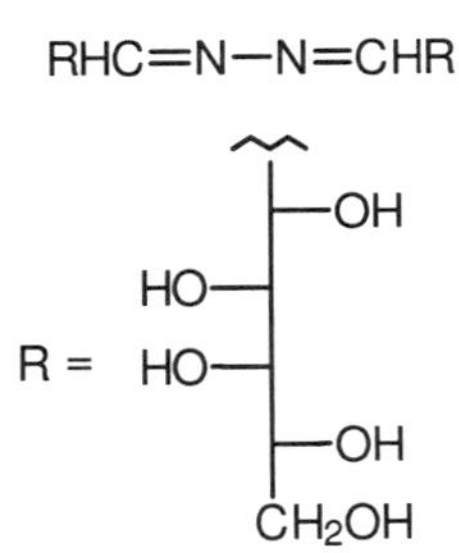	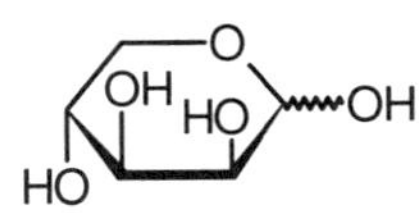 + 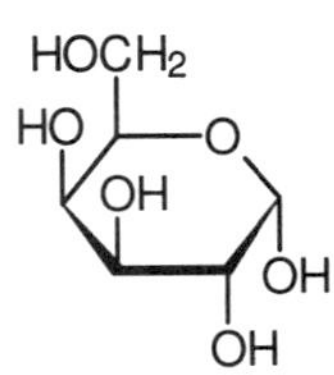	23[a] 14[b] / 37[a] 32[b]	20
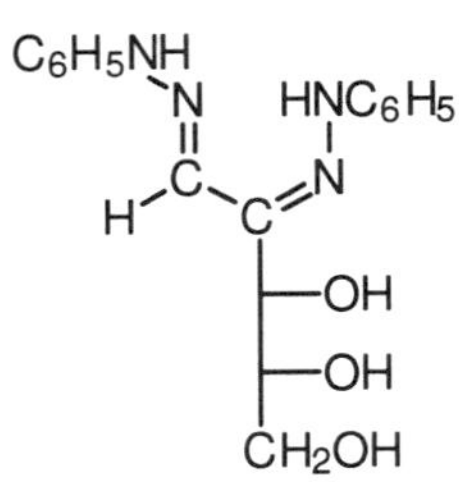	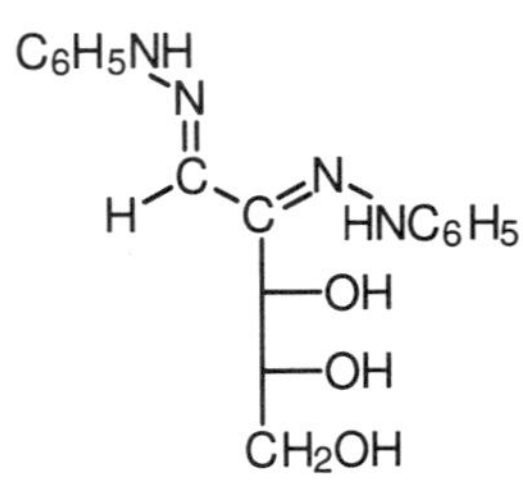	90	21, 22

[a]Reaction was conducted in the presence of oxygen.
[b]Reaction was conducted in the absence of oxygen.

hydroxylamines in that they involve formation of nitrogen radicals by homolytic cleavage of a nitrogen–oxygen or nitrogen–nitrogen bond. A typical example of this type of reaction was reported by Flechtner and Pohlid, who found that irradiation of the hydrazine **11** breaks the nitrogen–nitrogen bond to produce the corresponding aminodeoxy sugar **12** (eq 2).[19]

Flavins

Flavins are a group of biologically active compounds, each of which contains a characteristic, heterocyclic structure.[23] Some flavins, such as riboflavin (**13**), also have a carbohydrate portion of the molecule. These compounds experience a variety of photochemical processes, some of which involve the carbohydrate moiety (eq 3),[23–27] but the most important photochemistry of flavins centers on the nitrogen-containing heterocycle.

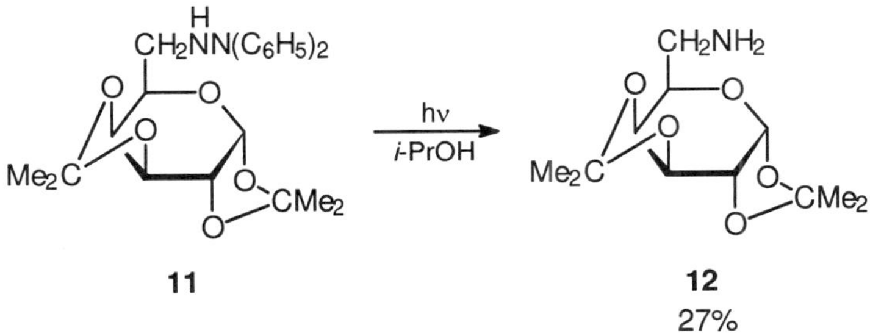

(2)

Equation 2

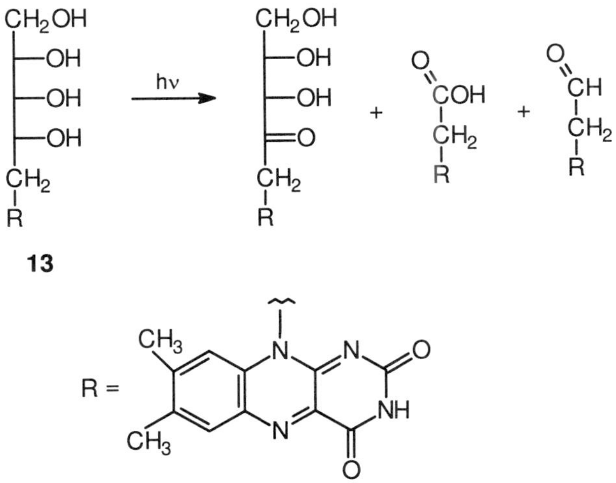

(3)

Equation 3

Summary

The variety of amine and amine derivatives that have been studied photochemically includes amines (eq 1), amides (Scheme 18.1), imides (Scheme 18.2), nucleosides (Scheme 18.3), and hydroxylamines (Scheme 18.4). None of the reactions of these compounds has found widespread use in synthesis.

References

1. Sowa, C. E.; Thiem, J. *Angew. Chem. Int. Ed. Engl.* **1994**, *33*, 1979.
2. Sowa, C. E.; Kopf, J.; Thiem, J. *J. Chem. Soc., Chem. Commun.* **1995**, 211.
3. Sowa, C. E.; Stark, M.; Heidelberg T.; Thiem, J. *Synlett* **1996**, 227.
4. Kolar, C; Dehmel, K.; Knoedler, U.; Paal, M.; Hermentin, P.; Gerken, M. *J. Carbohydr. Chem.* **1989**, *8*, 295.
5. Hanessian, S.; Masse, R. *Carbohydr. Res.* **1977**, *54*, 142.
6. Zehavi, U. *J. Org. Chem.* **1977**, *42*, 2821.
7. Hermentin, P.; Raab, E.; Paal, M.; Gerken, M.; Kolar, C.; Boettger, D.; Berscheid, H. G. *J. Carbohydr. Chem.* **1990**, *9*, 235.
8. Johdo, O.; Tone, H.; Okamoto, R.; Yoshimoto, A.; Naganawa, H.; Sawa, T.; Takeuchi, T. *J. Antibiotics* **1992**, *45*, 1653.
9. Tanaka, H.; Yoshioka, T.; Shimauchi, Y.; Matsushita, Y.; Matsuzawa, Y.; Oki, T.; Ishikura, T. *J. Antibiotics* **1982**, *35*, 312.
10. Oki, T.; Kitamura, I.; Shibamoto, N.; Ogasawara, T.; Yoshimoto, A.; Inui, T.; Naganawa, H.; Takeuchi, T.; Umezawa, H. *J. Antiboitics* **1979**, *32*, 801.
11. Cadet. J.; Kan, L.-S.; Wang, S. Y. *J. Am. Chem. Soc.* **1978**, *100*, 6715.
12. Sako, M.; Shimada, K.; Hirota, K.; Maki, Y. *J. Chem. Soc., Chem. Commun.* **1986**, 1704.
13. Shimada, K.; Sako, M.; Hirota, K.; Maki, Y. *Tetrahedron Lett.* **1987**, *28*, 207.
14. Kitade, Y.; Makino, T.; Nakanishi, R.; Sako, M.; Hirota, K.; Maki, Y. *Nucleic Acids Symp. Ser.* **1991**, *25*, 59.
15. Sako, M.; Makino, T.; Kitade, Y.; Hirota, K.; Maki, Y. *J. Chem. Soc., Perkin Trans. 1* **1992**, 1801.
16. Binkley, W. W.; Binkley, R. W. *Tetrahedron Lett.* **1970**, 3439.
17. Binkley, R. W.; Binkley, W. W. *Carbohydr. Res.* **1972**, *23*, 283.
18. Yokoyama, M.; Irie, M.; Sujino, K.; Kagemoto, T.; Togo, H.; Funabashi, M. *J. Chem. Soc., Perkin Trans. 1* **1992**, 2127.
19. Flechtner, T. W.; Pohlid, H. *J. Carbohydr. Chem.* **1984**, *3*, 107.
20. Binkley, R. W.; Binkley, W. W. *Carbohydr. Res.* **1970**, *13*, 163.
21. Maciejewski, A.; Wojtczak, J. *J. Photochem.* **1983**, *21*, 351.
22. Maciejewski, A.; Wojtczak, J. *J. Photochem.* **1983**, *22*, 165.
23. Heelis, P. F. *Chem. Soc. Rev.* **1982**, *11*, 15.
24. Cairns, W. L.; Metzler, D. E. *J. Am. Chem. Soc.* **1971**, *93*, 2772.
25. Jorns, M. S.; Hemmerich, P. *Z. Naturforsch., B: Anorg. Chem., Org. Chem.* **1972**, *27*, 1040.
26. Jorns, M. S.; Schöllnhammer, G.; Hemmerich, P. *Eur. J. Biochem.* **1975**, *57*, 35.
27. Moore, W. M.; Baylor, C., Jr. *J. Am. Chem. Soc.* **1969**, *91*, 7170.

19

Halogenated Carbohydrates

*Photochemically Initiated, Chain Reactions**

Halogenation

Free-radical halogenation of carbohydrates is able to produce synthetic intermediates that are difficult to form by other reactions.[1] Replacement of H-5 in a pyranoid ring with a bromine atom provides an example of such a reaction because this substitution would be difficult were it not for the simple, one-step bromination shown in eq 1.[2] This and other photoinitiated halogenation reactions of carbohydrates have been investigated by several research groups, but particularly noteworthy is the pioneering effort and extensive studies of Ferrier and co-workers, who defined the standard reaction conditions and determined many of the critical factors that influence the regio- and stereoselectivity of photobromination.[1]

Hanessian–Hullar and Related Reactions

The Hanessian–Hullar reaction is a free-radical bromination reaction in which a hydrogen atom in a benzylidene acetal is replaced by a bromine atom derived from *N*-bromosuccinimide (NBS). The unstable bromo compound **1**, created by this replacement, undergoes ring opening to form the bromobenzoate **2** (Scheme 19.1).[3–6] This reaction usually is conducted by heating the reactants with a radical initiator, but sometimes photolysis is used to assist the initiation process. The Hanessian–Hullar reaction has had considerable

*We continue in this chapter the policy adopted in earlier chapters of discussing photochemically initiated, chain reactions of carbohydrates but not including tables of these reactions.

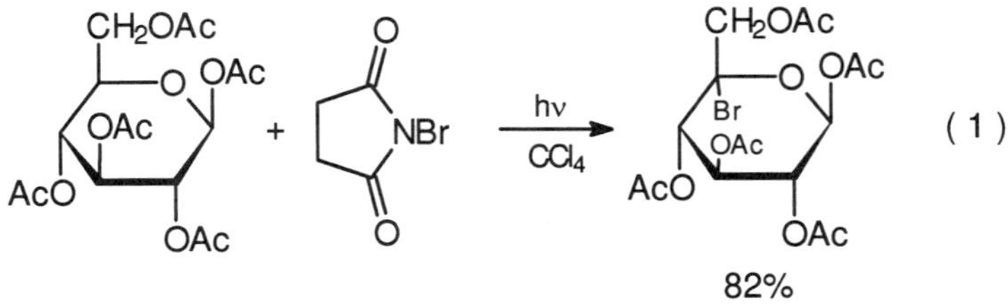

Equation 1

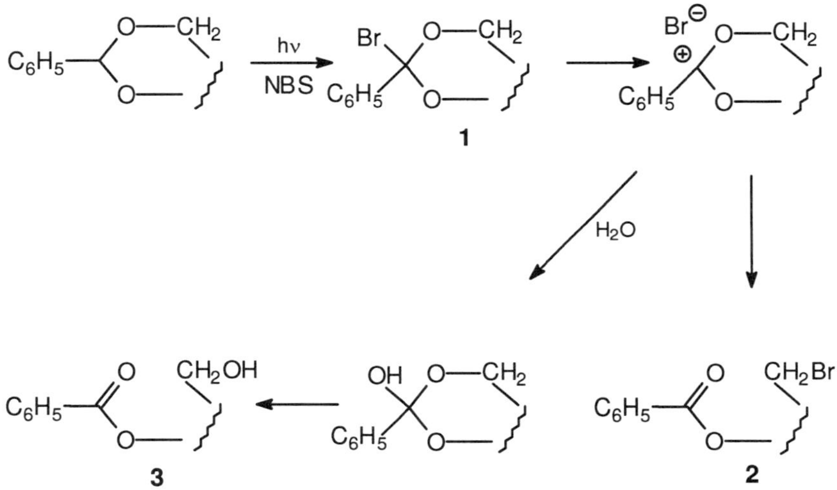

Scheme 19.1

impact on carbohydrate synthesis because benzylidene acetals are common protecting groups, and their ring opening following reaction with bromine produces useful synthetic intermediates in a process that (a) is simple, (b) typically occurs in good yield, and (c) is often highly regioselective.[7]

Several variations on the Hanessian–Hullar reaction have been investigated. One of these is the irradiation of benzylidene acetals in the presence of bromotrichloromethane (eq 2, X = $BrCCl_3$).[8–11] This reaction is sometimes highly regioselective in cases where reaction under typical Hanessian–Hullar conditions is not (eq 2, X = NBS).[8] Hanessian–Hullar ring opening also can be conducted in the presence of water, in which case a hydroxybenzoate (**3**) is formed rather than a bromobenzoate (**2**) (Scheme 19.1).[12–15] A reaction mechanistically similar to hydroxybenzoate formation is the irradiation of benzyl ethers in the presence of water and either bromine or NBS to cause removal of the benzyl group (eq 3).[16–19]

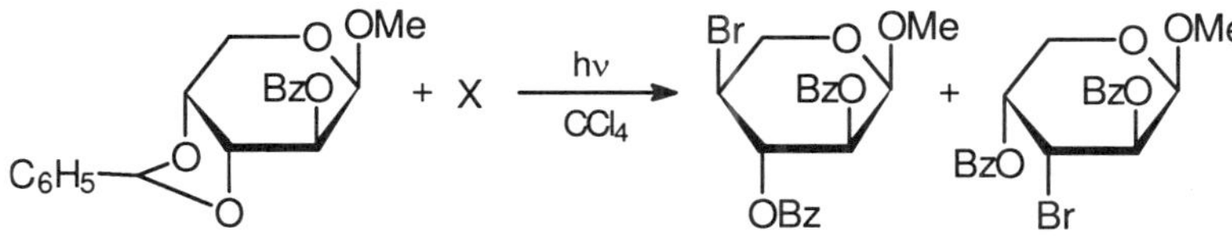

(2)

X		
$BrCCl_3$	86%	–
NBS	30%	60%

Equation 2

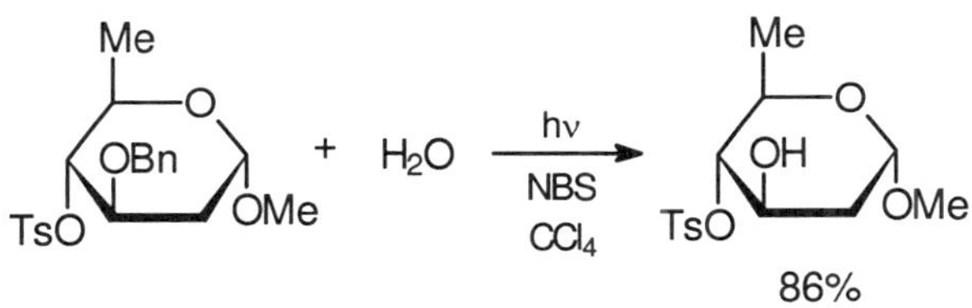

(3)

Equation 3

Reaction of Halogenated Carbohydrates with Organotin Compounds

Another chain reaction that is sometimes photochemically initiated is the reaction of halogenated carbohydrates in the presence of trialkyltin hydride, hexaalkyltin, or an allyltin derivative. These reactions produce organotin radicals that become chain-carrying agents for a variety of reactions, including synthesis of deoxy sugars (Scheme 19.2), deuterated deoxy sugars (eq 4),[20] compounds containing allyl groups (eq 5, the Keck reaction),[21] and, in the presence of electron-deficient alkenes, *C*-glycosides (eq 6).[22]

Deoxyhalogeno Sugars

Direct irradiation of a deoxyhalogeno sugar produces an (n,σ*) excited state.[23] This excited compound fragments at the carbon–halogen bond to give a halogen atom and a carbon radical. Photofragmentation is a logical reaction for an excited halogeno compound because the excitation energy greatly exceeds the energy needed for carbon–halogen bond homolysis. (For example, the excitation energy for methyl iodide is 111 kcal/mol, and the bond dissociation energy for the carbon–iodine bond is 53 kcal/mol.[23]) If reaction of a deoxyiodo sugar is conducted in a hydrogen-donating solvent (e.g., methanol or 2-propanol),

$$Bu_3Sn\cdot + RX \longrightarrow R\cdot + Bu_3SnX$$

$$R\cdot + Bu_3SnH \longrightarrow RH + Bu_3Sn\cdot$$

R = carbohydrate moiety

X = halogen

Scheme 19.2

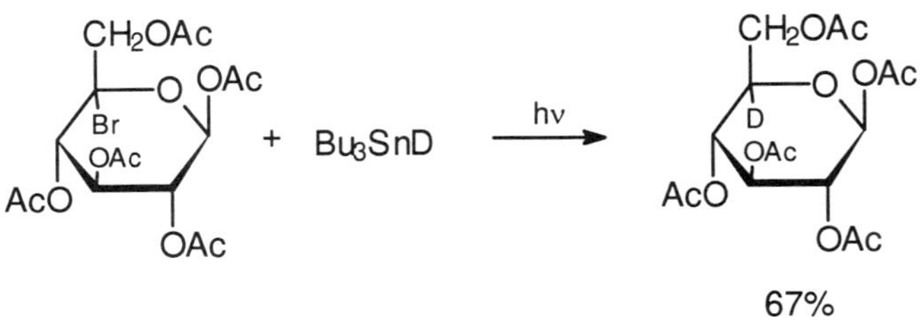

(4)

Equation 4

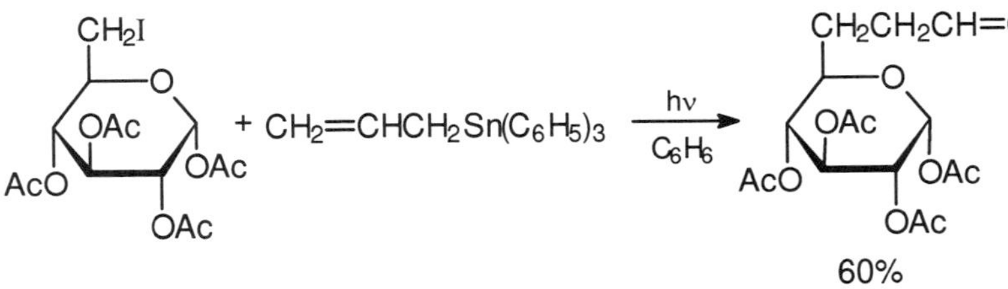

(5)

Equation 5

abstraction of a hydrogen atom from the solvent produces a deoxy sugar (Scheme 19.3). In general, better hydrogen-donating solvents produce higher product yields.[24] In molecules where internal hydrogen abstraction or addition to a π-system is possible, these reaction do occur, but by far the most common process is photosubstitution of a hydrogen atom for a halogen atom (Table 19.1). Deoxyiodo sugars are irradiated more frequently than other halogenated carbohydrates because they are often easier to synthesize and because a carbon–iodine bond absorbs light at longer wavelength and with a greater molar absorption coefficient than other carbon–halogen bonds.[23]

Roth and Binkley found that the epimeric deoxyiodo sugars **4** and **5** interconvert during photolysis (eq 7).[27] Such interconversion is not surprising because recombination of the radical pair formed by bond homolysis easily could cause epimerization (Scheme 19.3). The conversion of **4** into **5**, however, requires contribution from a chain reaction because the quantum yield for this process is 3.3.[27]

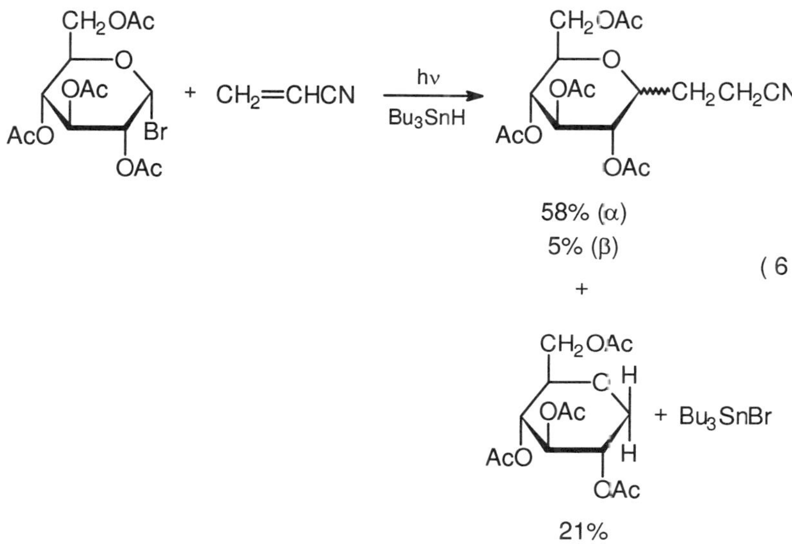

(6)

Equation 6

$$R{-}I \underset{}{\overset{h\nu}{\rightleftharpoons}} R\bullet \;\; \bullet I$$

$$R\bullet + (CH_3)_2CHOH \longrightarrow R{-}H + (CH_3)_2\dot{C}OH$$

6

R = carbohydrate moiety

Scheme 19.3

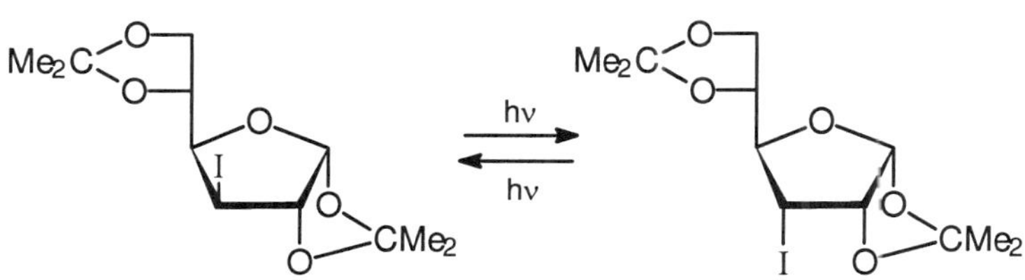

(7)

4 **5**

Equation 7

Table 19.1. Photochemical Reactions of Deoxyhalogeno Sugars

Reactant	*Product*	*Percent Yield*	*Ref.*
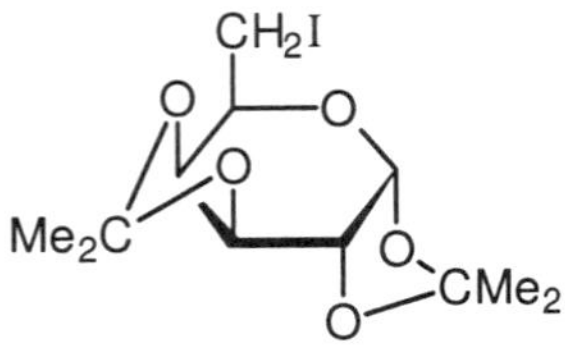	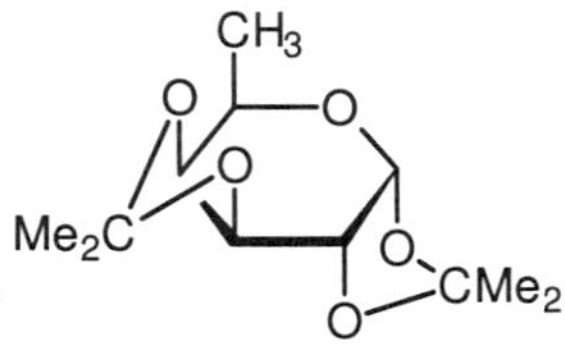	95 (*i*-PrOH) 83 (MeOH) 36 (*t*-BuOH)	24 25, 26[a] 27[b] 25, 26
	+		
	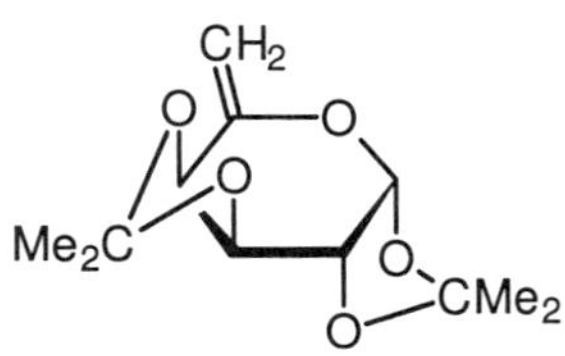	0 (*i*-PrOH) 0 (MeOH) 32 (*t*-BuOH)	24 25, 26[a] 27[b] 25, 26
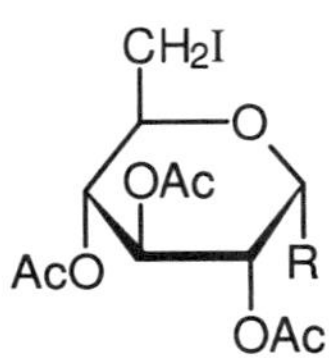	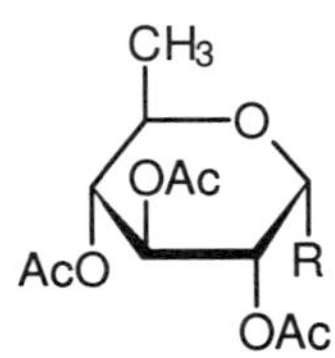	55 (MeOH)	28
R = 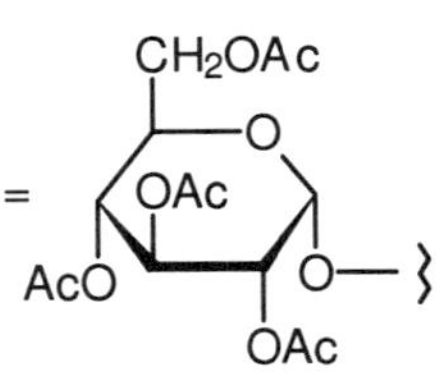			
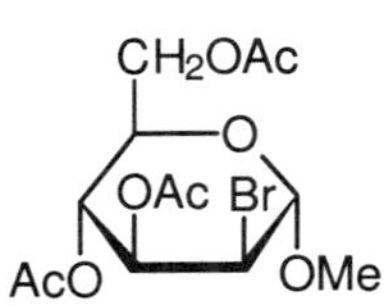	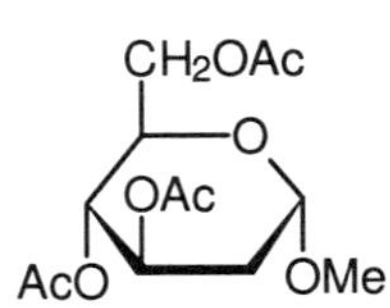	90 (*i*-PrOH)	29

Continued on next page

Table 19.1. Photochemical Reactions of Deoxyhalogeno Sugars—*Continued*

Reactant	*Product*	*Percent Yield*	*Ref.*
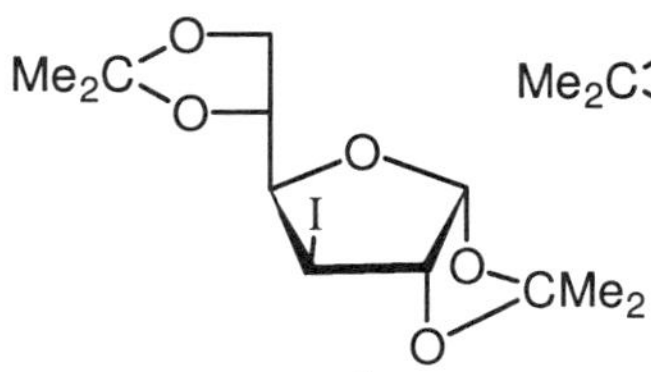 **A**	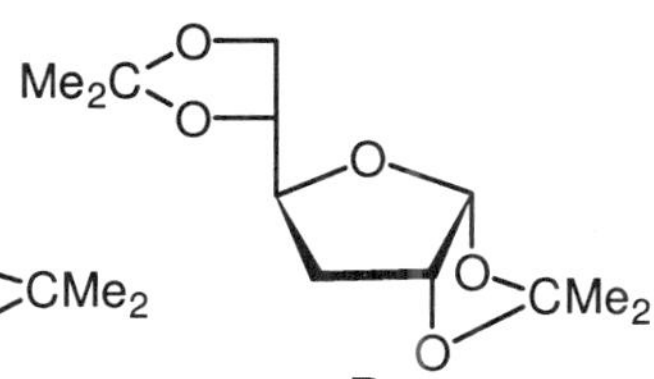 **B**	32 (MeOH) 100 (*i*-PrOH)	30 24
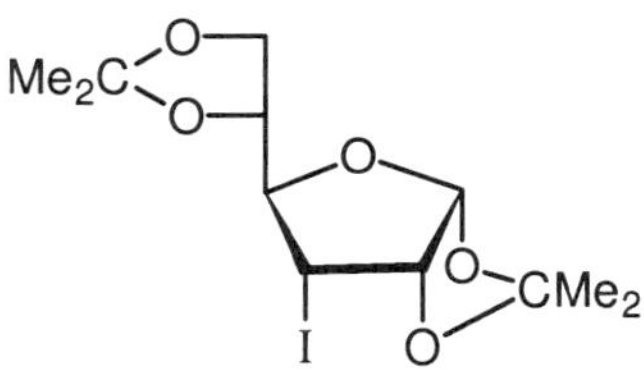 **C**	**B**	99 (*i*-PrOH)	24
A	**C**	99 (heptane)	27
C	**A**	98 (heptane)	27
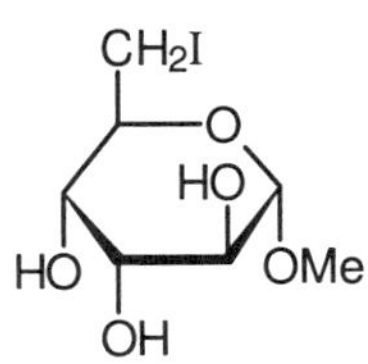 + H_2CO	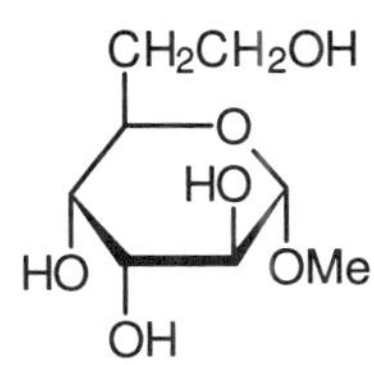 + 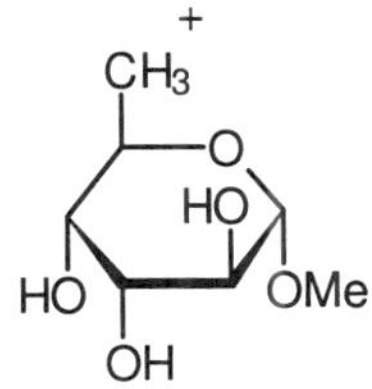	5 9	31
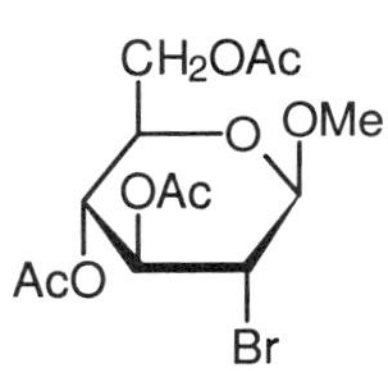	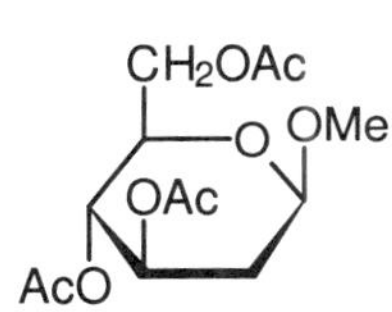	91 (*i*-PrOH)	29

Continued on next page

Table 19.1. Photochemical Reactions of Deoxyhalogeno Sugars—*Continued*

Reactant	*Product*	*Percent Yield*	*Ref.*
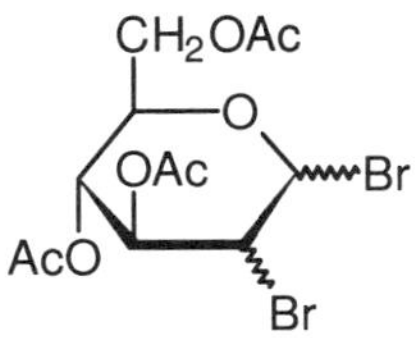	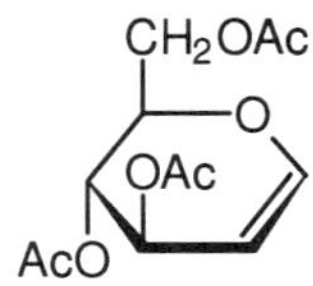	60	29
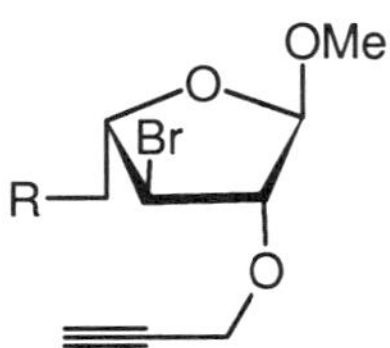	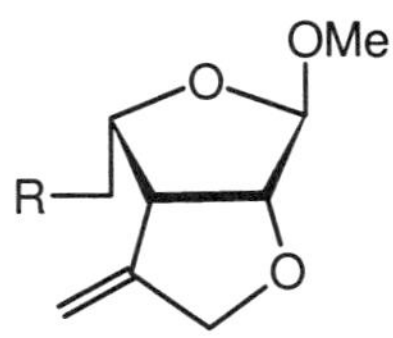		
R = $MsOCH_2$—		60 86	32 33
R = MeOC(=O)—		88	33
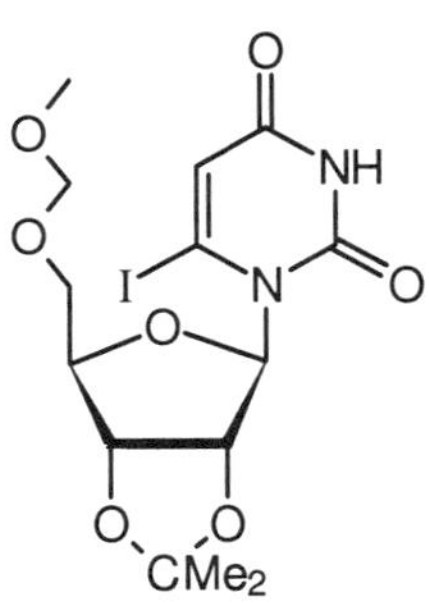	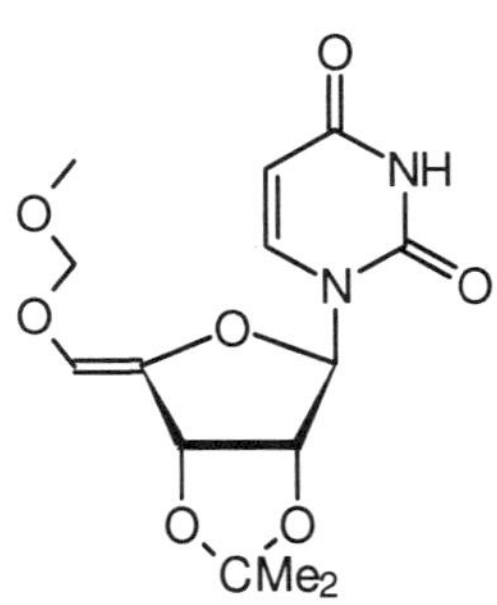	8[c]	34
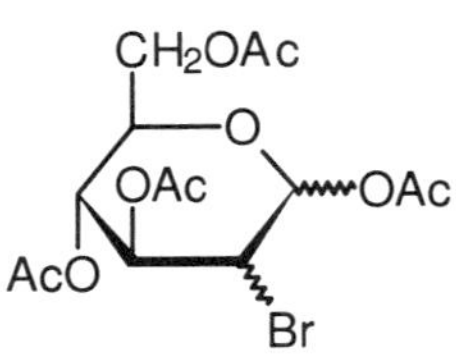	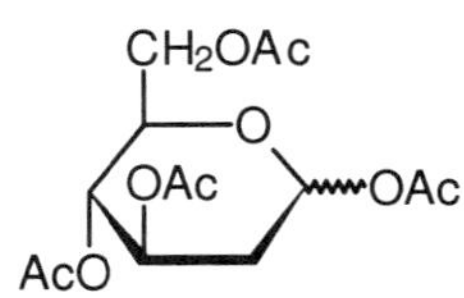	78	29

Continued on next page

Table 19.1. Photochemical Reactions of Deoxyhalogeno Sugars—*Continued*

Reactant	*Product*	*Percent Yield*	*Ref.*
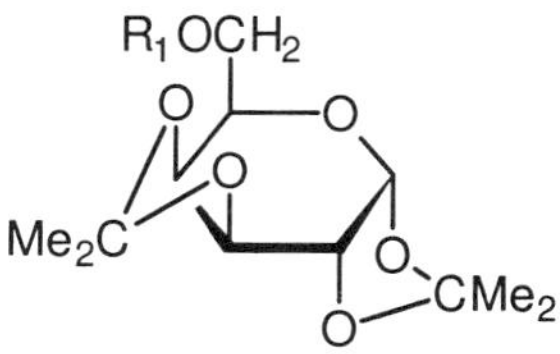	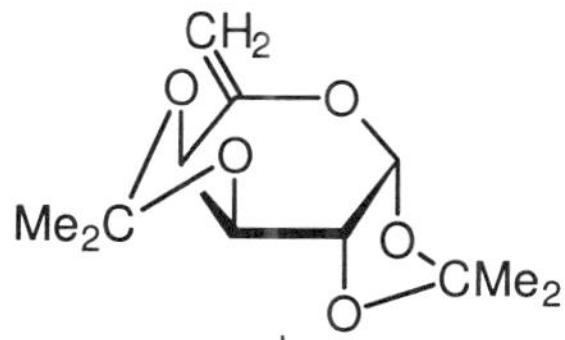	0 (*i*-PrOH) 80 (*t*-BuOH)	35
R_1 and R_2 given after next entry	+ 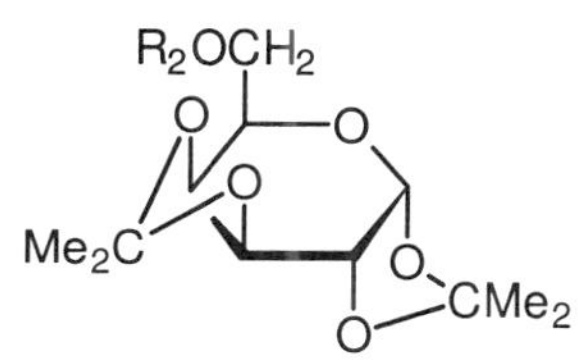	94 (*i*-PrOH) 14 (*t*-BuOH)	35
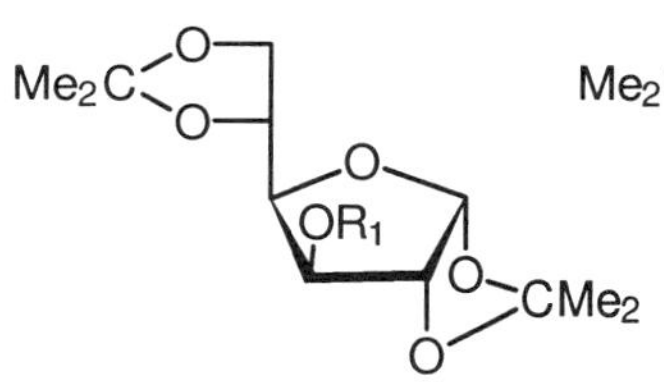	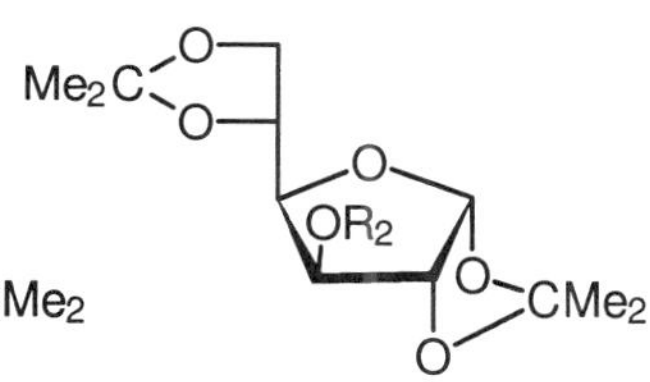	96	35
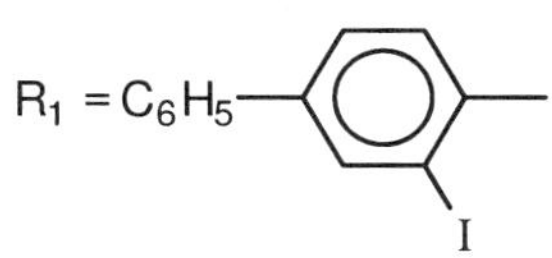	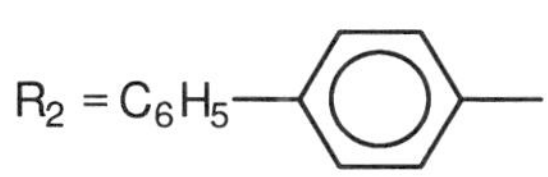		
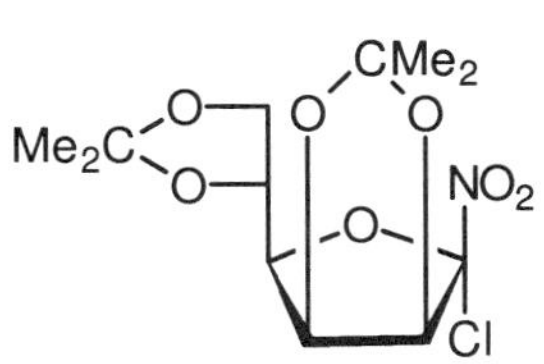 + $R_1R_2R_3C^{\ominus}$	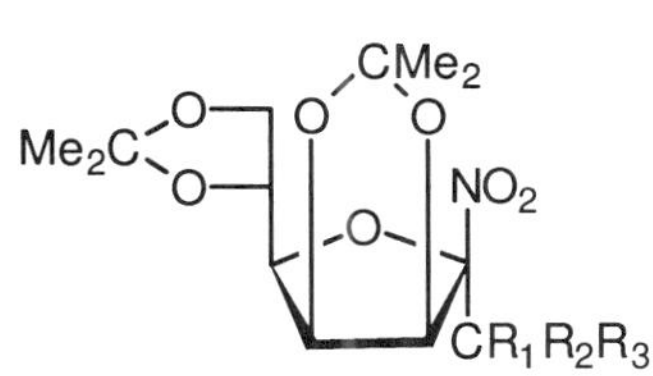 + $Cl^{\ominus}$		36
$R_1 = R_2 = CH_3$, $R_3 = NO_2$		81	
$R_1 = R_2 = H$, $R_3 = NO_2$		72	
$R_1 = R_2 = COOEt$, $R_3 = H$		83[d]	

Continued on next page

Table 19.1. Photochemical Reactions of Deoxyhalogeno Sugars—*Continued*

Reactant	*Product*	*Percent Yield*	*Ref.*
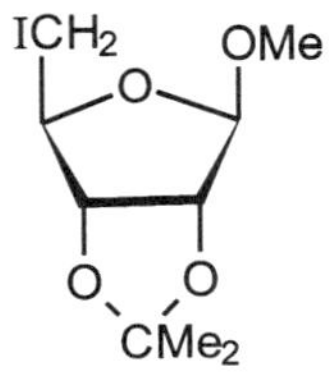	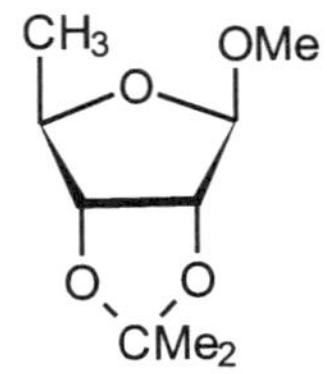	72	3
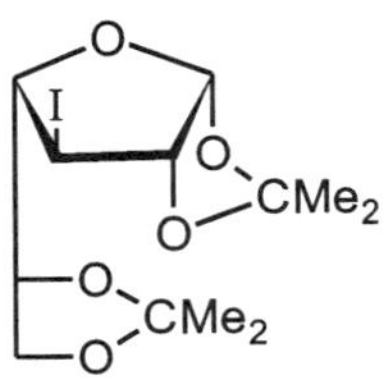	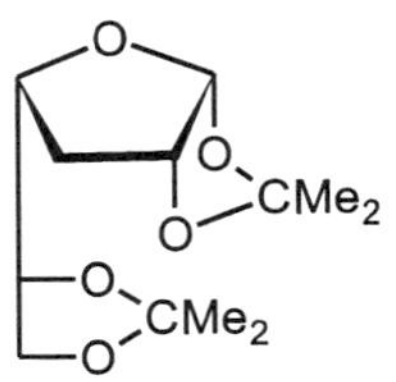	90	3
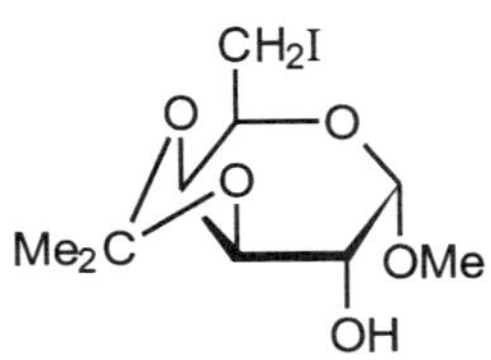	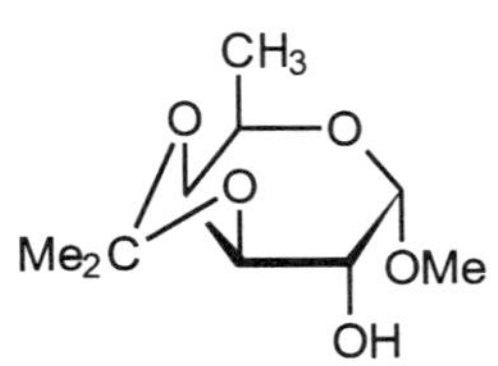	95	3
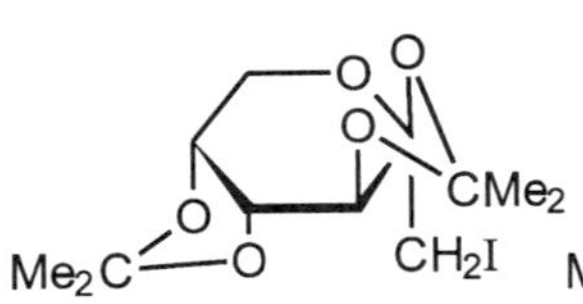	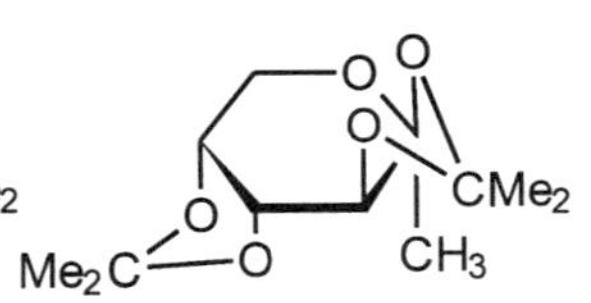	93	3

[a]Different yields were observed from irradiation in different solvents and with different light filters.

[b]A complex mixture formed when oxygen was present in the reaction mixture.

[c]A product was isolated in which substitution on the nitrogenous base had occurred.

[d]The product isolated had eliminated the elements of HNO_2.

Lemmes and von Sonntag proposed a modified version of the mechanism shown in Scheme 19.3 for the conversion of a deoxyhalogeno sugar into a deoxy sugar.[37] Based on γ-radiolysis of solutions of deoxyhalogeno sugars in 2-propanol containing acetone, they suggested that formation of deoxy sugars was taking place by a chain reaction involving the 2-hydroxypropyl radical **6**. Since both photolysis and γ-radiolysis in 2-propanol produce **6** and since γ-radiolysis caused a chain reaction, these authors concluded that a chain reaction was occurring photochemically also. The only quantum yields that have been determined for deoxy sugar formation from deoxyiodo sugar photolysis were measured in heptane and are less than unity; consequently, a chain reaction is not required to explain these results.[27] Such a finding does not exclude the possibility that a chain reaction is taking place in a solvent where the 2-hydroxypropyl radical **6** could function as a chain-carrying agent.

Photochemical electron-transfer reactions also can cause the dehalogenation of carbohydrates. Cossy and co-workers found that irradiation of a bromodeoxy sugar in the presence of an electron donor (triethylamine) converted the sugar into a radical anion (Scheme 19.4). After loss of bromide from the radical anion, the carbon radical that remained added internally to a triple bond to form a new ring system.[32,33]

Hypohalites

Hypohalites are thermally and photochemically unstable compounds that can be synthesized by reaction of alcohols with reagents such as a combination of iodine and (diacetoxyiodo)benzene. Irradiation of hypohalites generates alkoxy radicals and halogen atoms (Scheme 19.5). The most common situation in which carbohydrate hypohalites are irradiated is in the synthesis of spiro compounds in which the anomeric carbon atom is common to both rings

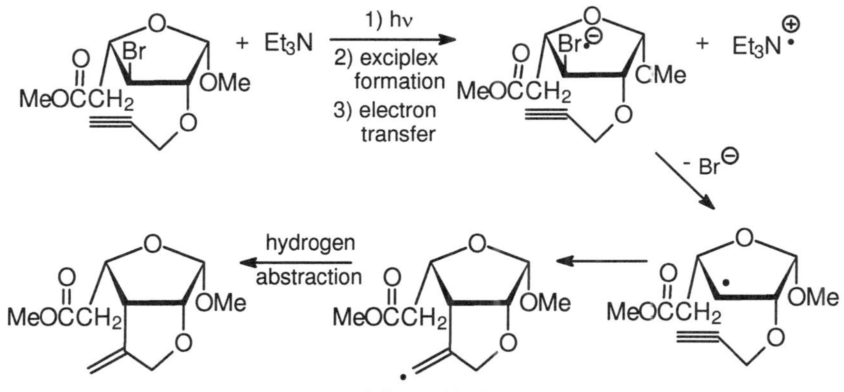

Scheme 19.4

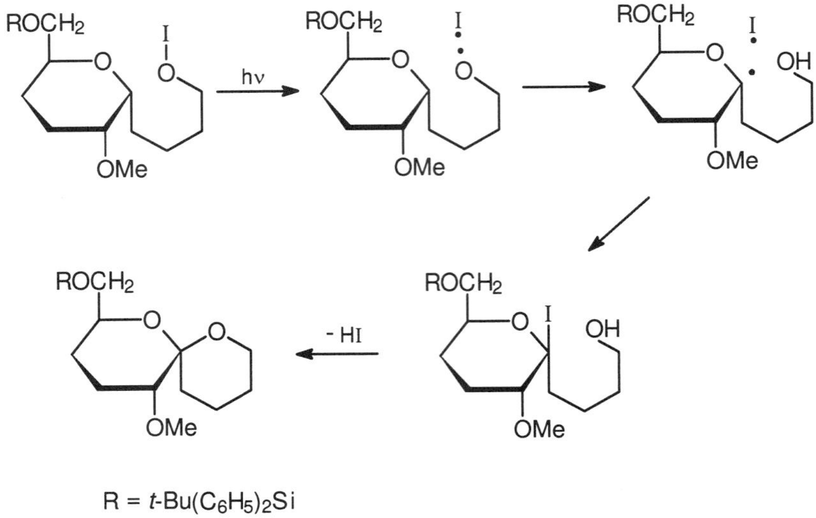

Scheme 19.5

(Table 19.2). Ring formation presumably comes from spontaneous elimination of the elements of HI from the halogenated alcohol formed by the photochemical reaction (Scheme 19.5).

Halogenated Nucleotides

Replacement of thymidine in DNA with 5-bromdeoxyuridine greatly increases the sensitivity of DNA to photochemical degradation. Photolysis of DNA modified in this manner creates alkali-labile sites, causes strand breaks, and promotes cross-linking with proteins.[46,47] Saito and co-workers have proposed the mechanism shown in Scheme 19.6 for photochemical reaction of polynucleotides containing 5-halogenodeoxyuridine. These workers studied a group of hexanucleotides in which thymidine was replaced by a 5-halogenodeoxyuridine (Table 19.3). They identified two fragmentation pathways that have a common beginning; that is, in each case internal transfer of an electron is followed by loss of bromide ion (Scheme 19.6). Once bromide ion has departed, the direction of the reaction is thought to be determined by whether the 5-uracilyl radical **7** abstracts a hydrogen atom from C-1′ or C-2′ in the neighboring adenosine. Cook and Greenberg also have studied the mechanism for photochemical reaction of 5-bromdeoxyuridine-containing polynucleotides.[51] They agree with Saito that electron transfer followed by hydrogen abstraction describes the basic reaction pathway, but they argue that hydrogen abstraction

Table 19.2. Hypohalite Photolysis

Reactant	*Product*	*Percent Yield*	*Ref.*
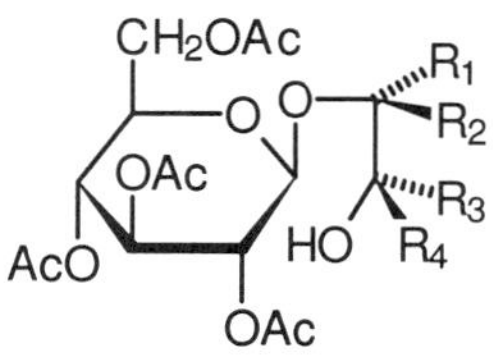	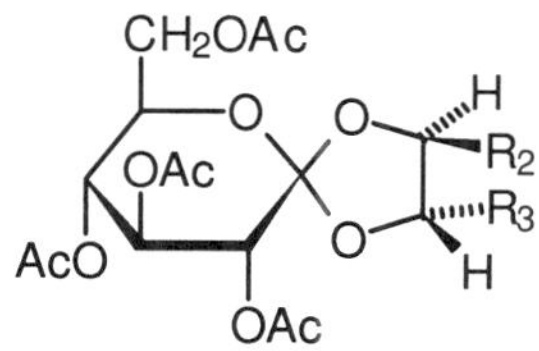		
+ I_2 + HgO			
$R_1 = R_2 = R_3 = R_4 = H$		68	38, 39
$R_1 = R_2 = R_3 = R_4 = H$ (α anomer)		60	39
$R_2, R_3 = (CH_2)_4$, $R_1 = R_4 = H$		38	38, 39
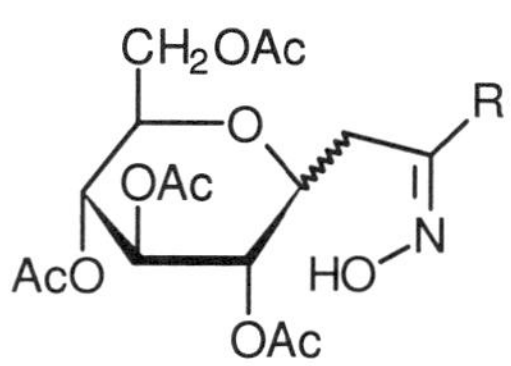	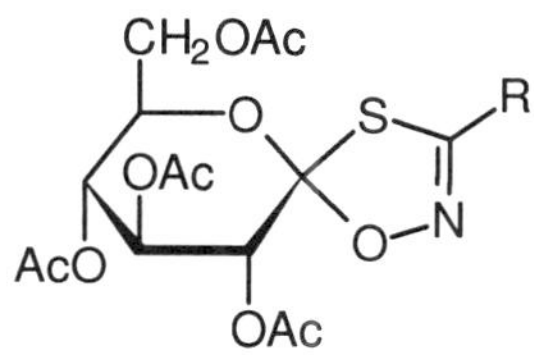 A		40, 41a
+			
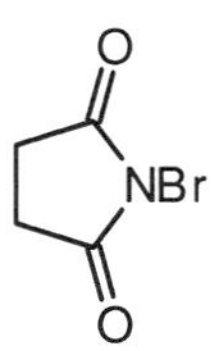	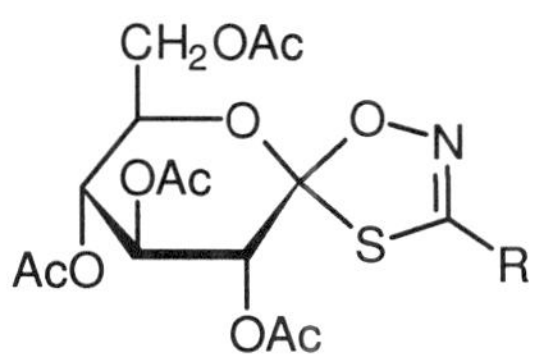 B		
		A B	
R = C_6H_5(β)		50 10	
R = C_6H_5(α)		44 0	

Continued on next page

Table 19.2. Hypohalite Photolysis—*Continued*

Reactant	*Product*	*Percent Yield*	*Ref.*
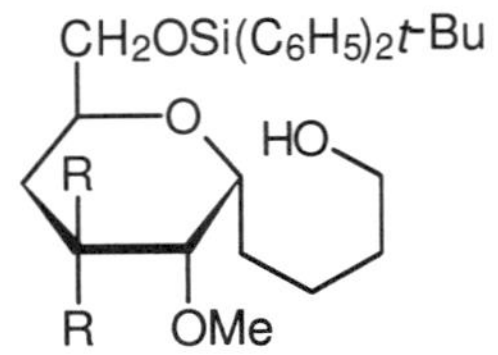 + I_2 + $C_6H_5I(OAc)_2$	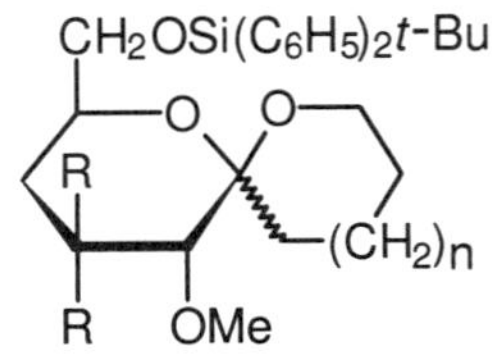		
R = H, n = 1		52	42
R, R = $-(CH_2)_4O-$, n = 0		56	42
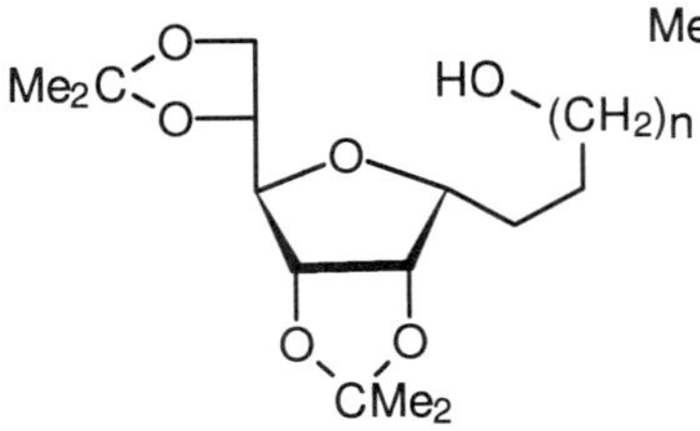 + I_2 + $C_6H_5I(OAc)_2$	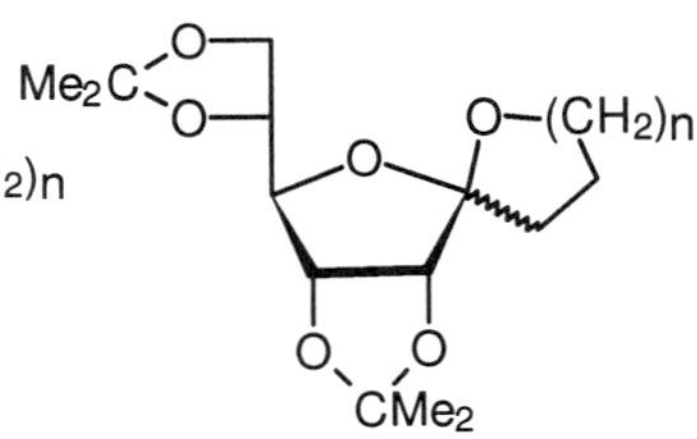		43
	n = 1	67	44, 45
	n = 2	33[b]	45
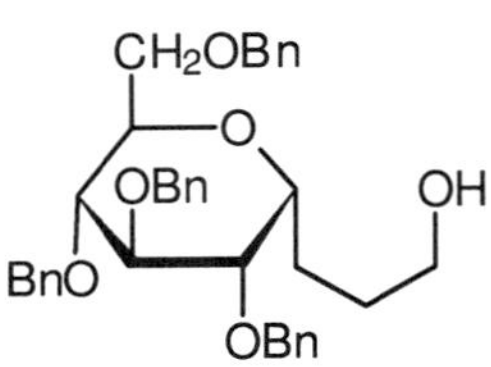 + I_2 + $C_6H_5I(OAc)_2$	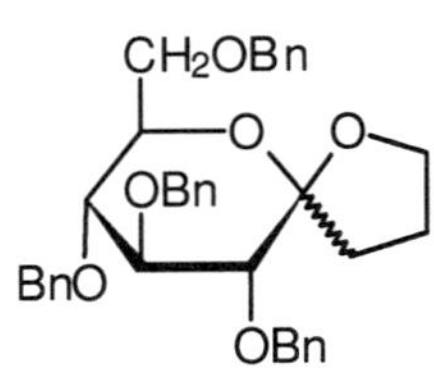	68	44, 45

Continued on next page

Table 19.2. Hypohalite Photolysis—*Continued*

Reactant	*Product*	*Percent Yield*	*Ref.*
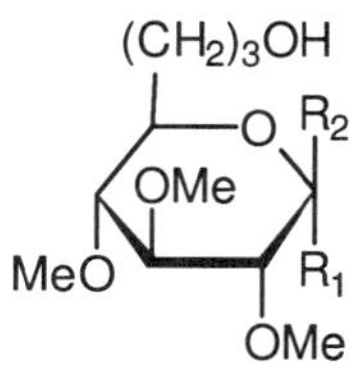 + I_2 + $C_6H_5I(OAc)_2$	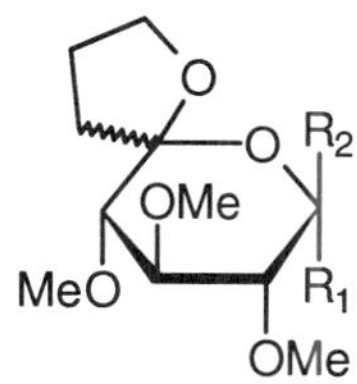		
R_1 = OMe, R_2 = H		75	44, 45
R_1 = H, R_2 = OMe		86	44, 45
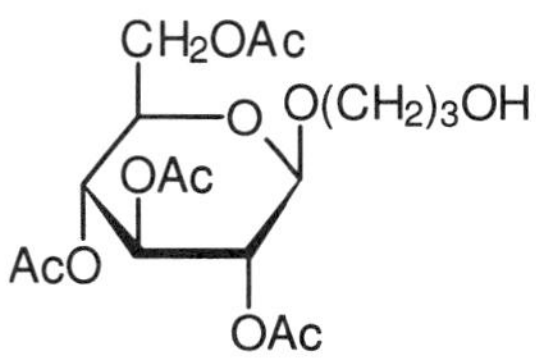 + I_2 + HgO	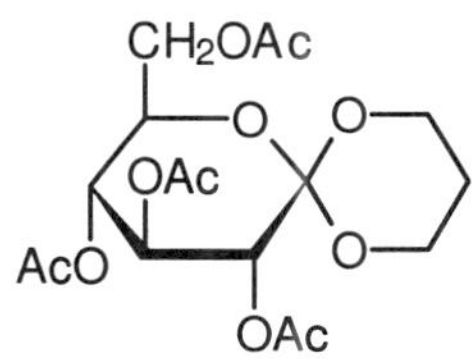	25	39
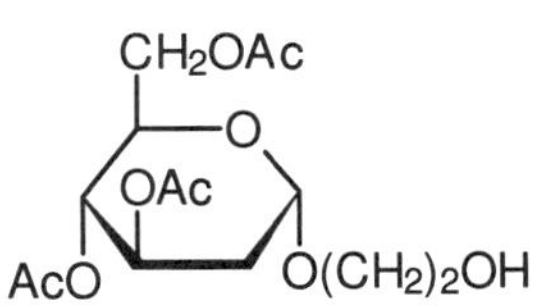 + I_2 + HgO	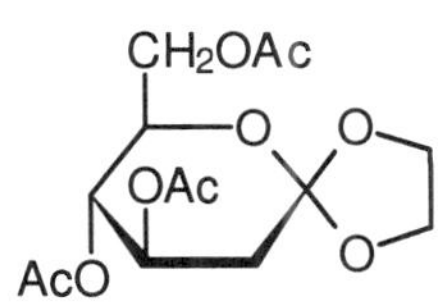	37	43

[a]Compounds with eight other "R" groups were irradiated.
[b]Small amounts of iodine-containing compounds were isolated.

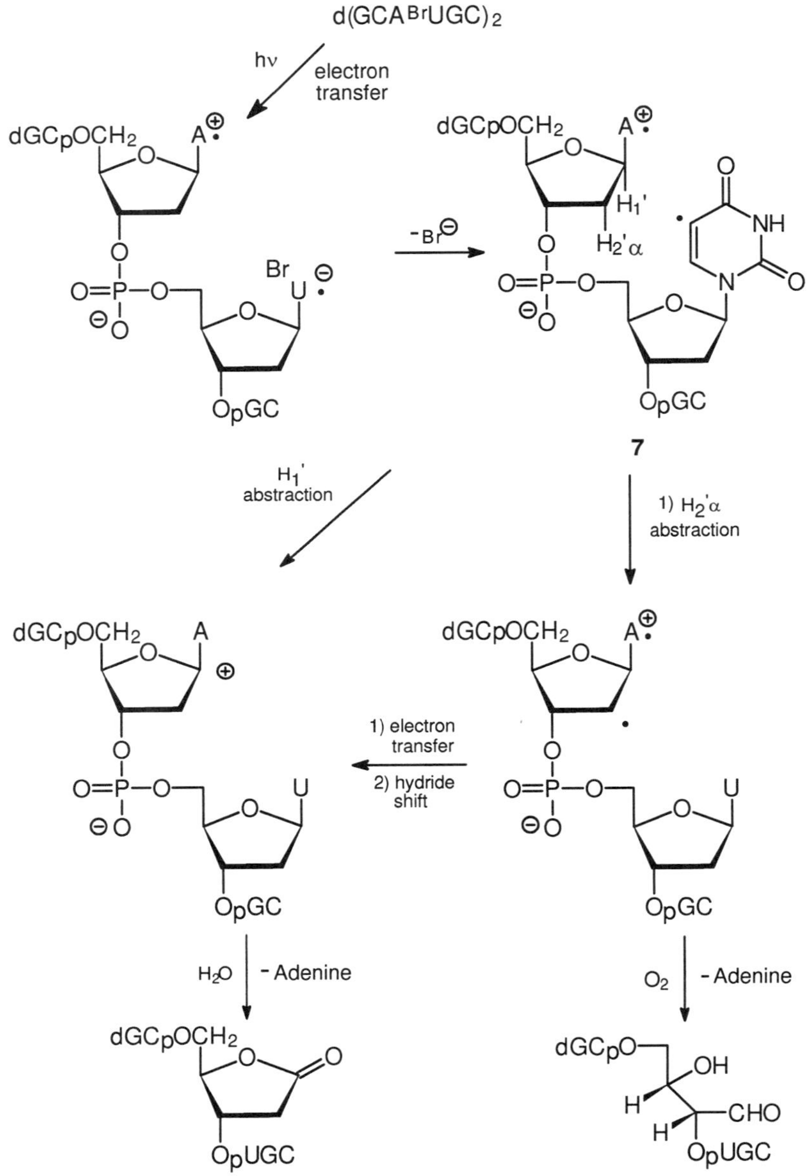
d(GCABrUGC)2
hν
electron transfer
-Br
7
H1' abstraction
1) H2'α abstraction
1) electron transfer
2) hydride shift
H2O
- Adenine
O2
- Adenine
dGCpOCH2
OpGC
OpUGC
dGCpO
CHO

Scheme 19.6

Table 19.3. Photochemical Reactions of Halogenated Nucleotides and Polynucleotides

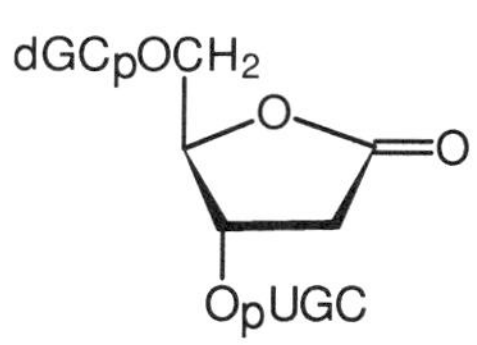

A

d(GCAUGC)

B

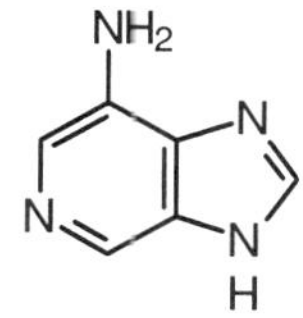

C

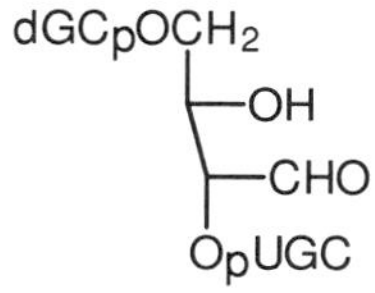

D

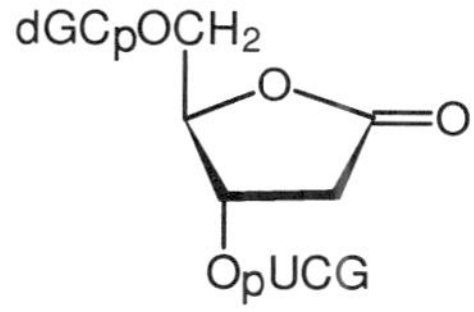

E

d(GCAUCG)

F

Reactant	Product	Percent Yield	Ref.
d(GCA^{Br}UGC)$_2$	**A**	47 38	46, 47 50
	B	10	46
	C	61	46, 47
	D	6.4	50
d(GCA^{I}UGC)$_2$	**A**	13 14	47-49 50
	C	29	47-49
	D	7 13	47-49 50

Continued on next page

Table 19.3. Photochemical Reactions of Halogenated Nucleotides and Polynucleotides—*Continued*

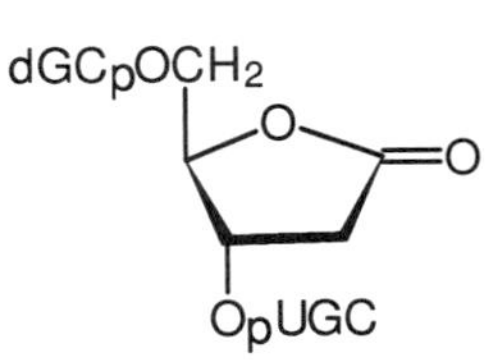

A

d(GCAUGC)

B

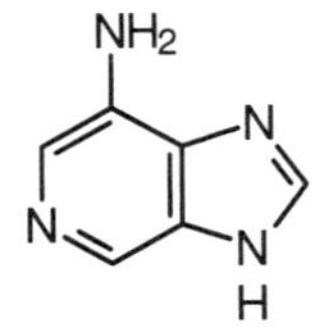

C

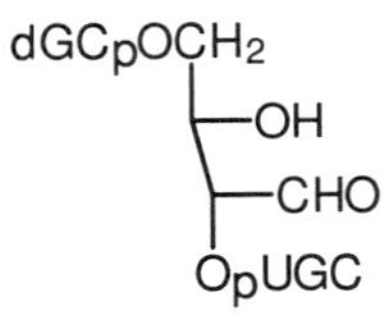

D

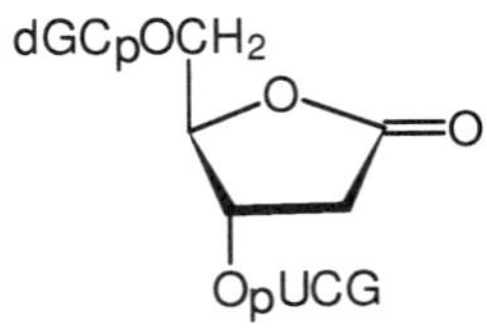

E

d(GCAUCG)

F

Reactant	Product	Percent Yield	Ref.
d(GCA^{Br}UCG) + d(CGTAGC)	E	29[a]	46
	F	8[a]	
	C	41[a]	
d(GC$^{2'I}$UUGC) + d(GCAAGC)	A	7.6	47
	D	9.5	

Continued on next page

Table 19.3. Photochemical Reactions of Halogenated Nucleotides and Polynucleotides—*Continued*

Reactant	*Product*	*Percent Yield*	*Ref.*
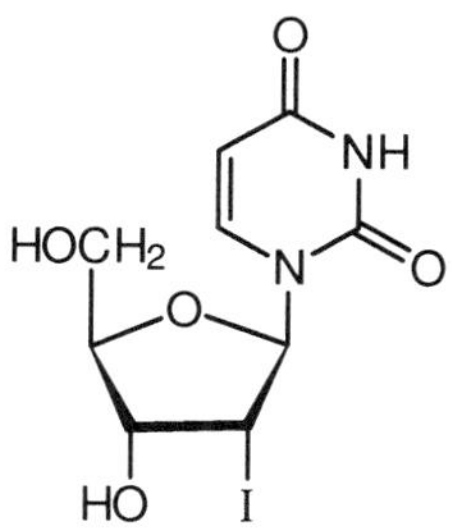	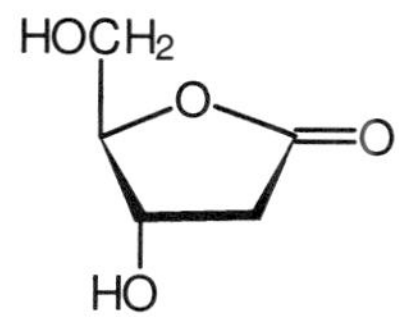	1.5	49[b]
	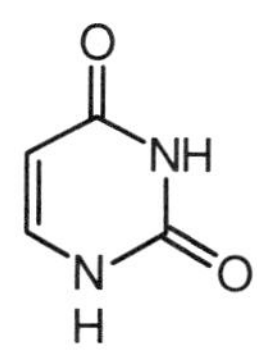	11	
	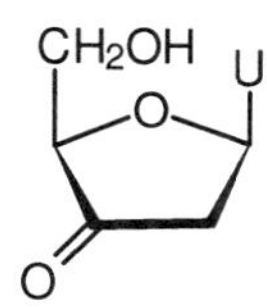	4	
	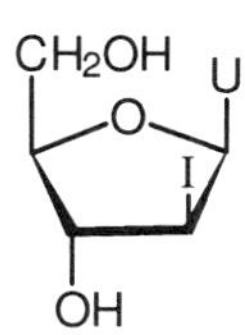	10	

[a]Without the d(CGTAGC) reaction was much slower.
[b]Seven other products were formed in yields of 2% or less.

only occurs from C-2′ and that the apparent abstraction from C-1′ is actually the result of a second electron transfer followed by a hydride shift (Scheme 19.6).

Summary

Many of the important, photochemical reactions of halogens with carbohydrates and of halogenated carbohydrates involve chain reactions in which light functions as an initiator. In reactions where an excited state of a carbohydrate actually is formed, the most frequently observed process is homolysis of a carbon–halogen bond. The carbon radicals thus produced typically abstract hydrogen atoms from the solvent. In polynucleotides, internal hydrogen abstraction by photochemically generated carbon radicals leads to compounds that are unstable and easily undergo further reaction resulting in cleavage of the polynucleotide chain.

References

1. Somsák, L.; Ferrier, R. J. *Adv. Carbohydr. Chem. Biochem.* **1991**, *49*, 37.
2. Blattner, R.; Ferrier, R. J. *J. Chem. Soc., Perkin Trans. 1* **1980**, 1523.
3. Hanessian, S.; Pleassas, N. R. *J. Org. Chem.* **1969**, *34*, 1035.
4. Hullar, T. L.; Siskin, S. B. *J. Org. Chem.* **1970**, *35*, 225.
5. Hanessian, S.; Pleassas, N. R. *J. Org. Chem.* **1969**, *34*, 1045.
6. Hanessian, S.; Pleassas, N. R. *J. Org. Chem.* **1969**, *34*, 1053.
7. Gelas, J. *Adv. Carbohydr. Chem. Biochem.* **1981**, *39*, 71.
8. Chana, J. S.; Collins, P. M.; Farnia, F.; Peacock, D. J. *J. Chem. Soc., Chem. Commun.* **1988**, 94.
9. Collins, P. M.; Manro, A.; Opara-Mottah, E. C.; Ali, M. H. *J. Chem. Soc., Chem. Commun.* **1988**, 272.
10. Collins, P. M.; Ali, M. H. *Tetrahedron Lett.* **1990**, *31*, 4517.
11. Myers, A. G.; Gin, D. Y.; Rogers, D. H. *J. Am. Chem. Soc.* **1993**, *115*, 2036.
12. Binkley, R. W.; Goewey, G. S; Johnston, J. C. *J. Org. Chem.* **1984**, *49*, 992.
13. Binkley, R. W. *J. Carbohydr. Chem.* **1985**, *4*, 227.
14. Adinolfi, M.; Barone, G.; Corsaro, M. M.; Lanzetta, R.; Mangoni, L.; Monaco, P. *J. Carbohydr. Chem.* **1995**, *14*, 913.
15. Han, O.; Liu, H.-W. *Tetrahedron Lett.* **1987**, *28*, 1073.
16. BeMiller, J. N.; Wing, R. E.; Meyers, C. Y. *J. Org. Chem.* **1968**, *33*, 4292.
17. Binkley, R. W.; Hehemann, D. G.; *J. Org. Chem.* **1990**, *55*, 378.
18. Binkley, R. W. *J. Carbohydr. Chem.* **1990**, *9*, 507.
19. Giuliano, R. M.; Villani, F. J., Jr. *J. Org. Chem.* **1995**, *60*, 202.
20. Korth, H.-G.; Sustmann, R.; Giese, B.; Rückert, B.; Gröninger, K. S. *Chem. Ber.* **1990**, *123*, 1891.
21. Praly, J.-P.; El Kharraf, Z.; Descotes, G. *Carbohydr. Res.* **1992**, *232*, 117.
22. Araki, Y.; Endo, T.; Tanji, M.; Nagasawa, J.; Ishido, Y. *Tetrahedron Lett.* **1987**, *28*, 5853.
23. Kopecký, J. *Organic Photochemistry, a Visual Approach;* VCH: New York, 1992; p 169.
24. Binkley, R. W.; Hehemann, D. G. *Carbohydr. Res.* **1979**, *74*, 337.
25. Binkley, W. W.; Binkley, R. W. *Carbohydr. Res.* **1968**, *8*, 370.
26. Binkley, W. W.; Binkley, R. W. *Carbohydr. Res.* **1969**, *11*, 1.

27. Roth, R. C.; Binkley, R. W. *J. Org. Chem.* **1985**, *50*, 690.
28. Guilloux, E. R.; Defaye, J.; Bell, R. H.; Horton, D. *Carbohydr. Res.* **1971**, *20*, 421.
29. Binkley, R. W.; Bankaitis, D. *J. Carbohydr. Chem.* **1982**, *1*, 1.
30. Bell, R. H.; Horton, D.; Williams, D. M.; Winter-Mihaly, E. *Carbohydr. Res.* **1977**, *58*, 109.
31. Garegg, P. J.; Hoffman, J.; Lindberg, B.; Samuelsson, B. *Carbohydr. Res.* **1978**, *67*, 263.
32. Cossy, J.; Ranaivosata, J.-L.; Bellosta, V. *Tetrahedron Lett.* **1994**, *35*, 3161.
33. Cossy, J.; Ranaivosata, J.-L.; Bellosta, V. *Tetrahedron* **1996**, *52*, 629.
34. Satoh, K.; Tanaka, H.; Andoh, A.; Miyasaka, T. *Nucleosides Nucleotides* **1986**, *5*, 461.
35. Binkley, R. W.; Koholic, D. J. *J. Org. Chem.* **1979**, *44*, 3357.
36. Aebischer, B.; Meuwly, R.; Vasella, A. *Helv. Chim. Acta* **1984**, *67*, 2236.
37. Lemmes, R.; von Sonntag, C. *Carbohydr. Res.* **1982**, *105*, 276.
38. Praly, J.-P.; Descotes, G. *Tetrahedron Lett.* **1982**, *23*, 849.
39. Praly, J.-P.; Descotes, G.; Grenier-Loustalot, M.-F., Metras, F. *Carbohydr. Res.* **1984**, *128*, 21.
40. Praly, J.-P.; Boyé, S.; Joseph, B.; Rollin, P. *Tetrahedron Lett.* **1993**, *34*, 3419.
41. Praly, J.-P.; Faure, R.; Joseph, B.; Kiss, L.; Rollin, P. *Tetrahedron* **1994**, *50*, 6559.
42. Dorta, R. L.; Martin, A.; Salazar, J. A.; Suárez, E. *Tetrahedron Lett.* **1996**, *37*, 6021.
43. Descotes, G.; Idmoumaz, H.; Praly, J.-P. *Carbohydr. Res.* **1984**, *128*, 341.
44. Martín, A.; Salazar, J. A.; Suárez, E. *Tetrahedron Lett.* **1995**, *36*, 4489.
45. Martín, A.; Salazar, J. A.; Suárez, E. *J. Org. Chem.* **1996**, *61*, 3999.
46. Sugiyama, H.; Tsutsumi, Y.; Saito, I. *J. Am. Chem. Soc.* **1990**, *112*, 6720.
47. Saito, I. *Pure Appl. Chem.* **1992**, *64*, 1305.
48. Sugiyama, H.; Tsutsumim Y.; Fujimoto, K.; Saito, I. *J. Am. Chem. Scc.* **1993**, *115*, 4443.
49. Sugiyama, H.; Fujimoto, K.; Saito, I. *J. Am. Chem. Soc.* **1995**, *117*, 2945.
50. Sugiyama, H.; Fujimoto, K., Saito, I.; Kawashima, E.; Sekine, T.; Ishido, Y. *Tetrahedron Lett.* **1996**, *37*, 1805.
51. Cook, G. P.; Greenberg, M. M. *J. Am. Chem. Soc.* **1996**, *118*, 10025.

20

Organometallic Compounds

Introduction

One of the aspects of photolysis of unprotected carbohydrates that was discussed in Chapter 11 is that irradiation of these compounds in the presence of metal ions changes their photochemistry. This change is attributed to complexation between the metal ion and oxygen atoms of the carbohydrate, an association that produces an organometallic complex that then undergoes photochemical reaction. Although spectroscopic evidence for the existence of these complexes in solution is convincing, the association is typically not sufficiently strong to allow them to be isolated. If, however, covalent bonding occurs between a carbon atom of a carbohydrate and metal ion, an isolable, organometallic compound often is formed. The photochemistry of carbohydrates covalently bonded to metal ions is the subject of the present chapter.

Cobaloximes

A carbon–cobalt bond readily cleaves homolytically upon photolysis; thus, a compound containing this bonding arrangement, such as a carbohydrate cobaloxime (**1**), is a good source of carbohydrate radicals (eq 1). A carbohydrate radical formed by cobaloxime photolysis can undergo typical radical reactions such as addition to an unsaturated compound[1] (Scheme 20.1), reaction with an ion[2] (eq 2), or combination with a species that has an unshared electron[3] (eq 3). In the absence of compounds that react readily with radicals, homolysis of a cobalt–carbon bond in a cobaloxime is followed by recombination to regenerate the starting material (or its epimer) or by elimination to give an unsaturated carbo-

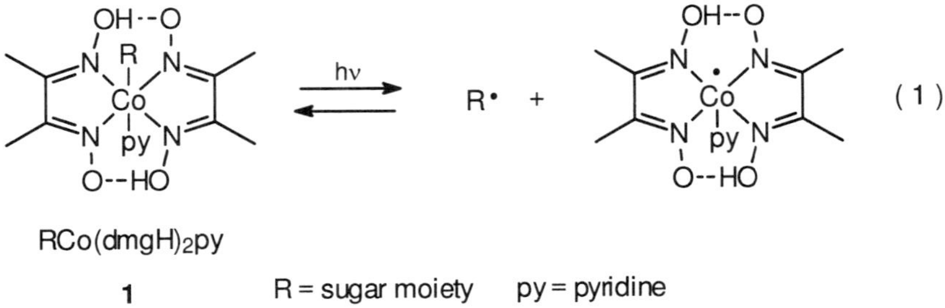

Equation 1

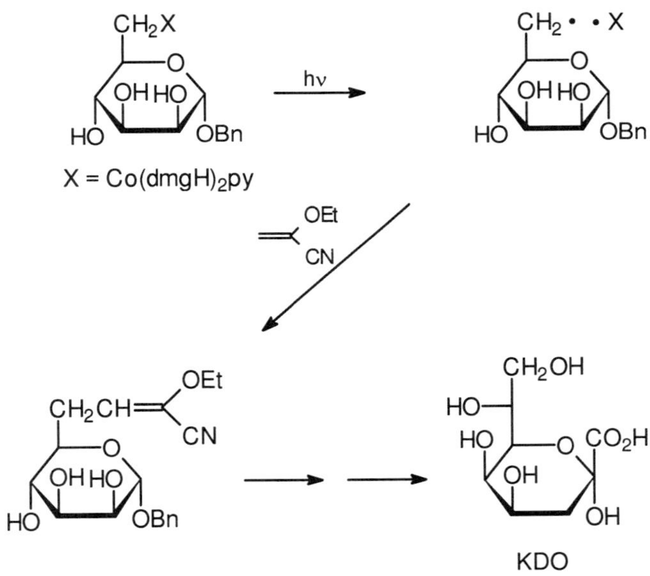

Scheme 20.1

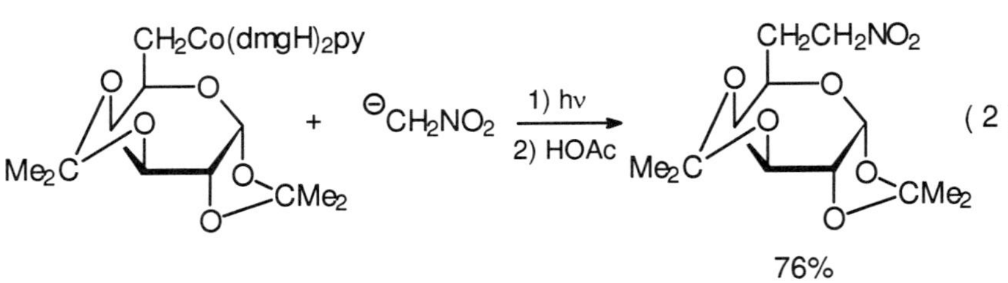

Equation 2

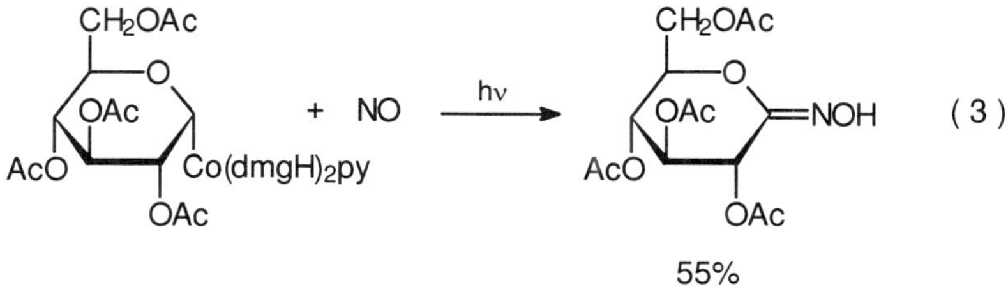

Equation 3

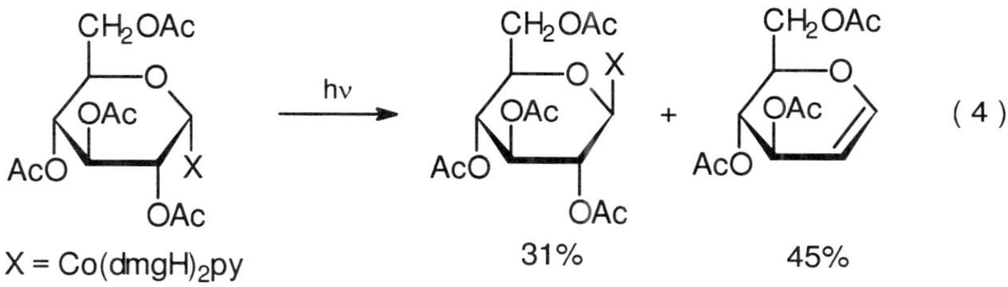

Equation 4

hydrate (eq 4).[3] A listing of the photochemical reactions of carbohydrate cobaloximes is given in Table 20.1.

Radical reactions brought about by photolysis of carbohydrate cobaloximes have some distinct synthetic advantages. Branchaud and Yu have pointed out that these reactions are run under conditions compatible with a wide variety of functional groups and that reactions can be conducted on unprotected sugars.[2] Since radicals do not easily eliminate a group on a carbon atom β to the radical center, they stand in contrast to carbanions, which readily eliminate β-substituents. The probability that an elimination reaction might occur reduces the usefulness of carbanions (but not radicals) in carbon–carbon bond construction reactions of highly functionalized compounds such as carbohydrates.

Undoubtedly, part of the motivation for investigating the synthetic possibilities of photolysis of cobaloximes came from knowledge of the photochemical reactivity of vitamin B_{12} coenzymes, that is, organocobalt compounds that generate carbon radicals upon photolysis (Scheme 20.2).[15, 16] (The structures of vitamin B_{12} and its coenzymes are sufficiently complicated that they will not be reproduced here. They are often represented symbolically, as is done in Figure 20.1, in a manner that emphasizes the presence of a carbon–cobalt bond.[16] For complete structures of these compounds see references 17 and 18.) The photochemical reactions of vitamin B_{12} coenzymes and related compounds are listed in Table 20.2.

Table 20.1. Reactions of Carbohydrate Cobaloximes[a]

Reactant	*Product and Percent Yield*	*Ref.*
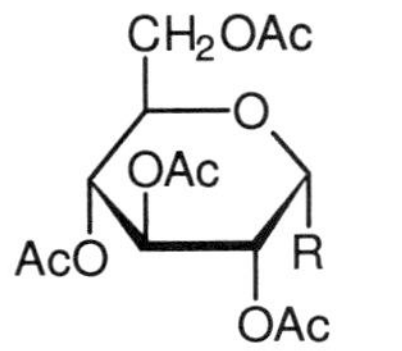 R = Co(dmgH)$_2$py **A**	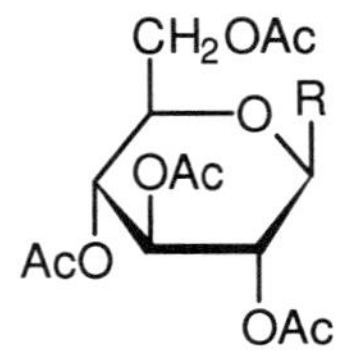 31% + 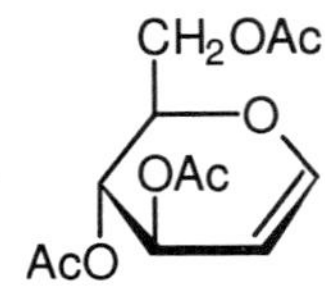 45%	3
A + NO	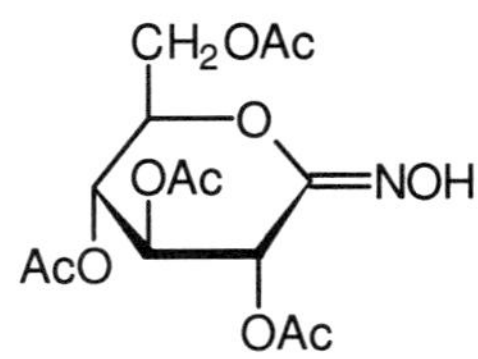 55%	3, 4
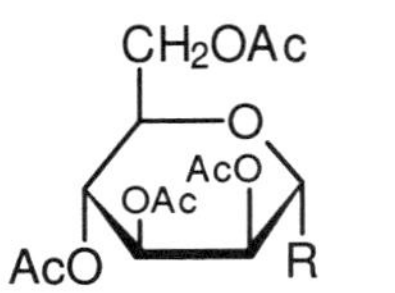 R = Co(dmgH)$_2$py **B**	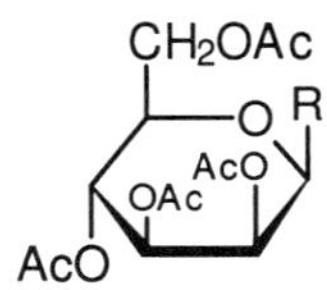 36% + 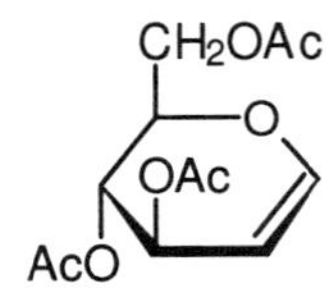 24%	3
B + NO	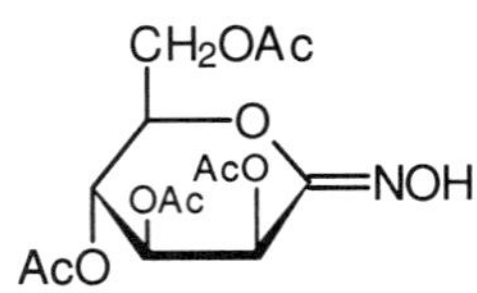 76%	3, 4, 5
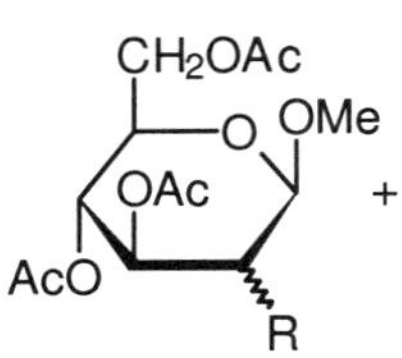 + NO R = Co(dmgH)$_2$py	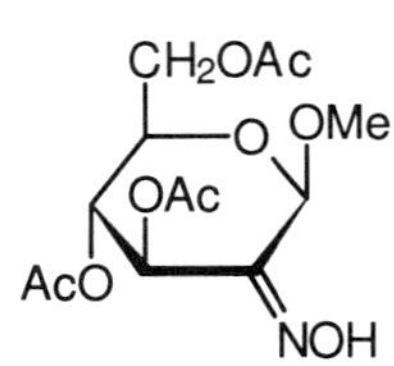 75%	4

Continued on next page

Table 20.1. Reactions of Carbohydrate Cobaloximes[a]—*Continued*

Reactant	*Product*	*Percent Yield*	*Ref.*
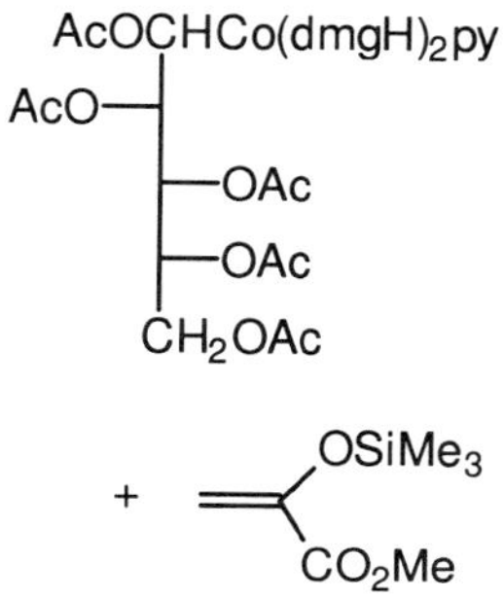	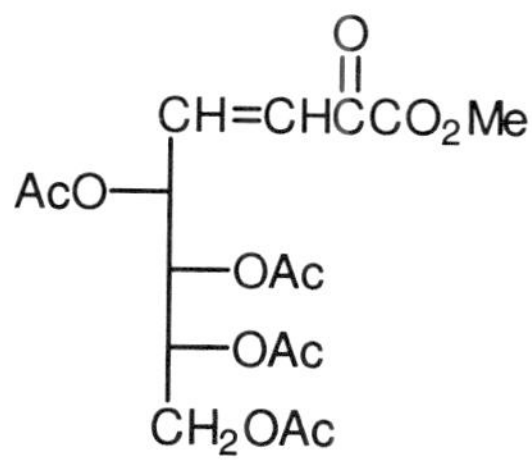	72	6
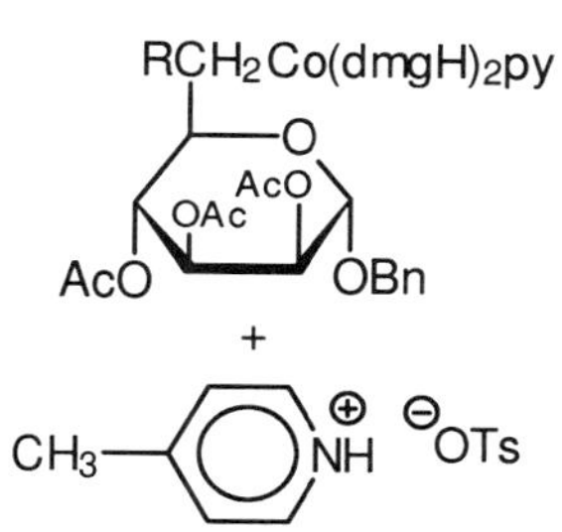	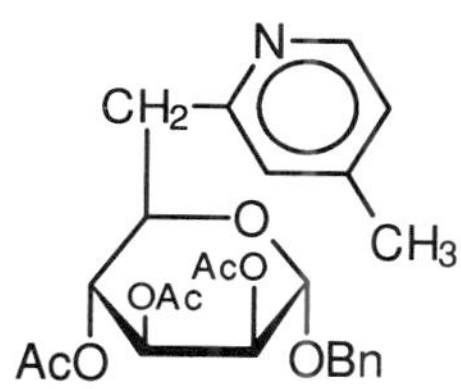	58-64	7
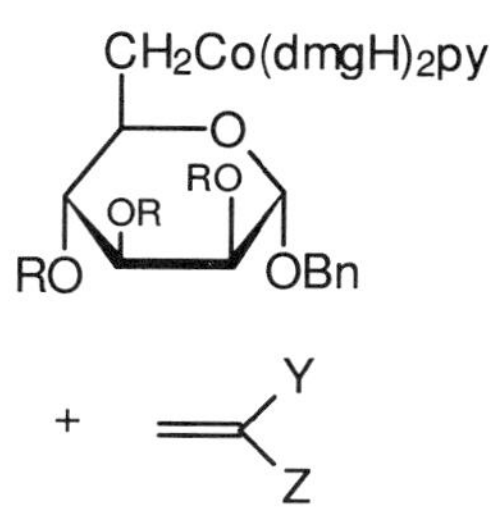	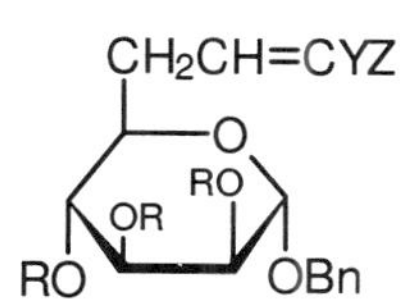		
	R = H, Y = H, Z = C_6H_5	63	8, 9
	R = Ac, Y = H, Z = C_6H_5	85	8, 9
	R = H, Y = CN, Z = OEt	71-81	1, 9
	R = Ac, Y = CN, Z = OEt	83	1, 9

Continued on next page

Table 20.1. Reactions of Carbohydrate Cobaloximes[a]—*Continued*

Reactant	*Product*	*Percent Yield*	*Ref.*
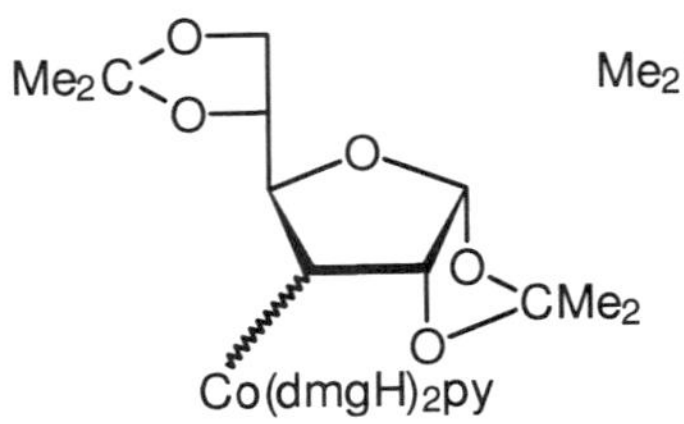 + 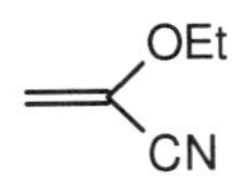	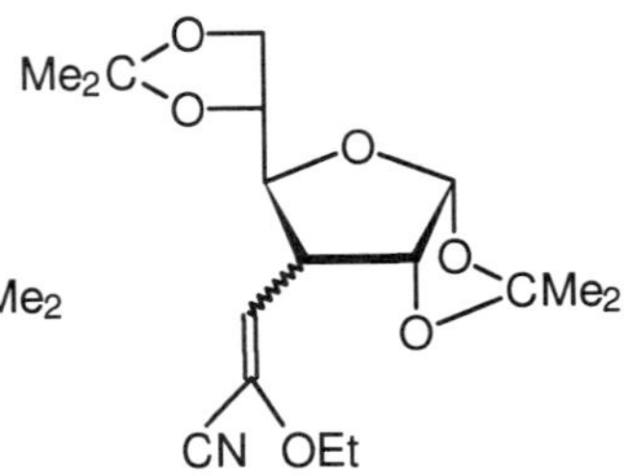	62	2
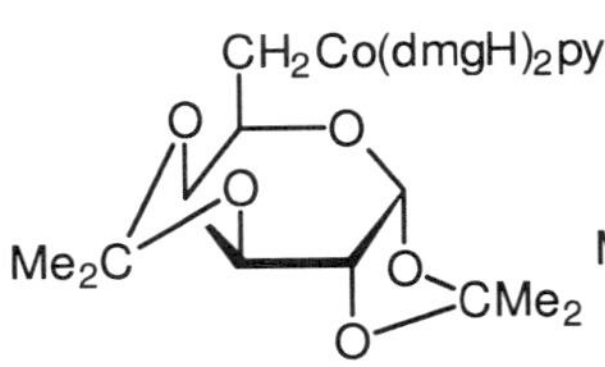 + X	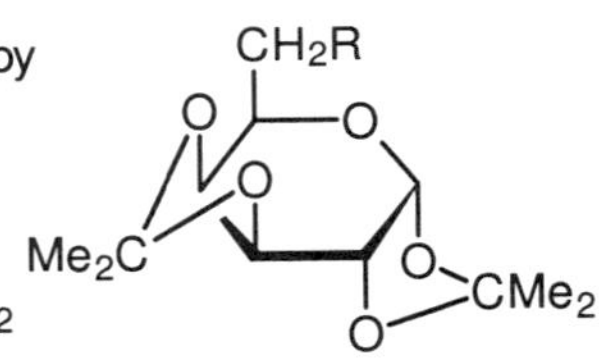 **C** 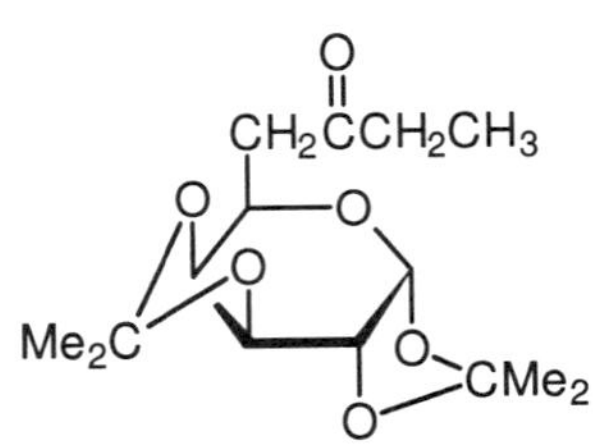 **D**		2
$X = {}^{\ominus}CH_2NO_2$, $R = -CH_2NO_2$		**C** = 76, **D** = 0	
$X = CH_3CH_2\overset{\ominus}{C}HNO_2$, $R = -CH_2(NO_2)CH_2CH_3$		**C** = 36, **D** = 32	

Continued on next page

Table 20.1. Reactions of Carbohydrate Cobaloximes[a]—*Continued*

Reactant	*Product*	*Percent Yield*	*Ref.*
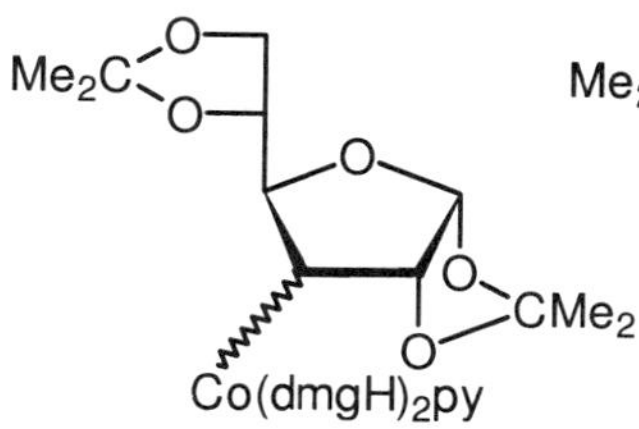 + $R\overset{\ominus}{C}HNO_2$	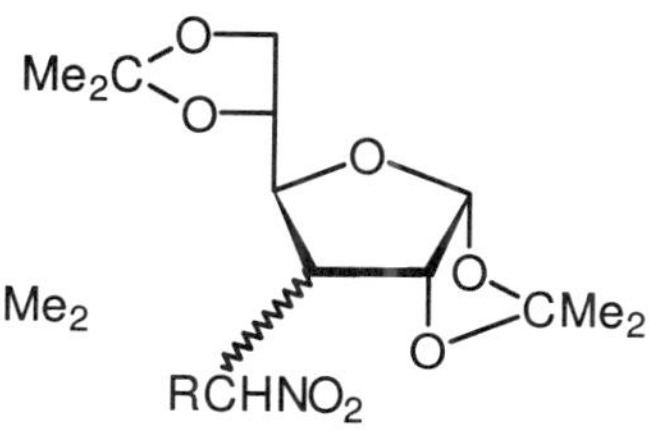		2
R = H		48	
R = CH_2CH_3		50	
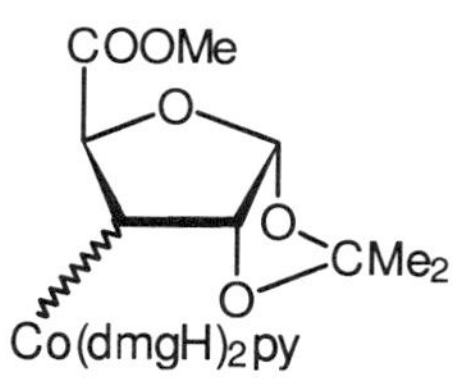 + $^{\ominus}CH_2NO_2$	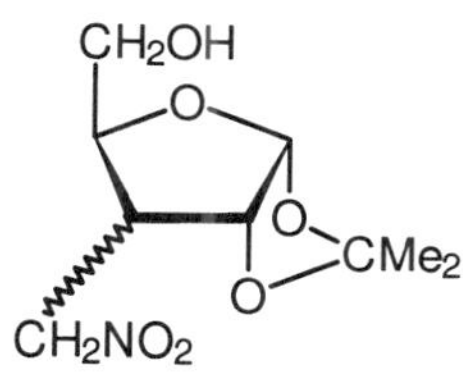	48	2
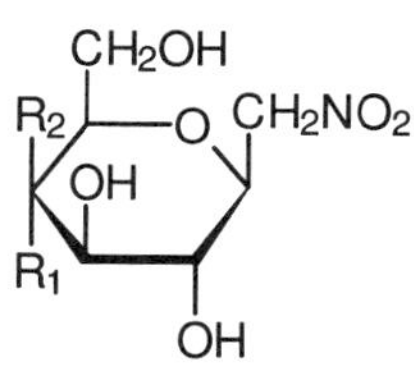 + $RCo(dmgH)_2py$	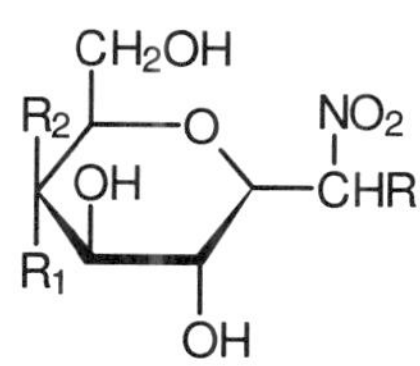		9
R = 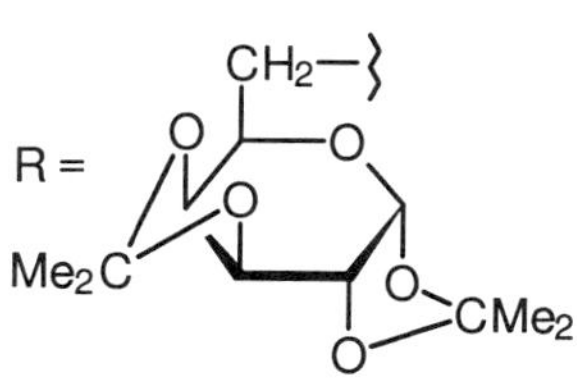	R_1 = OH, R_2 = H	45	
	R_1 = H, R_2 = OH	50	

Continued on next page

Table 20.1. Reactions of Carbohydrate Cobaloximes[a]—*Continued*

Reactant	*Product*	*Percent Yield*	*Ref.*
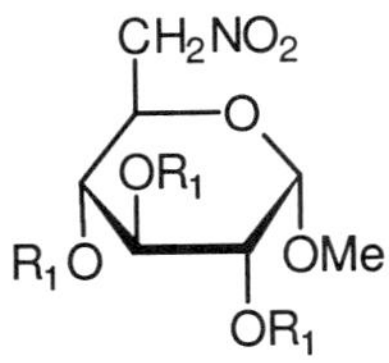 + $R_2Co(dmgH)_2py$	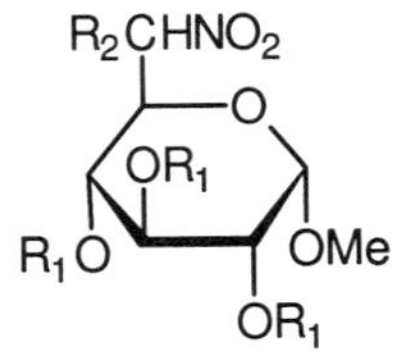 **E** + 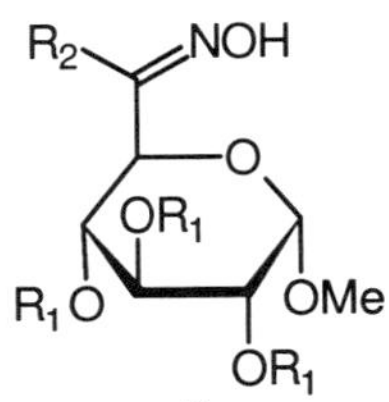 **F**		10
R_1 = H, R_2 = 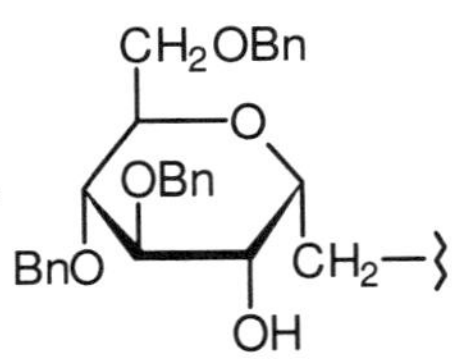		**E** = 30 **F** = 25	
R_1 = Bn, R_2 = 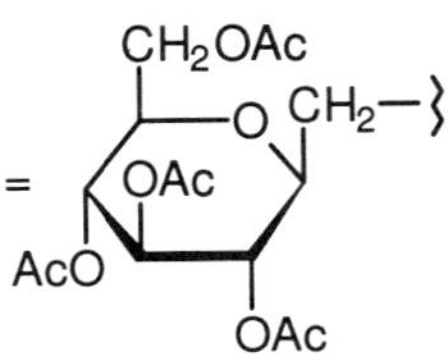		**E** = 51 **F** = 0	
5'-dCGAG + 3'-dGCTC + Co-PEM[b]	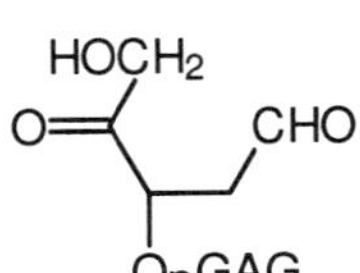		11[b]

Continued on next page

Table 20.1. Reactions of Carbohydrate Cobaloximes[a]—*Continued*

Reactant	*Product*	*Percent Yield*	*Ref.*
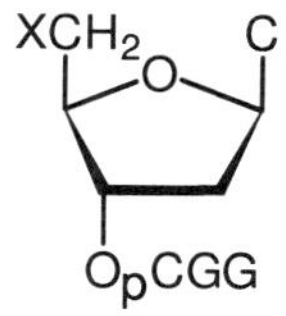 + Co-PEM	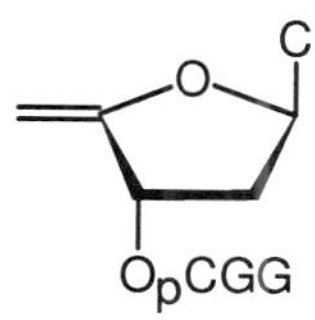		12b
X = I		19	
X = Br		14	
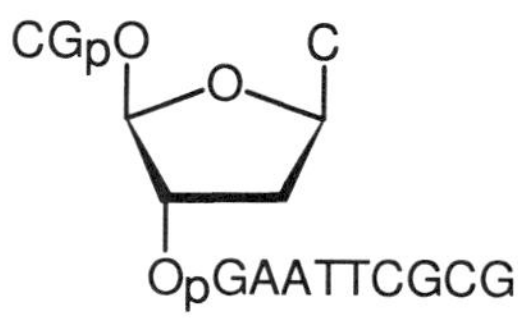 + O_2 + Co-PEM	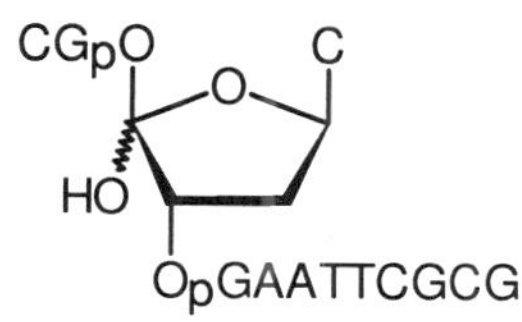		13b
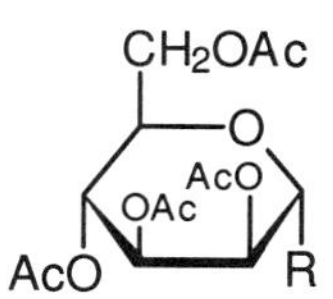 + $CH_2{=}CYZ$	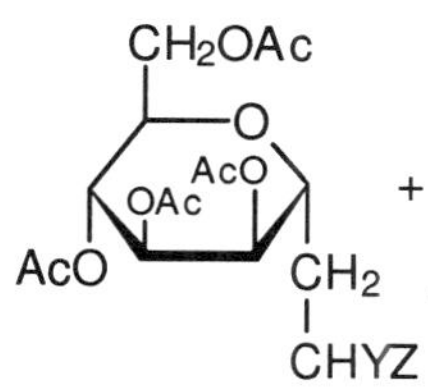 + 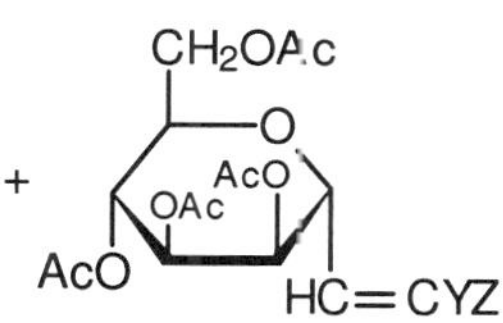		3, 5
R = $Co(dmgH)_2py$			
Y = H, Z = CN	50%	51%	
Y = H, Z = CO_2Me	4%	27%	
Y = H, Z = C_6H_5	1%	65%	
Y = OEt, Z = CN	1%	60%	

Continued on next page

Table 20.1. Reactions of Carbohydrate Cobaloximes[a]—*Continued*

Reactant	*Product*	*Percent Yield*	*Ref.*
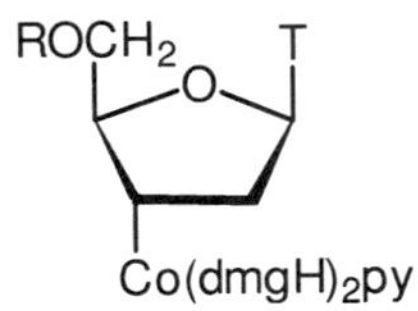 + $^{\ominus}CH_2NO_2$	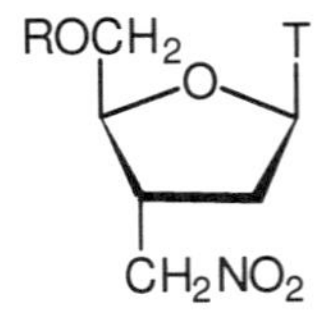	47	14
T = 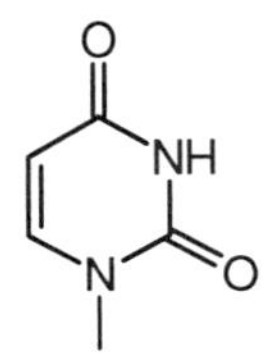	+ 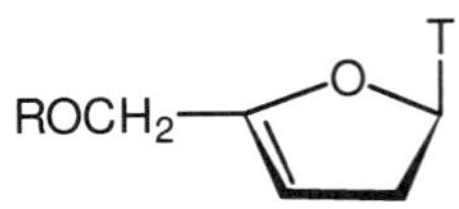	13	
R = *t*-BuMe$_2$Si	+ 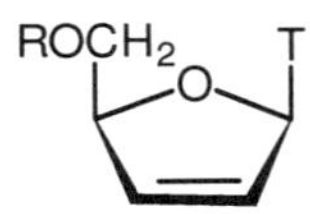	7	

[a]See equation 1 for the structure of Co(dmgH)$_2$py.
[b]Co-PEM is a cobalt complex of peplomycin.[14]

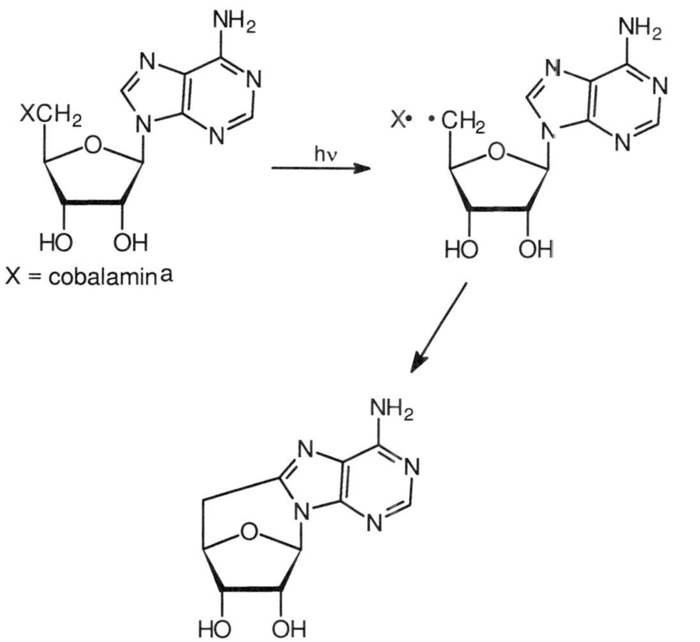

[a]The complete structure of cobalamin is given in references 17 and 18.

Scheme 20.2

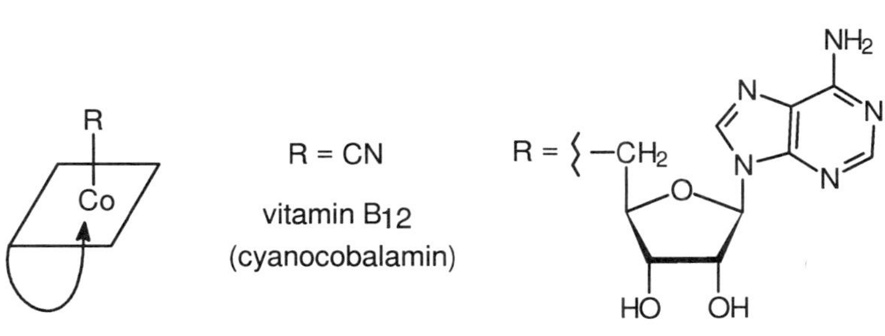

Figure 20.1. Vitamin B_{12} and a vitamin B_{12} coenzyme.

Table 20.2. Reactions of Vitamin B_{12} Coenzymes

Reactant	*Product*	*Percent Yield*	*Ref.*
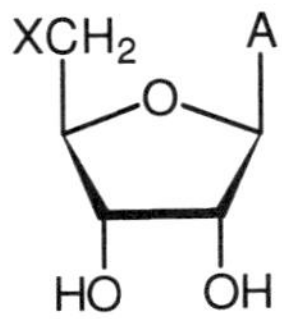 X = cobalamin[a] 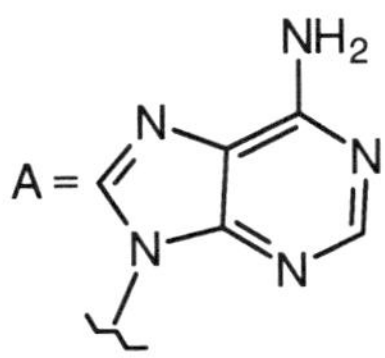	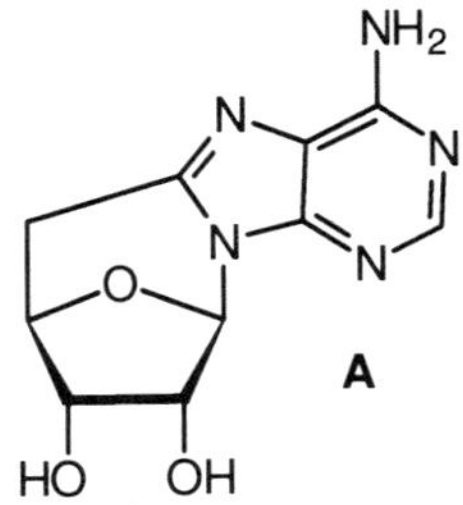 **A** + 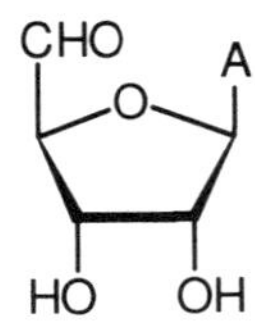 **B**	**A** + **B** = 70	19-22
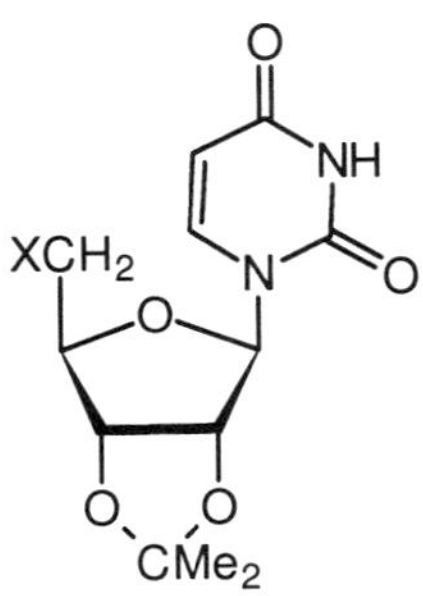	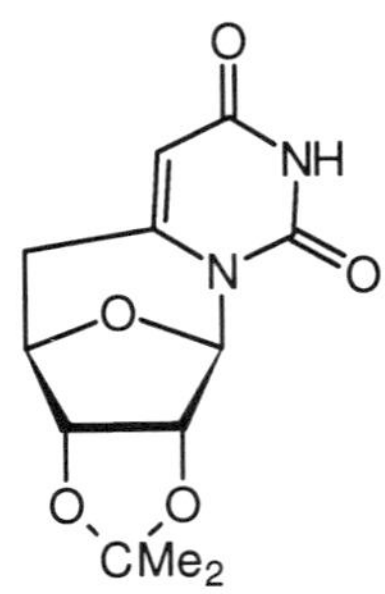	40	22
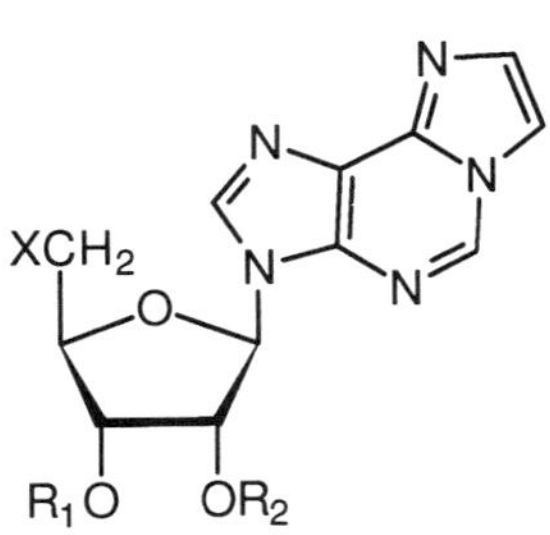 X = cobalamin[a]	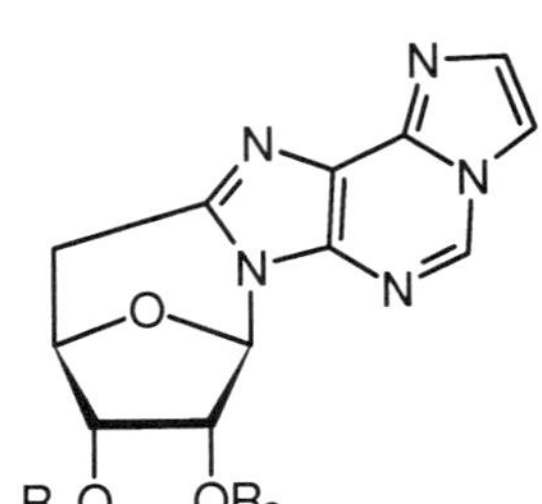 $R_1 = R_2 = H$ $R_1, R_2 = CMe_2$		23

[a]See reference 17 or 18 for the structure of cobalamin.

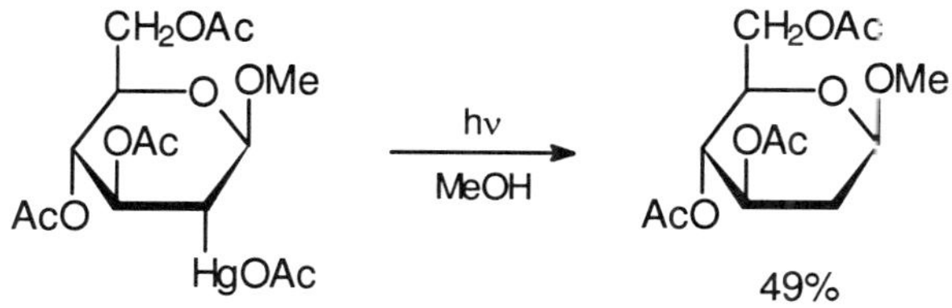

(5)

Equation 5

Organomercury Compounds

Organomercurial carbohydrate derivatives are converted into deoxy sugars by irradiation in hydrogen-donating solvents (eq 5).[24] Since the yield from this type of reaction is modest and since conduct of the reaction requires handling of toxic materials, photolysis of mercury-containing carbohydrates is a rare event.[24,25]

Compounds Containing Iron, Manganese, or Platinum

Isolated instances of photochemical reactions of carbohydrates covalently bonded to iron, manganese, or platinum have been reported. (Reactants and products for these reactions are given in Table 20.3). In one case chirality transfer to iron occurred during photosubstitution in a glucopyranosyliron compound.[26] In another type of reaction, photolysis caused the demetalation of manganese-containing *C*-glycosides.[27] In a third example, irradiation of a platinum complex of 1,2:5,6-di-*O*-isopropylidene-D-mannitol cleaved this sugar alcohol into two molecules of 1,2-*O*-isopropylidene-D-glyceraldehyde.[28]

Summary

Photolysis of carbohydrate cobaloximes produces carbon radicals that react with a variety of reagents. The general synthetic value of radical formation in this manner can be appreciated by noting the reactions listed in Table 20.1. A specific example of the synthetic application of cobaloxime photochemistry is found in the synthesis of the biologically important carbohydrate KDO (Scheme 20.1).

Table 20.3. Reactions of Carbohydrates Containing Iron, Platinum, and Manganese

Reactant	*Product*	*Percent Yield*	*Ref.*
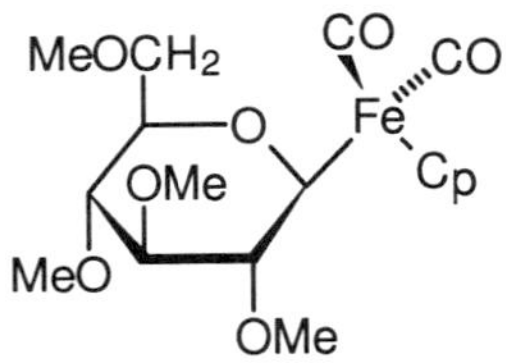 + $(C_6H_5)_3P$	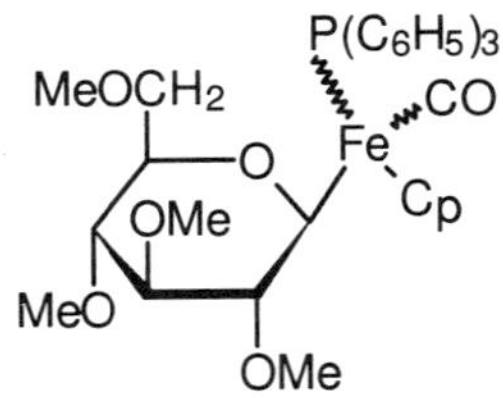	85	26
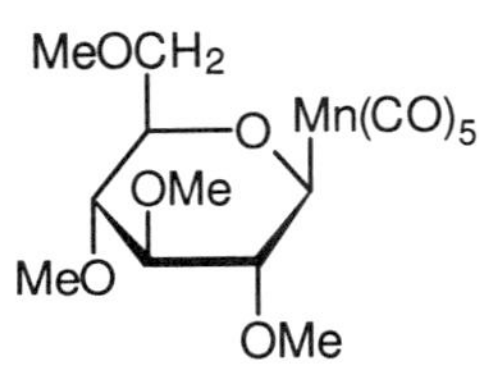 + $CH_2{=}CHCO_2Me$	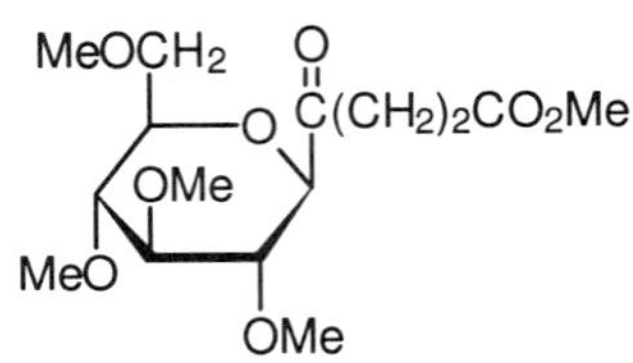	47	27
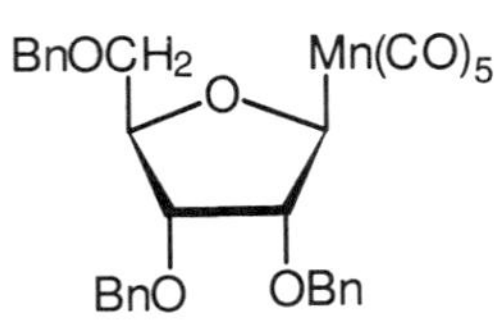 + $CH_2{=}CHCO_2Me$	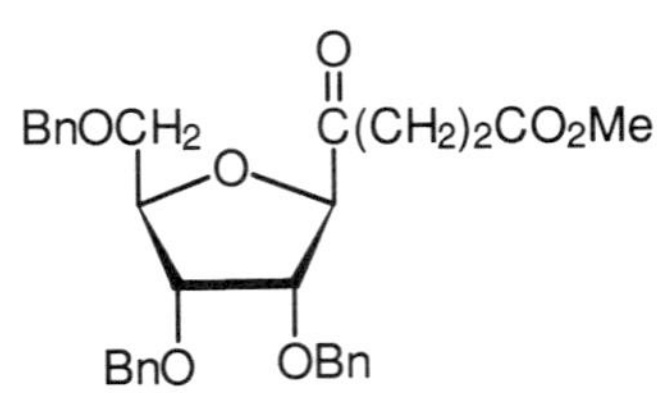	56	27
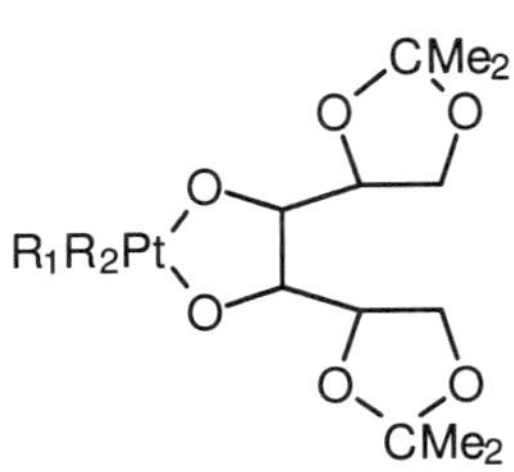 $R_1, R_2 = (C_6H_5)_2PCH_2CH_2P(C_6H_5)_2$	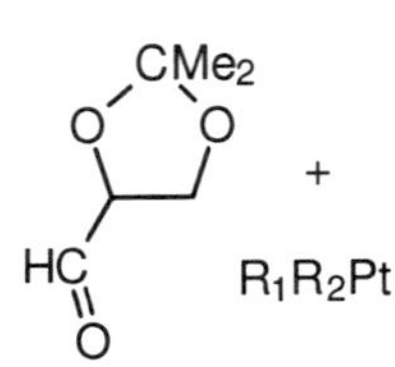	50-80	28

References

1. Branchaud, B. P; Meier, M. S. *Tetrahedron Lett.* **1988**, *29,* 3191.
2. Branchaud, B. P.; Yu, G.-X. *Tetrahedron Lett.* **1991**, *32,* 3639.
3. Ghosez, A.; Göbel, T., Giese, B. *Chem. Ber.* **1988**, *121,* 1807.
4. Veit, A.; Giese, B. *Synlett* **1990**, 166.
5. Giese, B. *Pure Appl. Chem.* **1988**, *60,* 1655.
6. Giese, B.; Carboni, B.; Göbel, T.; Muhn, R.; Wetterich, F. *Tetrahedron Lett.* **1992**, *33,* 2673.
7. Branchaud, B. P.; Choi, Y. L. *J. Org. Chem.* **1988**, *53,* 4641.
8. Branchaud, B. P.; Meier, M. S.; Choi, Y. *Tetrahedron Lett.* **1988**, *29,* 167.
9. Branchaud, B. P.; Meier, M. S. *J. Org. Chem.* **1989**, *54,* 1320.
10. Martin, O. R.; Xie, F.; Kakarla, R.; Benhamza, R. *Synlett* **1993**, 165.
11. Sugiyama, H.; Tashiro, T.; Dannoue, Y.; Miwa, T.; Matsuura, T.; Saito, I. *Tetrahedron Lett.* **1989**, *30,* 7213.
12. Sugiyama, H.; Ohmori, K.; Saito, I. *J. Am. Chem. Soc.* **1994**, *116,* 10326.
13. Saito, I.; Morii, T.; Sugiyama, H.; Matsuura, T.; Meares, C. F.; Hecht, S. M. *J. Am. Chem. Soc.* **1989**, *111,* 2307.
14. Takahashi, Y.; Honda, Y.; Tsuchiya, T. *Carbohydr. Res.* **1995**, *270,* 77.
15. Hecht, S. M. *Acc. Chem. Res.* **1986**, *19,* 383.
16. Joblin, K. N.; Johnson, A. W.; Lappert, M. F.; Nicholson, B. K. *J. Chem. Soc., Chem. Commun.* **1975**, 441.
17. Lott, W. B.; Chagovetz, A. M.; Grissom, C. B. *J. Am. Chem. Soc.* **1995**, *117,* 12194.
18. Cotton, F. A.; Wilkinson, G.; Gaus, P. L. *Basic Inorganic Chemistry;* John Wiley & Sons: New York, 1987; p 678.
19. Johnson, A. W.; Shaw, N. *J. Chem. Soc.* **1962**, 4608.
20. Hogenkamp, H. P. C. *J. Biol. Chem.* **1963**, *238*, 477.
21. Hogenkamp, H. P. C.; Ladd, J. N.; Barker, H. A. *J. Biol. Chem.* **1962**, *237,* 1950.
22. Johnson, A. W.; Oldfield, D.; Rodrigo, R.; Shaw, N. *J. Chem. Soc.* **1964**, 4080.
23. Gani, D.; Hollaway, M. R.; Johnson, A. W.; Lappert, M. F.; Wallis, O. C. *J. Chem. Res. (S)* **1981**, 190.
24. Horton, D.; Tarelli, J. M.; Wander, J. D. *Carbohydr. Res.* **1972**, *23,* 440.
25. Descotes, G.; Idmoumaz, H.; Praly, J.-P. *Carbohydr. Res.* **1984**, *128,* 341.
26. Trainor, G. L. *J. Organomet. Chem.* **1985**, *282,* C43.
27. DeShong, P.; Slough, G. A., Elango, V.; Trainor, G. L. *J. Am. Chem. Soc.* **1985**, *107,* 7788.
28. Andrews, M. A.; Gould, G. L. *Organometallics* **1991**, *10,* 387.

Index